An-Min Li, Udo Simon, Guosong Zhao, Zejun Hu
Global Affine Differential Geometry of Hypersurfaces

De Gruyter Expositions in Mathematics

—

Edited by
Victor P. Maslov, Moscow, Russia
Walter D. Neumann, New York City, New York, USA
Markus J. Pflaum, Boulder, Colorado, USA
Dierk Schleicher, Bremen, Germany
Raymond O. Wells, Bremen, Germany

Volume 11

An-Min Li, Udo Simon, Guosong Zhao, Zejun Hu

Global Affine Differential Geometry of Hypersurfaces

2nd revised and extended edition

DE GRUYTER

Mathematics Subject Classification 2010
Primary: 53A15, 53C21, 53C24, 53C40, 53C42; Secondary: 17C40, 26B20, 35J35, 35J60, 53B21, 53B25, 58E30, 58J32, 58J60.

Authors

Prof. Dr. An-Min Li
Sichuan University
School of Mathematics
Chengdu 610064, Sichuan
People's Republic of China
amli@scu.edu.cn

Prof. Guosong Zhao
Sichuan University
School of Mathematics
Chengdu 610064, Sichuan
People's Republic of China
gszhao@scu.edu.cn

Prof. Dr. Udo Simon
Technische Universität Berlin
Institut für Mathematik, MA 8–3
Straße des 17. Juni 136
D-10623 Berlin, Germany
simon@math.tu-berlin.de

Prof. Dr. Zejun Hu
Zhengzhou University
School of Mathematics and Statistics
Zhengzhou 450001, Henan
People's Republic of China
huzj@zzu.edu.cn

ISBN 978-3-11-026667-2
e-ISBN (PDF) 978-3-11-026889-8
e-ISBN (EPUB) 978-3-11-039090-2
Set-ISBN 978-3-11-026890-4
ISSN 0938-6572

Library of Congress Cataloging-in-Publication Data
A CIP catalog record for this book has been applied for at the Library of Congress.

Bibliographic information published by the Deutsche Nationalbibliothek
The Deutsche Nationalbibliothek lists this publication in the Deutsche Nationalbibliografie; detailed bibliographic data are available on the Internet at http://www.dnb.de.

© 2015 Walter de Gruyter GmbH, Berlin/Boston
Typesetting: PTP-Berlin, Protago T$_E$X-Production GmbH
Printing and binding: CPI books GmbH, Leck
♾ Printed on acid-free paper
Printed in Germany

www.degruyter.com

Contents

Introduction —— 1

1   Preliminaries and basic structural aspects —— 7
1.1   Affine spaces —— 7
1.1.1   Basic notations for affine spaces —— 7
1.1.2   Volume and orientation —— 8
1.1.3   Dual determinant forms —— 8
1.1.4   Duality and cross-product construction —— 9
1.1.5   Affine mappings and transformation groups —— 9
1.1.6   Affine invariance and duality —— 11
1.1.7   Directional derivatives —— 11
1.1.8   Frames —— 12
1.2   Euclidean spaces —— 15
1.2.1   Basic notations for Euclidean spaces —— 15
1.2.2   Normed determinant forms —— 15
1.2.3   Different Euclidean structures —— 15
1.2.4   Isometries and orthogonal transformations —— 16
1.3   Differential geometric structures of affine and Euclidean space —— 16
1.3.1   Differential geometric equiaffine structures —— 16
1.3.2   Differential geometric Euclidean structures —— 18
1.3.3 $\mathbb{R}^{n+1}$ as a standard space —— 18
1.4   Klein's Erlangen program —— 19
1.5   Motivation: A short sketch of the Euclidean hypersurface theory —— 20
1.6   Hypersurfaces in equiaffine space —— 24
1.6.1   Definition and notation —— 24
1.6.2   Bundles —— 24
1.7   Structural motivation for further investigations —— 25
1.8   Transversal fields and induced structures —— 26
1.8.1   Volume form —— 26
1.8.2   Weingarten structure equation —— 26
1.8.3   Gauß structure equation —— 26
1.8.4   Compatibility of the induced volume form and the connection —— 27
1.8.5   Relative normal fields —— 27
1.9   Conormal fields and induced structures —— 28
1.9.1   Volume form —— 28
1.9.2   Bilinear form —— 28
1.10   Relative normalizations —— 29
1.11   Nondegenerate hypersurfaces —— 29
1.12   Gauß structure equations for conormal fields —— 30

1.12.1   Compatibility of connections and volume forms in the conormal
            bundle —— 31
1.12.2   Compatibility of bilinear forms and connections; conjugate
            connections —— 31
1.12.3   Cubic form —— 32
1.12.4   Tchebychev form —— 32
1.12.5   Local notation —— 32
1.12.6   Relative Gauß maps —— 33
1.13   Affine invariance of the induced structures —— 35
1.14   A summary of relative hypersurface theory —— 36
1.14.1   Structure equations in terms of $\tilde{\nabla}$ —— 36
1.14.2   Integrability conditions in terms of $\tilde{\nabla}$ —— 37
1.15   Special relative normalizations —— 38
1.15.1   Euclidean normalization as a relative normalization —— 38
1.15.2   Equiaffine normalization —— 38
1.15.3   Centroaffine normalization —— 40
1.16   Comparison of relative normalizations —— 41
1.16.1   Comparison of corresponding geometric properties —— 42
1.16.2   Gauge invariance —— 43

2   **Local equiaffine hypersurface theory** —— 45
2.1   Blaschke–Berwald metric and structure equations —— 45
2.2   The affine normal and the Fubini–Pick form —— 48
2.2.1   Affine normal —— 49
2.2.2   Fubini–Pick form —— 52
2.2.3   Affine curvatures —— 54
2.2.4   Geometric meaning of the affine normal —— 56
2.3   The equiaffine conormal —— 59
2.3.1   Properties of the equiaffine conormal —— 59
2.3.2   The affine support function —— 62
2.4   Hyperquadrics —— 63
2.4.1   Hyperquadrics —— 63
2.4.2   Hypersurfaces with vanishing Pick invariant —— 68
2.5   Integrability conditions and the local fundamental theorem —— 72
2.5.1   Relations between the coefficients —— 73
2.5.2   Integrability conditions —— 73
2.5.3   The fundamental theorem —— 76
2.6   Euclidean boundary points of locally convex immersed
            hypersurfaces —— 80
2.7   Graph immersions —— 85
2.7.1   Graph immersions with equiaffine normalization —— 85
2.7.2   Graph Immersions with Calabi metric —— 89

3	**Affine hyperspheres** —— 93	
3.1	Definitions and basic results for affine hyperspheres —— 95	
3.1.1	Definition of affine hyperspheres —— 95	
3.1.2	Differential equations for affine hyperspheres —— 96	
3.1.3	Calabi compositions —— 100	
3.2	Affine hyperspheres with constant sectional curvature —— 104	
3.2.1	Examples —— 104	
3.2.2	Local classification of two-dimensional affine spheres with constant scalar curvature —— 107	
3.2.3	Generalization to higher dimensions —— 109	
3.3	Affine hypersurfaces with parallel Fubini–Pick form —— 116	
3.3.1	Implications from a parallel Fubini–Pick form —— 116	
3.3.2	Classification of affine hypersurfaces with parallel Fubini–Pick form —— 117	
3.4	Affine completeness, Euclidean completeness, and Calabi completeness —— 131	
3.4.1	Euclidean completeness and affine completeness —— 131	
3.4.2	The equivalence between Calabi metrics and Euclidean metrics —— 137	
3.4.3	Remarks —— 143	
3.5	Affine complete elliptic affine hyperspheres —— 144	
3.6	A differential inequality on a complete Riemannian manifold —— 147	
3.7	Estimates of the Ricci curvatures of affine complete affine hyperspheres of parabolic or hyperbolic type —— 151	
3.8	Qualitative classification of complete hyperbolic affine hyperspheres —— 154	
3.8.1	Euclidean complete hyperbolic affine hyperspheres —— 155	
3.8.2	Affine complete hyperbolic affine hyperspheres —— 159	
3.8.3	Proof of the second part of the Calabi conjecture —— 166	
3.9	Complete hyperbolic affine 2-spheres —— 172	
3.9.1	A splitting of the Levi–Civita operator —— 173	
4	**Rigidity and uniqueness theorems** —— 181	
4.1	Integral formulas for affine hypersurfaces and their applications —— 181	
4.1.1	Minkowski-type integral formulas for affine hypersurfaces —— 182	
4.1.2	Characterization of ellipsoids —— 183	
4.1.3	Some further characterizations of ellipsoids —— 186	
4.1.4	Global solutions of a differential equation of Schrödinger type —— 190	
4.1.5	Rigidity theorems for ovaloids —— 191	
4.1.6	Hypersurfaces with boundary —— 194	
4.2	The index method —— 201	
4.2.1	Fields of line elements and nets —— 201	

4.2.2	Vekua's system of partial differential equations —— **206**	
4.2.3	Affine Weingarten surfaces —— **207**	
4.2.4	An affine analogue of the Cohn–Vossen theorem —— **214**	

5	**Variational problems and affine maximal surfaces —— 217**	
5.1	Variational formulas for higher affine mean curvatures —— **218**	
5.2	Affine maximal hypersurfaces —— **223**	
5.2.1	Definitions and fundamental results —— **223**	
5.2.2	An affine analogue of the Weierstraß representation —— **226**	
5.2.3	Construction of affine maximal surfaces —— **227**	
5.2.4	Construction of improper (parabolic) affine spheres —— **229**	
5.2.5	Affine Bernstein problems —— **231**	
5.3	Differential inequalities for $\Delta^{(B)}J$, $\Delta^{(B)}\Phi$, and $\Delta^{(C)}\Phi$ —— **232**	
5.3.1	Notations from E. Calabi —— **232**	
5.3.2	Computation of $\Delta^{(B)}J$ —— **234**	
5.3.3	Computation of $\Delta^{(B)}(\|\hat{B}\|^2)$ —— **235**	
5.3.4	Estimations of $\Delta^{(C)}\Phi$ and $\Delta^{(B)}\Phi$ —— **236**	
5.4	Proof of Calabi's conjecture in dimension 2 —— **240**	
5.5	Chern's affine Bernstein conjecture —— **246**	
5.5.1	Some tools from p.d.e. theory —— **246**	
5.5.2	A partial result on Chern's conjecture in arbitrary dimensions —— **248**	
5.5.3	Proof of Chern's conjecture in dimension 2 —— **250**	
5.5.4	Estimates for the determinant of the Hessian —— **252**	
5.5.5	First proof of Lemma 5.29 —— **257**	
5.6	An analytic proof of Chern's conjecture in dimension 2 —— **259**	
5.6.1	Technical estimates —— **259**	
5.6.2	Estimates for $\sum f_{ii}$ —— **263**	
5.6.3	Estimates for the third order derivatives —— **266**	
5.6.4	Second proof of Lemma 5.29 —— **271**	
5.7	An affine Bernstein problem with respect to the Calabi metric —— **272**	

6	**Hypersurfaces with constant affine Gauß–Kronecker curvature —— 277**	
6.1	The affine Gauß–Kronecker curvature —— **277**	
6.1.1	Motivation —— **277**	
6.1.2	Main results —— **278**	
6.2	Splitting of the fourth order PDE S_n = const into two (second order) Monge–Ampère equations —— **280**	
6.3	Construction of Euclidean complete hypersurfaces with constant affine G-K curvature —— **282**	
6.4	Completeness with respect to the Blaschke metric —— **290**	

7   Geometric inequalities —— 303
7.1   The affine isoperimetric inequality —— 303
7.1.1   Steiner symmetrization —— 303
7.1.2   A characterization of ellipsoids —— 306
7.1.3   The affine isoperimetric inequality —— 308
7.2   Inequalities for higher affine mean curvatures —— 310
7.2.1   Mixed volumes —— 310
7.2.2   Integral inequalities for curvature functions —— 312
7.2.3   Total centroaffine area —— 314

A   Basic concepts from differential geometry —— 317
A.1   Tensors and exterior algebra —— 317
A.1.1   Tensors —— 317
A.1.2   Exterior algebra —— 319
A.2   Differentiable manifolds —— 321
A.2.1   Differentiable manifolds and submanifolds —— 321
A.2.2   Tensor fields on manifolds —— 324
A.2.3   Integration on manifolds —— 326
A.3   Affine connections and Riemannian geometry: Basic facts —— 328
A.3.1   Affine connections —— 328
A.3.2   Riemann manifolds —— 332
A.3.3   Manifolds of constant curvature, Einstein manifolds —— 335
A.3.4   Examples —— 336
A.3.5   Exponential mapping and completeness —— 336
A.4   Green's formula —— 338

B   Laplacian comparison theorem —— 341

Bibliography —— 345

Index —— 362

Introduction

Affine differential geometry has a long history: As far as we know it was Transon who published the first result in affine differential geometry in 1841; he considered the affine normal of a curve. But it needed more than 70 years before a systematic and intensive study of affine properties of curves and surfaces began. Following the ideas of Felix Klein, presented in his famous lecture at Erlangen in 1872, geometers like Pick (1906) [280], Tzitzeica (at the ICM congress 1912), and others proposed the study of curves and surfaces with respect to different transformation groups.

In 1907, Tzitzeica considered what are now called affine spheres. After another decade, in 1916, a larger group of geometers started a systematic study of properties of curves and surfaces with respect to the equiaffine (unimodular) transformation group: Berwald, Blaschke, Franck, Gross, König, Liebmann, Pick, Radon, Reidemeister, Salkowski, Thomsen, and Winternitz. The progress was so rapid that Blaschke, with Reidemeister as coauthor, published the first monograph about affine differential geometry in 1923 [20]. A commentary about this period can be found in [32] in Blaschke's collected works [22], while the surveys [333] and [238] also briefly describe later developments. Blaschke's monograph shows the attraction coming from global problems; this book contains an extensive chapter about global curve theory and several sections about surfaces in the large: affine isoperimetric inequalities, ovaloids with constant affine mean curvature and several characterizations of ellipsoids (including deformation results).

It was mainly a group of geometers in Japan who continued to study similar local and global problems in the equiaffine or in the relative context: Kubota, Nakajima, Su, Süss, and others. In particular, they studied the affine geometry of ovaloids. At that time Salkowski published the second monograph in this field: [292] in 1934, which contains the local equiaffine and centroaffine theory, but no global results. His presentation includes affine connections, but they do not appear as basic invariants in affine hypersurface theory. The book [311] of father and son Schirokow, published in 1957, is the next mark in the development of affine differential geometry. While it shows remarkable progress in the local theory (in particular the geometry of the induced connections) and in the local classification of special classes of surfaces, there is little progress in global results: in addition to the global uniqueness theorems which already appeared in Blaschke's book, [311] contains affine integral formulas of the generalized Minkowski-type, due to Grotemeyer.

It was about the same time that the monograph [311] appeared (1957) and was translated (1962), that global affine geometry began receiving new impulses from different directions (see the bibliography for references):
-   the work of Blaschke and his school was extended by Brickell, Süss, Grotemeyer, Schneider, Simon, Voss, and others;

- Laugwitz emphasized new aspects, in particular in centroaffine geometry (see also his book [153]); later Heil continued his research;
- Münzner extended the index method and gave new affine applications;
- Calabi made several major contributions about locally strongly convex affine spheres in the large; his contributions again inspired R. Schneider, Pogorelov, Cheng, and Yau; the progress in these contributions is mainly due to eminent progress in p.d.e. theory, in particular results on Monge–Ampère equations;
- Calabi and Chern independently posed different versions of an "affine Bernstein problem" concerning affine maximal surfaces.

Parallel to this development in global affine geometry, there was a significant influence from Cartan's ideas: the schools in Russia (in particular Kazan) and Romania, from 1935–1955, investigated the role of induced connections in affine hypersurface theory, and Barthel and his school (1965–1980) developed an affine geometry on manifolds. Contributions of Calabi and the lecture of Nomizu at Münster in 1982, continued by research by him and Pinkall, finally led to a new approach to affine geometry. This more structural approach emphasizes the role of the induced volume forms and two induced conjugate connections. It allows a geometric interpretation of the integrability conditions of the structure equations. For these more structural considerations we refer the reader to the introductory notes [239, 341] or the research publications [72, 245].

The developments which we briefly described led to an increasing interest in affine differential geometry. We note a burst of research in affine differential geometry, reflected by the following facts:

- there were two conferences on the topic at Oberwolfach in 1986 and 1991, documented in the proceedings volume [337] and [252]; other meetings, like the Leuven conferences in 1991 and 1992, contain major sections devoted to affine differential geometry (see [74, 75]);
- since 1986 there have been several hundred relevant research papers published in affine differential geometry;
- new research groups have appeared in several places, in particular in China.

We recall our intention for the first edition:

It was our original plan to write an extensive monograph about important developments, but the overflow of material made us change our plans. Thus for the first edition we started on the level for students with a basic knowledge in Euclidean differential geometry (curves and hypersurfaces) and Riemannian geometry. To be on the safe side we collected basic definitions and facts from differential geometry on manifolds in Appendix A; we added references for proofs. Appendix B contains more advanced topics (Laplacian comparison theorem). For the most part we have retained this concept for the second edition.

The first edition of this monograph appeared in 1993. Since then important new contributions to the field have appeared. Examples are:

(i) Calabi and Chern posed two different versions of an *affine Bernstein conjecture* in dimension two, respectively. For both versions there have been proofs during the last two decades.

(ii) Regarding locally strongly convex Blaschke hypersurfaces, the class of affine hyperspheres is a subclass of the class of such hypersurfaces with constant Gauß–Kronecker curvature. New results about this larger class appeared in the 1990s.

(iii) It was already known that a Blaschke hypersurface, with its Fubini–Pick form being parallel with respect to the Levi–Civita connection of the Blaschke metric, is an affine hypersphere. For the case where the Blaschke hypersurfaces of this subclass are locally strongly convex, an explicit classification was given about ten years ago. This class is very large.

To include such significant developments into our monograph has been the main motivation to write this second edition.

We now give an outline of the contents of the book. Preliminaries and elementary structural aspects about the affine theory of hypersurfaces are collected in Chapter 1. This chapter contains a structurally oriented review of the Euclidean hypersurface theory (Section 1.5) and preparatory material about transversal and conormal fields of hypersurfaces in Sections 1.8–1.15 to develop two central concepts: the regularity (nondegeneracy) of hypersurfaces and the notion of a relative normalization. As motivation and guideline for this we wrote Section 1.7 from a didactical point of view. Chapter 1 closes with Sections 1.14–1.16 introducing the three most important relative normalizations: the Euclidean, the equiaffine, and the centroaffine.

From there on we concentrate, with a few exceptions, on equiaffine hypersurface theory: a student at the level mentioned should be able to follow from the elementary beginnings up to recent research.

In Chapter 2 we develop the local theory and use the invariant and the local calculus together with Cartan's moving frames. Chapters 3–6 contain mainly global results. For this reason, we restrict to locally strongly convex hypersurfaces, as in this case the equiaffine metric is definite.

Chapter 3 centers around problems which were solved during the last decades. One knows from the p.d.e.'s for affine spheres that they admit many solutions, so the class of affine spheres is very large. Thus one tries to classify subclasses.

(i) In Section 3.2 we present the local classification of all locally strongly convex affine spheres with equiaffine metric of constant sectional curvature. The analogous problem for indefinite metrics with nonvanishing Pick invariant is also known, but it is still open for other cases in dimension $n > 2$.

(ii) All equiaffine hypersurfaces with a parallel Fubini–Pick form are affine hyperspheres. In case the hypersurfaces are locally strongly convex, they are either hyperquadrics or hyperbolic hyperspheres, which have been completely and explicitly classified by Z. Hu, H. Li, and L. Vrancken [133]. The proofs for this spectacular classification were given in recent years, and in Section 3.3 we outline this development.

(iii) From Section 3.4 on, we investigate affine spheres with complete Blaschke metric. The first global result is due to Blaschke, while important later contributions are due to Calabi, R. Schneider, Pogorelov, Cheng and Yau, Sasaki, Gigena, and A.-M. Li.

This classification requires subtle estimates; thus this section is an excellent demonstration of strong p.d.e. methods.

Chapter 4 additionally presents other methods: the application of integral formulas and the index method. We derive and apply several integral formulas of Minkowski and Lichnerowicz-type, respectively. Due to a result by Li we are able to prove that the constancy of certain affine curvature functions on an ovaloid implies that the ovaloid is an ellipsoid. First results of this type are due to Blaschke and Süss, but so far one needed additional positivity assumptions.

Another application is given by results about the global solutions of an important Schrödinger-type equation. The index method is restricted to dimension 2. We present an outline of the method in Sections 4.2.1–4.2.2 and give its exemplary applications in two cases:

(i) we prove uniqueness results for affine Weingarten surfaces;

(ii) we generalize the classical Cohn–Vossen theorem about isometric ovaloids in Euclidean space and show that adequate analogues in equiaffine space are uniqueness results for the induced connections.

In Chapter 5 we study two variational problems:

(i) we prove variational formulas for the affine curvature integrals;

(ii) we present recent results about what were classically called "affine minimal surfaces", but – following Calabi – nowadays are called "affine maximal surfaces".

Our exposition in Sections 5.4–5.6 leads to recent solutions of two versions of an affine Bernstein problem in two dimensions: This includes the solution of Calabi's conjecture by A.-M. Li and F. Jia; and the solution of Chern's conjecture, first by N. S. Trudinger and X.-J. Wang, then by A.-M. Li and F. Jia, who gave a completely different analytic proof. To explore such an analytic proof, Li and his collaborators developed a method, called the real (complex) affine technique, which was also applied successfully to the study of both Abreu's equation and the existence of extremal Kähler metrics. Finally, in Section 5.7 we describe another version of the affine Bernstein problem: namely in dimensions $n = 2$ and $n = 3$ we consider locally strongly convex, affine maximal hypersurfaces that are complete with respect to the so-called Calabi metric.

In Chapter 6 we study hypersurfaces with constant affine Gauß–Kronecker curvature. The contents of this chapter in this second edition are completely new; we make an important observation in Corollary 6.4; as a result it admits an extension of well-known results on affine hyperspheres.

Finally, Chapter 7 gives an introduction to affine problems related to the theory of convex bodies: we prove the affine isoperimetric inequality and some integral inequalities for affine curvature integrals.

Our selection of topics for this book has been made under two view points: to present interesting new results and important global methods. In the end a student should know about recent topics of research. Naturally, we also made our selection according to personal taste and interest.

Finally, we again include a bibliography in this second edition. In our first edition, the purpose of the bibliography was twofold: first, of course, to include the necessary references, and secondly, to continue and update a series of bibliographies, starting with that in the monograph by A. P. and P. A. Schirokow [311] and giving a reference system in the field as complete as possible. The first edition of our monograph appeared about two decades ago. Meanwhile databases like Zentralblatt MATH and MathSciNet offer an excellent bibliographical service, and thus there is no need any more to include the complete reference system of a field in a monograph. For this reason the references in the present edition have been reduced. Meanwhile, in case that some journals or proceedings have only a local/regional distribution, we also provide a link to a ZMATH-review (Zbl.) or an ISBN-number.

Acknowledgements. We were financially supported by different institutions: the first author by the Alexander von Humboldt-Stiftung, Tian-Yuan Foundation of China, TU Berlin, and the GADGET program of the European Union; the Chinese authors by the National Natural Science Foundation of China and the Science Foundation of the Educational Committee of China; U. Simon by an academy grant of Stiftung Volkswagenwerk in 1988/89, followed by partial support of the GADGET program of the EU and a fellowship of the Japan Society for the Promotion of Science, combined with partial support by DAAD in 1990; he had the pleasure to stay for several weeks as a guest professor at Science University of Tokyo in 1990 and Catholic University of Leuven in 1991 and 1992. Since 1992, the project was partially supported by the Deutsche Forschungs Gemeinschaft and Alexander von Humboldt-Foundation.

During the last decades we had discussions about affine differential geometry with many of our colleagues and friends; we learned much from this and would like to thank in particular: S. S. Chern, H. Li, E. Lutwak, A. Mihai, K. Nomizu, B. Opozda, B. Palmer, L. Sheng, A. P. Shirokov, and C. P. Wang; the "affine" research groups at Leuven (F. Dillen, L. Verstraelen, L. Vrancken), Granada (A. Martinez, F. Milan, F. G. Santos) and the research group at SUT (N. Abe, S. Yamaguchi and their students).

It is a pleasure to thank the financing institutions, the host institutions, and our colleagues there for their support as well as their hospitality.

In particular, we would like to thank AvH, Stiftung Volkswagenwerk, DFG and NSFC, Dierks von Zweck Foundation in Essen (Germany), and finally Sichuan University and TU Berlin; due to their support we were able to work together at the TU Berlin and at Sichuan University for longer periods, from 1986 on.

Berlin, Chengdu, and Zhengzhou, February 2015

An-Min Li
Udo Simon
Guosong Zhao
Zejun Hu

1 Preliminaries and basic structural aspects

We recall basic notions from affine and Euclidean spaces in Sections 1.1 and 1.2 of this chapter, including the notion of frames in affine space in Section 1.1.8. We summarize elementary facts from a structural viewpoint of differential geometry in Section 1.3. The basic material for Sections 1.1 and 1.2 is contained in books on linear algebra, analytic geometry, and calculus. The standard references are [94, 105, 106, 124, 237, 346].

1.1 Affine spaces

We recall the basic structure of $\mathbb{R}^{n+1}$:
  – an algebraic structure as a vector space over $\mathbb{R}$;
  – a geometric (affine) structure as a set of points;
  – the canonical topological structure;
  – the canonical differentiable structure.

For the study of the affine differential geometry of hypersurfaces in real affine space, it is sometimes convenient to introduce different notations for different structural aspects of $\mathbb{R}^{n+1}$. Therefore we introduce the following notations.

1.1.1 Basic notations for affine spaces

(i) Denote by A^{n+1} an $(n + 1)$-dimensional affine space, by V the real vector space associated with A^{n+1}, and by

$$\pi : A^{n+1} \times A^{n+1} \;\to\; V$$

the mapping relating A^{n+1} and V such that
(a) for any three points $p, q, r \in A^{n+1}$ we have

$$\pi(p, q) + \pi(q, r) = \pi(p, r)$$

(compatibility of π with addition and scalar multiplication on V);
(b) for any $p \in A^{n+1}$ and $v \in V$ there exists a unique $q \in A^{n+1}$ such that $\pi(p, q) = v$.
The mapping π allows to define the *affine structure* of A^{n+1} from the structure of V. It is very suggestive to write $\overrightarrow{pq} := \pi(p, q) = v$.
(ii) Denote by V^* the dual space of V and by

$$\langle\,,\,\rangle : V^* \times V \;\to\; \mathbb{R}$$

the *standard scalar product*.

(iii) Relation (b) in (i) is equivalent to the following: for any $p \in A^{n+1}$ the mapping

$$\pi_p := \pi(p, \ ) : \ A^{n+1} \to V, \qquad q \mapsto \pi_p(q) = \pi(p, q)$$

is bijective. Thus we get the *topological* and *differentiable structure* on A^{n+1} from V via pull back of π_p for an arbitrary $p \in A^{n+1}$.

(iv) Recall that the tangent space $T_p A^{n+1}$ at $p \in A^{n+1}$ is canonically isomorphic to V, which justifies the usual *identification* of $T_p A^{n+1}$ and V (cf. [120, p. 11]).

1.1.2 Volume and orientation

(i) The set of determinant forms over V is a real vector space of dimension one. If we fix a nontrivial determinant form Det we can define the nonoriented volume of the $(n + 1)$-dimensional parallelepiped $P(p_0, p_1, \ldots, p_{n+1})$ by

$$\mathrm{Vol}\, P(p_0, p_1, \ldots, p_{n+1}) := |\mathrm{Det}(\overrightarrow{p_0 p_1}, \ldots, \overrightarrow{p_0 p_{n+1}})|$$

where $p_0, \ldots, p_{n+1} \in A^{n+1}$. Therefore determinant forms are also called *volume forms*.

Any two nontrivial determinant forms Det and $\widehat{\mathrm{Det}}$ differ by a nonzero constant $\beta \in \mathbb{R}$:

$$\widehat{\mathrm{Det}} = \beta \cdot \mathrm{Det}.$$

As an obvious consequence we derive the fact that the definition of the volume depends on the choice of the determinant form, but the ratio of two volumes is independent of this choice.

(ii) Fix an arbitrary nontrivial determinant form Det over V. There are two equivalence classes,

$$\{D \mid D = \beta \cdot \mathrm{Det}, \ \beta > 0\}, \quad \text{and} \quad \{D \mid D = \beta \cdot \mathrm{Det}, \ \beta < 0\},$$

in the set of nontrivial determinant forms. Each class is called an *orientation* of V.

1.1.3 Dual determinant forms

The sets of determinant forms over V and V^* are related via duality: the determinant form Det^* over V^* is called the *dual form* of Det if

$$\mathrm{Det}^*(v^{*1}, \ldots, v^{*n+1}) \cdot \mathrm{Det}(v_1, \ldots, v_{n+1}) = \det(\langle v^{*i}, v_j \rangle),$$

where "det" on the right-hand side denotes the determinant of a matrix with coefficients $\langle v^{*i}, v_j \rangle$ for $v^{*i} \in V^*, v_j \in V$.

Fixing a volume form Det over V to define a volume, this automatically fixes a volume form Det^* over V^*. Moreover, if Det represents an orientation-class then Det^* defines an orientation of V^* induced by the orientation of V.

1.1.4 Duality and cross-product construction

(i) Let $w_1, \ldots, w_n \in V$. If these vectors are linearly independent, and that the span $W := \mathrm{span}(w_1, \ldots, w_n)$ has dimension n. Via duality there is a one-dimensional subspace $W^* \subset V^*$ such that

$$w^*(w_i) = 0$$

for any $w^* \in W^*$. A basis vector of W^* can be explicitly calculated from the cross-product construction: For a fixed nontrivial determinant form Det the *cross product* of $w_1, \ldots, w_n$

$$[\ \ ]: \times_n V \to V^*$$

satisfies

$$\langle [w_1, \ldots, w_n], z \rangle = \mathrm{Det}(w_1, \ldots, w_n, z)$$

for any $z \in V$.

It is obvious how the definition of the cross product depends on the choice of the determinant form.

(ii) When we restrict to the *equiaffine space* A^{n+1} with a fixed determinant form as volume form, the cross product induces a mapping

$$i: \Lambda^n(V) \to V^*$$

from the vector space of exterior n-forms to V^*. It is easy to see that the mapping i is an isomorphism. Thus we can identify $\Lambda^n(V)$ and V^* via this isomorphism.

(iii) Consider a hyperplane $H \subset A^{n+1}$ given by $n + 1$ affinely independent points $p_0, \ldots, p_n$.

We describe H by its position vector $\overrightarrow{Oq}$ for arbitrary $q \in H$; here O denotes the *origin*:

$$\overrightarrow{Oq} = \overrightarrow{Op_0} + \sum_{i=1}^{n} h^i \overrightarrow{p_0 p_i}.$$

The set $\{h^1, \ldots, h^n\}$ gives the *affine coordinates* of q with respect to $\{p_0, \ldots, p_n\}$. The cross-product construction allows to give another representation of H, the so-called *Hesse equation*:

$$\langle w^*, \overrightarrow{p_0 q} \rangle = 0,$$

where $w^* = [\overrightarrow{p_0 p_1}, \ldots, \overrightarrow{p_0 p_n}]$. In the classical terminology w^* is called a *conormal (vector)* of the hyperplane H.

1.1.5 Affine mappings and transformation groups

In order to define the affine structure of a space, one uses the structure of the associated vector space. Correspondingly, affine mappings (which keep the affine structure) are defined via linear mappings between the associated vector spaces.

1. Affine mappings

(i) Let A_1, A_2 be affine spaces and V_1, V_2, respectivly, their associated vector spaces, and let $\pi_i : A_i \times A_i \to V_i$. A mapping $\alpha : A_1 \to A_2$ is called *affine*, if there exists a linear mapping L_α such that

$$\pi_2(\alpha(p), \alpha(q)) = L_\alpha(\pi_1(p, q)) \quad \text{for all } p, q \in A_1,$$

i.e. the following diagram is commutative:

$$
\begin{array}{ccc}
A_1 \times A_1 & \overset{(\alpha,\alpha)}{\longrightarrow} & A_2 \times A_2 \\
\pi_1 \downarrow & & \pi_2 \downarrow \\
V_1 & \overset{L_\alpha}{\longrightarrow} & V_2.
\end{array}
$$

L_α is uniquely determined. We call L_α the *linear mapping associated with α*.

(ii) An affine mapping is injective (surjective) if and only if L_α is injective (surjective). We call bijective affine mappings *regular*. For $A_1 = A_2$ in (i) above, define the determinant of α to be

$$\det \alpha := \det L_\alpha.$$

(iii) Fix an origin $O \in A^{n+1}$. Then the affine mapping $\alpha : A^{n+1} \to A^{n+1}$ has the following representation

$$\overrightarrow{O\alpha(p)} = L_\alpha(\overrightarrow{Op}) + b,$$

where $b \in V$.

2. Affine transformation groups

It is well known that the set of all automorphisms of a vector space V of dimension $n + 1$ forms a group. We use the following standard notations for this group and its subgroups:

$$GL(n + 1, \mathbb{R}) := \{L : V \to V \mid L \text{ isomorphism}\};$$
$$SL(n + 1, \mathbb{R}) := \{L \in GL(n + 1, \mathbb{R}) \mid \det L = 1\}.$$

Correspondingly, for an affine space A^{n+1} we have the following *affine transformation groups*:

$\mathfrak{a}(n + 1) := \{\alpha : A^{n+1} \to A^{n+1} \mid L_\alpha \text{ regular}\}$, regular affine group

$\mathfrak{s}(n + 1) := \{\alpha \in \mathfrak{a}(n + 1) \mid \det \alpha = 1\}$, unimodular (equiaffine) group

$\mathfrak{z}_p(n + 1) := \{\alpha \in \mathfrak{a}(n + 1) \mid \alpha(p) = p\}$, centroaffine group with center
$\quad p \in A^{n+1}$

$\tau(n + 1) := \{\alpha : A^{n+1} \to A^{n+1} \mid \text{there exists } b(\alpha) \in V \text{ s.t. } \overrightarrow{p\alpha(p)} = b(\alpha)$
$\quad \text{for all } p \in A^{n+1}\}$, group of translations on A^{n+1}.

Let g be one of the groups above and $S_1, S_2 \subset A^{n+1}$ subsets. Then S_1 and S_2 are called *equivalent modulo g* if there exists $\alpha \in g$ such that

$$S_2 = \alpha S_1.$$

3. Properties of affine mappings

(i) The standard properties of affine mappings come from the properties of the asso-
     ciated linear mappings. Recall in particular:
     (a) *parallelism* is invariant under affine mappings;
     (b) the *partition ratio of 3 points* is invariant under affine mappings;
     (c) the *ratio of the volumes* of two parallelepipeds is affinely invariant.

(ii) **Theorem.** $\alpha : A^{n+1} \to A^{n+1}$ *is a regular affine transformation if and only if α is bijective, continuous, and preserves convexity.*

     In the sense of this theorem convexity is an affine property; for a more general
     result see [400].

(iii) **Theorem.** *Let $\alpha : A^{n+1} \to A^{n+1}$ be an affine transformation. Then the nonoriented volume of a parallelepiped is invariant under α if*

$$\det \alpha = 1.$$

1.1.6 Affine invariance and duality

Many differential geometric quantities of hypersurfaces are defined via the scalar
product $\langle \ , \ \rangle : V^* \times V \to \mathbb{R}$. Thus affine invariance depends on the associated linear
mapping and its dual mapping $L^* : V^* \to V^*$; as an immediate consequence of the
dual mapping they satisfy

$$\langle (L^*)^{-1} v^*, Lv \rangle = \langle v^*, v \rangle = \langle L^* v^*, L^{-1} v \rangle.$$

1.1.7 Directional derivatives

We consider an $\mathbb{R}$-vector space with the canonical differentiable structure. As usual
we identify the tangent space $T_u V$ at $u \in V$ with the vector space itself.

(i) We denote the directional derivative of a function $\tau \in C^\infty(V)$ in the direction of
     the vector v by [124, p. 95]

$$\bar{\nabla}_v \tau := v(\tau) := v\tau := d\tau(v).$$

For vector fields v, w the derivative $\bar{\nabla}_v w$ of w in the direction of v is again a vector
field. $\bar{\nabla}$ obeys the following rules for the differentiation of vector fields w, w_1, w_2,
and a differentiable function $\tau : V \to \mathbb{R}$, in the direction of the tangent vectors

v, v_1, v_2 at $u \in V$:

$$\bar{\nabla}_v(w_1 + w_2) = \bar{\nabla}_v w_1 + \bar{\nabla}_v w_2,$$
$$\bar{\nabla}_v(\tau w) = \tau \bar{\nabla}_v w + (\bar{\nabla}_v \tau)w,$$
$$\bar{\nabla}_{v_1 + v_2} w = \bar{\nabla}_{v_1} w + \bar{\nabla}_{v_2} w,$$
$$\bar{\nabla}_{\tau v} w = \tau \bar{\nabla}_v w.$$

In the language of differential geometry the directional derivative is a torsion-free flat affine connection on V. By the pull back of $\pi_0 : A^{n+1} \to V$ we can define a flat connection without torsion on A^{n+1}. Thus we consider A^{n+1} to be a differentiable manifold with a flat affine connection, which we again denote by $\bar{\nabla}$.

(ii) Recall that $\bar{\nabla}$ can be extended uniquely as a tensor derivation over A^{n+1}. This implies, in particular, the following relation for any (differentiable) vector field $v : A^{n+1} \to V$, 1-form $v^* : A^{n+1} \to V^*$, and tangent vector z:

$$z\langle v^*, v \rangle = \bar{\nabla}_z \langle v^*, v \rangle = \langle \bar{\nabla}_z v^*, v \rangle + \langle v^*, \bar{\nabla}_z v \rangle.$$

(iii) The rule to differentiate a determinant form $\mathrm{Det}(v_1, \ldots, v_{n+1})$, applied to $n + 1$ tangent vector fields, in direction of z can be reformulated as covariant differentiation:

$$(\bar{\nabla}_z \mathrm{Det})(v_1, \ldots, v_{n+1}) = \bar{\nabla}_z(\mathrm{Det}(v_1, \ldots, v_{n+1}))$$
$$- \sum_{B=1}^{n+1} \mathrm{Det}(v_1, \ldots, \bar{\nabla}_z v_B, \ldots, v_{n+1})$$
$$\equiv 0.$$

This in particular implies the fact that the volume of a parallelepiped is invariant under parallel translation along a given curve.

(iv) In the case of directional derivatives it is standard to denote the tangent vectors in direction of the coordinate axes by $\partial_B := \frac{\partial}{\partial x^B}$, where x^B denote the coordinate-functions ($B = 1, \ldots, n + 1$) and the partial derivatives $\partial_B f$ of a differentiable function (vector field) f by

$$\bar{\nabla}_B f := \bar{\nabla}_{\partial_B} f := \partial_B f.$$

The notation with components was used in classical textbooks on analytic geometry, differential geometry, and physics. This calculus was called tensor calculus. In many situations it is still useful to calculate with components instead of tensors.

1.1.8 Frames

In differential geometry three different "languages" are used: the classical tensor calculus in local notation, the invariant calculus (Koszul calculus), and the method of moving frames (Cartan). For beginners we now give an elementary introduction to the calculus of moving frames on an affine space A^{n+1}, considered as a differentiable manifold. In each calculus we use a standard notation.

A *coordinate system* on A^{n+1} is given by a fixed point O, called the *origin*, and a *basis* $\{\eta_1, \ldots, \eta_{n+1}\} \in V$, such that the *position vector* x of the point $x \in A^{n+1}$ has the representation

$$x = \sum_{B=1}^{n+1} x^B \eta_B.$$

Thus, for fixed origin O, we can identify $x \in A^{n+1}$ and $\overrightarrow{Ox} \in V$, and in this way we can identify A^{n+1} and V. This identification explains that one uses the simpler notation x for $\overrightarrow{Ox}$.

Considering A^{n+1} as a differentiable manifold, we can take $(x^1, \ldots, x^{n+1})$ as (local) coordinates for a chart of A^{n+1}. Then a *Gauß basis* is given by $\partial_1, \ldots, \partial_{n+1}$, where, as above, $\partial_B = \frac{\partial}{\partial x^B}$, and its dual basis by $dx^1, \ldots, dx^{n+1}$. With respect to the chart, for a tangent vector field $X : A^{n+1} \to V$, we have the (local) representation

$$X = \sum a^B(x) \partial_B,$$

while the representation of 1-form reads

$$\omega = \sum b_B(x) dx^B.$$

The differential dX of a vector field $X : A^{n+1} \to V$ is given by

$$dX = \sum da^B(x) \partial_B.$$

In particular, for the position vector field we get

$$dx = \sum dx^B \partial_B.$$

Let x be a point and $\{e_1, \ldots, e_{n+1}\}$ a basis satisfying $\mathrm{Det}(e_1, \ldots, e_{n+1}) = 1$. An ordered $(n + 1)$-tuple of differentiable mappings $e_i : U \to V$ on an open set $U \subset A^{n+1}$ is called a *unimodular affine frame*, if at each $x \in U$ we have $\mathrm{Det}(e_1, \ldots, e_{n+1}) = 1$. We denote such a frame by $\{x; e_1, \ldots, e_{n+1}\}$.

We express the relation between the canonical Gauß basis field and an arbitrary frame field by

$$e_A = \sum a_A^B(x) \partial_B, \tag{1.1}$$

where

$$a = (a_A^B) : A^{n+1} \to \mathbb{R}^{(n+1)^2}$$

is a matrix-function, and $b = a^{-1}$ its inverse, $b = (b_A^B)$,

$$\partial_A = \sum b_A^B(x) e_B. \tag{1.2}$$

We describe dx by 1-forms ω^A with respect to the frame

$$dx = \sum \omega^A e_A \tag{1.3}$$

and also

$$de_A = \sum w_A^B e_B. \tag{1.4}$$

The comparison $\sum dx^A b_A^B e_B = dx = \Sigma w^B e_B$ gives

$$w^A = \sum b_B^A(x) dx^B. \tag{1.5}$$

Analogously, considering the differential de_A of e_A and (1.4), we get

$$w_A^B = \sum b_C^B(x) da_A^C(x). \tag{1.6}$$

Obviously, the vector fields $\{e_1, \ldots, e_{n+1}\}$ and the forms $\{w^1, \ldots, w^{n+1}\}$ are dual, namely (1.1) and (1.5) imply $w^A(e_C) = \delta_C^A$. We want to point out the relation between the directional derivative $\bar{\nabla}$ and the differential of the mapping $e_A : A^{n+1} \to V$. As (1.4) describes the differential of the frame field with respect to the frame, the 1-forms w_A^B contain all information about the directional derivative of the frame. Therefore the forms w_A^B are called *connection one-forms* of the canonical connection $\bar{\nabla}$. As the frame is unimodular, differentiation gives

$$\begin{aligned} 0 &= d\,1 \\ &= d\,(\text{Det}\,(e_1, \ldots, e_{n+1})) \\ &= \sum_A \text{Det}\,(e_1, \ldots, de_A, \ldots, e_{n+1}) \\ &= \sum w_A^A. \end{aligned} \tag{1.7}$$

Finally, exterior differentiation of (1.3) and (1.4) gives

$$\begin{aligned} 0 &= dd\,x \\ &= dw^A e_A - w^A \wedge de_A \\ &= \sum \left(dw^B - \sum w^A \wedge w_A^B\right) e_B \end{aligned}$$

and

$$\begin{aligned} 0 &= dd\,e_A \\ &= \sum dw_A^B e_B - \sum w_A^B \wedge de_B \\ &= \sum \left(dw_A^C - \sum w_A^B \wedge w_B^C\right) e_C. \end{aligned}$$

As a frame is linearly independent along A^{n+1}, the coefficients in both equations must vanish. Thus the equations

$$dw^A = \sum w^B \wedge w_B^A, \tag{1.8}$$
$$dw_A^B = \sum w_A^C \wedge w_C^B \tag{1.9}$$

completely describe an arbitary frame $\{e_1, \ldots, e_{n+1}\}$ of A^{n+1} by integration; for this reason these equations are called *structure equations* of A^{n+1}.

1.2 Euclidean spaces

1.2.1 Basic notations for Euclidean spaces

Let V be a real vector space of dimension $(n + 1)$. We call a positive definite, symmetric bilinear form

$$\phi : V \times V \to \mathbb{R}$$

an *inner product*; (V, ϕ) is then called *Euclidean*. We simplify the notation and write brackets for the inner product

$$\phi(v, w) =: (v, w).$$

Recall the fact that one usually identifies V and V^* for Euclidean vector spaces (Theorem of Riesz). An affine space is called a *Euclidean space* if the associated vector space V is Euclidean; we denote such spaces by E.

1.2.2 Normed determinant forms

An inner product allows us to introduce a *norm* $\| \ \| : V \to \mathbb{R}$. From the inner product we get the notions of *length* and *angle*. Using the determinant of the Gram matrix we can also introduce, independently of the use of a determinant form, the *volume* of a parallelepiped $P(P_0, P_1, \ldots, P_{n+1})$ by

$$\mathrm{Vol}\, P(P_0, \ldots, P_{n+1}) := \left\{ \det((\overrightarrow{P_0 P_i}, \overrightarrow{P_0 P_j})) \right\}^{1/2}.$$

This definition is compatible with the definition of the volume via a determinant form (as in Section 1.1.2) if and only if the inner product and the determinant form Det are compatible, that means Det satisfies

$$\mathrm{Det} = \pm\, \mathrm{Det}^*.$$

In this case one calls the determinant form Det *normed* with respect to the inner product. Via the norm we can introduce the notion of *distance* on E; thus the inner product on V as a tangent space induces a *Riemannian structure* on E.

1.2.3 Different Euclidean structures

We recall the situation of two different inner products on V:

$$\Phi : V \times V \to \mathbb{R} \ \text{ and } \ \Psi : V \times V \to \mathbb{R}.$$

Then there exists a unique isomorphism $L : V \to V$ such that

$$\Psi(v, w) = \Phi(Lv, w) \ \text{ for all } \ v, w \in V,$$

and L is necessarily self-adjoint with respect to Φ because of the symmetry. Obviously we can construct from Φ all inner products on V using the set of all (with respect to Φ) self-adjoint isomorphisms of V. We have

$$\det(\Psi(v_i, v_j)) = \det L \cdot \det(\Phi(v_i, v_j));$$

therefore, Ψ and Φ define the same (nonoriented) volume if and only if L satisfies $\det L = \pm 1$. An equivalence class of inner products inducing the same nonoriented volume is called an *equiform structure*. We use this terminology with respect to the vector space V and the Euclidean space E.

1.2.4 Isometries and orthogonal transformations

Let (V, Φ) and (W, Ψ) be two Euclidean vector spaces of dimension n and m, respectively, where

$$\Phi : V \times V \to \mathbb{R}; \quad \Psi : W \times W \to \mathbb{R}.$$

A linear mapping $L : V \to W$ is an *isometry* if $\dim V = \dim W$ and L satisfies

$$\Psi(Lv, Lw) = \Phi(v, w)$$

for all $v, w \in V$. The identification $V = V^*$ implies $L^* = L^{-1}$ for an isometry. The isometries on V are called *orthogonal mappings*, and the group of these mappings is called the *orthogonal group*. The corresponding *group of motions* on E is compatible with the Euclidean structure of E.

1.3 Differential geometric structures of affine and Euclidean space

In this section we summarize the basic differential geometric structures for different spaces according to F. Klein's Erlangen program. This enables the beginner to recognize the geometry of hypersurfaces as induced from the structure of the ambient space. To emphasize this point of view we describe the structure of the ambient space in a differential geometric terminology.

1.3.1 Differential geometric equiaffine structures

We consider the affine space A^{n+1} as a differentiable manifold of dimension $(n + 1)$. At $p \in A^{n+1}$ we identify the tangent space $T_p A^{n+1}$ with the vector space V and the cotangent space $T_p^* A^{n+1}$ with V^*. The *differential geometric equiaffine structure* is given by three properties, namely:

At each point $p \in A^{n+1}$ we have:

(AS-1) the canonical scalar product

$$\langle \, , \, \rangle : V^* \times V \to \mathbb{R}.$$

(AS-2) A fixed volume form Det over V and its dual Det* over V^*; this allows us to define an orientation and a volume on A^{n+1}. We fix an orientation.

(AS-3) $\bar{\nabla}$ is a flat affine connection without torsion on A^{n+1}.

The structures are related by the following *affine compatibility conditions*:

(AC-1) The form Det and its dual Det* are related by

$$\mathrm{Det}^*(v^{*1}, \ldots, v^{*n+1}) \cdot \mathrm{Det}(v_1, \ldots, v_{n+1}) = \det(\langle v^{*i}, v_j \rangle),$$

where $v^{*i} \in V^*$ and $v_j \in V$.

(AC-2) Any differentiable 1-form $v^* : A^{n+1} \to V^*$ and any differentiable vector field $v : A^{n+1} \to V$ satisfy, for $z \in V$,

$$z\langle v^*, v \rangle = \bar{\nabla}_z\langle v^*, v \rangle = \langle \bar{\nabla}_z v^*, v \rangle + \langle v^*, \bar{\nabla}_z v \rangle.$$

The beginner should note that we use, as usual in the literature, the same notation for the differentiation in V and V^* (see 1.1.7.(ii)).

(AC-3) The volume forms Det and Det* are parallel with respect to the connection $\bar{\nabla}$:

$$\bar{\nabla}\mathrm{Det} = 0 \quad \text{and} \quad \bar{\nabla}\mathrm{Det}^* = 0,$$

that means they satisfy

$$(\bar{\nabla}\mathrm{Det})(v_1, \ldots, v_{n+1}; z) := (\bar{\nabla}_z\mathrm{Det})(v_1, \ldots, v_{n+1}) = 0$$

for $z \in V$ and vector fields $v_i : A^{n+1} \to V$, and the analogous formula holds for Det*.

The structures we considered are invariant with respect to the equiaffine transformation group. In particular, for any two vector fields $v : A^{n+1} \to V$ and $v^* : A^{n+1} \to V^*$ and an affine transformation $\alpha : A^{n+1} \to A^{n+1}$, from Section 1.1.6. we get in $p \in A^{n+1}$ and $q = \alpha(p)$:

$$\langle (L^*)^{-1}v^*, Lv \rangle_q = \langle v^*, v \rangle_p.$$

Moreover, for dual bases $v_1, \ldots, v_{n+1} \in V$ and $v^{*1}, \ldots, v^{*n+1} \in V^*$, we have

$$\mathrm{Det}(v_1, \ldots, v_{n+1}) = 1 \iff \mathrm{Det}^*(v^{*1}, \ldots, v^{*n+1}) = 1, \tag{1.10}$$

and this property is invariant under unimodular mappings in the following sense: if equation (1.10) is satisfied and $L : V \rightarrow V$ is linear then $Lv_1, \ldots, Lv_{n+1} \in V$ and $(L^*)^{-1}v^{*1}, \ldots, (L^*)^{-1}v^{*n+1} \in V^*$ are again dual bases, and

$$\mathrm{Det}(Lv_1, \ldots, Lv_{n+1}) = 1 \Leftrightarrow \det L = 1$$
$$\Leftrightarrow \det (L^*)^{-1} = 1$$
$$\Leftrightarrow \mathrm{Det}^*((L^*)^{-1}v^{*1}, \ldots, (L^*)^{-1}v^{*n+1}) = 1.$$

1.3.2 Differential geometric Euclidean structures

We consider an $(n + 1)$-dimensional Euclidean space as a differentiable manifold. We identify both, the tangent space T_pE and its dual T_p^*E at $p \in E$, with V. The *differential geometric structure* of E again is given by three structures:

(ES-1) the inner product $(,): V \times V \rightarrow \mathbb{R}$;

(ES-2) a normed determinant form Det;

(ES-3) a flat affine connection $\bar{\nabla}$ without torsion.

The three *Euclidean compatibility conditions* read:

(EC-1) $\{\det((b_i, b_j))\}^{1/2} = \mathrm{Det}(b_1, \ldots, b_{n+1})$
  for a positively oriented orthonormal basis $b_1, \ldots, b_{n+1} \in V$;

(EC-2) $z(v, w) = \bar{\nabla}_z(v, w) = (\bar{\nabla}_z v, w) + (v, \bar{\nabla}_z w)$
  for differentiable vector fields $v, w : E \rightarrow V$;

(EC-3) $\bar{\nabla}\mathrm{Det} = 0$.

The beginner should compare these conditions with the foregoing affine compatibility conditions (AC-1-3) and realize that the Euclidean conditions are specializations of the affine ones. In a differential geometric terminology they state the following facts:

The inner product induces a Riemann structure on E. The Riemannian volume form coincides with the volume form Det, see (EC-1). The flat connection $\bar{\nabla}$ is the Levi–Civita connection of the Riemannian metric, see (EC-2), and the volume form Det is parallel with respect to $\bar{\nabla}$, see (EC-3).

The given structures are invariant under the group of motions on E.

1.3.3 $\mathbb{R}^{n+1}$ as a standard space

As pointed out in Section 1.1, the space $\mathbb{R}^{n+1}$ carries different structures. We summarized these different structures and their compatibility conditions. Calculating examples and exercises we will frequently use $\mathbb{R}^{n+1}$ as the ambient space, recalling that every $(n + 1)$-dimensional affine space A^{n+1} is isomorphic (with respect to the algebraic structure) and diffeomorphic (with respect to the differentiable structure) to $\mathbb{R}^{n+1}$. Then we will point out which of the structures (equiaffine, Euclidean) we are going to use.

1.4 Klein's Erlangen program

In his inaugural address at Erlangen in 1872, Felix Klein developed a principle to classify geometries.

Starting from the elementary point of view, Euclidean geometry is the study of properties such as length, area, angle, volume, etc; these properties are invariants of the group of Euclidean motions. In order to classify the geometric objects in Euclidean space, one defines equivalence classes with respect to the group of motions; this expresses, in particular, that such geometric properties of an object are independent of its position in Euclidean space. The main tools for the classification are equivalence theorems which give necessary and sufficient conditions in terms of Euclidean invariants. Thus the aim of Euclidean geometry is the study of invariants with respect to the group of motions.

In Klein's view, by considering a larger group, one obtains a more general geometry. Thus Euclidean geometry is a special case of equiaffine geometry, the latter a special case of projective geometry (obviously one can extend such considerations to more general geometries like pseudo-Euclidean geometry etc).

We already listed some transformation groups of affine space in Section 1.1.5-2. Now we list some groups and the corresponding invariants, and give some examples of equivalent objects below.

Transformation group	Invariants
$\mathfrak{a}(n+1) = \{\alpha : A^{n+1} \to A^{n+1} \mid L_\alpha \text{ regular}\}$ regular affine group	parallelism partition ratio of 3 points ratio of volumes
$\mathfrak{s}(n+1) = \{\alpha \in \mathfrak{a}(n+1) \mid \det \alpha = 1\}$ equiaffine (unimodular) group	parallelism partition ratio of 3 points volume orientation
A^{n+1} is Euclidean; $\mathfrak{M}(n+1) = \{\alpha \in \mathfrak{a}(n+1) \mid L_\alpha \text{ orthogonal}\}$ group of motions	parallelism partition ratio of 3 points volume length angle

The equivalence of objects depends on the group: all parallelograms in A^2 (so in particular, squares and rectangles) are equivalent under the affine transformation group; all triangles are equivalent; the same is true for all ellipses (in particular circles).

If we restrict to the equiaffine group, the classification is finer: now parallelograms (triangles, respectively) might be nonequivalent; parallelograms (triangles, respectively) are equivalent if and only if they have the same area (two-dimensional volume); a corresponding statement holds true for ellipses.

Finally, with respect to the Euclidean group of motions, parallelograms (triangles, resp.) are equivalent if and only if the lengths of the edges and the angles coincide; circles and ellipses are nonequivalent. For the classification of those classes of objects we need classification theorems in this theory.

Thus, the smaller the group of transformations, the larger the set of invariants and the finer the classification of objects. As a general principle, we have that to each transformation group on a given space there is a particular geometry. It is the aim of the corresponding theory to study the invariants of the given group and to classify the geometric objects in terms of these invariants. Special classes of objects are classified by equivalence theorems, and it is obvious that one tries to minimize the system of invariants describing the objects of such a class completely.

Klein's program concerned, in particular, the elementary geometries; later all geometries satisfying Klein's principle were called *Klein geometries*.

It took nearly 20 years before Klein's program was published, first in Italian (1890), then in French (1891), and finally in German (1893) (see [139]). It essentially influenced the development of geometry. Its application to differential geometry started at the beginning of the twentieth century. Later, Elie Cartan and others extended Klein's program to include geometries other than Klein spaces (Riemannian geometry, affine connections, etc).

1.5 Motivation: A short sketch of the Euclidean hypersurface theory

In our treatise we confine the discussion to the equiaffine transformation group, as up to now most of the recent important results in the local and global classification of hypersurfaces concern the equiaffine theory. Thus our approach will be a compromise between the historic way of Blaschke and his school, and more recent structural developments.

It is our experience in teaching that a short review of the Euclidean hypersurface theory gives a good motivation for students to understand the equiaffine approach, to see analogies, and to understand differences.

As usual we describe the hypersurface by an immersion

$$x : M \to E^{n+1}$$

of an n-dimensional, connected, and oriented C^∞-manifold M into Euclidean space. To develop the Euclidean hypersurface theory, there are two independent fundamental steps:

(a) the introduction of the metric (first fundamental form) for the study of the *intrinsic geometry* of the hypersurface;

(b) the introduction of the unit normal field for the study of the *extrinsic geometry* of the hypersurface.

All intrinsic and extrinsic geometric structures are invariant with respect to the group of motions.

(a) Intrinsic geometry. The first fundamental form, which we denote by I, induces a Riemannian structure on the hypersurface; in particular this induces the Riemannian volume form $\omega(I)$, the Levi–Civita connection $\nabla(I)$, and the intrinsic curvatures. The volume form is defined by

$$\omega(I)(e_1,\ldots,e_n) := (\det(I(e_i, e_j)))^{1/2}$$

for a positively oriented frame $\{e_1,\ldots,e_n\}$ on M, while the Levi–Civita connection is the only torsion-free affine connection satisfying the Ricci identity.

The volume form $\omega(I)$ and the connection $\nabla(I)$ are compatible in the sense that $\omega(I)$ is parallel with respect to $\nabla(I)$:

$$\nabla(I)\omega(I) = 0.$$

(b) Normal field and extrinsic geometry. The unit normal field, which we denote by μ, is a nowhere vanishing section of the normal bundle, which is a line bundle. The unit normal field has three important properties:

(i) The *conormal property*: the equation

$$(\mu, dx) = 0$$

allows us to interpret μ at $x(p)$ as an element of the dual space V^*, where we denote $V = T_{x(p)}E^{n+1}$, $p \in M$. The cross product gives an explicit construction of μ from a local oriented frame $\{e_1,\ldots,e_n\}$ on M:

$$\mu = \frac{[dx(e_1),\ldots,dx(e_n)]}{\|[dx(e_1),\ldots,dx(e_n)]\|}.$$

(ii) *Transversal property*: μ is transversal to $x(M)$ and extends any basis of $dx(T_p(M))$ to a basis of V.

(iii) *Gauß map property*: we interpret μ as a mapping $\mu : M \to S^n(1) \subset V$, the so-called *Gauß map*. The relation $(\mu, d\mu) = 0$ implies that μ is transversal to $d\mu(T_pM)$ at $\mu(p)$.

The introduction of the normal field induces the following structure on M:

(b.1) The transversal field μ induces a volume form $\omega(\mu)$ on M by

$$\omega(\mu)(e_1,\ldots,e_n) := \mathrm{Det}(dx(e_1),\ldots,dx(e_n),\mu)$$

which coincides with the Riemannian volume form $\omega(I)$.

(b.2) The transversal field μ gives rise to the *structure equations of Gauß and Weingarten*:

$$\bar{\nabla}_w dx(z) = dx(\nabla_w z) + II(w,z)\mu,$$
$$\bar{\nabla}_w \mu = d\mu(w) = dx(-B(w)).$$

Here $B := B(\mu)$ denotes the Weingarten operator, $\nabla := \nabla(\mu)$ is an affine connection without torsion and $II := II(\mu)$ is the second fundamental form. It is obvious that these coefficients depend on μ (which explains our marks), but this fact generally is not accentuated. As it is well known, the connection $\nabla(\mu)$ coincides with the Levi–Civita connection $\nabla(I)$:

$$\nabla(I) = \nabla(\mu).$$

Thus, we can state the following:

Compatibility Theorem. *Let* $x : M \to E^{n+1}$ *be a hypersurface immersion. The three structures of the ambient space are*

– *the inner product,*
– *the volume form* Det, *and*
– *the connection* $\bar{\nabla}$,

which are compatible as in Section 1.3.2, via x and μ induce three corresponding structures on M:

– *the inner product I,*
– *the volume form $\omega(\mu)$, and*
– *the connection $\nabla(\mu)$,*

which are compatible by analogous relations:

(i) $\omega(I) = \omega(\mu)$,
(ii) $\nabla(I) = \nabla(\mu)$, *and*
(iii) $\nabla(\mu)\omega(\mu) = 0$.

Emphasizing such structural view points in Euclidean geometry will help us find adequate structures in affine geometry, which is the more recent approach to affine hypersurface theory.

It is easy to prove that the compatibility relations in the theorem characterize the Euclidean normal, namely:

Characterization of the Euclidean normal. *Let* $x : M \to E^{n+1}$ *be a hypersurface with nonzero Gauß–Kronecker curvature on M. Then a transversal field ξ coincides with the*

Euclidean normal μ if and only if the three induced structures
- the volume form $\omega(\xi)$,
- the affine connection $\nabla(\xi)$,
- the first fundamental form I,

satisfy three compatibility conditions, namely:
(i) $\omega(I) = \omega(\xi)$,
(ii) $\nabla(I) = \nabla(\xi)$,
(iii) $\nabla(\xi)\,\omega(\xi) = 0$ *(parallelity).*

The procedure for developing a Euclidean hypersurface theory then is the following:

The Weingarten and the Gauß equations together give a linear system of first-order p.d.e.'s; integrating once gives a local frame $dx(e_1), \ldots, dx(e_n), \mu$, and a second integration gives the hypersurface x itself from dx. From the integration theory of such systems one knows that for given initial values the frame field (as the solution of the first integration) depends uniquely on the coefficients B, ∇ and II. Furthermore, it can be easily seen that the coefficients are invariant under motions of the hypersurface in E^{n+1}. Both facts imply that the coefficients must contain all geometric information about x. This is the reason to study their geometric properties and their relations. Some of the relations are already contained in the compatibility theorem. The most important additional relations are summarized in a lemma.

Lemma 1.1. *The coefficients of the structure equations satisfy the following relations:*
(i) $I(B(z), w) = II(z, w)$;
(ii) *integrability conditions of Codazzi and Gauß.*

Corollary 1.2. *The first and second fundamental forms, I and II, determine all coefficients of the structure equations.*

In Lemma 1.1, relation (i) is the main one for the extrinsic curvature theory: from (i) the Weingarten operator is self-adjoint with respect to I, and thus all its eigenvalues are real. Geometrically they describe the curvature of the normal sections in eigendirections. As roots of a polynomial they are not necessarily differentiable, but the coefficients of the characteristic polynomial are differentiable. These (normalized) coefficients are appropriate curvature functions of x.

The Gauß integrability conditions and, as a consequence, the theorema egregium, are the basic relations to introduce the notion of "intrinsic" curvatures (sectional curvature, scalar curvature). The above corollary and the structure equations together with the integrability conditions are the tools for the fundamental existence and uniqueness theorem of Bonnet:

Theorem 1.3 (Fundamental theorem). *Let M be a connected, simply connected and oriented differentiable manifold of dimension n, and E^{n+1} a Euclidean space of dimension $n + 1$.*

(i) *Uniqueness: if for two immersion $x, x' : M \to E^{n+1}$ the first and the second fundamental forms coincide: $I = I'$, $II = II'$ then $x(M)$ and $x'(M)$ are equivalent modulo a Euclidean motion.*

(ii) *Existence: if two symmetric bilinear forms I, II are given on M, if I is positive definite and the integrability conditions of Gauß and Codazzi are satisfied then there exists a hypersurface immersion $x : M \to E^{n+1}$ with I as the first and II as the second fundamental form.*

Thus, the first and the second fundamental forms define a complete system of geometric quantities in Euclidean hypersurface theory, which are invariant with respect to the group of motions and describe the geometry of the hypersurface. Both forms are naturally induced on M from the immersion x and the structure of the ambient space: the form I via the restriction of the inner product structure, the form II via the unit normal field. All other invariant quantities can be derived from I and II.

1.6 Hypersurfaces in equiaffine space

1.6.1 Definition and notation

(i) A *hypersurface-immersion* $x : M \to A^{n+1}$ into an equiaffine space is a C^∞-differentiable map of maximal rank, where M again is a connected, oriented C^∞-manifold and $\dim A^{n+1} = \dim M + 1 = n + 1 \geq 3$. We describe the hypersurface by its position vector, denoted by x, with respect to a fixed origin $O \in A^{n+1}$. The differential mapping dx_p is injective; this allows to identify both tangent spaces $T_p M$ and $dx(T_p M)$ by this mapping.

(ii) A differentiable vector field $Z : M \to V$ is called a *transversal* (or sometimes *normal) field* along x if, at any point $p \in M$ and for any frame $\{e_1, \ldots, e_n\}$ on M, the $(n + 1)$-tuple of vectors $\{dx_p(e_1), \ldots, dx_p(e_n), Z_p\}$ is a basis of $T_{x(p)} A^{n+1}$.

1.6.2 Bundles

There are two fundamental vector-bundles over M:
(i) the tangent bundle TM;
(ii) the conormal bundle CM.

From the affine cross-product construction, applied to the canonical scalar product and to an arbitrary frame field $\{e_1, \ldots, e_n\}$, we get a *conormal* of x at $p \in M$:

$$U(p) := [dx(e_1), \ldots, dx(e_n)]. \tag{1.11}$$

U is a nowhere vanishing section $U : M \to CM$ and fixes the tangent plane $dx(T_pM)$ at $x(p)$:

$$\langle U, dx \rangle = 0, \tag{1.12}$$

which means $\langle U_p, dx_p(v) \rangle = 0$ for any $v \in T_pM$. We call such a nowhere vanishing section a *conormal field* of the hypersurface x; each such section defines a basis of the *conormal line bundle*. If $\beta \in C^\infty(M)$ is nowhere vanishing then the field βU is again a conormal field.

1.7 Structural motivation for further investigations

To study the geometry of a hypersurface with respect to the equiaffine transformation group it is quite natural to try to develop an affine differential geometry of hypersurfaces analogously to the Euclidean one. We recall the differential geometric structure of the equiaffine space in Section 1.3.1 and ask:

(i) How does a hypersurface inherit the structure of the ambient space?
(ii) Find a (if possible "minimal") system of geometric quantities which completely describes the geometry of the hypersurface; these quantities should be invariant with respect to the equiaffine group.

To search the appropriate affine quantities we recall the Euclidean situation in Section 1.5.

There we started with two fundamental steps (a) and (b), namely:

(a) The introduction of the Euclidean first fundamental form. In the equiaffine space there is no inner product on V, so there is no affine analogue to this Euclidean step.

(b) The introduction of an appropriate normalization. In the affine case there are many possible choices of conormal fields and transversal fields; thus one tries to find normalizations which are distinguished with respect to the affine structure.

The Euclidean situation suggests that we study the induced quantities under structural considerations:

We consider the structures induced on M from the structures of the ambient space by an arbitrary normalization, and their compatibility; we distinguish those normalizations such that the compatibility conditions formally look like the compatibility conditions of the structures in the ambient space.

With respect to the affine transformation group this principle leads to the following concepts:

(1) the notion of a *nondegenerate (regular) hypersurface* (see Sections 1.9 and 1.11);
(2) the notion of a *relative normalization* (see Sections 1.8.5 and 1.10).

1.8 Transversal fields and induced structures

Let $x : M \to A^{n+1}$ be a given hypersurface in equiaffine space A^{n+1} and $Z : M \to V$ be an arbitrary vector field transversal to x; with $\{e_1, \ldots, e_n\}$ we denote an arbitrary frame field. The field Z induces the following structures in Sections 1.8.1–1.8.3:

1.8.1 Volume form

Z induces a nontrivial volume form $\omega(Z)$ via

$$\omega(Z)(e_1, \ldots, e_n) := \mathrm{Det}(dx(e_1), \ldots, dx(e_n), Z).$$

1.8.2 Weingarten structure equation

Via the so-called *Weingarten structure equations*, Z induces an operator $B = B(Z)$ and a 1-form $\theta = \theta(Z)$, namely: the first derivative dZ admits a unique decomposition into a tangential and a transversal part:

$$\bar{\nabla}_v Z = -dx(B(v)) + \theta(v)Z,$$

where the coefficients are implicitly well defined and depend on Z; the sign for B is a convention. As in the Euclidean hypersurface theory it easily follows that B is a linear operator on each tangent space $B_p : T_pM \to T_pM$, and θ is a 1-form.

1.8.3 Gauß structure equation

Z induces a bilinear form $G = G(Z)$ and a connection $\nabla = \nabla(Z)$ on M, via the so-called *Gauß structure equations*: namely, the second derivative $\bar{\nabla}_v dx(w)$ has the following unique decomposition:

$$\bar{\nabla}_v dx(w) = dx(\nabla_v w) + G(v, w)Z.$$

The basic properties of the coefficients of G and ∇ coincide with properties of the coefficients of the Euclidean Gauß structure equations:

(i) G is a symmetric (0,2) tensor field;
(ii) ∇ is an affine connection without torsion.

Proof. As the decomposition is unique, G is uniquely determined. This immediately gives the differentiability of G, and also the tensorial property of G in the first variable. The Lie-bracket satisfies $[dx(v), dx(w)] = dx([v, w])$, and the connection $\bar{\nabla}$ is torsion-free. This implies

$$0 = \bar{\nabla}_v dx(w) - \bar{\nabla}_w dx(v) - [dx(v), dx(w)]$$
$$= dx(\nabla_v w - \nabla_w v) - dx([v, w]) + (G(v, w) - G(w, v))Z.$$

The linear independence of the tangential and the transversal parts gives first the symmetry of G and secondly $\nabla_v w - \nabla_w v - [v, w] = 0$ because of the injectivity of dx. To verify that ∇ is a connection we point out that all $\mathbb{R}$-linear properties of ∇ follow from corresponding properties of $\bar{\nabla}, G$, and dx, while the Leibniz rule for $\tau \in C^\infty(M)$ follows from

$$\nabla_v(\tau w) = (dx)^{-1}(\bar{\nabla}_v dx(\tau w) - G(v, \tau w)Z)$$
$$= (dx)^{-1}(v(\tau)dx(w) + \tau \bar{\nabla}_v dx(w) - \tau G(v, w)Z)$$
$$= v(\tau)w + \tau \nabla_v w. \qquad \square$$

1.8.4 Compatibility of the induced volume form and the connection

Following our structural principle motivated from the Euclidean hypersurface theory, we state the compatibility of the induced structures $\nabla = \nabla(Z)$ and $\omega = \omega(Z)$. The proofs are elementary and can be used as exercises. The reader also can find them in the Lecture Notes [341].

Proposition 1.4. *Let Z be a transversal field of the hypersurface x. Then the invariants $\omega = \omega\,(\mathrm{Det}, Z)$, $\nabla = \nabla(Z)$, $\theta = \theta\,(Z)$ satisfy*

$$\nabla \omega = \theta \omega.$$

For the proof see Proposition 3.3.3.1 in [341].

1.8.5 Relative normal fields

The compatibility condition between $\bar{\nabla}$ and Det in the ambient space reads (see Section 1.3.1)

$$\bar{\nabla}\,\mathrm{Det} = 0.$$

Thus, the induced structures ∇ and ω formally satisfy analogous compatibility conditions if and only if

$$\nabla \omega = 0,$$

which is equivalent to $\theta = \theta\,(Z) = 0$ in the Weingarten equation. Following our structural principle we consider such normalizations as distinguished with respect to the affine transformation group.

Definition 1.5. A transversal field $Z : M \to V$ is called a *relative normal field* of x if the Weingarten equation reads

$$dZ(v) = -dx(B(v))$$

for any tangent v on M. Nowadays the minus sign is a standard convention in the literature.

Lemma 1.6. *Let $Z : M \to V$ be a relative normal field of x. Then there is a unique conormal field $U : M \to V^*$ satisfying $\langle U, Z \rangle = 1$.*

Proof. The system $\langle U, dx(e_i) \rangle = 0$, $\langle U, Z \rangle = 1$ is a linear system of maximal rank, where $\{e_1, \ldots, e_n\}$ is an arbitrary frame at $p \in M$. □

1.9 Conormal fields and induced structures

Now we investigate which structures are induced on M from an arbitrary conormal field $U : M \to V^*$ of the hypersurface x.

1.9.1 Volume form

U induces a volume form

$$\omega^*(U)(e_1, \ldots, e_n) := \mathrm{Det}^*(dU(e_1), \ldots, dU(e_n), U).$$

A conormal field U is called *regular* if $\omega^*(U)$ is a nontrivial volume form.

1.9.2 Bilinear form

U induces, via the scalar product, a bilinear form on M

$$G(U)(v, w) := -\langle dU(v), dx(w) \rangle;$$

the sign of G is a convention.

Lemma 1.7 (Nondegeneracy). *Let $U, U^\# : M \to V^*$ be two conormal fields. Then*

$$\mathrm{Det}^*(dU(e_1), \ldots, dU(e_n), U) \neq 0 \Leftrightarrow \mathrm{Det}^*(dU^\#(e_1), \ldots, dU^\#(e_n), U^\#) \neq 0.$$

Proof. There exists a nowhere vanishing function $q \in C^\infty(M)$ such that $U^\# = q\,U$. The proof is now obvious. □

Definition 1.8. From the foregoing result the condition

$$\text{Det}^*(dU(e_1), \ldots, dU(e_n), U) \neq 0$$

is satisfied for all conormal fields if it is true for one; that means that this condition only depends on the hypersurface itself. A hypersurface is called *nondegenerate* (or *regular*) if there exists a regular conormal field.

1.10 Relative normalizations

Definition and Lemma 1.9. (i)   A pair (U, Z) is called a normalization of the immersion $x : M \to A^{n+1}$ if U is a conormal field and Z a normal field, and $\langle U, Z \rangle = 1$; the normalization is called a relative normalization if Z is a relative normal field.

(ii) Let (U, Z) be a relative normalization of x. Then $G(U) = G(Z) =: G$ is a symmetric $(0,2)$ tensor field. $G(Z)$ is called the relative metric of the normalization (U, Z).

(iii) $\omega^*(U)(e_1, \ldots, e_n) \cdot \omega(Z)(e_1, \ldots, e_n) = (-1)^n \det(G(e_i, e_j))$, where the determinant on the right-hand side is the Gram determinant (see [105, p. 192]).

For the proof see Proposition 3.4.8.3 in [341]. Compare the preceding relation (iii) with (AC-1) in Section 1.3.1 above.

1.11 Nondegenerate hypersurfaces

In Sections 1.1–1.10 we considered structures on M induced from the immersion x and a normalization (U, Z). It is of particular interest to investigate the case when $G(Z)$ and $\omega^*(U)$ are nondegenerate.

Theorem 1.10. *Let (U, Z) be a relative normalization of $x : M \to A^{n+1}$. Then the following statements are equivalent:*
(i) *x is nondegenerate;*
(ii) *G is nondegenerate;*
(iii) *$\omega^*(U)$ is nondegenerate;*
(iv) *the conormal mapping $U : M \to V^*$ is a hypersurface immersion and its position vector U is transversal to $U(M)$.*

Proof. Exercise.

As a consequence, the nondegeneracy of x does not depend on the normalization. Moreover, for a relative normalization (U, Z) the compatibility condition in Definition and Lemma 1.9. (iii) corresponds to (AC-1) in Section 1.3, and thus it looks formally like the compatibility condition of the ambient space. This way gives a structural view point for the definition of a relative normalization in Section 1.8.5.

Another approach is described in [340].

For a given nondegenerate hypersurface, consider the class $\mathcal{T}$ of all transversal fields. We have two possibilities of studying equivalences in the class $\mathcal{T}$, namely:

- Z and $Z^\sharp$ are *tangentially equivalent* if there exists a conormal U such that $\langle U, Z \rangle = \langle U, Z^\sharp \rangle$; if there exists one such conormal field U then obviously the latter relation is satisfied for any conormal field;
- Z and $Z^\sharp$ are *transversally equivalent* if there exists $q \in C^\infty(M)$, s.t. $Z^\sharp = q \cdot Z$.

The tangential equivalence geometrically is the interesting one, as in each equivalence class there is exactly one relative normal. To any equivalence class there is exactly one conormal U satisfying $\langle U, Z \rangle = 1$ for all transversal Z in this class. This bijective correspondence simplifies the relative theory.

Definition 1.11. Let x be a nondegenerate hypersurface and (U, Z) a relative normalization of x. Then (x, U, Z) is called a *relative hypersurface*.

Exercise 1.1. Prove the following:
On any nondegenerate hypersurface there exists infinitely many different relative normalizations.

1.12 Gauß structure equations for conormal fields

On a nondegenerate hypersurface x, any conormal field U can be considered as an immersion $U : M \to V^*$ (see Theorem 1.10). For this immersion with transversal field U the Gauß structure equations have the following form.

Proposition and Definition 1.12. On a nondegenerate hypersurface, for any conormal field U there is a unique decomposition

$$\bar{\nabla}_v dU(w) = dU(\nabla^*_v w) - \hat{B}(U)(v, w)U.$$

Here $\hat{B} := \hat{B}(U)$ is a symmetric (0.2)-tensor field, called the relative Weingarten form, and $\nabla^* := \nabla^*(U)$ is an affine connection without torsion, called the conormal connection.

Proof. The proof follows the lines of Section 1.8.3. Verify that $\hat{B}$ and ∇^* are well defined by U, and proceed now as before. □

1.12.1 Compatibility of connections and volume forms in the conormal bundle

As ω^* is trivial on a degenerate hypersurface, we assume x to be nondegenerate. We immediately can verify that

$$\nabla^* \omega^* = 0$$

for a given conormal field U, where $\nabla^* = \nabla^*(U)$ and $\omega^* = \omega^*(U)$.

1.12.2 Compatibility of bilinear forms and connections; conjugate connections

Now we start to find an analogue for the hypersurface to the compatibility condition (AC-2) (see Section 1.3.1) in the ambient space. Obviously the nondegeneracy of the hypersurface is needed again.

Proposition 1.13. *Let x be nondegenerate and (U, Z) a relative normalization. Then the three invariants G, $\nabla := \nabla(Z)$ and $\nabla^* := \nabla^*(U)$ satisfy the relation*

$$wG(v_1, v_2) = G(\nabla^*_w v_1, v_2) + G(v_1, \nabla_w v_2).$$

Proof. See Proposition 3.4.8.3 in [341]. ∎

Notes. (i) To compare the above relation with the compatibility condition (AC-2) in the ambient space (see Section 1.3.1), the beginner should realize again that we use the same notation $\bar{\nabla}$ for the differentiation in both V and V^*. With this in mind, we get a formal analogy between (AC-2) and the relation in Proposition 1.13 as $\langle dU, Z \rangle = 0$ for a relative normalization.

(ii) As usual we write the local components of ∇ and ∇^* as *Christoffel symbols* Γ^k_{ij} and Γ^{*k}_{ij}, respectively.

Definition 1.14. Two affine connections ∇, ∇^* on a differentiable manifold M are called *conjugate* relative to the symmetric nondegenerate $(0, 2)$-field G if

$$wG(v_1, v_2) = G(\nabla_w v_1, v_2) + G(v_1, \nabla^*_w v_2)$$

for all tangent fields v_1, v_2, w. In a modified terminology we call (∇, G, ∇^*) a *conjugate triple*.

The notion of conjugate connections, which already appeared in Norden's presentation (see the appendix in [311]), can be used for a different structural approach to relative hypersurface theory; this approach is presented in more detail in Chapters 3–4 of [341]. The conference report [261] gives an overall view of the topics that center around the concept of conjugate connections.

1.12.3 Cubic form

In relative hypersurface theory the *difference tensor* between both connections

$$K := \tfrac{1}{2}(\nabla - \nabla^*)$$

plays an important role. The *associated cubic form* is defined by

$$A(u, v, w) := G(u, K(v, w)).$$

We list properties of K and A and refer to Section 4.4 in [341].

Proposition 1.15. *Let x be nondegenerate. Then K and A satisfy:*
(i) *K is a symmetric (1,2) tensor field, $K(u, v) = K(v, u)$;*
(ii) *the Levi–Civita connection $\tilde{\nabla} := \nabla(G)$ of G satisfies*

$$\tilde{\nabla} := \nabla(G) = \tfrac{1}{2}(\nabla + \nabla^*);$$

(iii) *the cubic form A is totally symmetric.*

In the early unimodular theory of Blaschke and his school, the invariants G and A were the most important ones, while ∇ and ∇^* became of interest by the work of Norden and the monograph [311].

1.12.4 Tchebychev form

Consider the linear map $w \mapsto K(v, w)$. Its trace T defined by

$$nT(v) := \operatorname{tr} \{w \mapsto K(v, w)\}$$

is called the *Tchebychev form*, while the associated vector field $T^\sharp$ implicitly defined by

$$G(T^\sharp, v) := T(v);$$

it is called the *Tchebychev vector field*.

Lemma 1.16. *For any affine frame $\{e_1, \ldots, e_n\}$ we have*

$$2nT(e_i) = e_i \left(\ln \frac{|\omega(e_1, \ldots, e_n)|}{|\omega^*(e_1, \ldots, e_n)|} \right).$$

Proof. Exercise.

1.12.5 Local notation

As already mentioned in the introduction, there are different possibilities for calculations in differential geometry. A notation with components for tensor fields and connections is called a *local notation*.

Consider a differentiable manifold M of dimension $n \geq 2$. In the appendix we re-call the notions of a frame field $\{e_1, \ldots, e_n\}$, a local coordinate system $\{x^1, \ldots, x^n\}$, and an associated *Gauß basis* (or *coordinate frame*) $\{\partial_1, \ldots, \partial_n\}$, as well as the notion of components of tensor fields with respect to a local basis. In the following we summa-rize the local notation of the basic invariants and formulas for relative hypersurfaces.

Let $f : M \to \mathbb{R}$ be a (vector valued) differentiable function and ∇ an affine con-nection. With respect to a local (coordinate) frame we write partial derivatives of f by $f_i := \partial_i f$, and $f_{ij} = \partial_j \partial_i f$, etc. For a covariant differentiation, we write $f_{,i} := \nabla_i f = \nabla_{\partial_i} f$ and $f_{,ij} := \nabla_j f_{,i}$, etc.

As usual, with respect to a local frame field $\{e_1, \ldots, e_n\}$, the components of an affine connection are written as *Christoffel symbols*, i.e.

$$\nabla_{e_i} e_j =: \sum \Gamma^k_{ji} e_k,$$

and analogously for a coordinate frame.

With the foregoing local notation, and by denoting $x_i = e_i(x)$, we can summarize the structure equations of a hypersurface with relative normalization (U, Z) as follows:

$$\partial_j x_i = \sum \Gamma^k_{ij} x_k \quad + G_{ij} \cdot Z, \qquad \text{Gauß equation for } x$$

$$\nabla_j x_i = \qquad\qquad G_{ij} \cdot Z.$$

$$\partial_i Z = -\sum B^k_i x_k, \qquad\qquad\qquad \text{Weingarten equation}$$

$$\partial_j U_i = \sum \Gamma^{*k}_{ij} U_k \quad -B_{ij} \cdot U, \qquad \text{Gauß equation for } U,$$

$$\nabla^*_j x_i = \qquad\qquad -B_{ij} \cdot U.$$

The components of the difference tensor are denoted by $\sum K^k_{ij} e_k := K(e_i, e_j)$, the components of the cubic form by $A_{ijk} := A(e_i, e_j, e_k)$, and raising and lowering of indices with the relative metric identifies the components $K^k_{ij} = A^k_{ij}$.

Concerning the Tchebychev vector field, for an affine frame $\{e_1, \ldots, e_n\}$ we have

$$2nT_i = \sum (\Gamma^k_{ik} - \Gamma^{*k}_{ik}) = e_i \left(\ln \left| \frac{w(e_1, \ldots, e_n)}{w^*(e_1, \ldots, e_n)} \right| \right).$$

1.12.6 Relative Gauß maps

In the Euclidean theory of hypersurfaces the notion of "curvature" and the geometry of the Gauß map $\mu : M \to S^n(1) \subset V$ are closely related. In this subsection we introduce analogous notions in the relative theory of hypersurfaces.

1. Conormal indicatrix. The regularity statement in Definition 1.8 allows us to con-sider $U : M \to V^*$ as a hypersurface with transversal field U, which defines a relative normal $U' := -U$. This immediately gives $Y' := -Y$ as conormal of the hypersurface

$U : M \to V^*$, and thus (Y', U') is a relative normalization of U; the prime-notation is a standard convention in the literature.

The Gauß structure equation of this hypersurface $U : M \to V^*$ with relative normal $U' = -U$ is given by

$$\bar{\nabla}_v dU(w) = dU(\nabla_v^* w) + \hat{B}(v, w)U' = dU(\nabla_v^* w) + G(U')(v, w)U'.$$

This hypersurface immersion U is itself regular if and only if $G(U') = \hat{B}$.

The mapping U is called the *conormal indicatrix* or the *Gauß conormal map* of x. In case $\hat{B}$ is nondegenerate, $\hat{B}$ is called the *relative metric of the conormal map*. $\hat{B}$ is (positive) definite if and only if $U : M \to V^*$ is a locally strongly convex immersion.

2. Relative indicatrix. The relative normal Y defines the *relative normal indicatrix* or *Gauß normal map*. This map is not necessarily an immersion.

We will study the case where Y is an immersion; then rank $(Y, dY) = n + 1$. In this case let us (as in the Euclidean geometry) consider $x(M)$ and $Y(M)$ to be hypersurfaces in the affine space A (identifying A and V via π_0); both hypersurfaces have parallel tangent hyperplanes at corresponding points $x(p)$ and $Y(p)$. This correspondence is called *Peterson correspondence* (cf. [311]). Moreover, for an immersion Y the pair (U, Y) is a *dual pair*.

Historically the special case that $Y : M \to V$ is an immersion, to be considered as spherical indicatrix of x, was the beginning of relative geometry. E. Müller considered such a pair $\{x, Y\}$ generalizing thus the geometric situation of a Euclidean surface together with its Gauß mapping of maximal rank $\mu : M_2 \to S^2(1) \subset E^3$ as spherical indicatrix.

In the relative setting, $Y : M \to A$ was called *gauge surface (Eichfläche)* of the given surface x.

Proposition 1.17. *Let (U, Y) be a relative normalization of the regular hypersurface x.*
(i) *The following conditions are equivalent:*
    (a) *the conormal indicatrix $U : M \to V^*$ is a regular immersion;*
    (b) *the normal indicatrix $Y : M \to V$ is a regular immersion;*
    (c) *rank $(dY) = n$, that means Y is an immersion;*
    (d) *the relative Weingarten operator B has maximal rank.*
(ii) *Under the assumption rank $(dY) = n$, the relative metrics $G(U') = \hat{B}$ and $G' = G(Y')$ of the two indicatrices coincide: $\hat{B} = G'$.*
(iii) *U' is locally strongly convex if and only if Y' is so.*

Proof. Exercise.

1.13 Affine invariance of the induced structures

To describe geometric properties of a hypersurface in Euclidean space we use geometric quantities which are invariant with respect to the group of Euclidean motions. Correspondingly we have to prove that the affine-geometric quantities defined so far in this chapter – i.e. the quantities of x depending on transversal or conormal fields – are invariant with respect to the group of regular affine transformations in A^{n+1}. For that purpose we consider a hypersurface immersion $x : M \to A^{n+1}$ with a normalization (U, Z).

As above, the normalization induces on M two symmetric bilinear forms, namely $G = G(U) = G(Z)$ and $\hat{B} = \hat{B}(U)$, additionally the Weingarten operator $B = B(Z)$, the 1-form $\theta = \theta(Z)$, and the two connections ∇, ∇^*.

Let $\alpha : A^{n+1} \to A^{n+1}$ be a regular affine mapping and $L_\alpha : V \to V$ the associated linear mapping in Section 1.1.5. Then $\alpha x : M \to A^{n+1}$ is again a hypersurface immersion with the pair $(L_\alpha^*)^{-1}U : M \to V^*, L_\alpha Z : M \to V$ as normalization. We verify this using Sections 1.1.5 and 1.1.6, the relation $dL_\alpha = L_\alpha$, and the chain rule which implies $d(\alpha x) = L_\alpha dx$:

$$\langle (L_\alpha^*)^{-1}U, L_\alpha Z \rangle = \langle U, Z \rangle = 1;$$
$$\langle (L_\alpha^*)^{-1}U, d(\alpha x) \rangle = \langle (L_\alpha^*)^{-1}U, L_\alpha dx \rangle = \langle U, dx \rangle = 0.$$

In particular if Y is a relative normal, the same is true for $L_\alpha Y$, namely

$$\langle d(L_\alpha^*)^{-1}U, (L_\alpha Y) \rangle = \langle (L_\alpha^*)^{-1}dU, L_\alpha Y \rangle = \langle dU, Y \rangle = 0.$$

We denote the geometric quantities of the hypersurface αx by $G_\alpha, B_\alpha, \dots$, etc.

The affine invariance of the induced structures means that the hypersurface x with the normalization (U, Z) and the hypersurface αx with the normalization $((L_\alpha^*)^{-1}U, L_\alpha Z)$ define the same structure on M, which means $G_\alpha = G$, etc. For the proof of the affine invariance we do not need the regularity of x.

Proposition 1.18. *Let x be a hypersurface and $\alpha : A^{n+1} \to A^{n+1}$ be an affine transformation. Then*
(i) *x is nondegenerate if and only if αx is nondegenerate;*
(ii) *the structures defined via the Weingarten equation are affinely invariant and*
$$B_\alpha = B \text{ and } \theta_\alpha = \theta;$$
(iii) *if x is nondegenerate the structures defined via the Gauß equation for x are affinely invariant and*
$$\nabla_\alpha = \nabla \text{ and } G_\alpha = G;$$
   (M, G) is a semi-Riemannian manifold;
(iv) *if x is nondegenerate, the structures defined via the Gauß equation for the conormal field U are well defined and*
$$\nabla_\alpha^* = \nabla^* \text{ and } \hat{B}_\alpha = \hat{B}.$$

Proof. We give an exemplary proof for (iii). The rest is left as an exercise.

For tangent vector field v, w we have

$$
\begin{aligned}
-G_\alpha(v, w) : &= \langle d(L_\alpha^*)^{-1} U(v), d(\alpha x)(w) \rangle \\
&= \langle (L_\alpha^*)^{-1}(dU(v)), (L_\alpha(dx))w \rangle \\
&= \langle dU(v), dx(w) \rangle \\
&= -G(v, w).
\end{aligned}
$$

We show that $\nabla_\alpha = \nabla$:

$$
\begin{aligned}
(\nabla_\alpha)_v w &= (d(\alpha x))^{-1} (\bar\nabla_v d(\alpha x)(w) - G_\alpha(v, w) L_\alpha Z) \\
&= (L_\alpha dx)^{-1} L_\alpha(\bar\nabla_v dx(w) - G(v, w)Z) \\
&= (dx)^{-1}(\bar\nabla_v dx(w) - G(v, w)Z) \\
&= \nabla_v w.
\end{aligned}
$$

Now we are really able to call the quantities above "geometric quantities of affine geometry". Analogously one can prove the affine invariance of other quantities, or the invariance with respect to the equiaffine subgroup. ☐

1.14 A summary of relative hypersurface theory

It is well known that nowadays there are three different types of calculus used in differential geometry, namely
- the local tensor calculus;
- Cartan's moving frames;
- Koszul's invariant calculus.

Graduate students should be familiar with all three types. For this reason we present all three in this book, namely the first two types in the following chapter where we treat the basics of Blaschke's unimodular affine hypersurface theory. Here, in this section, we summarize the basics of relative hypersurface theory, using Koszul's invariant calculus.

1.14.1 Structure equations in terms of $\tilde\nabla$

We already presented the structure equations of Gauß and Weingarten in Sections 1.8 and 1.12, using the induced and the conormal connection, respectively. For many applications it is necessary to consider instead the structure equations written in terms of the Levi–Civita connection $\tilde\nabla$ of the relative metric G. From Section 1.12.4 we recall the relations between the three connections ∇, ∇^*, $\tilde\nabla$, the difference tensor K, and the cubic form A. Then we can rewrite the structure equations in terms of $\tilde{\nabla}$ as follows:

$$\tilde{\nabla}_v dx(w) = dx(K(v, w)) + G(v, w) \cdot Y, \tag{1.13}$$

$$dY(v) = -dx(B(v)), \tag{1.14}$$

$$\tilde{\nabla}_v dU(w) = dU(K(v, w)) - \hat{B}(v, w) \cdot U. \tag{1.15}$$

1.14.2 Integrability conditions in terms of $\tilde{\nabla}$

We add a summary of the integrability conditions for relative hypersurfaces, using Koszul's calculus; for proofs we refer to [341]. In the special case of a Blaschke normalization we state and prove the corresponding integrability conditions in Section 1.5, using moving frames and a local calculus.

(i) The $(0,4)$ Riemannian curvature tensor $R := R(G)$ satisfies

$$R(w, v, z, u) = G(K(w, u), K(v, z)) - G(K(v, u), K(w, z))$$
$$+ \frac{1}{2} \left(\hat{B}(v, u) \, G(w, z) - \hat{B}(w, u) \, G(v, z) \right. \tag{1.16}$$
$$\left. + \hat{B}(w, z) \, G(v, u) - \hat{B}(v, z) \, G(w, u) \right).$$

(ii) The covariant derivation of the Weingarten form reads

$$\tilde{\nabla}_v \hat{B}(w, z) - \tilde{\nabla}_w \hat{B}(v, z) = G(K(w, z), B(v)) - G(K(v, z), B(w)). \tag{1.17}$$

(iii) The covariant derivation of the cubic form reads

$$(\tilde{\nabla}_w A)(v, u, z) - (\tilde{\nabla}_v A)(w, u, z) =$$
$$\frac{1}{2} \left(\hat{B}(w, u) G(v, z) - \hat{B}(v, u) G(w, z) + \hat{B}(w, z) \, G(v, u) - \hat{B}(v, z) \, G(w, u) \right). \tag{1.18}$$

Contraction of the Riemannian curvature tensor with the metric gives the Ricci tensor Ric and the normalized scalar curvature κ. We use an orthonormal frame $\{e_1, \ldots, e_n\}$:

$$Ric(e_i, e_j) = \sum R(e_i, e_k, e_j, e_k)$$
$$= \sum \left(A(e_i, e_r, e_s)A(e_j, e_r, e_s) - A(e_i, e_j, e_s)A(e_r, e_r, e_s) \right)$$
$$+ \frac{1}{2}(n - 2)\hat{B}(e_i, e_j) + \frac{n}{2}L_1 G(e_i, e_j). \tag{1.19}$$

$$\kappa := \frac{1}{n(n - 1)} \sum Ric\,(e_i, e_i) = J + L_1 - \frac{n}{n - 1} \cdot G(T, T), \tag{1.20}$$

where the *Pick invariant* J is defined by $n(n-1)J := \|A\|^2$, the norm is taken with respect to G, and $L_1 := \frac{1}{n} \operatorname{tr} B$ is the relative mean curvature.

1.15 Special relative normalizations

We sketch some information on special relative normalizations. Details can be found in Chapter 6 of the Lecture Notes [341].

1.15.1 Euclidean normalization as a relative normalization

It is immediate from the Weingarten equation that the Euclidean normalization defines a relative normalization on a nondegenerate hypersurface, and that the bilinear form induced on M from the Euclidean normalization via the Gauß structure equation is the Euclidean second fundamental form.

1.15.2 Equiaffine normalization

For a given nondegenerate hypersurface x we want to find a relative normalization (U, Y) that is invariant under the equiaffine (unimodular) transformation group. The basic ideas of this approach were given by Flanders [93], Calabi [46] and Nomizu [235].

1. Construction of an equiaffinely invariant normalization. Volume forms are essential for equiaffine geometry, namely: for a nondegenerate hypersurface x with given relative normalization (U, Z), we consider the volume forms ω and ω^* and study their properties under the unimodular transformation group. Recall that the induced volume forms $\omega = \omega(Z)$ and $\omega^* = \omega^*(U)$ satisfy, for any frame $\{e_1, \ldots, e_n\}$,

(i) $\omega(e_1, \ldots, e_n) \cdot \omega^*(e_1, \ldots, e_n) = (-1)^n \det(G(e_i, e_j))$;

(ii) for a frame $\{e_1, \ldots, e_n\}$ which is orthonormal with respect to G at $p \in M$, we can state:

$$\omega(e_1, \ldots, e_n) \cdot \omega^*(e_1, \ldots, e_n) = (-1)^n \det(G(e_i, e_j))$$

$$= \prod_{i=1}^{n} (-\varepsilon_i),$$

where $G(e_i, e_j) = \varepsilon_i \delta_{ij}$ and $\varepsilon_i = \pm 1$;

(iii) the bases $\{dx(-\varepsilon_1 e_1), \ldots, dx(-\varepsilon_n e_n), Z\}$ and $\{dU(e_1), \ldots, dU(e_n), U\}$ are dual in V and V^*, respectively.

We try to determine a relative normalization (U, Y) such that

$$\mathrm{Det}(dx(-\varepsilon_1 e_1), \ldots, dx(-\varepsilon_n e_n), Y) = 1,$$

in analogy to Section 1.3.1. If such a normalization exists, U must satisfy

$$\mathrm{Det}^*(dU(e_1), \ldots, dU(e_n), U) = 1.$$

In this case, both volume forms are invariant under unimodular linear mappings, and the volume forms $\omega(Y) =: \omega(e)$ and $\omega^*(U) := \omega^*(e)$ (we use the mark "e" for *equiaffine*) have to satisfy

$$|\omega(e)(e_1,\ldots,e_n)| = |\omega^*(e)(e_1,\ldots,e_n)|;$$

but if ω and ω^* coincide on a basis (up to sign), we have $\omega = \pm\omega^*$. From the foregoing this property is invariant under equiaffine transformations of the hypersurface x and the relative normalization (U, Y) considered. We are now going to prove the existence and uniqueness of such a relative normalization.

Theorem 1.19. *Let $x : M \to A^{n+1}$ be a nondegenerate hypersurface. Then:*
(i) *There exists a relative normalization (U, Y) with the following properties:*
    (a) $|\omega| = |\omega^*|$;
    (b) *the normalization is equiaffinely invariant;*
(ii) *Up to orientation there is only one relative normalization of x satisfying (a) and (b) in (i).*
(U, Y) is called the equiaffine (or Blaschke) normalization of x, U the equiaffine (or Blaschke) conormal, Y the equiaffine (or Blaschke) normal.

Proof. The reader should give the proof as an exercise: consider two different relative normalizations (U, Y) and $(U^\sharp, Y^\sharp)$ on x and study how they are related, e.g. $U^\sharp = q \cdot U$ for some nowhere vanishing $q \in C^\infty$ (see Section 5.1 in [341]). Calculate the volume forms; then the equation $\omega = \pm\omega^*$ gives a unique solution for q. $\qquad\square$

Definition 1.20. A nondegenerate hypersurface is *locally strongly convex* at a point $p \in M$ if an arbitrary relative metric (and then any) is definite at p. If a nondegenerate hypersurface is *locally strongly convex* at any $p \in M$ we call the hypersurface to be *locally strongly convex*.

Remark 1.21. (i) The relative metric that corresponds to the above choice of (U, Y) is called the *Blaschke metric*. From Chapter 3 on, the notation G will only be used to denote the Blaschke metric. In the context of a Blaschke normalization the induced geometry nowadays is called a *Blaschke hypersurface*.
(ii) We will fix the orientation of (U, Y) on locally strongly convex surfaces such that G is positive definite.

Corollary 1.22. *Let x be a nondegenerate hypersurface. A relative normalization (U, Z) coincides (up to orientation) with the equiaffine normalization if and only if it satisfies $|\omega(Z)| = |\omega(G)| = |\omega^*(U)|$, which means that the Riemannian volume $\omega(G)$ coincides (up to sign) with the volume forms induced by Z and U, respectively.*

2. Equiaffine structure equations. We summarize the results of this introductory subsection on the equiaffine normalization. For a nondegenerate hypersurface there exists, up to orientation, a unique relative normalization (U, Y) satisfying

$$|\omega^*(U)| = |\omega(Y)| = |\omega(G)|,$$

this normalization is equiaffinely invariant.
Via the structure equations

(Weingarten) $dY(v) = -dx(B(v)),$

(Gauß – x) $\bar{\nabla}_v dx(w) = dx(\nabla_v w) + G(v,w)Y,$

(Gauß – U) $\bar{\nabla}_v dU(w) = dU(\nabla_v^* w) - \hat{B}(v,w)U,$

the normalization induces the following structures on M:
-   the (1,1)-tensor field B which is the *affine shape operator*;
-   the bilinear, symmetric forms G (the *Blaschke metric*) and $\hat{B}$ (the *Weingarten form*);
-   the affine connections ∇ and ∇^*, which are conjugate relative to G; both connections are torsion free and Ricci symmetric (that means their Ricci tensors are symmetric); the difference of the connections gives the difference tensor K and finally the cubic form A.

The volume forms $\omega = \omega(Y)$ and $\omega^* = \omega^*(U)$ are parallel:

$$\nabla\omega = 0 \text{ and } \nabla^*\omega^* = 0.$$

3. Apolarity for Blaschke hypersurfaces. On a nondegenerate hypersurface the unimodular normalization can be characterized by $T = 0$ within the class of all relative normalizations.

1.15.3 Centroaffine normalization

Now we consider the geometry of a hypersurface $x : M \rightarrow A^{n+1}$ with respect to the centroaffine transformation group $\mathfrak{z}_p$ with center p (see Section 1.1.5-2). Without loss of generality we choose the origin $O \in A^{n+1}$ to be fixed as the center; thus we can identify A^{n+1} (with fixed origin) with the vector space V.

The following observation is obvious: on a nondegenerate hypersurface (see Section 1.11) the set $\{p \in M \mid x(p) \in dx(T_pM)\}$ is nowhere dense (the beginner should verify this as an exercise). It is a standard convention in the centroaffine literature to consider the position vector x to be always transversal to $x(M)$; we follow this convention from now on.

From the point of view of linear algebra, neither of the dual vector spaces V, V^* is distinguished as $(V^*)^* = V$. This suggests to consider the nondegenerate hypersurface

$x : M \rightarrow V$ itself and, if it exists, its centroaffinely invariant conormal immersion $U : M \rightarrow V^*$ as a dual pair:

$$\langle U, \pm x \rangle = 1, \quad \langle U, dx \rangle = 0, \quad \langle dU, \pm x \rangle = 0. \tag{1.21}$$

For x given, at any $p \in M$, this linear system has a unique solution $U(p)$, thus defining a mapping $U : M \rightarrow V^*$. From the nondegeneracy of x, the relations in (1.21) define a bijective correspondence, namely a duality, between $x : M \rightarrow V$ and $U : M \rightarrow V^*$. As the equations (1.21) are centroaffinely invariant, we can choose a centroaffinely invariant normalization $(U(c), Y(c))$ of $x : M \rightarrow V$, where $U(c) := U$ is a solution of (1.21) and $Y(c) := \pm x$. The pair $(U(c), Y(c))$ is called the *centroaffine normalization* of x.

For a locally strongly convex hypersurface the following choices are standard. Consider $x(p) \in x(M)$ and its tangent hyperplane $dx(T_p M)$.

Choose $Y(c) := -x$ if the hypersurface in $x(p)$ seperates $dx(T_p M)$ and the origin (so called *elliptic case*).

Choose $Y(c) :=+ x$ if $dx(T_p M)$ and the origin are on the same side of the hypersurface (so called *hyperbolic case*).

Now it is a trivial consequence of the Weingarten structure equations in Definition 1.5 that the Weingarten operator $B(c)$ of this normalization satisfies $B(c) = \pm \mathrm{id}$, and the centroaffine Weingarten form satisfies $\hat{B}(c) = \pm G(c)$ (verify this with Section 1.12.1).

One of the standard examples of a hypersurface with centroaffine normalization is the following: consider a nondegenerate hypersurface x with relative normalization (U, Y). Then $U : M \rightarrow V^*$ is a hypersurface as $\mathrm{rank}(dU) = n$ and U is transversal to $U(M)$. Thus one can consider $(\pm U)$ as *centroaffine normal* of the hypersurface $U : M \rightarrow V^*$.

It is important to realize that the invariants of centroaffine hypersurface theory essentially depend on the choice of the origin; this is immediately clear from the fact that the position vector is used as centroaffine normal (modulo sign).

We will use the centroaffine normalization for the study of the Gauß map for affine maximal surfaces in Section 5.2. The reader interested in a more detailed study of the centroaffine normalization is referred to Section 6.3 of [341].

1.16 Comparison of relative normalizations

From the structural point of view, within the class of all possible normalizations, the subclass of relative normalizations is distinguished with respect to the affine transformation group; namely, according to Section 1.11, a relative normalization is distinguished as a unique representative of a *tangential equivalence class*.

In this section we study relations between different geometric quantities induced from a relative normalization.

1.16.1 Comparison of corresponding geometric properties

Two conormal vector fields U, and $U^\sharp$ corresponding to a nondegenerate hypersurface $x : M \to A^{n+1}$ differ only by a nowhere vanishing C^∞-function $q : M \to \mathbb{R}$, that is $U^\sharp = q \cdot U$. If we restrict to conormal fields inducing a fixed orientation on M, q must be strictly positive; with $2\tau = \ln q$ we have the relation $U^\sharp = e^{2\tau} U$. Note that q (and τ resp.) can be expressed by the *relative support functions*:

$$\Lambda^\sharp = q \cdot \Lambda,$$

where $\Lambda := \langle U, \pm x \rangle$ and $\Lambda^\sharp := \langle U^\sharp, \pm x \rangle$; the sign can be chosen appropriately.

On the other hand, for a given conormal field U and any C^∞-function $\tau : M \to \mathbb{R}$, the definition $U^\sharp := e^{2\tau} U$ gives another conormal field, and the orientation is preserved. As the relative geometry of a nondegenerate hypersurface is uniquely determined by the choice of the conormal, we can express the fundamental quantities corresponding to $U^\sharp$ using the function τ.

Proposition 1.23. *Let U and $U^\sharp$ be two conormal fields of a nondegenerate hypersurface x with $U^\sharp = q \cdot U = e^{2\tau} U$. Then the corresponding relative normal field, the relative metric G and the connections ∇ and ∇^* are related by:*

(i) $Z^\sharp = e^{-2\tau}(Z + dx(2\mathrm{grad}_G\tau))$, *or locally* $Z^\sharp = e^{-2\tau}(Z + 2\tau_r G^{rk}\partial_k x)$, *where* $\tau_r = \partial_r\tau$ *and* $G^{rk}G_{ks} = \delta^r_s$;

   (as we use different metrics G and $G^\sharp$, we have to be careful in raising and lowering indices in the local calculus; we prefer the explicit use of the local representation (G^{rk}) of the inverse matrix of (G_{ks}));

(ii) $G^\sharp(v, w) = e^{2\tau} G(v, w)$, *locally* $G^\sharp_{ij} = e^{2\tau} G_{ij}$; *thus all relative metrics define a conformal structure on M;*

(iii) $\nabla^\sharp_v w = \nabla_v w - 2G(v, w)\mathrm{grad}_G\tau$, *locally:* $\Gamma^{\sharp\,k}_{ij} = \Gamma^k_{ij} - 2G_{ij}G^{rk}\tau_r$;

(iv) $\nabla^{*\sharp}_v w = \nabla^*_v w + 2 \cdot (d\tau(w)v + d\tau(v)w)$;

   this relation implies that the connections $\nabla^{*\sharp}$ and ∇^* are projectively equivalent (see [90]);

(v) $\omega(\mathrm{Det}, Z^\sharp) = e^{-2\tau}\omega(\mathrm{Det}, Z)$ *and* $\omega^*(\mathrm{Det}^*, U^\sharp) = e^{2(n+1)\tau}\omega^*(\mathrm{Det}^*, U)$.

Proof. Verification is left as an exercise. E.g. using the Gauß equation for x in two different forms we get

$$dx(\nabla_v w) + G(v, w)Z = \bar{\nabla}_v dx(w)$$
$$= dx(\nabla^\sharp_v w) + G^\sharp(v, w)Z^\sharp$$
$$= dx(\nabla^\sharp_v w) + G^\sharp(v, w)e^{-2\tau}[Z + dx(2\mathrm{grad}_G\tau)].$$

Compare the transversal and the tangential parts; this gives (ii) and (iii).

Corollary 1.24. *Let x be a nondegenerate hypersurface, $\{(U, Z)\}$ be the set of relative normalizations and $\{G\}$ be the conformal class of relative metrics. Then the correspondence*

$$(U, Z) \mapsto G(Z) = G(U)$$

is bijective.

1.16.2 Gauge invariance

The Proposition 1.18 lists how certain relative invariants behave under a change $(U, Z) \mapsto (U^{\sharp}, Z^{\sharp})$ of relative normalizations of the same hypersurface x. We draw the attention to the properties in (ii) and (iv) and add some consequences:

(1) All relative metrics $\{G\}$ of x define a conformal class. This implies that the conformal curvature tensor W of Weyl is invariant under the change of a relative normalization of x.

(2) All conormal connections $\{\nabla^*\}$ of x define a projective class of torsion free, Ricci symmetric connections. This implies that the projective curvature tensor P is invariant under the change of a relative normalization of x.

We call a change of the relative normalization of x a *gauge transformation*. Using this terminology, we can present a list of gauge invariants; for a proof we refer to [340].

Theorem 1.25. *Consider the class of relative normalizations $\{(U, Z)\}$ of a nondegenerate hypersurface x. The following invariants are gauge invariants under the change of relative normalizations of x:*

(i) *the Weyl conformal curvature tensor W;*

(ii) *the projective curvature tensor P; any conormal connection is projectively flat, thus $P \equiv 0$;*

(iii) *the traceless part $\tilde{K}$ of the difference tensor K of an arbitrary relative normalization (U, Z), given by*

$$\tilde{K}(u, v) := K(u, v) - \frac{n}{n + 2} \cdot \left(T(u)v + T(v)u + G(u, v)T^{\sharp} \right).$$

Here G denotes the relative metric and T ($T^{\sharp}$, respectively) the Tchebychev form (the Tchebychev vector field, respectively) of the same normalization (U, Z);

(iv) *fix an origin in the affine space A^{n+1} and assume that the position vector of x is always transversal along the hypersurface; then the support function $\Lambda := \langle U, \pm x \rangle$ is nowhere zero. We have a gauge invariant Tchebychev form*

$$\tilde{T} := T + \frac{n + 2}{2n} d \ln |\Lambda|.$$

2 Local equiaffine hypersurface theory

There exist different approaches to a hypersurface theory with respect to the unimodular affine transformation group.

As mentioned in the introduction, Blaschke and his school tried to develop the equiaffine theory in analogy to the Euclidean hypersurface theory. The two fundamental forms I and II of Euclidean hypersurface theory define a complete system of invariants describing the Euclidean differential geometry of a hypersurface; thus, in the beginning of a systematic study of equiaffine surface theory, the suggestion was to search for equiaffinely invariant forms. From this, it was plausible that Blaschke started the theory by introducing an equiaffinely invariant quadratic form as an "equiaffine metric" [20, § 39]. In the next step he derived the equiaffinely invariant normalization [20, § 40]. The same approach is used in §22 of [311].

In [42] and [47] Calabi used another approach: he considered an equiform (with respect to the given volume) Euclidean structure on A^{n+1} and used this to define an equiaffinely invariant conormal field of the hypersurface in terms of Euclidean invariants. From this he got a transversal field, structure equations, etc. In another paper [46] Calabi defined, in fact, an equiaffine normalization via the volume forms ω, ω^* as in Theorem 1.19, while in [235] Nomizu defined such a normalization as in Corollary 1.22. Another structural approach starts with a connection [245]. In the previous chapter we proved the equivalence of these different approaches. A more detailed investigation is given in [341]. We refer also to [311, 345, 352].

In this chapter we give an introduction to the local equiaffine hypersurface theory. We start, like Blaschke, by introducing the "equiaffine metric". The reason for using this classical concept comes from the fact that the equiaffine metric can be easily calculated from a parametric representation of a hypersurface; thus Blaschke's approach is still very useful for the investigation of many particular problems; this is one of the main goals of this volume.

Besides the invariant calculus we will use Cartan's moving frames. This exposition of the equiaffine theory is due to S. S. Chern [60] and H. Flanders [93]. The moving frame method allows to use progress in the solution of p.d.e.'s via exterior forms. For the convenience of the reader, we give an introduction to the local equiaffine theory in terms of moving frames from the beginning; the beginner will find the fundamentals of exterior forms in the first appendix.

2.1 Blaschke–Berwald metric and structure equations

Let A^{n+1} be a unimodular affine space of dimension $(n + 1)$; with respect to the coordinates $(x^1, \ldots, x^{n+1})$, the standard notation for the volume element in the exterior calculus is $dV = dx^1 \wedge dx^2 \wedge \cdots \wedge dx^{n+1}$. Then, on a Gauß basis, we have the relation

$dV(\partial_1,\ldots,\partial_{n+1}) = \det(dx^i(\partial_j)) = 1$. The unimodular linear group preserves this volume form. We recall Section 1.1.8 about frames.

Let M be a C^∞-manifold of dimension n, and let $x : M \to A^{n+1}$ be a C^∞-immersion. Since the investigation is local, we may identify M with $x(M)$. Then T_xM can be identified with a subspace of the vector space V. Choose a local unimodular affine frame field $\{x; e_1,\ldots,e_{n+1}\}$ on M such that $x \in M$ and $e_1,\ldots,e_n \in T_xM$. We call such a frame *a frame adapted to x (adapted frame)*. We can apply the so-called moving frame equations for A^{n+1}, (1.3) and (1.4):

$$dx = \sum_{A=1}^{n+1} w^A e_A,$$

$$de_A = \sum_{B=1}^{n+1} w_A^B e_B.$$

If we restrict the forms to M and T_xM we have

$$w^{n+1} = 0. \tag{2.1}$$

Taking the exterior differentiation of (2.1) and using (1.8) we get

$$\sum w^i \wedge w_i^{n+1} = 0. \tag{2.2}$$

Our convention for the range of indices is the following

$$1 \leq A, B, \cdots \leq n+1,$$
$$1 \leq i, j, \cdots \leq n,$$

and we shall follow the Einstein summation convention. We apply Cartan's Lemma (Appendix A, Theorem A.10), and from (2.2) we obtain that

$$w_i^{n+1} = \sum h_{ij} w^j, \tag{2.3}$$

where h_{ij} denote the coefficients of a bilinear symmetric form:

$$h_{ij} = h_{ji}. \tag{2.4}$$

We state the equations from Sections 1.8.2 and 1.8.3 in terms of moving frames. For an adapted frame field the structure equations read:

Weingarten: $\qquad de_{n+1} = \sum w_{n+1}^i e_i + w_{n+1}^{n+1} e_{n+1},$ \hfill (2.5)

Gauß: $\qquad de_i = \sum w_i^j e_j + \sum h_{ij} w^j e_{n+1}.$ \hfill (2.6)

As in the previous chapter (Section 1.8 and following) we are going to investigate the coefficients and their properties; moreover, we want to find the relations between the two presentations.

We consider the quadratic differential form

$$\sum \omega^i \omega_i^{n+1} = \sum h_{ij} \omega^i \omega^j. \qquad (2.7)$$

It is easy to see that the expression (2.7) is invariant under unimodular affine transformations in A^{n+1}, although it depends on the choice of the local frame fields. Let $\{x; \hat{e}_1, \ldots, \hat{e}_{n+1}\}$ be another adapted frame field. Then we have

$$\hat{e}_i = \sum a_i^k e_k, \quad i = 1, \ldots, n,$$
$$\hat{e}_{n+1} = \sum a_{n+1}^i e_i + a^{-1} e_{n+1},$$

or

$$e_i = \sum b_i^k \hat{e}_k,$$
$$e_{n+1} = \sum b_{n+1}^i \hat{e}_i + a \cdot \hat{e}_{n+1},$$

where $a = \det(a_i^j)$, and (b_i^j) is the inverse matrix of (a_i^j). In analogy to the above relations we have the following ones with respect to the frame $\{x; \hat{e}_1, \ldots, \hat{e}_{n+1}\}$:

$$dx = \sum \hat{\omega}^i \hat{e}_i,$$
$$d\hat{e}_A = \sum \hat{\omega}_A^B \hat{e}_B,$$
$$\hat{\omega}^{n+1} = 0,$$
$$\hat{\omega}_i^{n+1} = \sum \hat{h}_{ij} \hat{\omega}^j, \quad \hat{h}_{ij} = \hat{h}_{ji},$$

which gives

$$\omega^k = \sum a_i^k \hat{\omega}^i, \qquad (2.8)$$
$$\omega_i^{n+1} = \mathrm{Det}(e_1, \ldots, e_n, de_i) = a^{-1} \sum b_i^k \hat{\omega}_k^{n+1} \qquad (2.9)$$

and

$$\hat{h}_{il} = \sum a a_i^j h_{jk} a_l^k. \qquad (2.10)$$

This leads to the important conclusion that the rank of the quadratic differential form (2.7) is an affine invariant. Comparing the Gauß equation above with that in Section 1.8.3 we see that $h = G(e_{n+1})$. As we assumed the hypersurface to be nondegenerate, we have

$$H := \det(h_{ij}) \neq 0.$$

From (2.10) we get

$$\hat{H} := \det(\hat{h}_{ij}) = a^{n+2} H.$$

Assume that M is oriented. We shall restrict to frames such that a fame $\{e_1, \ldots, e_n\}$ is positively oriented. Then $a > 0$ and

$$|\hat{H}|^{1/(n+2)} = a|H|^{1/(n+2)}. \qquad (2.11)$$

(2.8), (2.9), and (2.11) give

$$|\hat{H}|^{-1/(n+2)} \sum \hat{\omega}^i \hat{\omega}_i^{n+1} = |H|^{-1/(n+2)} \sum \omega^i \omega_i^{n+1}.$$

This shows that the quadratic differential form

$$|H|^{-1/(n+2)} \sum \omega^i \omega_i^{n+1} = |H|^{-1/(n+2)} \sum h_{ij} \omega^i \omega^j$$

is independent of the choice of the local unimodular affine frame field. Therefore it is equiaffinely invariant.

Remark 2.1. Choosing e_{n+1} to be the Euclidean normal vector field we see that this quadratic form is an element of the conformal class of relative metrics. As each metric in this class is generated from a uniquely determined relative normalization (see Section 1.10 and the bijectivity statement of Corollary 1.24) and as the above quadratic form is equiaffinely invariant, it must coincide with the Blaschke metric $G = G(Y)$, where Y is the equiaffine normal vector field from Remark 1.21. We express the Blaschke metric by

$$G = \sum G_{ij} \omega^i \omega^j, \tag{2.12}$$

where

$$G_{ij} = |H|^{-1/(n+2)} h_{ij}. \tag{2.13}$$

When $x : M \to A^{n+1}$ is a locally strongly convex hypersurface, the quadratic form is definite. We recall Section 1.15, Proposition 1.23, and Remark 1.21: the equiaffinely invariant form G is conformally related to the second fundamental form of Euclidean hypersurface theory; however, for the second fundamental form the relation between the shape of the hypersurface and the definiteness of the second form is well known. (The beginner should recall that the locally strong convexity of a hypersurface is affinely invariant.)

On a nondegenerate hypersurface $x : M \to A^{n+1}$ the Blaschke metric defines a semi-Riemannian space (M, G). We use the well-known properties of this structure from now on (for the beginner we summarized these properties in Appendix A.3).

2.2 The affine normal and the Fubini–Pick form

In Section 1.15.2 we proved that there exists, up to orientation, a unique relative normalization which is invariant under the group of equiaffine transformations. From this, one can construct the induced structures as in Chapter 1 via the structure equations.

Vice versa, we now construct from the Blaschke metric G the equiaffine normal. Again we use the bijective relation in Corollary 1.24 between the normalization and the conformal class of relative metrics.

2.2.1 Affine normal

Let $x : M \to A^{n+1}$ be a nondegenerate hypersurface and let $x = (x^1, \ldots, x^{n+1})$ be its position vector field. Using the Blaschke metric we define a vector field $\tilde{Y} : M \to V$ by

$$\tilde{Y} := \frac{1}{n} \Delta x = \frac{1}{n} (\Delta x^1, \ldots, \Delta x^{n+1}), \tag{2.14}$$

where Δ denotes the Laplacian with respect to the Blaschke metric, acting on functions. From the construction, $\tilde{Y}$ is equiaffinely invariant.

Lemma 2.2. *Let $x : M \to A^{n+1}$ be a nondegenerate hypersurface. The canonical vector field $\tilde{Y} : M \to V$ from (2.14) coincides with the equiaffine normalization from Section 1.15.2 up to orientation.*

Proof. In the first step we prove that $\tilde{Y}$ is transversal, in the second step that $\tilde{Y}$ is a relative normalization. Thus $\tilde{Y}$ defines an equiaffinely invariant normalization, and from the bijectivity relation in Corollary 1.24 we get $\tilde{Y} = Y$.

Step 1. We choose a local adapted frame field $\{x; e_1, \ldots, e_{n+1}\}$ on M. Denote the Levi–Civita connection matrix of the Blaschke metric G by $(\tilde{\omega}_j^i)$. We recall the local notations from Section 1.12.6. From

$$dx = \sum \omega^i e_i,$$

the Gauß structure equation

$$de_i = \sum \omega_i^j e_j + \sum h_{ij} \omega^j e_{n+1},$$

and the definition of the covariant derivatives (in local notation) we get

$$x_{,ij} = x_{ij} - \sum \tilde{\Gamma}_{ij}^k e_k, \tag{2.15}$$

$$x_{,ij} = \sum (\Gamma_{ij}^k - \tilde{\Gamma}_{ij}^k) e_k + h_{ij} e_{n+1}, \tag{2.16}$$

where the Christoffel symbols Γ_{ij}^k and $\tilde{\Gamma}_{ij}^k$ and the connection forms are related by

$$\omega_i^k = \sum \Gamma_{ij}^k \omega^j, \tag{2.17}$$

$$\tilde{\omega}_i^k = \sum \tilde{\Gamma}_{ij}^k \omega^j. \tag{2.18}$$

The two relations

$$\sum \omega^j \wedge \omega_j^i = d\omega^i = \sum \omega^j \wedge \tilde{\omega}_j^i$$

imply

$$\sum (\omega_j^i - \tilde{\omega}_j^i) \wedge \omega^j = 0.$$

Then Cartan's lemma gives

$$\Gamma_{ij}^k - \tilde{\Gamma}_{ij}^k = \Gamma_{ji}^k - \tilde{\Gamma}_{ji}^k. \tag{2.19}$$

Now we compute $\tilde{Y}$:

$$\tilde{Y} = \frac{1}{n}\Delta x$$

$$= \frac{1}{n}\sum G^{ij}x_{,ij}$$

$$= \frac{1}{n}\sum |H|^{1/(n+2)}h^{ij}\left(\sum(\Gamma_{ij}^k - \tilde{\Gamma}_{ij}^k)e_k + h_{ij}e_{n+1}\right) \qquad (2.20)$$

$$= \frac{1}{n}\sum G^{ij}(\Gamma_{ij}^k - \tilde{\Gamma}_{ij}^k)e_k + |H|^{1/(n+2)}e_{n+1},$$

where (h^{ij}) denotes the inverse matrix of (h_{ij}). Hence

$$\text{Det}\,(e_1,\dots,e_n,\tilde{Y}) = |H|^{1/(n+2)} \neq 0. \qquad (2.21)$$

This shows that $\tilde{Y}$ is transversal. When $x : M \to A^{n+1}$ is locally strongly convex, from the above calculation one can easily see that $\tilde{Y}$ always points to the concave side of $x(M)$.

Step 2. To obtain the analogue of the Weingarten structure equation in Section 1.15.2-2, we now derive a necessary and sufficient condition for e_{n+1} to be parallel to $\tilde{Y}$. Exterior differentiation of $\omega_i^{n+1} = \sum h_{ij}\omega^j$ together with the structure equations gives

$$\sum\left(dh_{ij} + h_{ij}\omega_{n+1}^{n+1} - \sum h_{ik}\omega_j^k - \sum h_{kj}\omega_i^k\right)\wedge\omega^j = 0.$$

Cartan's Lemma implies that

$$\sum h_{ijk}\omega^k := dh_{ij} + h_{ij}\omega_{n+1}^{n+1} - \sum h_{ik}\omega_j^k - \sum h_{kj}\omega_i^k, \qquad (2.22)$$

where the coefficients h_{ijk}, implicitly defined from (2.22), satisfy

$$h_{ijk} = h_{jik} = h_{ikj}. \qquad (2.23)$$

Multiplying (2.22) by $|H|^{-1/(n+2)}$, we obtain

$$dG_{ij} + \frac{1}{n+2}G_{ij}d\ln|H| + G_{ij}\omega_{n+1}^{n+1} - \sum G_{ik}\omega_j^k - \sum G_{kj}\omega_i^k = |H|^{-1/(n+2)}\sum h_{ijk}\omega^k. \qquad (2.24)$$

On the other hand, since $\tilde{\omega}_i^j$ is the Levi–Civita connection of G, we have the Ricci lemma

$$dG_{ij} - \sum G_{ik}\tilde{\omega}_j^k - \sum G_{jk}\tilde{\omega}_i^k = 0. \qquad (2.25)$$

From (2.24) and (2.25) we get

$$\left[\omega_{n+1}^{n+1} + \frac{1}{n+2}d\ln|H|\right]G_{ij} - \sum G_{ik}(\Gamma_{jl}^k - \tilde{\Gamma}_{jl}^k)\omega^l - \sum G_{kj}(\Gamma_{il}^k - \tilde{\Gamma}_{il}^k)\omega^l$$
$$= |H|^{-1/(n+2)}\sum h_{ijk}\omega^k. \qquad (2.26)$$

As $\{\omega^1, \ldots, \omega^n\}$ is a basis, there exist $a_1, \ldots, a_n \in C^\infty(M)$ such that

$$\omega_{n+1}^{n+1} + \frac{1}{n+2} d \ln |H| = \sum a_k \omega^k. \tag{2.27}$$

We try to determine these coefficients such that e_{n+1} is parallel to $\tilde{Y}$. Insert (2.27) into (2.26) to obtain

$$a_k G_{ij} - \sum G_{il}(\Gamma_{jk}^l - \tilde{\Gamma}_{jk}^l) - \sum G_{lj}(\Gamma_{ik}^l - \tilde{\Gamma}_{ik}^l) = |H|^{-1/(n+2)} h_{ijk}. \tag{2.28}$$

By contraction of indices, from (2.28) we obtain that

$$na_k - 2 \sum (\Gamma_{jk}^j - \tilde{\Gamma}_{jk}^j) = |H|^{-1/(n+2)} \sum G^{ij} h_{ijk}, \tag{2.29}$$

$$a_j - \sum (\Gamma_{jk}^k - \tilde{\Gamma}_{jk}^k) - \sum G_{lj} G^{ik}(\Gamma_{ik}^l - \tilde{\Gamma}_{ik}^l) = |H|^{-1/(n+2)} \sum G^{ik} h_{ikj}. \tag{2.30}$$

Let (H^{ij}) be the adjoint matrix of (h_{ij}), so that

$$\sum H^{ij} h_{jk} = H \delta_k^i.$$

Differentiating $H = \det(h_{ij})$ and using (2.22), we obtain

$$dH = \sum H^{ij} dh_{ij} = -(n+2)H\omega_{n+1}^{n+1} + \sum H^{ij} h_{ijk} \omega^k,$$

i.e.

$$(n+2)a_k = |H|^{-1/(n+2)} \sum G^{ij} h_{ijk}. \tag{2.31}$$

From (2.20), e_{n+1} is parallel to $\tilde{Y}$ if and only if

$$\sum G^{ij}(\Gamma_{ij}^k - \tilde{\Gamma}_{ij}^k) = 0. \tag{2.32}$$

Combination of (2.29), (2.30), (2.31), and (2.22) shows that this is equivalent to

$$a_j = 0, \quad j = 1, 2, \ldots, n,$$

i.e.

$$\omega_{n+1}^{n+1} + \frac{1}{n+2} d \ln |H| = 0. \tag{2.33}$$

When e_{n+1} is parallel to $\tilde{Y}$, the last formula and

$$de_{n+1} = \sum \omega_{n+1}^i e_i + \omega_{n+1}^{n+1} e_{n+1}$$

give

$$d\tilde{Y} = |H|^{1/(n+2)} \sum \omega_{n+1}^i e_i. \tag{2.34}$$

Thus $\tilde{Y}$ is a relative normal vector field and coincides with the equiaffine normal vector field, which we denote by Y as before. $\qquad \square$

We will frequently need (2.33) for the explicit calculation of hypersurfaces; thus we state:

Proposition 2.3. *Let M be a nondegenerate hypersurface in A^{n+1}, and let the expression $\{x; e_1, \ldots, e_n, e_{n+1}\}$ be an adapted frame field on M. Then e_{n+1} is parallel to Y if and only if*

$$\omega_{n+1}^{n+1} + \frac{1}{n+2} d \ln |H| = 0.$$

Remark and Definition 2.4. From now on we shall choose an adapted frame field $\{x; e_1, \ldots, e_n, e_{n+1}\}$ such that e_{n+1} is parallel to Y. We call such a frame an *equiaffine frame*; so an equiaffine frame has the following three properties:
(i)   it is unimodular;
(ii) $e_1, \ldots, e_n$ are tangential;
(iii) e_{n+1} is parallel to the affine normal vector Y.

Under this choice we have

$$\sum G^{ij}(\Gamma_{ij}^k - \tilde{\Gamma}_{ij}^k) = 0, \tag{2.35}$$

$$Y = |H|^{1/(n+2)} e_{n+1}. \tag{2.36}$$

Relation (2.35) is called the *apolarity condition* in classical equiaffine geometry. The line through x in the direction of Y is called the *affine normal line* at x. While Y itself is equiaffinely invariant, the line is affinely invariant.

2.2.2 Fubini–Pick form

We recall from Section 1.8.3: for a given hypersurface $x : M \to A^{n+1}$, the affine connection $\bar{\nabla}$ of A^{n+1} induces on M a connection ∇ via the affine normal field Y, called the *induced connection* on M. The construction of ∇ follows from the Gauß structure equations in Section 1.8.3 in terms of Y; we repeat these equations in terms of moving frames on M using an equiaffine frame $\{x; e_1, \ldots, e_{n+1}\}$:

$$de_i = \sum \omega_i^j e_j + \omega_i^{n+1} e_{n+1}.$$

The assignment $\nabla : \{x; e_1, \ldots, e_n\} \to (\omega_i^j)$ describes the induced affine connection on M. Let

$$\omega_i^j - \tilde{\omega}_i^j =: \sum A_{ik}^j \omega^k, \tag{2.37}$$

i.e.

$$\Gamma_{ij}^k - \tilde{\Gamma}_{ij}^k =: A_{ij}^k, \tag{2.38}$$

where the Christoffel symbols Γ_{ij}^k and $\tilde{\Gamma}_{ij}^k$ were defined in (2.17) and (2.18). Obviously, A_{ij}^k is a symmetric (1,2)-tensor field on M. Define

$$A_{ijk} := \sum G_{il} A_{jk}^l. \tag{2.39}$$

We are going to prove that

$$A_{ijk} = -\frac{1}{2}|H|^{-1/(n+2)}h_{ijk}.$$ (2.40)

In the first step we state a consequence of (2.24) and (2.33):

$$dG_{ij} - \sum G_{ik}\omega_j^k - \sum G_{kj}\omega_i^k = |H|^{-1/(n+2)}\sum h_{ijk}\omega^k,$$ (2.41)

which means that $|H|^{-1/(n+2)}h_{ijk}$ is the covariant derivative of G_{ij} with respect to the induced connection ∇. Hence, both sides of (2.40) are tensors. Thus it is sufficient to prove the formula (2.40) with respect to a special frame field. We choose a local orthonormal frame field $\{x; e_1, \ldots, e_n\}$ with respect to the Blaschke metric G, with $\{x; e_1, \ldots, e_n, e_{n+1}\}$ equiaffine. Then

$$G_{ij} = \eta_{ij} = h_{ij}, \qquad |H| = 1,$$
$$\omega_i^{n+1} = \sum \eta_{ij}\omega^j, \qquad \omega^j = \sum \eta^{ji}\omega_i^{n+1}, \qquad \omega_{n+1}^{n+1} = 0,$$

where

$$(\eta_{ij}) = \begin{pmatrix} \pm 1 & & & 0 \\ & \pm 1 & & \\ & & \ddots & \\ 0 & & & \pm 1 \end{pmatrix}.$$

Exterior differentiation of $\omega^j = \sum \eta^{jk}\omega_k^{n+1}$ gives

$$\sum \omega^k \wedge \omega_k^j = d\omega^j = \sum \eta^{jm}\omega_m^k \wedge \omega_k^{n+1} = \sum \eta^{jm}\eta_{lk}\omega_m^l \wedge \omega^k.$$

It follows immediately that

$$d\omega^j = \sum \omega^i \wedge \frac{1}{2}\left(\omega_i^j - \sum \eta^{jl}\eta_{im}\omega_l^m\right).$$

From the fundamental theorem of pseudo-Riemannian geometry (see Appendix A, Section A.3.2.2) it follows that

$$\tilde{\omega}_i^j = \frac{1}{2}\left(\omega_i^j - \sum \eta^{jm}\eta_{il}\omega_m^l\right).$$ (2.42)

Hence

$$\sum A_{ijk}\omega^k = \frac{1}{2}\left(\sum \eta_{im}\omega_j^m + \sum \eta_{jm}\omega_i^m\right).$$ (2.43)

On the other hand, since $G_{ij} = \eta_{ij}$ from (2.41) we obtain that

$$-\sum \eta_{im}\omega_j^m - \sum \eta_{jm}\omega_i^m = |H|^{-1/(n+2)}\sum h_{ijk}\omega^k.$$ (2.44)

Combining (2.43) and (2.44) gives the assertion in (2.40). An immediate consequence of (2.40) and (2.23) is that the components A_{ijk} define a totally symmetric (0,3) tensor field, called the *cubic form* (tensor field):

$$A_{ijk} = A_{jik} = A_{ikj}.$$ (2.45)

From (2.35) and (2.38) we reformulate the apolarity condition in terms of the cubic form A_{ijk}:

$$\sum G^{ij} A_{ijk} = 0, \quad k = 1, 2, \ldots, n. \tag{2.46}$$

The simplest invariant of the quadratic form G and the cubic form A

$$A = \sum A_{ijk} \omega^i \omega^j \omega^k \tag{2.47}$$

is defined by

$$
\begin{aligned}
J &:= \frac{1}{n(n-1)} \sum G^{il} G^{jm} G^{kr} A_{ijk} A_{lmr} \\
&:= \frac{1}{n(n-1)} \|A\|_G^2,
\end{aligned}
\tag{2.48}
$$

where $\| \cdot \|_G$ denotes the tensor norm with respect to the metric G. The cubic form A is also called the *Fubini–Pick form*; it measures the deviation of the induced connection (ω_j^i) from the Levi–Civita connection $(\tilde{\omega}_j^i)$ of the Blaschke metric G. J is called the *Pick invariant*.

We rewrite the Gauß structure equation in local form, both in terms of the second order partial derivatives x_{ij} and the covariant derivatives $x_{,ij}$; we will need both versions:

$$x_{ij} = \sum \Gamma_{ij}^k e_k + G_{ij} Y, \tag{2.49}$$

$$x_{,ij} = \sum A_{ij}^k e_k + G_{ij} Y. \tag{2.50}$$

2.2.3 Affine curvatures

In Euclidean differential geometry, curvature intuitively is described by the deviation of the normal field. Thus, the geometric information about curvature is contained in the coefficients of the Weingarten structure equation. This leads to the definition of curvature functions using the Weingarten operator.

In the affine case the Weingarten equation looks similar to the Euclidean one. We stated the Weingarten structure equation for the equiaffine normal two times: in Section 1.15.2-2 in the invariant form, in (2.34) in terms of moving frames.

An essential property in the Euclidean theory is the self-adjointness of the Weingarten operator with respect to the first fundamental form; an analogue is true in the affine (relative) theory.

Lemma 2.5. *The Weingarten operator B is self-adjoint with respect to G. The associated form $\hat{B}(v, w) := G(B(v), w)$ is symmetric and equiaffinely invariant. $\hat{B}$ is called the equiaffine Weingarten form.*

Proof. Exterior differentiation of (2.33) gives

$$\sum \omega_{n+1}^i \wedge \omega_i^{n+1} = 0. \tag{2.51}$$

Since M is nondegenerate, $\omega_1^{n+1}, \ldots, \omega_n^{n+1}$ are linearly independent. Cartan's Lemma and (2.51) imply

$$\omega_{n+1}^i = -\sum l^{ij}\omega_j^{n+1}, \tag{2.52}$$

where the coefficients implicitly defined in (2.52) are symmetric:

$$l^{ij} = l^{ji}. \tag{2.53}$$

Inserting $\omega_i^{n+1} = \sum h_{ij}\omega^j$ into (2.52), we obtain the relation

$$\omega_{n+1}^i = -\sum l_j^i \omega^j, \tag{2.54}$$

where

$$l_j^i = \sum h_{jk} l^{ki}. \tag{2.55}$$

Comparing (2.54) with (2.34) we see that the coefficients B_j^i of the Weingarten operator with respect to the given frame are given by $|H|^{1/(n+2)} l_j^i = B_j^i$, and the coefficients of the associated equiaffine Weingarten form are given by

$$B_{ij} = \sum l^{kl} h_{ki} h_{lj}.$$

The lemma follows immediately from the symmetry of l^{ij}. $\qquad\square$

For locally strongly convex hypersurfaces, G is definite, and positive definite for an appropriate orientation of the affine normalization. From the self-adjointness of B its eigenvalues $\lambda_1, \ldots, \lambda_n$ are real (if G is indefinite, the eigenvalues might be complex, see Section 4.7 in [341]); they are called the *affine principal curvatures*, and the eigendirections are called the *directions of the affine principal curvatures*.

In the case of eigenvalues with higher multiplicities it might occur that the principal curvatures are continuous, but not everywhere differentiable on M. For this reason one considers the coefficients $L_r : M \to \mathbb{R}$ $(r = 0, 1, \ldots, n)$ of the associated characteristic polynomial of B

$$\det(B - \lambda\ \text{id}) = \sum_{r=0}^{n} \binom{n}{r} L_r(-\lambda)^{n-r}. \tag{2.56}$$

We have the following well-known relation (theorem of Vieta) between these coefficients and the eigenvalues as roots of the polynomial

$$\binom{n}{r} L_r = \sum_{1 \le i_1 < \cdots < i_r \le n} \lambda_{i_1} \cdots \lambda_{i_r}.$$

These coefficients are differentiable on M, and $L_1, \ldots, L_n$ are suitable functions to describe the equiaffine curvature of the hypersurface x. We call $L := L_1$ the *equiaffine mean curvature* and L_n the *equiaffine Gauß–Kronecker curvature*; obviously $L_0 = 1$.

Because of $\hat{B}(v, w) = G(B(v), w)$ it is obvious that one can calculate the curvature functions from $\hat{B}$ and G:

$$\det(\hat{B} - \lambda G) = \det(G) \cdot \det(B - \lambda\ \text{id}) = 0. \tag{2.57}$$

The quadratic Weingarten form $\hat{B}$ can be expressed in terms of moving frames by

$$\hat{B} = -\sum \omega_{n+1}^i \omega_i^{n+1} = \sum l^{ij} h_{ik} h_{jl} \omega^k \omega^l =: \sum B_{ij} \omega^i \omega^j. \qquad (2.58)$$

A point $p \in M$ is called an *umbilic* of the hypersurface x if the Weingarten operator satisfies $B = \lambda$ id at p, where $\lambda \in \mathbb{R}$. Obviously this is equivalent to $\hat{B} = \lambda G$ at p, which again is equivalent to $\lambda_1 = \cdots = \lambda_n = \lambda$ at p.

Exercise 2.1. Readers who so far do not quite feel at ease with the moving frame calculus should, as an exercise, reprove the equiaffine invariance of $\hat{B}$.

Hint: Consider another equiaffine frame and follow the idea of Section 2.1, where we proved the equiaffine invariance of G.

In Section 1.15.2-2 we summarized the structure equations coming from the equiaffine normalization, which with respect to an equiaffine frame read

(Weingarten) $\qquad\qquad Y_i = -\sum B_i^j e_j, \qquad\qquad\qquad\qquad (2.59)$

(Gauß – x) $\qquad\qquad x_{ij} = \sum \Gamma_{ij}^k e_k + G_{ij} Y, \text{ or} \qquad\qquad (2.60)$

(Gauß – x) $\qquad\qquad x_{,ij} = \sum A_{ij}^k e_k + G_{ij} Y. \qquad\qquad\quad (2.61)$

Note that the cubic form with coefficients A_{ijk} is totally symmetric:

$$A_{ijk} = A_{jki} = A_{kij},$$

and satisfies the apolarity condition.

2.2.4 Geometric meaning of the affine normal

In this subsection we discuss the geometric meaning of the affine normal; the result is due to Blaschke for $n = 2$ (cf. [20]) and Leichtweiß for $n \geq 2$ (cf. [159]). First we derive a canonical expansion of M in a neighborhood of a point. Let $p \in M$ be an arbitrary point. We can choose an equiaffine frame $\{p; e_1, \ldots, e_n, e_{n+1}\}$ such that, at the point p,

$$G_{ij} = \delta_{ij}, \quad e_{n+1} = Y.$$

By a unimodular affine transformation we can assume that the origin $O \in A^{n+1}$ coincides with p and that $e_1, e_2, \ldots, e_{n+1}$ have the coordinates

$$(1, 0, \ldots, 0), (0, 1, 0, \ldots, 0), \ldots, (0, \ldots, 0, 1),$$

respectively. Obviously, the hyperplane $x^{n+1} = 0$ is tangent to M at p. Assume that M is locally strongly convex and locally given by a strongly convex C^∞-function:

$$x^{n+1} = f(x^1, \ldots, x^n).$$

Then we have

$$f(0,\ldots,0) = 0,$$
$$\partial_i f(0,\ldots,0) = 0, \quad i = 1, 2, \ldots, n.$$

Hence the Taylor expansion of $f(x^1,\ldots,x^n)$ at p is given by

$$
\begin{aligned}
x^{n+1} &= f(x^1,\ldots,x^n) \\
&= \frac{1}{2}\sum a_{ij}x^i x^j + \frac{1}{6}\sum a_{ijk}x^i x^j x^k + \cdots,
\end{aligned}
\tag{2.62}
$$

where the coefficients in the expansion are symmetric:

$$a_{ij} = a_{ji}, \quad a_{ijk} = a_{jik} = a_{ikj}.$$

We choose the following local frame field in a neighborhood of p:

$$e_1 = (1, 0, \ldots, 0, \partial_1 f),$$
$$e_2 = (0, 1, 0, \ldots, 0, \partial_2 f),$$
$$\vdots$$
$$e_n = (0, \ldots, 0, 1, \partial_n f),$$

and e_{n+1} such that it is parallel to the affine normal vector field Y and satisfies $\mathrm{Det}\,(e_1,\ldots,e_n,e_{n+1}) = 1$. With this frame field, one verifies that

$$
\begin{aligned}
h_{ij} &= a_{ij} + \sum a_{ijk}x^k + \cdots, \\
a_{ij} &= \delta_{ij}.
\end{aligned}
\tag{2.63}
$$

It is not difficult to see that, at p, $w_i^j = 0, H = 1, w_{n+1}^{n+1} = \sum w_i^i = 0$. Hence from (2.62) and (2.63), at p we obtain

$$A_{ijk} = -\frac{1}{2}H^{-1/(n+2)}h_{ijk} = -\frac{1}{2}a_{ijk}.
\tag{2.64}$$

Inserting (2.64) into (2.62), we get the canonical expansion of $x(M)$ in a neighborhood of p:

$$x^{n+1} = \frac{1}{2}\sum (x^i)^2 - \frac{1}{3}\sum A_{ijk}(p)x^i x^j x^l + \cdots.
\tag{2.65}$$

We use the canonical expansion (2.65) to give a geometric interpretation of the affine normal. We shall show that the affine normal at p is the tangent line of the locus of the centers of gravity of slices parallel to the tangent hyperplane of M at p. Denote by S the intersection of $x(M)$ and the hyperplane

$$\Pi : x^{n+1} =: t^2 = \text{const.} > 0.$$

We denote by Ω the domain in Π bounded by S. Then the coordinates of the center of gravity of Ω, denoted by $(g^1, g^2, \ldots, g^{n+1})$, are given by

$$g^k = \frac{\int_\Omega x^k d\bar{V}}{\int_\Omega d\bar{V}}, \quad k = 1, 2, \ldots, n,$$

$$g^{n+1} = t^2,$$

where $d\bar{V}$ denotes the volume element of the subspace $\Pi = A^n$. By parallel translation of Π, the center of gravity of Ω describes a differentiable curve $g(t)$ in A^{n+1}. To compute the coordinates of the center of gravity, we introduce a cylindrical coordinate system in Π:

$$x^1 = r\theta^1,$$
$$x^2 = r\theta^2,$$
$$\vdots$$
$$x^n = r\theta^n,$$
$$x^{n+1} = t^2,$$

where $r \geq 0$, $(\theta^1)^2 + \cdots + (\theta^n)^2 = 1$; $\theta^1, \ldots, \theta^n$ are the homogeneous coordinates of the $(n-1)$-dimensional Euclidean unit hypersphere S^{n-1} in A^n. Then the canonical expansion (2.65) becomes

$$t^2 = \frac{r^2}{2} - \frac{C}{3}r^3 + \cdots, \tag{2.66}$$

where

$$C = \sum A_{ijk}\theta^i\theta^j\theta^k.$$

Hence we have

$$r = \sqrt{2}t + \frac{2C}{3}t^2 + \cdots. \tag{2.67}$$

The coordinates of the center of gravity are now given by

$$g^k = \frac{n}{n+1} \frac{\int_{S^{n-1}} r^{n+1}\theta^k dO_{n-1}}{\int_{S^{n-1}} r^n dO_{n-1}}, \quad k = 1, 2, \ldots, n,$$

$$g^{n+1} = t^2,$$

where dO_{n-1} denotes the volume element of S^{n-1}. We recall a formula of H. Weyl (cf. formula (12) of Section 3 in [402]):

$$\int_{S^{n-1}} (\theta^1)^{r_1}(\theta^2)^{r_2}\cdots(\theta^n)^{r_n} dO_{n-1}$$

$$= \begin{cases} \dfrac{(r_1 - 1)!!(r_2 - 1)!!\cdots(r_n - 1)!!}{n(n+2)\cdots(n+\sigma-2)}O_{n-1}, & \text{if all the numbers } r_i \text{ are even,} \\ 0, & \text{otherwise,} \end{cases}$$

where $\sigma = r_1 + r_2 + \cdots + r_n$ and O_{n-1} is the volume of S^{n-1}, and

$$(r-1)!! := 1 \cdot 3 \cdot 5 \cdot \cdots \cdot (r-1) \text{ for } r > 0 \text{ even and } 0!! := 1.$$

Using this formula we obtain

$$\int_{S^{n-1}} \theta^k dO_{n-1} = 0, \quad k = 1, 2, \ldots, n,$$

$$\int_{S^{n-1}} C\theta^k dO_{n-1} = \sum A_{ijl}(p) \int_{S^{n-1}} \theta^i \theta^j \theta^l \theta^k dO_{n-1}$$

$$= 3 \sum_{i \neq k} A_{iik} \int_{S^{n-1}} (\theta^i)^2 (\theta^k)^2 dO_{n-1} + A_{kkk} \int_{S^{n-1}} (\theta^k)^4 dO_{n-1}$$

$$= 3 \sum A_{iik} \frac{O_{n-1}}{n(n+2)} = 0, \quad k = 1, 2, \ldots, n,$$

because of the apolarity condition. It follows immediately that

$$\int_{S^{n-1}} r^n dO_{n-1} = 2^{n/2} t^n O_{n-1} + O(t^{n+1}),$$

$$\int_{S^{n-1}} r^{n+1} \theta^k dO_{n-1} = O(t^{n+3}), \quad k = 1, 2, \ldots, n.$$

Hence

$$g^k = O(t^3) \text{ for all } k = 1, 2, \ldots, n; \text{ and}$$
$$g^{n+1} = t^2,$$

and therefore

$$dg^1 : dg^2 : \cdots : dg^{n+1} = 0 : 0 : \cdots : 1,$$

i.e. the affine normal of M at p coincides with the tangent line of the curve of the centers of gravity at $t = 0$.

2.3 The equiaffine conormal

2.3.1 Properties of the equiaffine conormal

In Section 1.6.2 we defined conormal fields, and in Section 1.9 we listed their elementary properties. The equiaffine normal field Y is, in particular, a relative normal field (see Section 1.15.2), and therefore there exists a unique conormal field U of a hypersurface $x : M \to A^{n+1}$ satisfying

$$\langle U, dx(w) \rangle = 0, \quad w \in T_p M,$$
$$\langle U, Y \rangle = 1.$$

Recall that, in the calculus of moving frames, if we identify the tangent spaces T_pM and $dx(T_pM)$, then the first of the above equations reads

$$\langle U, w \rangle = 0, \quad w \in T_pM.$$

We summarize the following basic formulas (i)–(v) from Section 1.15.2 in the moving frame calculus; it is easy to verify (iv) and (v).

Lemma 2.6. *Let M be a nondegenerate hypersurface in A^{n+1},*

$$U : M \to V^*$$

be the equiaffine conormal field, $\{e_1, \ldots, e_n\}$ an arbitrary local frame field on M and $\{\omega^1, \ldots, \omega^n\}$ its dual coframe field. For $dU = \sum U_i \omega^i$, i.e.

$$dU(e_i) = U_*(e_i) =: U_i,$$

we have

(i) $\langle U, e_i \rangle = 0, \quad \langle U, Y \rangle = 1, \quad \langle U_i, Y \rangle = 0$;

(ii) $\langle U_i, e_j \rangle = -G_{ij}$, *where G denotes the Blaschke metric*;

(iii) *the volume forms $\omega := \omega(Y)$, $\omega^* := \omega^*(U)$ and the Riemannian volume form $\omega(G)$ satisfy*

$$\omega(e_1, \ldots, e_n) \cdot \omega^*(e_1, \ldots, e_n) = (-1)^n \det(G_{ij}),$$

and from $|\omega| = |\omega^*|$ in Theorem 1.19 we have

$$|\omega(e_1, \ldots, e_n)| = |\omega(G)(e_1, \ldots, e_n)| = |\det(G_{ij})|^{1/2} = |\omega^*(e_1, \ldots, e_n)|.$$

We simply use the notation ω, ω^* instead of $\omega(Y), \omega^*(U)$ from now on.

(iv) $\omega(G)(e_1, \ldots, e_n)U = e_1 \wedge \cdots \wedge e_n$,

$\quad \omega(G)(e_1, \ldots, e_n)U_i = \sum G_{ij} e_1 \wedge \cdots \wedge e_{j-1} \wedge Y \wedge e_{j+1} \wedge \cdots \wedge e_n$;

(v) $\omega^*(e_1, \ldots, e_n)e_i = \sum G_{ij} U_1 \wedge \cdots \wedge U_{j-1} \wedge U \wedge U_{j+1} \wedge \cdots \wedge U_n$,

$\quad \omega^*(e_1, \ldots, e_n)Y = U_1 \wedge \cdots \wedge U_n$.

As a corollary, from (v) and $dx = \sum \omega^i e_i$ we get a generalization of the Lelieuvre formulas from § 52 in [20]; obviously the moving frame calculus is the ideal calculus to write down these formulas.

Proposition 2.7. *A locally strongly convex hypersurface satisfies the following Lelieuvre formula*

$$x = \int_M [\omega^*(e_1, \ldots, e_n)]^{-1} \sum G_{ij} U_1 \wedge \cdots \wedge U_{j-1} \wedge U \wedge U_{j+1} \wedge \cdots \wedge U_n \omega^i.$$

Remark 2.8. As already stated in Blaschke's book [20, p. 140], these formulas allow us to determine the hypersurface from its conormal image and the conformal class of the Blaschke metric. For details the reader is referred to the paper [183], where nonconvex hypersurfaces are also investigated.

Finally, we recall the structure equations for conormal field in Section 1.15.2-2; the structure equation for the equiaffine conormal reads in terms of moving frames:

$$dU_i = \sum \omega_i^{*j} U_j + \omega_i^{*n+1} U, \tag{2.68}$$

where

$$\omega_i^{*j} = \sum \Gamma_{ik}^{*j} \omega^k, \tag{2.69}$$

$$\omega_i^{*n+1} = \sum B_{ij}^* \omega^j; \tag{2.70}$$

ω_i^{*j} is the connection form of ∇^*.

Proposition 2.9 (Conjugate connections). (i) *The connections ∇, ∇^* induced from the equiaffine normalization $\{U, Y\}$ of x are conjugate; that means their connection forms satisfy*

$$dG_{ij} = \sum \omega_i^{*k} G_{kj} + \sum \omega_j^k G_{ik}.$$

Locally:

$$\partial_k G_{ij} = \sum \Gamma_{ik}^{*l} G_{lj} + \sum \Gamma_{jk}^l G_{il}.$$

(ii) *The Levi–Civita connection $\tilde{\nabla}$ of G is the mean connection of ∇, ∇^*:*

(ii.1) $$\tilde{\nabla} = \frac{1}{2}(\nabla + \nabla^*),$$

or in local notation

(ii.2) $$\tilde{\Gamma}_{ij}^k = \frac{1}{2}(\Gamma_{ij}^k + \Gamma_{ij}^{*k}),$$

or in the moving frame calculus

(ii.3) $$\tilde{\omega}_i^k = \frac{1}{2}(\omega_i^k + \omega_i^{*k}).$$

Moreover

(ii.4) $$\nabla = \tilde{\nabla} + A \ or \ \Gamma_{ij}^k = \tilde{\Gamma}_{ij}^k + A_{ij}^k \ and$$

(ii.5) $$\nabla^* = \tilde{\nabla} - A \ or \ \Gamma_{ij}^{*k} = \tilde{\Gamma}_{ij}^k - A_{ij}^k.$$

Proof. (i) This is obtained by differentiation of (ii) in Lemma 2.6.

(ii) It follows from (i) that

$$dG_{ij} = \frac{1}{2}\sum(\omega_i^k + \omega_i^{*k})G_{kj} + \frac{1}{2}\sum(\omega_j^k + \omega_j^{*k})G_{ik}, \tag{2.71}$$

which shows that the connection $\frac{1}{2}(\nabla + \nabla^*)$ satisfies the Ricci lemma. As ∇ and ∇^* are torsion-free, the connection $\frac{1}{2}(\nabla + \nabla^*)$ is torsion-free. From (2.71) the connection $\frac{1}{2}(\nabla + \nabla^*)$ must be the Levi–Civita connection of G. $\qquad\square$

Let $dU_i = \sum U_{ij}\omega^j$, i.e. U_{ij} denote the components of the second order partial derivatives. We have

Proposition 2.10. (1) $U_{ij} = \sum \Gamma_{ij}^{*k}U_k - B_{ij}U,$
(2) $U_{,ij} = -\sum A_{ij}^k U_k - B_{ij}U.$

Proof. The assertion (1) follows from (2.68)–(2.70) and the fact that

$$B_{ij}^* = \langle U_{ij}, Y \rangle = -\langle U_i, Y_j \rangle = -B_{ij}.$$

Assertion (2) follows from (ii) of Proposition 2.9. $\qquad\square$

Corollary 2.11. *The conormal field U satisfies the vector valued Schrödinger-type (Laplacian plus potential) p.d.e.*

$$\Delta U + nL_1 U = 0.$$

Proof. From (ii) of Proposition 2.9 and the apolarity condition (2.46), we get

$$\Delta U = \sum G^{ij}U_{,ij} = \sum G^{ij}A_{ij}^k U_k - \sum G^{ij}B_{ij}U = -nL_1 U. \qquad\square$$

Note. The p.d.e. in Corollary 2.11 plays an important role in many global investigations in later chapters.

2.3.2 The affine support function

Definition 2.12. Let b be a vector in V. The function $\Lambda(x)$ on M defined by

$$\Lambda(x) := \langle U, b - x \rangle, \quad x \in M,$$

is called the *affine support function* relative to the vector b.

The support function satisfies p.d.e.'s, which play an important role for global investigations. We are going to compute the Laplacian of $\Lambda(x)$. Let us write $d\Lambda = \sum \Lambda_i \omega^i$. By Lemma 2.6 (i) we have

$$d\Lambda = \langle dU, b - x \rangle - \langle U, dx \rangle$$
$$= \sum \langle U_i, b - x \rangle \omega^i.$$

Hence

$$\Lambda_i = \langle U_i, b - x \rangle.$$

By the definition of the second covariant derivative (called the covariant Hessian of $\Lambda(x)$)

$$\sum \Lambda_{,ij}\omega^j = d\Lambda_i - \sum \tilde{\omega}_i^j \Lambda_j$$

$$= \langle dU_i, b - x \rangle - \langle U_i, dx \rangle - \sum \tilde{\omega}_i^j \Lambda_j$$

$$= \langle dU_i - \sum \tilde{\omega}_i^j U_j, b - x \rangle - \sum \langle U_i, e_j \rangle \omega^j$$

$$= \sum \left(\langle U_{,ij}, b - x \rangle - \langle U_i, e_j \rangle \right) \omega^j.$$

Therefore, the covariant Hessian of $\Lambda(x)$ satisfies

$$\Lambda_{,ij} = \langle U_{,ij}, b - x \rangle - \langle U_i, e_j \rangle, \tag{2.72}$$

or

$$\Lambda_{,ij} = -\sum A_{ij}^k \Lambda_k - \Lambda B_{ij} + G_{ij}. \tag{2.73}$$

It follows immediately that

$$\Delta \Lambda = \sum G^{ij} \Lambda_{,ij} = -nL_1 \Lambda + n. \tag{2.74}$$

Recall the related homogeneous equations for U in Proposition 2.10 and Corollary 2.11. The system (2.73) can be rewritten in terms of ∇^* and plays an important role in the geometry of projectively flat manifolds. For details see [341, §4.13].

2.4 Hyperquadrics

2.4.1 Hyperquadrics

Theorem 1.10 on nondegeneracy in Section 1.11 excludes hyperplanes, i.e. hypersurfaces which can be described by polynomials of first order. Thus the simplest class of surfaces in the affine theory consists of regular quadrics, which can be described by polynomials of second order.

As an example and exercise, we calculate the fundamental quantities such as the Blaschke metric, Fubini-Pick (cubic) form, etc.

Example 2.1. Elliptic paraboloid: we consider the graph

$$x^{n+1} = \frac{1}{2}((x^1)^2 + (x^2)^2 + \cdots + (x^n)^2)$$

and choose a local unimodular affine frame field as follows:

$$e_1 = (1, 0, \ldots, 0, x^1),$$
$$e_2 = (0, 1, 0, \ldots, 0, x^2),$$
$$\vdots$$
$$e_n = (0, \ldots, 0, 1, x^n),$$
$$e_{n+1} = (0, \ldots, 0, 1).$$

It is easy to verify the relations

$$h_{ij} = \delta_{ij} \text{ and } H = \det(h_{ij}) = 1,$$

which give the Blaschke metric G:

$$G_{ij} = \delta_{ij},$$
$$G = (dx^1)^2 + (dx^2)^2 + \cdots + (dx^n)^2.$$

Therefore the Riemann space (M, G) is flat. Obviously

$$\Delta x^1 = \Delta x^2 = \cdots = \Delta x^n = 0,$$
$$\Delta x^{n+1} = \frac{1}{2}\Delta\left((x^1)^2 + \cdots + (x^n)^2\right) = n.$$

It follows that the affine normal is given by a constant vector field:

$$Y = (0, \ldots, 0, 1) = e_{n+1}.$$

Since $de_{n+1} = 0$, we have

$$\omega^i_{n+1} = 0.$$

Thus, the affine principal curvatures vanish, that is,

$$\lambda_1 = \lambda_2 = \cdots = \lambda_n = 0. \tag{2.75}$$

As a consequence the quadratic form $\hat{B}$ vanishes identically. Moreover, it is easy to see that the connection ∇ is also a flat connection, thus

$$0 = \omega^j_i = \tilde{\omega}^j_i, \tag{2.76}$$

and so the difference tensor $K = \frac{1}{2}(\nabla - \nabla^*)$ and the corresponding cubic form vanish identically:

$$A_{ijk} = 0, \quad 1 \le i, j, k \le n, \tag{2.77}$$

therefore

$$J = 0. \tag{2.78}$$

Example 2.2. The ellipsoid

$$\frac{(x^1)^2}{(a^1)^2} + \frac{(x^2)^2}{(a^2)^2} + \cdots + \frac{(x^{n+1})^2}{(a^{n+1})^2} = 1,$$

where $a^1, a^2, \ldots, a^{n+1}$ are positive constants. By a unimodular affine transformation, the equation of M can be written in the equiaffine normal form

$$(x^1)^2 + (x^2)^2 + \cdots + (x^{n+1})^2 = r^2,$$

where r is a positive constant. We consider the canonical inner product

$$(\cdot, \cdot) : V \times V \to \mathbb{R},$$

given by

$$(x, y) := \sum_{A=1}^{n+1} x^A y^A \text{ for } x, y \in V.$$

This defines a Euclidean structure on A^{n+1}. Then the equation $\sum_{A=1}^{n+1}(x^A)^2 = r^2$ describes a Euclidean n-sphere, and the above transformation confirms the well-known fact that an ellipsoid is equiaffinely equivalent to a sphere of appropriate radius.

Choose a local-oriented orthonormal frame field $\{x; e_1, \ldots, e_{n+1}\}$ with respect to this inner product such that on M the vector fields $e_1, \ldots, e_n$ are tangent and e_{n+1} is the interior unit normal field. Then

$$\mathrm{Det}\,(e_1, \ldots, e_n, e_{n+1}) = 1,$$

and

$$e_{n+1} = -\frac{1}{r}x,$$

$$de_{n+1} = -\frac{1}{r}dx,$$

$$\omega_j^{n+1} = -\omega_{n+1}^j = \frac{1}{r}\omega^j,$$

$$h_{ij} = \frac{1}{r}\delta_{ij}, \quad H = \frac{1}{r^n},$$

$$G_{ij} = r^{-2/(n+2)}\delta_{ij}.$$

Let g denote the Euclidean first fundamental form induced from $(A^{n+1}, (\cdot, \cdot))$. We have

$$G = r^{-2/(n+2)}g, \tag{2.79}$$

$$\Delta_G = r^{2/(n+2)}\Delta_g, \tag{2.80}$$

where Δ_G and Δ_g denote the Laplace operators with respect to G and g, respectively. The Euclidean Gauß structure equations imply the well-known relation

$$\Delta_g x = \frac{n}{r}e_{n+1} = -\frac{n}{r^2}x. \tag{2.81}$$

It follows immediately that

$$Y = \frac{1}{n}\Delta_G x = \frac{1}{n}r^{2/(n+2)}\Delta_g x = -r^{-(2n+2)/(n+2)}x, \tag{2.82}$$

which shows that the affine normal vector is parallel to the position vector, and the equiaffine and Euclidean normals coincide up to a constant nonzero factor. From (2.82) we obtain that the affine Weingarten operator B is a constant multiple of the identity operator, and thus the Weingarten form satisfies

$$\hat{B} = \frac{1}{r^2}[(\omega^1)^2 + (\omega^2)^2 + \cdots + (\omega^n)^2],$$

and all affine principal curvatures are equal:

$$\lambda_1 = \lambda_2 = \cdots = \lambda_n = r^{-(2n+2)/(n+2)}. \tag{2.83}$$

Moreover, as the Euclidean and affine normalizations coincide (up to a factor), the induced connections ∇ and $\tilde{\nabla}$ coincide:

$$\omega_j^i = \tilde{\omega}_j^i, \tag{2.84}$$

and so the difference tensor and the cubic form, respectively, vanish:

$$A_{ijk} = 0; \quad 1 \le i, j, k \le n, \tag{2.85}$$

and hence

$$J = 0. \tag{2.86}$$

Example 2.3. The hyperboloid: we start with the equiaffine normal form (see Example 2.2). Consider

$$M : (x^1)^2 + (x^2)^2 + \cdots + (x^n)^2 - (x^{n+1})^2 = -c^2,$$

where c is a positive constant. Again we simplify the considerations introducing the Minkowski inner product $(\cdot, \cdot)_1 : V \times V \to \mathbb{R}$ as follows:

$$(x, y)_1 := x^1 y^1 + \cdots + x^n y^n - x^{n+1} y^{n+1}, \quad x, y \in V.$$

The above equation reads now: $(x, x)_1 = -c^2$, or

$$\left(\frac{x}{c}, \frac{x}{c}\right)_1 = -1. \tag{2.87}$$

Hence

$$(dx, x)_1 = 0. \tag{2.88}$$

Choose a local frame field $\{x; e_1, \ldots, e_n, e_{n+1}\}$ on M such that $e_1, \ldots, e_n \in T_x M$ and $e_{n+1} = \frac{x}{c}$. From (2.87) and (2.88) it follows that

$$(e_i, e_{n+1})_1 = 0, \quad (e_{n+1}, e_{n+1})_1 = -1. \tag{2.89}$$

The inner product on the Minkowski space induces a Euclidean metric g on M:

$$(e_i, e_j)_1 = g_{ij}. \tag{2.90}$$

It is easy to see that (g_{ij}) is positive definite from the inertial theorem in linear algebra. We consider an orthonormal frame $\{x; e_1, \ldots, e_n\}$ on M relative to g, such that

$$\mathrm{Det}\,(e_1, \ldots, e_n, e_{n+1}) = 1,$$

and the matrix with coefficients $(e_A, e_B)_1$ satisfies

$$((e_A, e_B)_1) = \begin{pmatrix} 1 & & & 0 \\ & \ddots & & \\ & & 1 & \\ 0 & & & -1 \end{pmatrix}. \tag{2.91}$$

Let $e_A = (\eta_A^1, \eta_A^2, \ldots, \eta_A^{n+1})$. Then the equation $\mathrm{Det}\,(e_1, \ldots, e_n, e_{n+1}) = 1$ implies that

$$\begin{vmatrix} \eta_1^1 & \cdots & \eta_1^{n+1} \\ \vdots & & \vdots \\ \eta_{n+1}^1 & \cdots & \eta_{n+1}^{n+1} \end{vmatrix} = 1.$$

From (2.89) and $(e_i, e_j)_1 = g_{ij} = \delta_{ij}$ it follows that

$$(w_i^{n+1} e_{n+1}, e_{n+1})_1 + (e_i, \sum w_{n+1}^j e_j)_1 = 0, \quad w_{n+1}^i = w_i^{n+1}. \tag{2.92}$$

A computation similar to Example 2.2 gives

$$w_i^{n+1} = w_{n+1}^i = \frac{1}{c} w^i, \tag{2.93}$$

$$G_{ij} = c^{-2/(n+2)} \delta_{ij}, \tag{2.94}$$

$$\Delta_G = c^{2/(n+2)} \Delta_g. \tag{2.95}$$

We compute $\Delta_g x$; since

$$de_i = \sum w_i^j e_j + \frac{1}{c} w^i e_{n+1},$$

we have

$$\Delta_g x = \frac{n}{c} e_{n+1} = \frac{n}{c^2} x; \tag{2.96}$$

consequently,

$$Y = c^{-(2n+2)/(n+2)} x. \tag{2.97}$$

As in Example 2.2, the affine normal vector is parallel to the position vector, and the Weingarten operator must be a constant multiple of the identity, or

$$\hat{B} = -\frac{1}{c^2}[(w^1)^2 + (w^2)^2 + \cdots + (w^n)^2], \tag{2.98}$$

and all affine principal curvatures are equal:

$$\lambda_1 = \lambda_2 = \cdots = \lambda_n = -c^{-(2n+2)/(n+2)}. \tag{2.99}$$

Again as in Example 2.2 we conclude that

$$A_{ijk} = 0, \quad 1 \le i, j, k \le n, \tag{2.100}$$

$$J = 0. \tag{2.101}$$

2.4.2 Hypersurfaces with vanishing Pick invariant

We have seen that all locally strongly convex hyperquadrics satisfy $J = 0$. In the following we prove that, within the class of all locally strongly convex hypersurfaces, the condition $J = 0$ characterizes hyperquadrics. In the locally strongly convex case this important classical result is due to Maschke (for analytic surfaces), Pick (for all surfaces) and Berwald (for hypersurfaces).

Theorem 2.13. *Let M be a locally strongly convex hypersurface in A^{n+1}. If the Pick invariant vanishes on M, i.e. $J = 0$, then M is locally a hyperquadric.*

Proof. Choose a local equiaffine frame field $\{x; e_1, \ldots, e_{n+1}\}$ on M, such that

$$e_{n+1} = Y, \quad G_{ij} = \delta_{ij},$$

and

$$\omega_i^{n+1} = \omega^i, \quad \omega_{n+1}^{n+1} = 0. \tag{2.102}$$

Since $J = 0$, we have

$$\omega_j^i + \omega_i^j = 0. \tag{2.103}$$

Exterior differentiation of (2.103) yields immediately

$$\omega_i^{n+1} \wedge \omega_{n+1}^j + \omega_j^{n+1} \wedge \omega_{n+1}^i = 0. \tag{2.104}$$

Inserting $\omega_{n+1}^i = -\sum l^{ij}\omega_j^{n+1} = -\sum l^{ij}\omega^j$ into (2.104) we get

$$\sum (\delta_{im} l^{jk} + \delta_{jm} l^{ik})\omega^m \wedge \omega^k = 0.$$

Hence

$$\delta_{im} l^{jk} + \delta_{jm} l^{ik} - \delta_{ik} l^{jm} - \delta_{jk} l^{im} = 0. \tag{2.105}$$

Contraction of indices gives

$$l^{jk} = a\delta^{jk},$$

where $a = \frac{1}{n}\sum l^{ii}$. Consequently the Weingarten operator satisfies $B = a \cdot id$ and all principal curvatures are equal: $\lambda_1 = \lambda_2 = \cdots = \lambda_n = a$. In forms

$$\omega_{n+1}^i = -a\omega^i, \quad i = 1, 2, \ldots, n, \tag{2.106}$$

and exterior differentiation implies

$$da \wedge \omega^i = 0, \quad i = 1, 2, \ldots, n.$$

Thus

$$a = \text{const.} \tag{2.107}$$

There are two cases: $a = 0$ and $a \neq 0$.

1) If $a = 0$, that is, $\omega^i_{n+1} = 0$, it follows immediately that

$$de_{n+1} = 0.$$

This means that e_{n+1} is a constant vector field. Then we can choose it to be in the form $e_{n+1} = (0, \ldots, 0, 1)$ so that M can be locally represented as a graph:

$$x^{n+1} = f(x^1, \ldots, x^n).$$

We consider the frame

$$\hat{e}_1 = (1, 0, \ldots, 0, \partial_1 f),$$
$$\hat{e}_2 = (0, 1, 0, \ldots, 0, \partial_2 f),$$
$$\vdots$$
$$\hat{e}_n = (0, \ldots, 0, 1, \partial_n f),$$
$$e_{n+1} = (0, \ldots, 0, 1),$$

and obtain

$$\hat{\omega}^j_i = 0, \quad \hat{h}_{ij} = \frac{\partial^2 f}{\partial x^i \partial x^j} = \partial_i \partial_j f.$$

Furthermore, since $e_{n+1} = Y = \text{const.}$,

$$\hat{H} = \det(\partial_i \partial_j f) = 1$$

and

$$\hat{G}_{ij} = \hat{h}_{ij}. \tag{2.108}$$

On the other hand, from (2.40), (2.41) and $\hat{\omega}^j_i = 0$ we have that

$$d\hat{G}_{ij} = d\hat{G}_{ij} - \sum \hat{\omega}^k_i \hat{G}_{kj} - \sum \hat{\omega}^k_j \hat{G}_{ik} = 0,$$

and so

$$\hat{G}_{ij} = \text{const.}$$

From (2.108), we obtain

$$\partial_i \partial_j f = \hat{h}_{ij} = \text{const.}$$

This shows that f is a quadratic polynomial. By a suitable unimodular affine transformation, we have

$$x^{n+1} = \rho\left((x^1)^2 + (x^2)^2 + \cdots + (x^n)^2\right),$$

where ρ is a positive constant. This shows that M is an elliptic paraboloid.

2) If $a \neq 0$, then it follows from (2.106) that

$$d\left(x + \frac{e_{n+1}}{a}\right) = 0.$$

Hence $x + \frac{e_{n+1}}{a}$ is a constant vector. Without loss of generality, we can assume

$$e_{n+1} = -ax. \tag{2.109}$$

Let $\{O; \eta_1, \ldots, \eta_{n+1}\}$ be a fixed unimodular affine frame and $x^A, e_1^A, e_2^A, \cdots, e_{n+1}^A$ be the A-th component of $x, e_1, e_2, \ldots, e_{n+1}$ relative to this frame, respectively. We recall that $de_{n+1} = -a \sum w^i e_i$, and so

$$-d\left(\frac{1}{a} e_{n+1}^A e_{n+1}^B\right) = \sum e_k^A e_{n+1}^B w^k + \sum e_k^B e_{n+1}^A w^k. \tag{2.110}$$

On the other hand, from $de_i = \sum w_i^j e_j + w^i e_{n+1}$ it follows that

$$d\left(\sum e_i^A e_i^B\right) = \sum e_k^A e_{n+1}^B w^k + \sum e_k^B e_{n+1}^A w^k. \tag{2.111}$$

Combining (2.110) and (2.111) gives

$$d\left(\sum e_i^A e_i^B\right) = -d\left(\frac{1}{a} e_{n+1}^A e_{n+1}^B\right).$$

This implies that

$$\sum e_i^A e_i^B + \frac{1}{a} e_{n+1}^A e_{n+1}^B =: C_{AB} = \text{const.} \tag{2.112}$$

Define

$$E := \begin{pmatrix} e_1^1 & e_2^1 & \cdots & e_{n+1}^1 \\ \vdots & & & \vdots \\ e_1^{n+1} & e_2^{n+1} & \cdots & e_{n+1}^{n+1} \end{pmatrix}, \quad F := \begin{pmatrix} 1 & & & 0 \\ & \vdots & & \\ & & 1 & \\ 0 & & & 1/a \end{pmatrix},$$

$$C := \begin{pmatrix} C_{11} & C_{12} & \cdots & C_{1,n+1} \\ \vdots & & & \vdots \\ C_{n+1,1} & C_{n+1,2} & \cdots & C_{n+1,n+1} \end{pmatrix}.$$

Then (2.112) can be written as

$$EFE^T = C,$$

where E^T denotes the transpose of the matrix E. For a fixed point $x_0 \in M$, we can choose a basis $\{\eta_1, \eta_2, \ldots, \eta_{n+1}\}$ for V such that

$$\eta_1 = e_1(x_0), \quad \eta_2 = e_2(x_0), \quad \cdots, \quad \eta_{n+1} = e_{n+1}(x_0).$$

Then at x_0, E is the identity matrix, and so

$$F = C.$$

Since C is a constant matrix on M, we have

$$EFE^\tau = F,$$

or

$$E^\tau F^{-1} E = F^{-1}. \tag{2.113}$$

Inserting the definition of E and F into (2.113) we are led to

$$\sum_i (e_{n+1}^i)^2 + a(e_{n+1}^{n+1})^2 = a.$$

This and (2.109) finally give

$$\sum_i (x^i)^2 + a(x^{n+1})^2 = \frac{1}{a}. \tag{2.114}$$

This shows that M is a hyperquadric.

If $a > 0$, the unimodular affine transformation

$$\begin{cases} \hat{x}^i = a^{-1/(2n+2)} x^i, \ i = 1, 2, \ldots, n, \\ \hat{x}^{n+1} = a^{n/(2n+2)} x^{n+1}, \end{cases}$$

transforms (2.114) into

$$(\hat{x}^1)^2 + (\hat{x}^2)^2 + \cdots + (\hat{x}^{n+1})^2 = a^{-(n+2)/(n+1)};$$

thus M is an ellipsoid.

If $a < 0$, the transformation

$$\begin{cases} \hat{x}^i = (-a)^{-1/(2n+2)} x^i, \ i = 1, 2, \ldots, n, \\ \hat{x}^{n+1} = (-a)^{n/(2n+2)} x^{n+1}, \end{cases}$$

of (2.114) gives

$$(\hat{x}^1)^2 + \cdots + (\hat{x}^n)^2 - (\hat{x}^{n+1})^2 = -(-a)^{-(n+2)/(n+1)},$$

which means that M is a hyperboloid. $\qquad\square$

Remark 2.14. From the proof of Theorem 2.13 it can be easily seen that the vanishing of the Fubini–Pick form $A = 0$ locally characterizes nondegenerate hyperquadrics. We point out the fact that, for locally strongly convex hypersurfaces in A^{n+1}, $A = 0$ is equivalent to $J = 0$.

2.5 Integrability conditions and the local fundamental theorem

As in Euclidean hypersurface theory, the structure equations of Weingarten and Gauß
for a hypersurface

$$de_{n+1} = \sum \omega_{n+1}^i e_i,$$

$$de_i = \sum \omega_i^j e_j + \omega_i^{n+1} e_{n+1}$$

give a linear system of first order p.d.e.'s for a local frame $\{e_1, \ldots, e_n, e_{n+1}\}$. The coefficients define linear forms which are related to the connections, quadratic and cubic
forms which we considered so far:

$$\omega_{n+1}^i = -\sum l^{ik}\omega_k^{n+1} = -\sum l_k^i \omega^k,$$

$$\omega_i^j = \sum \Gamma_{ik}^j \omega^k,$$

$$\omega_i^{n+1} = \sum h_{ij}\omega^j \quad \text{where } G_{ij} = |H|^{-1/(n+2)} h_{ij}.$$

From the integration theory for such linear systems, we know that there exists at most
one solution $e_1, \ldots, e_n, e_{n+1}$, for given coefficients and given initial values. Such a solution determines in particular

$$dx = \sum \omega^i e_i,$$

and a second integration locally gives the hypersurface x itself. Thus, roughly speaking, the coefficients must contain all geometric information about x.

The existence of a solution depends on the fact that the coefficients satisfy integrability conditions.

These facts motivate the investigations of this paragraph:

(1) in the first step we recall relations between the coefficients;
(2) in the second step we calculate the integrability conditions;
(3) in the third step we prove a fundamental existence and uniqueness theorem in
    analogy to Bonnet's fundamental theorem in Euclidean hypersurface theory.

Although the third structure equation, the Gauß equation for Y, is not essential for the
proof of the fundamental theorem, it is quite useful to include it in our investigations.
We will use the moving frame calculus and realize that it is adequate for the integration of the linear system considered. To date this approach has not yet appeared in
any of the monographs, and only in the two papers [60, 93].

As before, we restrict ourselves to a locally strongly convex hypersurface, i.e. to a
definite metric G. However, the results hold true for an indefinite metric. There exist
different versions of the fundamental theorem in the literature. We prove the classical version due to Radon, which so far is the most important version for many global
applications. The lecture notes [341, § 4] contains a detailed study and comparison of
the different versions; there the invariant calculus is used.

2.5.1 Relations between the coefficients

(i) From Lemma 2.5, the coefficients G, B and $\hat{B}$ satisfy

$$G(Bv, w) = \hat{B}(v, w).$$

(ii) The connections ∇, ∇^* and the Levi–Civita connection $\tilde{\nabla}$ are related by

$$\tilde{\nabla} = \frac{1}{2}(\nabla + \nabla^*),$$

while the cubic form satisfies

$$K = \frac{1}{2}(\nabla - \nabla^*), \quad A(u, v, w) = G(u, K(v, w)).$$

(iii) A and G satisfy the apolarity condition (2.46):

$$\text{tr}_G A = 0,$$

or

$$\sum G^{ij} A_{ijk} = 0, \quad \text{for } k = 1, 2, \ldots, n.$$

2.5.2 Integrability conditions

Choose a local equiaffine frame field $\{x; e_1, \ldots, e_n, e_{n+1}\}$ on M such that $e_{n+1} = Y$, $G_{ij} = \delta_{ij}$.

Proposition 2.15. *The integrability conditions of the system*
(i) $dx = \sum \omega^i e_i,$
(ii) $de_i = \sum \omega_i^j e_j + \omega_i^{n+1} e_{n+1},$
(iii) $de_{n+1} = \sum \omega_{n+1}^i e_i,$
(iv) $\omega_i^{n+1} = \omega^i, \quad \omega_{n+1}^{n+1} = 0$

read

(v) $\sum \omega_i^i = 0,$
(vi) $d\omega^i = \sum \omega^j \wedge \omega_j^i,$
(vii) $d\omega_i^j = \sum \omega_i^k \wedge \omega_k^j + \omega_i^{n+1} \wedge \omega_{n+1}^j,$
       $\omega_i^{n+1} = \omega^i,$
(viii) $d\omega_{n+1}^i = \sum \omega_{n+1}^j \wedge \omega_j^i.$

The equations (v)–(viii) between the linear differentiable forms $\omega^i, \omega_i^j, \omega_{n+1}^i$ are sufficient for the integration of (i)–(iv). In the terminology of moving frames the equations (v)–(viii) are called structure equations, which means that they are necessary and sufficient for the existence of the hypersurface structure.

Proof. Since $\omega^{n+1} = 0$, the proof is obvious from (i)–(iv); use the rules in the Appendix A.2.2. □

We are going to express these conditions in terms of the quadratic and cubic forms $G, \hat{B}$, and A.

With the above frame field, we have

$$\tilde{\omega}_i^j = \frac{1}{2}(\omega_i^j - \omega_j^i), \tag{2.115}$$

$$\omega_i^j - \tilde{\omega}_i^j = \frac{1}{2}(\omega_i^j + \omega_j^i) = \sum A_{ijk}\omega^k, \tag{2.116}$$

$$\omega_{n+1}^i = -\sum l^{ik}\omega_k^{n+1} = -\sum l^{ik}\omega^k. \tag{2.117}$$

Exterior differentiation of (2.116) gives

$$\sum \omega_i^k \wedge \omega_k^j + \sum \omega_j^k \wedge \omega_k^i + \omega_i^{n+1} \wedge \omega_{n+1}^j + \omega_j^{n+1} \wedge \omega_{n+1}^i$$
$$= 2\sum dA_{ijk} \wedge \omega^k + 2\sum A_{ijk}\omega^l \wedge \omega_l^k. \tag{2.118}$$

Inserting (2.116) into (2.118), we get

$$2\sum \left(dA_{ijk} - \sum A_{ijl}\omega_k^l - \sum A_{ilk}\omega_j^l - \sum A_{ljk}\omega_i^l\right) \wedge \omega^k = \omega_i^{n+1} \wedge \omega_{n+1}^j + \omega_j^{n+1} \wedge \omega_{n+1}^i.$$

Using $\omega_i^j = \tilde{\omega}_i^j + \sum A_{ijk}\omega^k$ and the symmetry of A_{ijk}, one verifies that

$$2\sum \left(dA_{ijk} - \sum A_{ljk}\tilde{\omega}_i^l - \sum A_{ilk}\tilde{\omega}_j^l - \sum A_{ijl}\tilde{\omega}_k^l\right) \wedge \omega^k = \omega_i^{n+1} \wedge \omega_{n+1}^j + \omega_j^{n+1} \wedge \omega_{n+1}^i,$$

or

$$2\sum A_{ijk,l}\omega^l \wedge \omega^k = -\sum (\delta_{il}l^{jk} + \delta_{jl}l^{ik})\omega^l \wedge \omega^k,$$

where $A_{ijk,l}$ denotes the covariant derivative of A_{ijk} with respect to G. From the last equality above it follows that

$$A_{ijk,l} - A_{ijl,k} = \frac{1}{2}(\delta_{ik}l^{jl} + \delta_{jk}l^{il} - \delta_{il}l^{jk} - \delta_{jl}l^{ik}). \tag{2.119}$$

The components of the Riemannian curvature form and tensor of G are denoted by Ω_i^j and R_{ijkl}, respectively. From Riemannian geometry (see Appendix A.3.1.4 and A.3.1.5) we know that

$$\Omega_i^j = d\tilde{\omega}_i^j - \sum \tilde{\omega}_i^k \wedge \tilde{\omega}_k^j \tag{2.120}$$

and

$$\Omega_i^j = -\frac{1}{2}\sum R_{ijkl}\omega^k \wedge \omega^l, \quad R_{ijkl} + R_{ijlk} = 0. \tag{2.121}$$

We insert (2.116) into (2.120) and check directly that

$$\Omega_i^j = \omega_i^{n+1} \wedge \omega_{n+1}^j + \sum A_{iml}A_{mjk}\omega^l \wedge \omega^k$$
$$- \sum \left(dA_{ijk} - \sum A_{ijl}\tilde{\omega}_k^l - \sum A_{ilk}\tilde{\omega}_j^l - \sum A_{ljk}\tilde{\omega}_i^l\right) \wedge \omega^k,$$

or

$$\Omega_i^j = \omega_i^{n+1} \wedge \omega_{n+1}^j + \sum A_{iml}A_{mjk}\omega^l \wedge \omega^k - \sum A_{ijk,l}\omega^l \wedge \omega^k.$$

The curvature tensor (see Appendix A.3.2.3) is then given by

$$R_{ijkl} = A_{ijl,k} - A_{ijk,l} + \delta_{ik}l^{jl} - \delta_{il}l^{jk} + \sum(A_{iml}A_{mjk} - A_{imk}A_{mjl}). \tag{2.122}$$

From (2.119) we get

$$R_{ijkl} = \frac{1}{2}(\delta_{ik}l^{jl} + \delta_{jl}l^{ik} - \delta_{il}l^{jk} - \delta_{jk}l^{il}) + \sum(A_{iml}A_{mjk} - A_{imk}A_{mjl}). \tag{2.123}$$

Exterior differentiation of (2.117) gives

$$\sum\left(dl^{ik} + \sum l^{il}\omega_l^k + \sum l^{lk}\omega_l^i\right) \wedge \omega^k = 0.$$

It follows from $G_{ij} = \delta_{ij}, B_{ij} = l^{ij}$ and

$$\omega_l^k = 2\sum A_{lkm}\omega^m - \omega_k^l$$

that

$$\sum\left(dB_{ik} - \sum B_{il}\omega_k^l - \sum B_{lk}\omega_i^l\right) \wedge \omega^k = 2\sum B_{lk}A_{ilm}\omega^k \wedge \omega^m.$$

Inserting $\omega_i^j = \tilde{\omega}_i^j + \sum A_{ijk}\omega^k$ into the last equation, we conclude that

$$\sum\left(dB_{ik} - \sum B_{il}\tilde{\omega}_k^l - \sum B_{lk}\tilde{\omega}_i^l\right) \wedge \omega^k = \sum B_{lk}A_{ilm}\omega^k \wedge \omega^m,$$

or

$$\sum B_{ik,j}\omega^j \wedge \omega^k = \sum B_{lk}A_{ilj}\omega^k \wedge \omega^j.$$

It follows immediately that

$$B_{ik,j} - B_{ij,k} = \sum(B_{lj}A_{ilk} - B_{lk}A_{ilj}). \tag{2.124}$$

Formulas (2.119), (2.123), and (2.124) are derived by choosing a special frame field. Relative to an arbitrary unimodular affine frame field we state them in the following.

Proposition 2.16.

$$A_{ijk,l} - A_{ijl,k} = \frac{1}{2}(G_{ik}B_{jl} + G_{jk}B_{il} - G_{il}B_{jk} - G_{jl}B_{ik}), \tag{2.119'}$$

$$R_{ijkl} = \sum(A_{il}^m A_{mjk} - A_{ik}^m A_{mjl}) + \frac{1}{2}(G_{ik}B_{jl} + G_{jl}B_{ik} - G_{il}B_{jk} - G_{jk}B_{il}), \tag{2.123'}$$

$$B_{ik,j} - B_{ij,k} = \sum(B_{jl}A_{ik}^l - B_{lk}A_{ij}^l). \tag{2.124'}$$

While (2.123') is called an integrability condition of Gauß type, the other two systems are said to be of Codazzi type; these notions are analogous to the Euclidean theory.

Corollary 2.17. *The integrability conditions (2.119$'$), (2.123$'$), and (2.124$'$) imply by contraction*

(i) $\sum A^l_{jk,l} = \frac{n}{2}(L_1 G_{jk} - B_{jk})$,

(ii) $R_{ik} = \sum A^m_{il} A^l_{mk} + \frac{n-2}{2} B_{ik} + \frac{n}{2} L_1 G_{ik}$,

(iii) $\sum B^i_{k,i} = n(L_1)_{,k} + \sum B^i_l A^l_{ik}$,

where L_1 is the affine mean curvature and R_{ik} is the Ricci tensor of (M, G). From (ii) we obtain, by another contraction, the so-called equiaffine theorema egregium which relates three fundamental invariants:

$$\chi = J + L_1, \tag{2.125}$$

where

$$\chi = \frac{1}{n(n-1)} R = \frac{1}{n(n-1)} \sum G^{ik} G^{jl} R_{ijkl}, \tag{2.126}$$

R is the scalar curvature and χ the normalized scalar curvature of the metric G.

Corollary 2.18. *The form $\hat{B}$ can be expressed in terms of $G, \hat{A}$, and their derivatives:*

$$B_{jk} = (\chi - J)G_{jk} - \frac{2}{n} \sum A^l_{jk,l}.$$

Proof. This follows from (i) of Corollary 2.17 and (2.125). □

Exercise 2.2. Use (2.123$'$) and the results from Section 2.4 to prove that for any quadric the Blaschke metric G defines a Riemannian space of constant sectional curvature. Prove that the sectional curvature is: (1) positive, for the ellipsoid; (2) zero, for the paraboloid; (3) negative, for the hyperboloid.

2.5.3 The fundamental theorem

First we prove the uniqueness part of the fundamental theorem, secondly the existence part.

Theorem 2.19 (Uniqueness). *Let $M, \bar{M}$ be two locally strongly convex hypersurfaces in A^{n+1} and let $f : M \to \bar{M}$ be a diffeomorphism such that*

$$G = f^* \bar{G}, \quad A = f^* \bar{A}.$$

Then M differs from $\bar{M}$ by a unimodular affine transformation; that means both hypersurfaces are equiaffinely equivalent.

Proof. Choose a local equiaffine frame field $\{x; e_1, \ldots, e_{n+1}\}$ on M such that $G_{ij} = \delta_{ij}$. Set

$$\bar{e}_i = f_*(e_i), \quad i = 1, 2, \ldots, n.$$

At $f(x)$ choose $\bar{e}_{n+1}$ such that $\bar{e}_{n+1}$ is parallel to the affine normal vector $\bar{Y}$ and $\mathrm{Det}\,(\bar{e}_1,\ldots,\bar{e}_n,\bar{e}_{n+1}) = 1$. Then

$$\bar{\omega}^i = \omega^i, \quad \bar{G}_{ij} = G_{ij}, \quad \bar{A}_{ijk} = A_{ijk},$$

where we identify $f^*\bar{\omega}^i$ with $\bar{\omega}^i$. Consequently,

$$\bar{Y} = \bar{e}_{n+1}.$$

The conditions $G = \bar{G}$ and $A = \bar{A}$ give

$$\tilde{\bar{\omega}}_i^j = \tilde{\omega}_i^j, \quad \bar{\omega}_i^j = \omega_i^j,$$

where $\tilde{\bar{\omega}}_i^j$ denote the Levi–Civita connection forms of the metric $\bar{G}$. Moreover, from Corollary 2.18 we get

$$\bar{B}_{jk} = B_{jk}, \quad \bar{\omega}_{n+1}^i = \omega_{n+1}^i.$$

Assume that $x_0 \in M$ is a fixed point. Applying a unimodular affine transformation σ (its determinant is equal to 1 or −1) to $\bar{M}$, we can assume that

$$x_0 = \sigma f(x_0), \quad e_i(x_0) = \sigma\bar{e}_i(f(x_0)), \quad i = 1,\ldots,n,$$
$$e_{n+1}(x_0) = \sigma\bar{e}_{n+1}(f(x_0)).$$

Now let x be an arbitrary point on M, let $x(t)$ be a curve connecting x_0 with x. Along the curve, Proposition 2.15 (i)–(iv) becomes a system of ordinary differential equations. Since the coefficients of this system coincide, $\omega_A^B = \bar{\omega}_A^B$ and $\omega^i = \bar{\omega}^i$, it follows from the uniqueness of the solution of such a system that $x(t)$ and $\sigma(\bar{x}(t))$ coincide. Then our theorem follows from the fact that x is arbitrary. □

Next we consider the existence problem. Let M be an n-dimensional smooth Riemannian manifold with a positive definite metric $G = \sum G_{ij}\omega^i\omega^j$ and a symmetric cubic covariant tensor field $A = \sum A_{ijk}\omega^i\omega^j\omega^k$ which satisfies the apolarity condition $\sum G^{ij}A_{ijk} = 0$ for $k = 1,\ldots,n$. We want to find conditions for the existence of a locally strongly convex immersion $x : M \to A^{n+1}$ such that G_{ij} and A_{ijk} are the Blaschke metric and the Fubini–Pick form of the immersion, respectively.

Choose an orthonormal co-frame field $\{\omega^1,\omega^2,\ldots,\omega^n\}$ relative to G on M. Then

$$G_{ij} = \delta_{ij}, \quad G = (\omega^1)^2 + \cdots + (\omega^n)^2.$$

Denote by $\tilde{\omega}_j^i$ the connection forms of the Levi–Civita connection of the Riemann manifold (M, G), they satisfy

$$d\omega^i = \sum \omega^j \wedge \tilde{\omega}_j^i, \quad \tilde{\omega}_i^j + \tilde{\omega}_j^i = 0. \tag{2.127}$$

Define an equiaffine symmetric connection ∇ with connection forms ω_i^j by

$$\omega_i^j := \tilde{\omega}_i^j + \sum A_{ijk}\omega^k.$$

The symmetry of A_{ijk} and (2.127) give

$$\omega_i^j + \omega_j^i = 2 \sum A_{ijk} \omega^k,$$

$$d\omega^i = \sum \omega^j \wedge \omega_j^i,$$

$$\sum \omega_i^i = 0 \quad \text{(apolarity)}.$$

The last equation above is due to the condition $\sum G^{ij} A_{ijk} = 0$. The remaining key problem is whether or not there exist n differential forms ω_{n+1}^i which satisfy (vii) and (viii) in Proposition 2.15. First we treat the case $n = 2$. In this case, we will prove that there exist unique forms ω_3^1 and ω_3^2 (without any further condition being required) such that

$$d\omega_i^j = \sum_k \omega_i^k \wedge \omega_k^j + \omega^i \wedge \omega_3^j, \quad i, j = 1, 2. \tag{2.128}$$

To see this, we set

$$\theta_i^j := d\omega_i^j - \sum \omega_i^k \wedge \omega_k^j = a_i^j \omega^1 \wedge \omega^2.$$

The coefficients satisfy

$$a_1^1 + a_2^2 = 0.$$

Set

$$\omega_3^1 = -B_{11} \omega^1 - B_{12} \omega^2, \quad \omega_3^2 = -B_{21} \omega^1 - B_{22} \omega^2 \tag{2.129}$$

with appropriate coefficients B_{ij}. Inserting (2.129) into (2.128), we see that these coefficients have to satisfy

$$B_{11} = a_2^1, \quad B_{12} = B_{21} = -a_1^1 = a_2^2, \quad B_{22} = -a_1^2. \tag{2.130}$$

Thus ω_3^1 and ω_3^2, defined by (2.129) with B_{ij} from (2.130), satisfy (2.128).

In our second step we will prove that ω_3^1, ω_3^2 satisfy the integrability conditions. We already know that (vi) and (vii) in Proposition 2.15 imply (i) in Corollary 2.17 and (2.126). These last formulas give

$$B_{ij} = (\chi - J) G_{ij} - \sum A_{ijl,l}. \tag{2.131}$$

Furthermore from our preceding argument it is easy to see that (viii) in Proposition 2.15 is equivalent to (2.124)', which in the case $n = 2$ is expressed by the two independent equations

$$B_{12,1} - B_{11,2} = \sum B_{l1} A_{l12} - \sum B_{l2} A_{l11},$$

and

$$B_{21,2} - B_{22,1} = \sum B_{l2} A_{l21} - \sum B_{l1} A_{l22}.$$

They are equivalent to the following system:

$$2(L_1)_{,k} = \sum B_{ik,i} - \sum B_{jl} A_{jlk}, \quad k = 1, 2.$$

Now it follows, with (2.131), that

$$(\chi - J)_{,k} = \sum A_{jlm,m} A_{jlk} - \sum A_{jkl,lj}. \qquad (2.132)$$

Thus when G_{ij} and A_{ijk} satisfy (2.132), all integrability conditions (v)–(viii) in Proposition 2.15 are satisfied, and thus the system (i)–(iv) is integrable. Consequently, there exists an immersion $x : M \rightarrow A^3$. In addition, it is not difficult to see that $\{e_1, e_2, e_3\}$ is a unimodular affine frame field over M and G_{ij} and A_{ijk} are the Blaschke metric and the Fubini–Pick form for the immersion, respectively. We have proved the following theorem due to J. Radon [289].

Theorem 2.20 (Existence and uniqueness for $n = 2$). *Let (M, G) be a 2-dimensional smooth Riemannian manifold with $G = \sum G_{ij}\omega^i\omega^j$ a positive definite metric. Assume that $A = \sum A_{ijk}\omega^i\omega^j\omega^k$ is a cubic symmetric covariant tensor field on M satisfying the apolarity condition*

$$\sum G^{ij} A_{ijk} = 0, \quad k = 1, 2$$

and the integrability conditions (2.132). Then there exists a locally strongly convex immersion $x : M \rightarrow A^3$ such that G and A are the Blaschke metric and the Fubini–Pick form of the immersion, respectively. The immersion is uniquely determined modulo unimodular transformations of A^3.

To extend Theorem 2.20 to general dimensions $n \geq 2$, first we notice that, in Proposition 2.15, (viii) is a consequence of (vi) and (vii). In fact, by exterior differentiation of (vii) we get

$$(d\omega^j_{n+1} - \sum \omega^k_{n+1} \wedge \omega^j_k) \wedge \omega^i = 0 \text{ for } i, j = 1, 2, \ldots, n. \qquad (2.133)$$

Put

$$a^j := d\omega^j_{n+1} - \sum \omega^k_{n+1} \wedge \omega^j_k =: \sum a^j_{kl}\omega^k \wedge \omega^l,$$

where

$$a^j_{kl} = -a^j_{lk}.$$

Then (2.133) implies

$$a^j \wedge \omega^3 \wedge \omega^4 \wedge \cdots \wedge \omega^n = 0.$$

It follows immediately that

$$a^j_{12} = 0, \quad j = 1, 2, \ldots, n.$$

Similarly, for all k and l, one verifies

$$a^j_{kl} = 0, \quad j = 1, 2, \ldots, n.$$

Thus

$$a^j = 0,$$

and so

$$d\omega_{n+1}^j = \sum \omega_{n+1}^k \wedge \omega_k^j, \ j = 1, 2, \ldots, n.$$

Finally we need only to find ω_{n+1}^i satisfying (vii). Inserting the formula from Corollary 2.18 into formula (2.122), where we put $B_{ij} := l^{ij}$, we get as a necessary condition

$$
\begin{aligned}
R_{ijkl} &= A_{ijl,k} - A_{ijk,l} + (\chi - J)(\delta_{ik}\delta_{jl} - \delta_{il}\delta_{jk}) \\
&+ \frac{2}{n}\sum(\delta_{il}A_{jkm,m} - \delta_{ik}A_{jlm,m}) + \sum(A_{iml}A_{mjk} - A_{imk}A_{mjl}).
\end{aligned}
\tag{2.134}
$$

Conversely, if G_{ij} and A_{ijk} satisfy the equation (2.134), we define B_{ij} via Corollary 2.18, and $\omega_{n+1}^i := -\sum B_j^i \omega^j$. Inverting the argument above, one verifies (vii) of Proposition 2.15. Therefore we have proved the following fundamental theorem of L. Berwald ([15], also in [20, p. 167]), that holds for all dimensions $n \geq 2$:

Theorem 2.21 (Existence and uniqueness for $n \geq 2$). *Let (M, G) be an n-dimensional smooth Riemannian manifold with $G = \sum G_{ij}\omega^i\omega^j$ a positive definite metric. Assume that $A = \sum A_{ijk}\omega^i\omega^j\omega^k$ is a cubic symmetric covariant tensor field on M. If G and A satisfy the apolarity condition $\sum G^{ij}A_{ijk} = 0$ and the integrability conditions (2.134), then there exists a locally strongly convex immersion $x : M \to A^{n+1}$ such that G and A are the Blaschke metric and the Fubini-Pick form for the immersion, respectively. The immersion is uniquely determined modulo unimodular transformations of A^{n+1}.*

2.6 Euclidean boundary points of locally convex immersed hypersurfaces

Let M be a connected oriented n-dimensional C^∞-manifold without boundary, and $x : M \to \mathbb{R}^{n+1}$ be a locally strongly convex immersed hypersurface.

Definition 2.22. Let $\overline{x(M)}$ be the closure of $x(M)$ with respect to the Euclidean topology in $\mathbb{R}^{n+1}$, then an element of $\overline{x(M)} \setminus x(M)$ is called a *Euclidean boundary point* of $x(M)$. The set of all Euclidean boundary points of $x(M)$ is denoted by $\partial_E(M) := \overline{x(M)} \setminus x(M)$.

The geometric and topological behavior of locally strongly convex immersed hypersurfaces is very complicated. In [394], B. Wang constructed many examples to illustrate this. In this section we will present some of his examples.

For any fixed $p \in M$, let $\{x(p); e_1, \ldots, e_n, e_{n+1}\}$ be an equiaffine frame on $\mathbb{R}^{n+1}$ such that $e_1, \ldots, e_n \in x_*(T_pM)$, where as above $x_*(T_pM)$ denotes the tangent hyperplane of $x(M)$ at $x(p)$. Locally, we have a neighborhood $U_p \subset M$ such that $x(U_p)$ is given as a graph of some smooth strongly convex function $x^{n+1} = f(x^1, \ldots, x^n)$, defined in a domain $\Omega \subset \mathbb{R}^n$. Denote

$$S(x(p), r) : = \{(x^1, \ldots, x^n) \in \Omega \mid x^{n+1} = f(x^1, \ldots, x^n) < r\},$$

$$M(x(p), r) : = \{(x^1, \ldots, x^n, f(x^1, \ldots, x^n)) \mid (x^1, \ldots, x^n) \in S(x(p), r)\}.$$

Definition 2.23. For $q \in \partial_E(M)$, if there exists $p \in M$ and a constant $r > 0$ such that
$$q \in \overline{M(x(p), r)},$$
we will call q a *first class Euclidean boundary point*. Otherwise, we will call q a *second class Euclidean boundary point*.

Example 2.4. Let $\alpha > 0$ be a constant, $I = (0, +\infty)$. Consider the immersion $x : I \to \mathbb{R}^2$ given by
$$x(t) = \left(\frac{1}{t^\alpha} \cos t, \frac{1}{t^\alpha} \sin t \right).$$
A direct calculation shows that this is a locally strongly convex immersion, and if t tends to $+\infty$, then $x(t) \to (0, 0)$. It is easy to see that $q = (0, 0)$ is the unique Euclidean boundary point of the immersion, and it is a second class Euclidean boundary point (see Fig. 2.1).

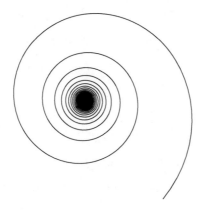

Fig. 2.1: An illustration of the curve in Example 2.4.

Remark 2.24. If we take $I = (a, b)$, $0 < a < b < +\infty$ in the above example, then $x(a)$ and $x(b)$ are Euclidean boundary points of the first class.

We pose the following problem:

Problem and Remark 2.25. For $n \geq 2$, does there exist a locally strongly convex immersed hypersurface of dimension n, which has only Euclidean boundary points of the second class?

If an immersed hypersurface is not assumed to be locally strongly convex, the following examples show that there are hypersurfaces which have only Euclidean boundary points of the second class.

Example 2.5. Let $M = (0, +\infty) \times \mathbb{R}$. Define $x : M \to \mathbb{R}^3$ by
$$x(u, v) = \left(\frac{1}{u^\alpha} \cos u, \frac{1}{u^\alpha} \sin u, v \right).$$

The set of the Euclidean boundary points of $x(M)$ is

$$\partial_E(M) = \{(0,0,v) \mid -\infty < v <+ \infty\}.$$

It can be easily seen that, although the immersed surface $x(M)$ is not "locally strongly convex", its Euclidean boundary points all are of the second class.

Example 2.6. Let $M = (0,+\infty) \times \mathbb{R}$. Consider the immersion $x : M \to \mathbb{R}^3$ given by

$$x(u,v) = (uf(v)\cos v, uf(v)\sin v, h(u)), \tag{2.135}$$

where $h(u)$ and $f(v)$ are smooth functions to be explicitly chosen later.
    Denote $h_u := \frac{\partial h}{\partial u}$, $f_v := \frac{\partial f}{\partial v}$, $\cdots$. By a direct calculation we have

$$x_u = (f\cos v, f\sin v, h_u),$$

$$x_v = u(f_v \cos v - f\sin v, f_v \sin v + f\cos v, 0),$$

$$[x_u, x_v] = -u(h_u(f_v \sin v + f\cos v), -h_u(f_v \cos v - f\sin v), -f^2),$$

$$|[x_u, x_v]|^2 = u^2\left[(h_u)^2((f_v)^2 + f^2) + f^4\right] =: u^2 A,$$

and

$$x_{uu} = (0, 0, h_{uu}),$$

$$x_{uv} = (f_v \cos v - f\sin v, f_v \sin v + f\cos v, 0),$$

$$x_{vv} = u(f_{vv}\cos v - 2f_v \sin v - f\cos v, f_{vv}\sin v + 2f_v \cos v - f\sin v, 0).$$

Let $II = L(du)^2 + 2Mdudv + N(dv)^2$ denote the Euclidean second fundamental form, where

$$L = \frac{1}{|[x_u, x_v]|}\mathrm{Det}(x_u, x_v, x_{uu}) = A^{-1/2}f^2 h_{uu},$$

$$M = \frac{1}{|[x_u, x_v]|}\mathrm{Det}(x_u, x_v, x_{uv}) = 0, \tag{2.136}$$

$$N = \frac{1}{|[x_u, x_v]|}\mathrm{Det}(x_u, x_v, x_{vv}) = A^{-1/2}v(f^2 + 2(f_v)^2 - ff_{vv})h_u.$$

Now we choose $h(u) = \exp\{u\}$ and $f(v) = \exp\{\frac{1}{4}v^2\}$, then

$$h_u(u) = h_{uu}(u) = \exp\{u\} > 0, \tag{2.137}$$

$$f^2 + 2(f_v)^2 - ff_{vv} = \left(\frac{1}{2} + \frac{v^2}{4}\right)\exp\left\{\frac{1}{2}v^2\right\} > 0. \tag{2.138}$$

Thus, $x : M \to \mathbb{R}^3$ is a locally strongly convex immersed surface.

To study the Euclidean boundary points of $x(M)$, let $q = (\bar{x}^1, \bar{x}^2, \bar{x}^3)$ be one of them. Let $(u(t), v(t))$, $0 \le t <+ \infty$, be a curve in M such that

$$x(u(t), v(t)) \to (\bar{x}^1, \bar{x}^2, \bar{x}^3) \text{ as } t \to+ \infty,$$

and at least one of the following conditions holds as $t \to+ \infty$:

$$(1) \; u(t) \to 0; \quad (2) \; v(t) \to+ \infty.$$

It follows that $u(t)$ must tend to 0 as $t \to+ \infty$. We conclude that all Euclidean boundary points of $x(M)$ lie on the hyperplane $\Pi := \{(x^1, x^2, x^3) \in \mathbb{R}^3 \,|\, x^3 = 1\}$.

On the other hand, it is easy to show that for any point $(\bar{x}^1, \bar{x}^2, 1)$, we can find a curve $(u(t), v(t))$ such that when $t \to+ \infty$, it holds $u(t) \to 0$, $v(t) \to+ \infty$, and $u(t)\exp\{\frac{1}{4}(v(t))^2\} \to \sqrt{(\bar{x}^1)^2 + (\bar{x}^2)^2}$. Thus the set of Euclidean boundary points of $x(M)$ is exactly the hyperplane Π.

We notice that all points of $\Pi \setminus (0, 0, 1)$ are Euclidean boundary points of the second class, and that $x : M \to \mathbb{R}^3$ is not globally convex; see Figure 2.2.

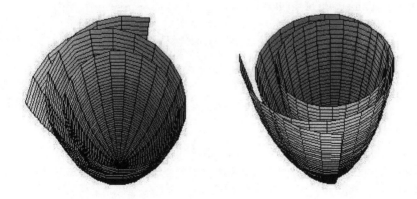

Fig. 2.2: Surface with Euclidean boundary points forming a hyperplane.

Example 2.7. Let $M = (-\frac{\pi}{2}, \frac{\pi}{2}) \times \mathbb{R}$. Consider the immersion $x : M \to \mathbb{R}^3$ given by

$$x(u, v) = \left(\cos u \cos v \exp\left\{-\frac{v^2}{a}\right\}, \; \cos u \sin v \exp\left\{-\frac{v^2}{a}\right\}, \; \sin u\right),$$

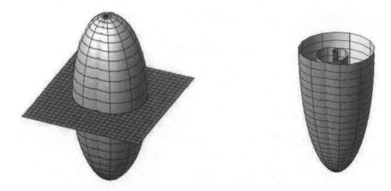

Fig. 2.3: An illustration of the surface in Example 2.7.

where $a = \text{const.} > 0$. Direct calculations show that the second fundamental form of $x(M)$ is given by

$$II = A^{-1}\left[\exp\left\{-\frac{2v^2}{a}\right\}(du)^2 + \left(1 + \frac{2}{a} + \frac{4v^2}{a^2}\right)\exp\left\{-\frac{2v^2}{a}\right\}\cos^2 u\,(dv)^2\right],$$

where

$$A = \sqrt{\left(1 + \frac{4v^2}{a^2}\right)\exp\left\{-\frac{2v^2}{a}\right\}\cos^2 u + \exp\left\{-\frac{4v^2}{a}\right\}\sin^2 u}.$$

Therefore, $x : M \to \mathbb{R}^3$ is a locally strongly convex immersed surface.

One can easily see (cf. also Figure 2.3) that $x(M)$ is (globally) convex at many of its points, e.g. at any point of the half circle

$$\{(\cos u,\ 0, \sin u)\ |\ -\frac{\pi}{2} < u < \frac{\pi}{2}\} \subset x(M).$$

Moreover, the set of Euclidean boundary points of $x(M)$, which all are of the second class, is a line segment lying on the x^3-axis:

$$\partial_E(M) = \left\{(0,\ 0,\ \sin u)\ |\ -\frac{\pi}{2} < u < \frac{\pi}{2}\right\}.$$

However, $x : M \to \mathbb{R}^3$ obviously is not globally convex; see also Figure 2.3.

Two pictures of $x(M)$ are shown above, where in Figure 2.3 the left picture is for $a = 2$ while the right one is its cross-section.

Remark 2.26. Example 2.7 provides a counterexample to Lemma 2.1 of [369], which states that if a locally strongly convex hypersurface M is convex at some point and that its boundary ∂M lies on a hyperplane then M is convex.

2.7 Graph immersions

2.7.1 Graph immersions with equiaffine normalization

1. Notations and formulas in terms of $\{x^i\}$

We introduce coordinates $(x^1, \ldots, x^{n+1})$ in $\mathbb{R}^{n+1}$. Let us consider the graph hypersurface

$$M := \{(x^1, \ldots, x^{n+1}) \mid x^{n+1} = f(x), \ x = (x^1, \ldots, x^n) \in \Omega\},$$

where f is a convex smooth function defined on a domain $\Omega \subset \mathbb{R}^n$.

Let $\boldsymbol{x} := (x^1, \ldots, x^n, f(x^1, \ldots, x^n))$ denote the position vector of the hypersurface M. First we choose the following unimodular affine frame field:

$$e_i = \frac{\partial \boldsymbol{x}}{\partial x^i} \ \text{ for } \ i = 1, \ldots, n \ \text{ and } \ e_{n+1} = (0, \ldots, 0, 1).$$

Then the Blaschke metric G is given by

$$G = \sum G_{ij} \, dx^i \, dx^j, \ G_{ij} = \left[\det\left(\frac{\partial^2 f}{\partial x^j \partial x^i} \right) \right]^{-1/(n+2)} \cdot \frac{\partial^2 f}{\partial x^j \partial x^i}. \tag{2.139}$$

In the following, we write out basic formulas with respect to the Blaschke metric for a graph hypersurface, which will be used in later chapters.

First of all, we calculate the affine normal $Y^{(B)} = \frac{1}{n} \Delta^{(B)} \boldsymbol{x}$, where, as above, $\Delta^{(B)}$ denotes the Laplacian with respect to the Blaschke metric, given by

$$\Delta^{(B)} := \frac{1}{\sqrt{\det(G_{kl})}} \sum \frac{\partial}{\partial x^i} \left(\sqrt{\det(G_{kl})} G^{ij} \frac{\partial}{\partial x^j} \right), \quad (G^{ij}) = (G_{ij})^{-1};$$

as already stated we mark invariants of the Blaschke geometry by the superscript "B". Denote

$$\rho := \left[\det\left(\frac{\partial^2 f}{\partial x^i \partial x^j} \right) \right]^{-1/(n+2)}.$$

Then we have $\sqrt{\det(G_{kl})} = \frac{1}{\rho}$, and $\Delta^{(B)}$ can be written as

$$\Delta^{(B)} = \sum G^{ij} \frac{\partial^2}{\partial x^i \partial x^j} - \frac{2}{\rho^2} \sum f^{ij} \frac{\partial \rho}{\partial x^j} \frac{\partial}{\partial x^i} + \frac{1}{\rho} \sum \frac{\partial f^{ij}}{\partial x^i} \frac{\partial}{\partial x^j}, \tag{2.140}$$

where (f^{ij}) denotes the inverse matrix of (f_{ij}) and $f_{ij} := \frac{\partial^2 f}{\partial x^i \partial x^j}$. Differentiation of the equation $\sum f^{ik} f_{kj} = \delta^i_j$ gives

$$\sum \frac{\partial f^{ik}}{\partial x^i} f_{kj} = -\sum f^{ik} \frac{\partial f_{kj}}{\partial x^i} = \frac{n+2}{\rho} \frac{\partial \rho}{\partial x^j}.$$

It follows that

$$\sum \frac{\partial f^{ik}}{\partial x^i} = \frac{n+2}{\rho} \sum f^{jk} \frac{\partial \rho}{\partial x^j}. \tag{2.141}$$

Inserting (2.141) into (2.140), we obtain

$$\Delta^{(B)} = \frac{1}{\rho} \sum f^{ij} \frac{\partial^2}{\partial x^i \partial x^j} + \frac{n}{\rho^2} \sum f^{ij} \frac{\partial \rho}{\partial x^j} \frac{\partial}{\partial x^i}. \tag{2.142}$$

Therefore

$$Y^{(B)} = \frac{1}{n} \Delta^{(B)} x = \frac{1}{\rho^2} \sum f^{ij} \frac{\partial \rho}{\partial x^j} e_i + \frac{1}{\rho} e_{n+1}. \tag{2.143}$$

Accordingly, the affine conormal vector field $U^{(B)}$ can be identified with

$$U^{(B)} = \left[\det \left(\frac{\partial^2 f}{\partial x^j \partial x^i} \right) \right]^{\frac{-1}{n+2}} \left(-\frac{\partial f}{\partial x^1}, \dots, -\frac{\partial f}{\partial x^n}, 1 \right). \tag{2.144}$$

From Corollary 2.11 we know that a locally strongly convex hypersurface with affine mean curvature L_1 satisfies $\Delta^{(B)} U^{(B)} = -nL_1 U^{(B)}$. It follows that f satisfies the following p.d.e.:

$$\Delta^{(B)} \left\{ \left[\det \left(\frac{\partial^2 f}{\partial x^i \partial x^j} \right) \right]^{\frac{-1}{n+2}} \right\} = -nL_1 \left[\det \left(\frac{\partial^2 f}{\partial x^i \partial x^j} \right) \right]^{\frac{-1}{n+2}}, \tag{2.145}$$

i.e.

$$\Delta^{(B)} \rho = -nL_1 \rho. \tag{2.146}$$

Convention on notations for the gradient of a function. In this book, we will use the same notation grad $*$ to denote the gradient vector of a function $*$ with respect to different metrics; it will be indicated in the context which metric we use. For example, when we write $G(\text{grad} \ln \rho, \text{grad} f)$ below, the gradient is taken with respect to the Blaschke metric G: grad $* = \sum G^{ij} \frac{\partial *}{\partial x^i} \frac{\partial}{\partial x^j}$; whereas when we write $\mathcal{H}(\text{grad} \ln \rho, \text{grad} f)$ in later subsections, the gradient is assumed to be taken with respect to the Calabi metric $\mathcal{H}$: grad$* = \sum f^{ij} \frac{\partial *}{\partial x^i} \frac{\partial}{\partial x^j}$.

By (2.142) and the above notations, we also have

$$\Delta^{(B)} f = \frac{n}{\rho} + nG (\text{grad} \ln \rho, \text{grad} f), \tag{2.147}$$

$$\Delta^{(B)} \left(\sum (x^k)^2 \right) = \frac{2}{\rho} \sum f^{kk} + nG(\text{grad} \ln \rho, \text{grad} \sum (x^k)^2), \tag{2.148}$$

and

$$\Delta^{(B)} x_k = nG (\text{grad} \ln \rho, \text{grad} x_k). \tag{2.149}$$

Next, to calculate the equiaffine Weingarten form $\hat{B}$ of the hypersurface M, we assume

$$\bar{e}_i = e_i, \ 1 \le i \le n; \ \bar{e}_{n+1} := \rho Y^{(B)} = e_{n+1} + \sum f^{ij} \frac{\partial \ln \rho}{\partial x^j} e_i.$$

Then $\{\bar{e}_1, \dots, \bar{e}_n, \bar{e}_{n+1}\}$ is still a unimodular affine frame field. Setting

$$dx = \sum \bar{\omega}^A \bar{e}_A, \ d\bar{e}_A = \sum \bar{\omega}_A^B \bar{e}_B,$$

direct calculations show that

$$
\begin{cases}
\bar{\omega}^i = dx^i, \quad \bar{\omega}^{n+1} = 0, \quad \bar{\omega}_{n+1}^{n+1} = \sum \dfrac{\partial \ln \rho}{\partial x^i} \bar{\omega}^i, \\[2mm]
\bar{\omega}_i^j = -\sum f_{ie} f^{kj} \dfrac{\partial \ln \rho}{\partial x^k} \bar{\omega}^\ell, \quad \bar{\omega}_i^{n+1} = \sum f_{ij} \bar{\omega}^j, \\[2mm]
\bar{\omega}_{n+1}^i = \sum \left[\dfrac{\partial}{\partial x^j} \left(f^{ki} \dfrac{\partial \ln \rho}{\partial x^k} \right) - f^{ki} \dfrac{\partial \ln \rho}{\partial x^k} \cdot \dfrac{\partial \ln \rho}{\partial x^j} \right] \bar{\omega}^j.
\end{cases}
$$

Therefore we obtain $\hat{B} = -\sum \bar{\omega}_{n+1}^i \bar{\omega}_i^{n+1} = \sum B_{ij} dx^i dx^j$, where

$$
\begin{aligned}
B_{ij} &= \sum \left[-\dfrac{\partial}{\partial x^i} \left(f^{lk} \dfrac{\partial \ln \rho}{\partial x^l} \right) + f^{lk} \dfrac{\partial \ln \rho}{\partial x^l} \dfrac{\partial \ln \rho}{\partial x^i} \right] f_{kj} \\[2mm]
&= -\dfrac{1}{\rho} \dfrac{\partial^2 \rho}{\partial x^i \partial x^j} + \dfrac{2}{\rho^2} \dfrac{\partial \rho}{\partial x^i} \dfrac{\partial \rho}{\partial x^j} + \dfrac{1}{\rho} \sum f^{kl} f_{ijk} \dfrac{\partial \rho}{\partial x^l}.
\end{aligned}
\tag{2.150}
$$

Finally, a direct computation shows that, with respect to the above basis $\{\bar{e}_i\}$, the Levi–Civita connection of G satisfies

$$
\bar{\omega}_i^j = \dfrac{1}{2} \sum \left(\dfrac{\partial \ln \rho}{\partial x^l} \delta_i^j + \dfrac{\partial \ln \rho}{\partial x^i} \delta_l^j - f^{jk} f_{il} \dfrac{\partial \ln \rho}{\partial x^k} + f^{jk} f_{ikl} \right) dx^l.
$$

Hence, using the relations $\bar{\omega}_i^j - \omega_i^j = \sum A_{il}^j dx^l$ and $A_{ikl} = \sum G_{jk} A_{il}^j$, we obtain the Fubini–Pick form $A = \sum A_{ijk} dx^i dx^j dx^k$, where

$$
A_{ijk} = -\dfrac{1}{2} \left(f_{ij} \dfrac{\partial \rho}{\partial x^k} + f_{jk} \dfrac{\partial \rho}{\partial x^i} + f_{ik} \dfrac{\partial \rho}{\partial x^j} + \rho f_{ijk} \right).
\tag{2.151}
$$

2. The gradient map and the Legendre transformation

Given a strongly convex smooth function f defined on a domain $\Omega \subset \mathbb{R}^n$, the Hessian $(\partial_i \partial_j f)$ is positive definite. Consider the mapping

$$
\begin{aligned}
F : \; &\Omega \to \mathbb{R}^n, \\
&(x^1, \ldots, x^n) \mapsto (\xi_1, \xi_2, \ldots, \xi_n),
\end{aligned}
\tag{2.152}
$$

where

$$
\xi_i = f_i := \partial_i f = \dfrac{\partial f}{\partial x^i}; \quad i = 1, \ldots, n.
\tag{2.153}
$$

The Jacobian of F satisfies

$$
\det \left(\dfrac{\partial \xi_i}{\partial x^j} \right) = \det \left(\dfrac{\partial^2 f}{\partial x^j \partial x^i} \right) > 0.
$$

This shows that F locally is a diffeomorphism. Assume that $\mathcal{U} \subset \Omega$ is an open set such that $\mathcal{U}$ is diffeomorphic to $F(\mathcal{U})$. Define a function u on $F(\mathcal{U})$ by

$$
\begin{aligned}
u(\xi_1, \ldots, \xi_n) &= \sum x^i \cdot \partial_i f(x^1, \ldots, x^n) - f(x^1, \ldots, x^n), \\
\xi_i &= \partial_i f(x^1, \ldots, x^n), \quad \text{for } i = 1, 2, \ldots, n.
\end{aligned}
\tag{2.154}
$$

Definition 2.27. The mapping F defined by (2.152) and (2.153) is called the *gradient map relative to f*, and the function u defined by (2.154) is called the *Legendre transformation function of f (relative to the origin O)*. $F(\Omega)$ is called the *domain of the Legendre transformation*.

Lemma 2.28. *Let $\Omega \subset \mathbb{R}^n$ be a convex domain and f be a strongly convex C^∞-function on Ω. Let F be the gradient map relative to f. Assume that x and y are two different points in Ω and that $\xi = F(x), \eta = F(y)$. Then*

$$(y - x, \eta - \xi) > 0,$$

where the brackets denote the canonical inner product of $\mathbb{R}^n$.

Proof. Define $\varphi : [0, 1] \to \mathbb{R}$ by

$$\varphi(t) = f(ty + (1 - t)x), \quad 0 \le t \le 1.$$

Differentiating $\varphi(t)$ with respect to t, we have

$$\varphi'(t) = \sum [\partial_i f(ty + (1 - t)x)](y^i - x^i),$$

$$\varphi''(t) = \sum [\partial_j \partial_i f(ty + (1 - t)x)](y^i - x^i)(y^j - x^j).$$

It follows from the convexity of f that $\varphi''(t) > 0$, and so $\varphi'(t)$ is a monotone increasing function. Thus

$$\varphi'(1) > \varphi'(0),$$

and

$$\sum \eta_i(y^i - x^i) > \sum \xi_i(y^i - x^i).$$

This proves the Lemma. $\qquad\square$

Lemma 2.28 shows that, when Ω is convex, $F : \Omega \to F(\Omega)$ is a diffeomorphism.

3. Formulas in terms of $\{\xi_i\}$

Denote by Ω^* the Legendre transformation domain of f so that the Legendre transform function $u : \Omega^* \to \mathbb{R}$ is defined on

$$\Omega^* = \{(\xi_1(x), \ldots, \xi_n(x)) \mid x \in \Omega\}.$$

Considering a locally strongly convex graph hypersurface, it is an advantage that we can express the basic formulas in terms of the x-coordinates as well in terms of the ξ-coordinates. In terms of the coordinates $(\xi_1, \ldots, \xi_n)$ and of $u(\xi)$, the Blaschke metric G is given by

$$G = \sum \bar{G}_{ij} d\xi_i d\xi_j, \quad \bar{G}_{ij} = \rho \frac{\partial^2 u}{\partial \xi_i \partial \xi_j}, \quad \rho = \left[\det\left(\frac{\partial^2 u}{\partial \xi_i \partial \xi_j}\right)\right]^{1/(n+2)}.$$

Note that $\frac{\partial u}{\partial \xi_i} = x^i$, $\left(\frac{\partial^2 u}{\partial \xi_i \partial \xi_j}\right)$ is the inverse matrix of $\left(\frac{\partial^2 f}{\partial x^i \partial x^j}\right)$ and $\sqrt{\det(\bar{G}_{kl})} = \rho^{n+1}$; from the expression

$$\Delta^{(B)} = \frac{1}{\sqrt{\det(\bar{G}_{kl})}} \sum \frac{\partial}{\partial \xi_i} \left(\sqrt{\det(\bar{G}_{kl})} \bar{G}^{ij} \frac{\partial}{\partial \xi_j} \right),$$

and by a similar calculation as in deriving (2.142), we can show that

$$\Delta^{(B)} = \frac{1}{\rho} \sum u^{ij} \frac{\partial^2}{\partial \xi_i \partial \xi_j} - \frac{2}{\rho^2} \sum u^{ij} \frac{\partial \rho}{\partial \xi_j} \frac{\partial}{\partial \xi_i}. \tag{2.155}$$

In particular, we have

$$\Delta^{(B)} u = \frac{n}{\rho} - 2G\left(\operatorname{grad}\ln\rho, \operatorname{grad}u\right), \tag{2.156}$$

$$\Delta^{(B)} \left(\sum(\xi_k)^2\right) = \frac{2}{\rho} \sum u^{ii} - 2G\left(\operatorname{grad}\ln\rho, \operatorname{grad}\sum(\xi_k)^2\right), \tag{2.157}$$

$$\Delta^{(B)} \xi_k = -2G\left(\operatorname{grad}\ln\rho, \operatorname{grad}\xi_k\right). \tag{2.158}$$

2.7.2 Graph Immersions with Calabi metric

In Section 2.7.1 we considered a graph immersion equipped with a Blaschke geometry. Calabi [40] considered a different normalization as follows (we use the superscript "C" to mark invariants of this geometry).

Let f be a strongly convex C^∞-function defined on a domain $\Omega \subset \mathbb{R}^n$, and consider the graph hypersurface

$$M \equiv M_f := \left\{ (x^1, \dots, x^{n+1}) \mid x^{n+1} = f(x), \ x = (x^1, \dots, x^n) \in \Omega \right\}.$$

For M we choose the constant relative normalization given by

$$Y := Y^{(C)} := (0, \dots, 0, 1).$$

Then the associated conormal field $U^{(C)}$ satisfies

$$U^{(C)} = (-f_1, \dots, -f_n, 1);$$

here the reader should recall the notation for partial derivatives from Sections 1.12.5 and 2.7.1. We consider the Riemannian metric $\mathcal{H}$ on M, defined by the Hessian of the graph function f:

$$\mathcal{H} \equiv \mathcal{H}_f := \sum f_{ij} dx^i dx^j,$$

where as before $f_{ij} = \partial_i \partial_j f$. Then $\mathcal{H}$ is the relative metric with respect to the relative normalization defined by $Y^{(C)}$. This metric is very natural for a convex graph; we call it the *Calabi metric*, in the literature one also finds the terminology *Hessian metric*.

Using the conventions in a local notation, as before we denote the inverse matrix of the matrix (f_{ij}) by (f^{jk}), thus $\sum f_{ij} f^{jk} = \delta_i^k$.

As before, denote by $\mathbf{x} = (x^1, \ldots, x^n, f(x^1, \ldots, x^n))$ the position vector of M. Then, in covariant form, the *Gauß structure equation* reads

$$\mathbf{x}_{,ij} = \sum A_{ij}^{(C)k} \mathbf{x}_k + f_{ij} Y^{(C)}.$$

One calculates the following relations in terms of the coordinates $(x^1, \ldots, x^n)$ (see e.g. [284]): The Levi–Civita connection with respect to the metric $\mathcal{H}$ is determined by its Christoffel symbols

$$\Gamma_{ij}^k = \frac{1}{2} \sum f^{kl} f_{ijl}. \tag{2.159}$$

The Fubini–Pick tensor $A_{ijk}^{(C)} = A_{ij}^{(C)l} f_{lk}$ and the Weingarten tensor satisfy

$$A_{ijk}^{(C)} = -\frac{1}{2} f_{ijk}, \qquad B_{ij}^{(C)} = 0.$$

The relative Tchebychev vector field is given by

$$T := \frac{1}{n} \sum f^{ij} A_{ij}^{(C)k} \partial_k.$$

Consequently, for the relative Pick invariant, we have

$$J^{(C)} = \frac{1}{4n(n-1)} \sum f^{il} f^{jm} f^{kr} f_{ijk} f_{lmr}. \tag{2.160}$$

The *integrability conditions* and the Ricci tensor read

$$R_{ijkl}^{(C)} = \sum f^{mh} \left[A_{jkm}^{(C)} A_{hil}^{(C)} - A_{ikm}^{(C)} A_{hjl}^{(C)} \right], \tag{2.161}$$

$$A_{ijk,l}^{(C)} = A_{ijl,k}^{(C)}, \tag{2.162}$$

$$R_{ik}^{(C)} = \sum f^{mh} f^{lj} \left[A_{iml}^{(C)} A_{hjk}^{(C)} - A_{imk}^{(C)} A_{hlj}^{(C)} \right]. \tag{2.163}$$

The scalar curvature satisfies

$$\begin{aligned} R^{(C)} &= n(n-1) \chi^{(C)} \\ &= \sum f^{ik} f^{jlj} f^{mh} \left[A_{jkm}^{(C)} A_{hil}^{(C)} - A_{ikm}^{(C)} A_{hjl}^{(C)} \right]. \end{aligned} \tag{2.164}$$

In terms of the Calabi metric $\mathcal{H}$ of a graph hypersurface we calculate its Laplacian. We consider the Legendre transformation function u of f and define a function Φ on M, which will be used in later chapters, by

$$\Phi := \|\operatorname{grad} \ln \rho\|_{\mathcal{H}}^2 = \rho \|\operatorname{grad} \ln \rho\|_G^2, \tag{2.165}$$

where

$$\rho = \left[\det \left(\frac{\partial^2 u}{\partial \xi_i \partial \xi_j} \right) \right]^{\frac{1}{n+2}}.$$

Recalling the involutionary character by using different coordinates x and ξ, and noting that $\mathcal{H} = \sum f_{ij} dx^i dx^j = \sum u_{ij} d\xi_i d\xi_j$ for $u_{ij} = \frac{\partial^2 u}{\partial \xi_i \partial \xi_j}$, we have

$$
\begin{aligned}
\Delta^{(C)} &= \sum f^{ij} \frac{\partial^2}{\partial x^i \partial x^j} + \frac{n+2}{2\rho} \sum f^{ij} \frac{\partial \rho}{\partial x^j} \frac{\partial}{\partial x^i} \\
&= \sum u^{ij} \frac{\partial^2}{\partial \xi_i \partial \xi_j} - \frac{n+2}{2\rho} \sum u^{ij} \frac{\partial \rho}{\partial \xi_j} \frac{\partial}{\partial \xi_i} .
\end{aligned}
$$

In particular, we have

$$\Delta^{(C)} f = n + \frac{n+2}{2} \mathcal{H} (\operatorname{grad} \ln \rho, \operatorname{grad} f), \tag{2.166}$$

$$\Delta^{(C)} u = n - \frac{n+2}{2} \mathcal{H} (\operatorname{grad} \ln \rho, \operatorname{grad} u), \tag{2.167}$$

and

$$\Delta^{(C)} \left(\sum_k (x^k)^2 \right) = 2 \sum_i f^{ii} + \frac{n+2}{2} \mathcal{H} \left(\operatorname{grad} \ln \rho, \operatorname{grad} \sum_k (x^k)^2 \right), \tag{2.168}$$

$$\Delta^{(C)} \left(\sum_k (\xi_k)^2 \right) = 2 \sum_i u^{ii} - \frac{n+2}{2} \mathcal{H} \left(\operatorname{grad} \ln \rho, \operatorname{grad} \sum_k (\xi_k)^2 \right), \tag{2.169}$$

$$\Delta^{(C)} x_k = \frac{n+2}{2} \mathcal{H} (\operatorname{grad} \ln \rho, \operatorname{grad} x_k), \tag{2.170}$$

$$\Delta^{(C)} \xi_k = -\frac{n+2}{2} \mathcal{H} (\operatorname{grad} \ln \rho, \operatorname{grad} \xi_k). \tag{2.171}$$

The above formulas will be used frequently later, see particularly in the proof of Proposition 5.36.

3 Affine hyperspheres

In Section 2.4 we studied locally strongly convex quadrics and proved that the affine normal has the following property: either the affine normal is a constant vector field (elliptic paraboloid) or the affine normal is parallel to the position vector of the (central) quadric with respect to the center as origin (ellipsoid, hyperboloid).

These properties remind one of the plane and the sphere in Euclidean geometry, which can be defined by the behaviour of the Euclidean normal.

It suggests itself to investigate whether the quadrics are the only hypersurfaces with the above properties of the affine normal. The answer is negative, and we will learn in this chapter that the two classes of hypersurfaces, defined by these properties, are both called affine spheres, and both classes are very large; this might be surprising at first glance, if one compares the affine situation with the Euclidean one. After deriving the p.d.e.'s for these hypersurfaces we will have a better understanding of the situation.

Historically, G. Tzitzeica was the first to study affine spheres [370, 371]. He used Euclidean invariants to define affine spheres, while Blaschke was the first to study these properties of the affine normal systematically in the affine context. Blaschke's monograph contains a series of local and global results (see [20, Chaps. 6–7]). Affine spheres attracted many geometers again during the last decades; we mention E. Calabi, A. V. Pogorelov, L. Nirenberg, S. Y. Cheng and S. T. Yau, R. Schneider, T. Sasaki, S. Gigena, U. Simon, K. Nomizu, U. Pinkall, A.-M. Li, L. Vrancken, F. Dillen, Z. Hu, H. Li, Ch. Scharlach, and others.

The affine hyperspheres are exactly the umbilical (see Section 2.2.3) affine hypersurfaces. They are divided into three types: the elliptic, the parabolic, and the hyperbolic affine hyperspheres, according to whether the affine mean curvature is positive, zero, or negative, respectively.

The study of the parabolic affine hyperspheres is equivalent to the study of the convex solutions of the equation (cf. Section 3.1 below)

$$\det(\partial_j \partial_i f) = 1. \tag{3.1}$$

By a Legendre transformation, the study of the elliptic and the hyperbolic affine hyperspheres is equivalent to the study of the equation

$$\det(\partial_j \partial_i u) = (L_1 u)^{-n-2}. \tag{3.2}$$

These facts indicate why, unlike in the case of the Euclidean hypersphere, there are many different types of affine hyperspheres. In Section 3.2 we will give some examples of affine hyperspheres and prove a local classification theorem of affine hyperspheres with constant sectional curvature. The second and main part of this chapter is devoted to the classification of locally strongly convex complete affine hyperspheres. The first result in this direction is due to W. Blaschke, who proved that every closed affine sphere in A^3 is an ellipsoid. Later, Deicke [65] generalized the result of Blaschke to higher dimensions. Calculating the Laplacian ΔJ of the Pick invariant J and estimating J, Calabi proved in 1958 that if a parabolic affine hypersphere is complete with respect to the Blaschke metric, then it is an elliptic paraboloid [40]. Furthermore, in 1972, E. Calabi proved that if a parabolic or hyperbolic affine hypersphere is complete with respect to the Blaschke metric then its Ricci curvature is negative semidefinite [41]. As a consequence, a parabolic affine hypersphere must be an elliptic paraboloid if its Blaschke metric is complete. In Sections 3.5–3.6 we will give a simpler proof of Calabi's result, extending a method of Cheng and Yau [57] to a differential inequality in a weak sense.

From an analytical view point it is natural to study convex solutions of the p.d.e. (3.1) which are defined as a graph $x : \mathbb{R}^n \to A^{n+1}$. In [40], Calabi proved that if a parabolic affine hypersphere of dimension $n \leq 5$ is such a graph, then its Blaschke metric is complete, therefore it is a paraboloid. This is a generalization of a theorem by K. Jörgens [138]. In 1971 A. V. Pogorelov was able to extend Calabi's theorem to arbitrary dimensions [284]. Later, Cheng and Yau gave a simpler and more analytic proof [57]. In fact, their theorem is a little more general than Pogorelov's theorem. They proved that if an affine hypersphere is complete relative to a Euclidean structure in A^{n+1}, then its Blaschke metric is complete. A similar result was obtained by E. Calabi and L. Nirenberg [233]. In Section 3.4 we shall give a generalization of this theorem. We will prove that if a locally strongly convex hypersurface M in A^{n+1} is complete relative to the Euclidean topology, and if the affine principal curvatures satisfy the condition that $\lambda_1^2 + \cdots + \lambda_n^2$ is bounded above by a constant, then the Blaschke metric of M is complete (cf. [168]).

Section 3.8 is devoted to the study of the Calabi conjecture on hyperbolic affine hyperspheres. In [57], Cheng and Yau proved the first part of the Calabi conjecture, but their proof was only valid under the additional condition that the affine hyperspheres of hyperbolic type are of Euclidean completeness. Later in [169, 170], A.-M. Li finally solved the problem showing that if a locally strongly convex hyperbolic affine hypersphere is complete with respect to the Blaschke metric, then it is complete with respect to the Euclidean topology. Combining Li's result and the result of Cheng and Yau, we get a complete solution of the first part of the Calabi conjecture. The second part of the Calabi conjecture is equivalent to the existence of the solution of the p.d.e. (3.2) on a bounded convex domain with zero boundary data. Using a result of Cheng and Yau on Monge–Ampère equations [56], T. Sasaki [295] and S. Gigena [98] solved the second part of the Calabi conjecture.

Altogether, Sections 3.4–3.8 give a qualitative classification of all (affinely or Euclidean) complete locally strongly convex affine hyperspheres. For a subclass of affine hyperspheres, namely all locally strongly convex Blaschke hypersurfaces with parallel cubic form (parallel with respect to Levi–Civita connection of the Blaschke metric), there is a new explicit classification in Section 3.3 (cf. [133]). In particular, this classification shows that the subclass of hyperbolic affine hyperspheres is very large.

In Section 3.9 we add some references from [395] about other results on complete hyperbolic affine spheres.

3.1 Definitions and basic results for affine hyperspheres

3.1.1 Definition of affine hyperspheres

Definition 3.1. A hypersurface in A^{n+1} is called an *affine hypersphere* if all the affine normal lines through each point of the hypersurface either intersect at one point, called its *center*, or else are mutually parallel. An affine hypersphere with center is called *proper*, and an affine hypersphere with parallel affine normals is called *improper*. In case $n = 2$, traditionally one also uses the terminology *affine sphere*.

Proposition 3.2. *Let M be a locally strongly convex hypersurface in A^{n+1} with the affine principal curvatures $\lambda_1, \ldots, \lambda_n$. Then:*
(i) *M is an affine hypersphere if and only if $\lambda_1 = \cdots = \lambda_n =$ const. on M;*
(ii) *M is an affine hypersphere if and only if $B = L_1 \cdot$ id or $\hat{B} = L_1 \cdot G$.*

Proof. (i) Choose an equiaffine frame field $\{x; e_1, \ldots, e_n, e_{n+1}\}$ on M such that it satisfies $G_{ij} = \delta_{ij}$, $e_{n+1} = Y$.

1) Assume that the affine normal lines are mutually parallel. Then

$$e_{n+1} = Y = f\eta,$$

where η is a constant vector and f is a C^∞-function on M. The last equation and the Weingarten equation give

$$df\eta = de_{n+1} = \sum \omega_{n+1}^i e_i,$$

thus

$$df = 0 \text{ and } \omega_{n+1}^i = 0,$$

and equation (2.54) implies

$$\lambda_1 = \cdots = \lambda_n = 0.$$

Conversely, if $\lambda_1 = \cdots = \lambda_n = 0$ on M, then

$$\omega_{n+1}^i = 0, \quad i = 1, \ldots, n,$$

and so

$$e_{n+1} = Y = \text{constant vector.}$$

2) Assume that all affine normal lines intersect at one point. By a unimodular affine transformation, we may assume that the center is at the origin, and then

$$e_{n+1} = fx,$$

where x is the position vector of M and f is a C^∞-function on M with $f \neq 0$ everywhere. Again we use the Weingarten equation

$$\sum \omega_{n+1}^i e_i = de_{n+1} = df \cdot x + f \sum \omega^i e_i,$$

which implies

$$df = 0, \quad \omega_{n+1}^i = f\omega^i.$$

Therefore

$$\lambda_1 = \cdots = \lambda_n = -f = \text{const.} \neq 0.$$

Conversely, if $\lambda_1 = \cdots = \lambda_n = -f = \text{const.} \neq 0$, we have

$$de_{n+1} = \sum \omega_{n+1}^i e_i = f \sum \omega^i e_i = f \cdot dx.$$

Thus

$$\frac{e_{n+1}}{f} - x = \text{constant vector.}$$

This means that all affine normal lines of the hypersurface intersect at one point, the center of the proper affine hypersphere.

(ii) is obviously equivalent to (i). □

Remark 3.3. One can easily prove that if $\lambda_1 = \cdots = \lambda_n = \lambda$ on a hypersurface M, then λ is constant on M. In fact, choosing a local frame field as above, we have $\omega_{n+1}^i = -\lambda\omega^i$. Taking exterior derivatives and using the structure equations we get $d\lambda \wedge \omega^i = 0$ for $i = 1, \ldots, n$, which implies that λ is constant.

Definition 3.4. *Let M be an affine hypersphere in A^{n+1}. Then:*
- *if $\lambda_1 = \cdots = \lambda_n = \text{const.} > 0$, we call M an elliptic affine hypersphere;*
- *if $\lambda_1 = \cdots = \lambda_n = 0$, we call M a parabolic affine hypersphere;*
- *if $\lambda_1 = \cdots = \lambda_n = \text{const.} < 0$, we call M a hyperbolic affine hypersphere.*

Obviously, the parabolic affine hyperspheres are exactly the improper affine hyperspheres.

It is easy to see that, for an elliptic affine hypersphere M, the center is on the concave side of M. For a parabolic affine hypersphere M, we may consider the center to be at infinity, while for a hyperbolic affine hypersphere M, the center is on the convex side of M.

The prototypes of affine hyperspheres of the three types are given by the locally strongly convex quadrics, namely the ellipsoids, convex paraboloids and hyperboloids, after which the types of affine hyperspheres are named.

3.1.2 Differential equations for affine hyperspheres

We will now derive the equations which are satisfied by locally strongly convex affine hyperspheres. First we consider parabolic affine hyperspheres.

Let M be a parabolic affine hypersphere in A^{n+1}. By a unimodular affine transformation, we can assume that

$$Y = (0, \ldots, 0, 1)$$

and that M is locally described by a strongly convex function defined on a domain $D \subset A^n$

$$x^{n+1} = f(x^1, \ldots, x^n).$$

Choose a local unimodular affine frame field on M as follows:

$$e_1 = (1, 0, \ldots, 0, \partial_1 f),$$
$$e_2 = (0, 1, \ldots, 0, \partial_2 f),$$

$$\vdots \tag{3.3}$$

$$e_n = (0, \ldots, 1, \partial_n f),$$
$$e_{n+1} = (0, \ldots, 0, 1).$$

The Gauß structure equations (2.49) and $\langle U, e_{n+1} \rangle = 1$, $\langle U, e_i \rangle = 0$ give

$$h_{ij} = \partial_j \partial_i f. \tag{3.4}$$

From $H^{1/(n+2)} e_{n+1} = Y = e_{n+1}$, we get

$$\det(\partial_j \partial_i f) = H = 1. \tag{3.5}$$

On the other hand, assume that M is locally given by $x^{n+1} = f(x^1, \ldots, x^n)$ and that $f(x^1, \ldots, x^n)$ satisfies (3.5). Considering the frame field (3.3), we have $d \ln H = 0$ and $de_{n+1} = 0$. Hence

$$\omega_{n+1}^{n+1} + \frac{1}{n+2} d \ln H = 0. \tag{3.6}$$

Now from Proposition 2.3 it follows that $e_{n+1} = (0, \ldots, 0, 1)$ is the affine normal vector at each point of M. This shows that M is a parabolic affine hypersphere.

Next we consider a locally strongly convex elliptic or hyperbolic affine hypersphere. We may assume that M is locally given as a graph of a strongly convex C^∞-function

$$x^{n+1} = f(x^1, \ldots, x^n), \quad (x^1, \ldots, x^n) \in D \subset A^n.$$

Consider the gradient map relative to f

$$F : D \to \mathbb{R}^n \tag{3.7}$$
$$(x^1, \ldots, x^n) \mapsto (\xi_1, \xi_2, \ldots, \xi_n)$$

where

$$\xi_i = \partial_i f = \frac{\partial f}{\partial x^i}; \quad i = 1, \ldots, n. \tag{3.8}$$

Let u be the Legendre transformation function of f:

$$u(\xi_1, \ldots, \xi_n) = \sum x^i \cdot \partial_i f(x^1, \ldots, x^n) - f(x^1, \ldots, x^n), \tag{3.9}$$

Then the hypersurface can be represented in terms of $\xi_1, \ldots, \xi_n$ as follows:

$$x = (x^1, \ldots, x^n, f(x^1, \ldots, x^n)) = \left(\frac{\partial u}{\partial \xi_1}, \ldots, \frac{\partial u}{\partial \xi_n}, -u + \sum \xi_i \frac{\partial u}{\partial \xi_i} \right). \tag{3.10}$$

From the representation of the hypersurface as a graph

$$\bar{x} := x \circ F^{-1} : F(D) \to A^{n+1}$$

in (3.10), and from the Gauß structure equation (2.6) we get

$$h_{ij} = \mathrm{Det}(\partial_1 \bar{x}, \ldots, \partial_n \bar{x}, \partial_j \partial_i \bar{x}),$$

where the partial derivatives read $\partial_i \bar{x} := \frac{\partial \bar{x}}{\partial \xi_i}$ etc. Using the rules for calculating determinants one verifies the following relations:

$$h_{ij} = [\det(\partial_k \partial_l u)] \cdot (\partial_j \partial_i u), \tag{3.11}$$

$$H = \det(h_{ij}) = [\det(\partial_j \partial_i u)]^{n+1}; \tag{3.12}$$

$$G_{ij} = \rho \frac{\partial^2 u}{\partial \xi_i \partial \xi_j}, \tag{3.13}$$

where ρ satisfies (see (2.13))

$$\rho = \left[\det \left(\frac{\partial^2 u}{\partial \xi_i \partial \xi_j} \right) \right]^{1/(n+2)}. \tag{3.14}$$

From (3.13) we have

$$\sqrt{G} := [\det(G_{ij})]^{1/2} = \rho^{n+1}.$$

Therefore

$$\Delta x^1 = \sum \frac{1}{\sqrt{G}} \frac{\partial}{\partial \xi_i} \left(G^{ij} \sqrt{G} \frac{\partial}{\partial \xi_j} \left(\frac{\partial u}{\partial \xi_1} \right) \right)$$

$$= \rho^{-n-1} \frac{\partial}{\partial \xi_1} (\rho^n) = n\rho^{-2} \frac{\partial \rho}{\partial \xi_1}.$$

Similarly, we have

$$\Delta x^2 = n\rho^{-2} \frac{\partial \rho}{\partial \xi_2},$$

$$\vdots$$

$$\Delta x^n = n\rho^{-2} \frac{\partial \rho}{\partial \xi_n},$$

$$\Delta x^{n+1} = \sum \rho^{-n-1} \frac{\partial}{\partial \xi_i} (\rho^n \xi_i)$$

$$= n\rho^{-1} + n\rho^{-2} \sum \xi_i \frac{\partial \rho}{\partial \xi_i}.$$

This gives

$$Y = \frac{1}{n}\Delta x = \left(\rho^{-2}\frac{\partial\rho}{\partial\xi_1},\ldots,\rho^{-2}\frac{\partial\rho}{\partial\xi_n},\rho^{-1} + \rho^{-2}\sum_i \xi_i\frac{\partial\rho}{\partial\xi_i}\right). \tag{3.15}$$

From (3.10) and (3.15) we see that the necessary and sufficient condition for $Y = -L_1 x$ is given by the relation

$$\rho = \frac{1}{L_1 u}, \quad L_1 = \text{const.} \neq 0, \tag{3.16}$$

i.e.

$$\det\left(\frac{\partial^2 u}{\partial\xi_i\partial\xi_j}\right) = (L_1 u)^{-n-2}, \tag{3.16'}$$

where L_1 is a constant.

We summarize the foregoing results:

Theorem 3.5. *Let M be an immersed hypersurface in A^{n+1} which is locally given as the graph of a strongly convex C^∞-function $x^{n+1} = f(x^1,\ldots,x^n)$.*
(i) *M is an elliptic or hyperbolic affine hypersphere with center at the origin if and only if the Legendre transformation function u of f satisfies the differential equation (3.16').*
(ii) *M is a parabolic affine hypersphere, with constant affine normal vector $(0,\ldots,0,1)$, if and only if f satisfies the equation (3.5).*

As a consequence of the last theorem we can state the following:

Any solution of the Monge–Ampère equation (3.5) defines locally an improper affine hypersphere given as the graph of the solution. Any solution of the Monge–Ampère equation (3.16') similarly defines a proper affine hypersphere. Thus both classes of affine hyperspheres are large.

Example 3.1. The equation

$$x^{n+1} = \frac{c}{x^1 x^2 \cdots x^n} =: f(x^1,\ldots,x^n),$$

where $c = \text{const.} \neq 0, x^1 > 0, x^2 > 0,\ldots,x^n > 0$, obviously defines a locally strongly convex hypersurface in A^{n+1}. Following [164] and [167] we denote this hypersurface by $Q(c,n)$. We will show that it is a hyperbolic affine hypersphere. To prove this, we put

$$\xi_i = \frac{\partial f}{\partial x^i} = -\frac{f}{x^i}.$$

The Legendre transformation function u of f is

$$u(\xi_1,\ldots,\xi_n) = -(n+1)f$$
$$= -(n+1)c^{1/(n+1)}[(-1)^n\xi_1\xi_2\cdots\xi_n]^{1/(n+1)}.$$

By a direct computation, one can verify that

$$\frac{\partial u}{\partial \xi_i} = \frac{u}{(n+1)\xi_i},$$

$$\frac{\partial^2 u}{\partial \xi_i \partial \xi_j} = \frac{u}{(n+1)^2 \xi_i \xi_j} - \frac{u}{(n+1)\xi_i^2}\delta_{ij},$$

$$\det\left(\frac{\partial^2 u}{\partial \xi_i \partial \xi_j}\right) = (L_1 u)^{-n-2},$$

where

$$L_1 = -(n+1)^{-(n+1)/(n+2)} c^{-2/(n+2)} < 0.$$

Thus, the hypersurface is a hyperbolic affine hypersphere. Note that it is not a quadric.

Exercise 3.1. Consider the hyperbolic affine hypersphere M in Example 3.1. The Blaschke metric G defines a Riemannian manifold (M, G). Prove that the sectional curvature of the metric vanishes identically; that means (M, G) is a flat space; see [184].

Lemma 3.6. *A nondegenerate hypersurface is an affine hypersphere if and only if the cubic form satisfies the covariant p.d.e.*

$$A_{ijk,l} = A_{ijl,k}$$

with respect to the Levi–Civita connection.

Proof. Use Proposition 2.16. If the p.d.e. is satisfied, contraction gives $\hat{B} = L_1 G$ (see Proposition 3.2). The converse is trivial. □

3.1.3 Calabi compositions

E. Calabi gave a construction whereby any two locally strongly convex hyperbolic affine hyperspheres can be composed to yield a new one, and he stated further results for similar other compositions (cf. [41]).

1. Composition of two hyperbolic affine hyperspheres
Let $x' : M' \to A^{p+1}$ and $x'' : M'' \to A^{q+1}$ be two locally strongly convex hyperbolic affine hyperspheres of dimension p and q, respectively, with Blaschke metrics G', G'' and constant negative affine mean curvatures L_1', L_1'', respectively. We shall endow A^{p+1} and A^{q+1} with the structure of unimodular vector spaces, additionally we fix in each an origin, namely the centers of $x'(M')$ and $x''(M'')$.

Now we shall construct a $(p + q + 1)$-dimensional hyperbolic affine hypersphere in $A^{p+1} \times A^{q+1} = A^{p+q+2}$ as follows: the abstract manifold M is the Cartesian product $\mathbb{R} \times M' \times M''$ with the orientation induced by the indicated ordering of the factors. To define the immersion $x : M \to A^{p+q+2}$, we let t denote the canonical parameter of $\mathbb{R}$, choose two positive constants C' and C'' (arbitrary for the moment) and set, for $u \in M'$, $v \in M''$, $t \in \mathbb{R}$,

$$x(t, u, v) = \left(C'x'(u) \exp\{\frac{-t}{p+1}\}, \ C''x''(v) \exp\{\frac{t}{q+1}\} \right). \tag{3.17}$$

A direct computation shows that $x(M)$ is locally strongly convex, and the Blaschke metric of $x(M)$ is

$$G = C\left\{ \frac{p+q+2}{(p+1)(q+1)}dt^2 \oplus (p+1)(-L_1')G' \oplus (q+1)(-L_1'')G'' \right\}, \tag{3.18}$$

where $C \in \mathbb{R}$ is given by

$$C^{p+q+3} = \frac{(C')^{2p+2}(C'')^{2q+2}}{(p+q+2)(p+1)^{p+1}(q+1)^{q+1}(-L_1')^{p+2}(-L_1'')^{q+2}}. \tag{3.19}$$

The affine normal vector Y on M satisfies

$$
\begin{aligned}
(p+q+1)Y &= \Delta_G x \\
&= \frac{1}{C}\left(\frac{C'}{p+1} \exp\left\{\frac{-t}{p+1}\right\} \left(-\frac{1}{L_1'}\Delta_{G'}x' + \frac{q+1}{p+q+2}x' \right), \right. \\
&\qquad \left. \frac{C''}{q+1} \exp\left\{\frac{t}{q+1}\right\} \left(-\frac{1}{L_1''}\Delta_{G''}x'' + \frac{p+1}{p+q+2}x'' \right) \right) \\
&= \frac{p+q+1}{(p+q+2)C}x.
\end{aligned}
\tag{3.20}
$$

Here, Δ_* denotes the Laplacian with respect to the metric $*$. By definition, equation (3.20) implies that $x(M)$ is an affine hypersphere; its affine mean curvature equals a negative constant and so the hypersphere is hyperbolic:

$$L_1 = -\frac{1}{(p+q+2)C}. \tag{3.21}$$

If we choose $C' = C'' = 1$, then L_1 is given by

$$(p+q+2)^{p+q+2}(-L_1)^{p+q+3} = (p+1)^{p+1}(q+1)^{q+1}\left(-L_1'\right)^{p+2}\left(-L_1''\right)^{q+2}. \tag{3.22}$$

2. Composition of a hyperbolic affine hypersphere and a point

Let $x' : M' \to A^{p+1}$ be a hyperbolic affine hypersphere of dimension p with mean curvature L_1' and affine metric G'; we assume that the center of $x'(M')$ is fixed as origin of A^{p+1}. Then for positive constants C', C'' and a, according to [41] and [132], one can also define the composition of M' with a point by

$$x(t, u) = \left(C' \exp\left\{\frac{at}{\sqrt{p+1}}\right\}x'(u), \ C'' \exp\left\{-\sqrt{p+1}\,at\right\} \right), \quad u \in M', \ t \in \mathbb{R}.$$

Then a direct computation shows that the composition x produces also a new locally strongly convex hyperbolic affine hypersphere. Moreover, its affine metric G is given by

$$G = \Lambda \left(\frac{p + 2}{(p + 1)(-L_1')} a^2 dt^2 \oplus G' \right),$$

where Λ is a positive constant such that

$$\Lambda^{p+3} = -L_1' \cdot \frac{(p + 1)^2}{p + 2} \cdot (C')^{2(p+1)}(C'')^2.$$

To calculate the affine normal vector of x, we get

$$Y = \frac{1}{p + 1} \Delta_G x = \frac{1}{(p + 1)\Lambda} \left[\Delta_{G'} x + \frac{(p + 1)(-L_1')}{(p + 2)a^2} x_{tt} \right] = \frac{(p + 1)(-L_1')}{(p + 2)\Lambda} x.$$

It follows that the composition x of the hyperbolic affine hyperspheres x' with mean curvature L_1' and a point is a hyperbolic affine hypersphere with mean curvature satisfying $L_1 = \frac{(p+1)L_1'}{(p+2)\Lambda}$.

3. The difference tensor of the Calabi compositions

Now we consider properties of the Calabi compositions in terms of the difference tensor K. Here, we recall that, by Section 1.12.3 and (2.39), the difference tensor K and the Fubini–Pick form A are equivalently related modulo the metric G via

$$G(K(e_i, e_j), e_l) := A(e_i, e_j, e_l).$$

(i) Let us begin with the hyperbolic hypersphere immersion

$$x : M = \mathbb{R} \times M' \times M'' \to A^{p+q+2}$$

as stated in Section 3.1.3-1. Let $\{u_i\}$ and $\{v_\alpha\}$ be local coordinates of M' and M'', respectively. Then $x_*(TM)$ has a G-direct product decomposition

$$x_*(TM) = \mathrm{Span}\{V\} \oplus \mathrm{Span}\{\frac{\partial x}{\partial u_i}\} \oplus \mathrm{Span}\{\frac{\partial x}{\partial v_\alpha}\},$$

where $V = \sqrt{\frac{(p+1)(q+1)}{(p+q+2)C}} \frac{\partial x}{\partial t}$ denotes a G-unit vector field.

By a direct calculation, we easily see that

$$K(V, V) = \lambda_1 V, \quad K(V, \frac{\partial x}{\partial u_i}) = \lambda_2 \frac{\partial x}{\partial u_i}, \quad K(V, \frac{\partial x}{\partial v_\alpha}) = \lambda_3 \frac{\partial x}{\partial v_\alpha}, \quad K(\frac{\partial x}{\partial u_i}, \frac{\partial x}{\partial v_\alpha}) = 0,$$

where

$$\lambda_1 = \frac{p - q}{(p + 1)(q + 1)} \sqrt{\frac{(p + 1)(q + 1)}{(p + q + 2)C}}, \quad \lambda_2 = -\frac{1}{p + 1} \sqrt{\frac{(p + 1)(q + 1)}{(p + q + 2)C}},$$

$$\lambda_3 = \frac{1}{q + 1} \sqrt{\frac{(p + 1)(q + 1)}{(p + q + 2)C}}.$$

It follows that

$$\lambda_2 + \lambda_3 = \lambda_1, \ \lambda_2 \lambda_3 = L_1.$$

(ii) Next we consider the hyperbolic hypersphere immersion

$$x : \ M = \mathbb{R} \times M' \to A^{p+2}$$

as stated in Section 3.1.3-2. Let $\{u_i\}$ be local coordinates of M'. Then $x_*(TM)$ has a G-direct product decomposition

$$x_*(TM) = \text{Span}\{V\} \oplus \text{Span} \left\{ \frac{\partial x}{\partial u_i} \right\},$$

where

$$V = \frac{1}{a} \sqrt{\frac{-(p+1)L_1'}{(p+2)\Lambda}} \frac{\partial x}{\partial t}$$

denotes a G-unit vector field. By a direct calculation, we easily see that

$$K(V, V) = \lambda_1 V, \ K\left(V, \frac{\partial x}{\partial u_i}\right) = \lambda_2 \frac{\partial x}{\partial u_i},$$

where

$$\lambda_1 = -p \sqrt{\frac{-L_1'}{(p+2)\Lambda}}, \ \lambda_2 = \sqrt{\frac{-L_1'}{(p+2)\Lambda}}.$$

It follows that

$$\lambda_1 \lambda_2 - \lambda_2^2 = \frac{(p+1)L_1'}{(p+2)\Lambda} = L_1.$$

4. Remarks
(i) Dillen and Vrancken [81] gave many other constructions for proper and improper affine hyperspheres.
(ii) Straightforward calculations show that the Calabi composition of two hyperbolic affine hyperspheres has a parallel Fubini–Pick form (with respect to the Levi–Civita connection) if and only if both original hyperbolic affine hyperspheres have parallel Fubini–Pick forms; similarly one has that the Calabi composition of a hyperbolic affine hypersphere and a point has a parallel Fubini–Pick form if and only if the original affine hypersphere has a parallel Fubini–Pick form; cf. [81, 132].
(iii) In [132], it was shown that the Calabi compositions can be characterized by the above mentioned properties of the difference tensor. It is worthy to point out that such characterizations are extremely important for the complete classification of locally strongly convex affine hypersurfaces with parallel Fubini–Pick form, given in [133] (cf. [131]). See also the Classification Theorem 3.23 below.

3.2 Affine hyperspheres with constant sectional curvature

From the last section we know that there are many affine hyperspheres, and that one is far from a local classification. In such a situation, it is a natural mathematical procedure to try to classify subclasses, at least. From Exercises 2.2 and 3.1 we know examples of affine hyperspheres with constant sectional curvature: all quadrics and the affine hypersphere $Q(c, n)$ from Example 3.1. When Li and Penn [184] realized that $Q(c, n)$ has a flat Blaschke metric [41], the problem was formulated to classify all affine hyperspheres with constant sectional curvature metric. For locally strongly convex affine hyperspheres the solution was given in [387], based on results in [167, 184, 411]. This classification is studied in this section.

For affine hyperspheres with indefinite Blaschke metric (as already mentioned, hypersurfaces with indefinite Blaschke metric are not treated in this monograph) the problem is still open: based on results in [210], the paper [338] gives a complete classification in dimension $n = 2$, while a partial classification is given in [211] for dimension $n = 3$, and in [396] the author classified all hyperbolic affine hyperspheres with parallel cubic form and flat Lorentzian metric; the paper [386] finally gave a full classification of all indefinite affine hyperspheres with constant sectional curvature for all dimensions under the additional condition that the Pick invariant is not zero.

We start with examples.

3.2.1 Examples

In the following we recall examples of affine hyperspheres with constant sectional curvature. We also give two examples of affine hyperspheres with constant Pick invariant J, constant mean curvature L_1, and constant normalized scalar curvature χ.

Example 3.2. Elliptic paraboloid:

$$x^{n+1} = \frac{1}{2}\left((x^1)^2 + (x^2)^2 + \cdots + (x^n)^2\right), \quad (x^1, \ldots, x^n) \in A^n.$$

We have

$$J = 0, \quad L_1 = 0, \quad \chi = 0.$$

Since (see Section 2.4.1, Example 2.1)

$$G = (dx^1)^2 + \cdots + (dx^n)^2$$

we have

$$R_{ijkl} = 0.$$

Example 3.3. Ellipsoid: any ellipsoid is equiaffinelly equivalent to a hypersphere

$$(x^1)^2 + \cdots + (x^{n+1})^2 = r^2.$$

We have

$$J = 0, \ L_1 = \chi = r^{-(2n+2)/(n+2)}.$$

For a standard hypersphere, the Blaschke metric and the induced metric differ by a constant factor (see Section 2.4.1, Example 2.2). Hence

$$R_{ijkl} = r^{-(2n+2)/(n+2)}(G_{ik}G_{jl} - G_{il}G_{jk}).$$

Example 3.4. Hyperboloid $H(c, n)$:

$$x^{n+1} = (c^2 + (x^1)^2 + \cdots + (x^n)^2)^{1/2}, \ (x^1, \ldots, x^n) \in A^n.$$

We have

$$J = 0, \ L_1 = \chi = -c^{-(2n+2)/(n+2)}.$$

Similar to Example 3.3, the Riemannian curvature tensor satisfies

$$R_{ijkl} = -c^{-(2n+2)/(n+2)}(G_{ik}G_{jl} - G_{il}G_{jk}).$$

Example 3.5. The hypersurface $Q(c, n)$:

$$x^1 x^2 \cdots x^{n+1} = c, \ c > 0, \ (x^1, \ldots, x^{n+1}) \in A^{n+1}.$$

We know (cf. Example 3.1) that $Q(c, n)$ is a hyperbolic affine hypersphere with affine mean curvature

$$L_1 = -(n + 1)^{-(n+1)/(n+2)} c^{-2/(n+2)}.$$

We now compute its Riemannian curvature tensor and Pick invariant. Consider the connected component

$$x^{n+1} = \frac{c}{x^1 x^2 \cdots x^n}, \ x^1 > 0, \ldots, \ x^n > 0.$$

We introduce the new parameters $u^1, u^2, \ldots, u^n$:

$$\begin{cases} x^1 = e^{u^1}, \\ x^2 = e^{u^2}, \\ \vdots \\ x^n = e^{u^n}. \end{cases}$$

Then $Q(c, n)$ can be represented in terms of $u^1, \ldots, u^n$:

$$\begin{cases} x^1 = e^{u^1}, \\ x^2 = e^{u^2}, \\ \vdots \\ x^n = e^{u^n}, \\ x^{n+1} = c e^{-u^1 - u^2 - \cdots - u^n}. \end{cases}$$

Take

$$e_1 = (e^{u^1}, 0, \ldots, 0, -ce^{-u^1-u^2-\cdots-u^n}),$$

$$e_2 = (0, e^{u^2}, 0, \ldots, 0, -ce^{-u^1-\cdots-u^n}),$$

$$\vdots$$

$$e_n = (0, \ldots, 0, e^{u^n}, -ce^{-u^1-\cdots-u^n}).$$

By a direct computation one verifies

$$(h_{ij}) = \begin{pmatrix} 2c & c & \cdots & c \\ c & 2c & \cdots & c \\ & & \cdots & \\ c & c & \cdots & 2c \end{pmatrix}.$$

Since (G_{ij}) is a constant matrix we have

$$R_{ijkl} = 0, \quad \chi = 0, \quad J = -L_1 = (n+1)^{-(n+1)/(n+2)} c^{-2/(n+2)}.$$

In the following we give two examples of affine hyperspheres with constant data J, L_1, and χ. We use Calabi's composition formula. Let

$$x' : M' \to A^{p+1} \quad \text{and} \quad x'' : M'' \to A^{q+1}$$

be two hyperbolic affine hyperspheres with affine mean curvatures L_1', L_1'', and normalized scalar curvatures χ', χ'', respectively. Using the expression (3.17) in the last section, we obtain a new hyperbolic affine hypersphere $x : M \to A^{p+q+2}$ with the following values for the normalized scalar curvature and the Pick invariant:

$$\chi = \frac{1}{(p+q+1)(p+q)C} \left[\frac{p(p-1)\chi'}{(p+1)(-L_1')} + \frac{q(q-1)\chi''}{(q+1)(-L_1'')} \right],$$

$$J = \chi - L_1 = \chi + \frac{1}{(p+q+2)C},$$

where C is given by (3.19).

Example 3.6. Taking $x'(M') = H(1, p)$ and $x''(M'') = H(1, q)$, we obtain the hyperbolic affine hypersphere

$$[(x^{p+1})^2 - ((x^1)^2 + \cdots + (x^p)^2)]^{(p+1)/2} \cdot [(x^{p+q+2})^2 - ((x^{p+2})^2 + \cdots + (x^{p+q+1})^2)]^{(q+1)/2} = 1.$$

Its affine mean curvature, Pick invariant and normalized scalar curvature are given, respectively, by

$$L_1 = \frac{-1}{(p+q+2)c},$$

$$J = \frac{1}{(p+q+2)c} - \frac{1}{(p+q+1)(p+q)c} \left[\frac{p(p-1)}{p+1} + \frac{q(q-1)}{q+1} \right],$$

$$\chi = -\frac{1}{(p+q+1)(p+q)c} \left[\frac{p(p-1)}{p+1} + \frac{q(q-1)}{q+1} \right],$$

where

$$c = [(p + q + 2)(p + 1)^{p+1}(q + 1)^{q+1}]^{-1/(p+q+3)}.$$

Example 3.7. Taking $x'(M') = H(1, p)$ and $x''(M'') = Q(1, q)$, we obtain the hyperbolic affine hypersphere

$$[(x^{p+1})^2 - ((x^1)^2 + \cdots + (x^p)^2)]^{(p+1)/2} x^{p+2} \cdots x^{p+q+2} = 1.$$

Its affine mean curvature, Pick invariant and normalized scalar curvature, respectively, are given by

$$L_1 = -\frac{1}{(p + q + 2)c},$$

$$J = \frac{1}{(p + q + 2)c} - \frac{p(p - 1)}{(p + q + 1)(p + q)(p + 1)c},$$

$$\chi = \frac{-p(p - 1)}{(p + q + 1)(p + q)(p + 1)c},$$

where

$$c = [(p + q + 2)(p + 1)^{p+1}]^{-1/(p+q+3)}.$$

3.2.2 Local classification of two-dimensional affine spheres with constant scalar curvature

First we derive a differential equation for the Pick invariant J on 2-dimensional affine spheres; this equation was first derived by W. Blaschke [20, §76].

Let $x : M \to A^3$ be an affine sphere. Choose a local unimodular affine frame field $\{x; e_1, e_2, e_3\}$ on M such that $e_1, e_2 \in T_x M$, $G_{ij} = \delta_{ij}$ and $e_3 = Y$. Since $x(M)$ is an affine sphere, we have

$$A_{ijk,l} = A_{ijl,k}.$$

Using this and the Ricci identity of Section A.3.1.8 we get

$$\Delta J = 6\chi J + \sum (A_{ijk,l})^2. \tag{3.23}$$

We will prove a corresponding formula in Section 3.2.3 for any dimension, so we omit the details here.

To apply (3.23) we compute $\sum (A_{ijk,l})^2$ in the case $J \neq 0$. We can choose e_1, e_2 such that $A_{111} = 0$ at a fixed point $x \in M$. In fact, if $A_{111} \neq 0$ at x, by a change of frame

$$\begin{cases} \bar{e}_1 = \cos \theta\ e_1 + \sin \theta\ e_2, \\ \bar{e}_2 = -\sin \theta\ e_1 + \cos \theta\ e_2, \end{cases}$$

we get

$$\bar{A}_{111} = (\cos^3 \theta - 3 \cos \theta \sin^2 \theta) A_{111} + (\sin^3 \theta - 3 \cos^2 \theta \sin \theta) A_{222}.$$

Choosing θ such that

$$\frac{\cos^3 \theta - 3 \cos \theta \sin^2 \theta}{\sin^3 \theta - 3 \cos^2 \theta \sin \theta} = -\frac{A_{222}}{A_{111}},$$

we arrive at $\bar{A}_{111} = 0$. Thus in the following we assume that $A_{111} = 0$ at the point x. Since $J = 2((A_{111})^2 + (A_{222})^2)$, we have

$$4(A_{111}A_{111,1} + A_{222}A_{222,1}) = J_{,1},$$

$$4(A_{111}A_{111,2} + A_{222}A_{222,2}) = J_{,2}.$$

Then $A_{111}|_x = 0$ implies that, at x,

$$A_{222,1} = \frac{J_{,1}}{4A_{222}}, \quad A_{222,2} = \frac{J_{,2}}{4A_{222}}.$$

The apolarity condition 2.5.1 (iii) and Lemma 3.6 together imply

$$A_{111,1} = -A_{221,1} = -A_{112,2} = A_{222,2},$$

$$A_{111,2} = -A_{221,2} = -A_{222,1} = A_{112,1}.$$

Thus we obtain (see (A.31) for the definition of the gradient operator)

$$\sum (A_{ijk,l})^2 = 8\left((A_{222,1})^2 + (A_{222,2})^2\right) = \frac{\|\operatorname{grad} J\|^2}{J}. \tag{3.24}$$

Inserting (3.24) into (3.23) we get

$$\Delta J = 6\chi J + \frac{\|\operatorname{grad} J\|^2}{J}, \tag{3.25}$$

or

$$\Delta \ln J^{1/6} = \chi. \tag{3.26}$$

As an application of formula (3.26) we obtain the following theorem (cf. [184]).

Theorem 3.7. *Let $x : M \to A^3$ be an affine sphere with $\chi = \text{const}$. Then $x(M)$ is contained either in a quadric or in an affine image of the surface $Q(c, 2)$ (for the definition of $Q(c, 2)$ see Example 3.5 in Section 3.2.1).*

Proof. $\chi = \text{const.}$ implies $J = \text{const.}$

1) Case $J \neq 0$. From (3.26) we have $\chi = 0$, then $x(M)$ must be locally a flat space with respect to the Blaschke metric. From (3.23) we have

$$A_{ijk,l} = 0. \tag{3.27}$$

We can choose coordinates (u, v) on $x(M)$ such that the Blaschke metric is

$$G = (du)^2 + (dv)^2.$$

Then (3.27) implies

$$A_{ijk} = \text{const.}$$

If $A_{111} \neq 0$, we make an orthogonal transformation:

$$\begin{cases} \bar{u} = \cos\theta\; u + \sin\theta\; v, \\ \bar{v} = -\sin\theta\; u + \cos\theta\; v. \end{cases}$$

We choose θ such that $\bar{A}_{111} = 0$ and $\bar{A}_{222} = \sqrt{\frac{I}{2}}$.

We choose a constant c such that $Q(c, 2)$ and $x(M)$ have the same Pick invariant. As above, we find coordinates $(\tilde{u}, \tilde{v})$ on $Q(c, 2)$ such that the Blaschke metric of $Q(c, 2)$ is $\tilde{G} = (d\tilde{u})^2 + (d\tilde{v})^2$, and $\tilde{A}_{111} = 0, \tilde{A}_{222} = \sqrt{\frac{I}{2}}$. Since both $x(M)$ and $Q(c, 2)$ are locally Euclidean spaces, there is a local isometry

$$f : x(M) \rightarrow Q(c, 2).$$

Since $\bar{A}_{ijk} = \tilde{A}_{ijk}$, by Theorem 2.19, $Q(c, 2)$ and $x(M)$ differ by an equiaffine transformation.

2) Case $J = 0$. It is well known that a locally strongly convex surface with $J = 0$ is contained in a quadric. $\qquad\square$

Theorem 3.7 gives the following classification: in dimension $n = 2$ the ellipsoids, the elliptic paraboloids, the hyperboloids and the surfaces $Q(c, 2)$ are the only affine spheres with constant scalar curvature.

3.2.3 Generalization to higher dimensions

We now generalize Theorem 3.7 to higher dimensional locally strongly convex affine hyperspheres. First, we give a local characterization of the affine hypersphere $Q(c, n)$ due to A.-M. Li (cf. [167]).

Theorem 3.8. Let $x : M \rightarrow A^{n+1}$ be a hyperbolic affine hypersphere. If $\chi = 0$ on M, then, up to an affine transformation, $x(M)$ lies on the hypersurface $Q(c, n)$.

Proof. Choose a local equiaffine frame field $\{x; e_1, \ldots, e_{n+1}\}$ on M (cf. Remark and Definition 2.4) such that $G_{ij} = \delta_{ij}$. Since $x(M)$ is an affine hypersphere, we have (see Lemma 3.6 and Proposition 2.16)

$$A_{ijk,l} = A_{ijl,k}, \tag{3.28}$$

$$R_{ijkl} = \sum(A_{iml}A_{jmk} - A_{imk}A_{jml}) - L_1(\delta_{il}\delta_{jk} - \delta_{ik}\delta_{jl}). \tag{3.29}$$

Applying the Ricci identity (see Section A.3.1.8), the symmetry relation (3.28) and the apolarity condition (2.46), we get

$$
\begin{aligned}
\Delta A_{ijk} &= \sum A_{ijk,ll} \\
&= \sum A_{ijl,kl} \\
&= \sum A_{ijl,lk} + \sum A_{ijr} R_{rlkl} + \sum A_{irl} R_{rjkl} + \sum A_{rjl} R_{rikl} \\
&= \sum A_{ijr} R_{rlkl} + \sum A_{irl} R_{rjkl} + \sum A_{rjl} R_{rikl}.
\end{aligned}
\tag{3.30}
$$

The Laplacian of the Pick invariant J then is given by

$$
\begin{aligned}
\Delta J &= \frac{1}{n(n-1)} \Delta \left(\sum (A_{ijk})^2 \right) \\
&= \frac{2}{n(n-1)} \left[\sum (A_{ijk,l})^2 + \sum A_{ijk} A_{ijk,ll} \right] \\
&= \frac{2}{n(n-1)} \left[\left(\sum A_{ijk,l} \right)^2 + \sum A_{ijk} A_{ijr} R_{rlkl} + \sum (A_{ijk} A_{irl} - A_{ijl} A_{irk}) R_{rjkl} \right].
\end{aligned}
\tag{3.31}
$$

Inserting (3.29) into (3.31) we get

$$
\frac{n(n-1)}{2} \Delta J = \sum (A_{ijk,l})^2 + \sum (R_{ij})^2 + \sum (R_{ijkl})^2 - (n+1) R L_1.
\tag{3.32}
$$

Since $R = 0 = \chi$ and $L_1 = \text{const.} < 0$, we have $J = \chi - L_1 = \text{const.} > 0$. From (3.32) it follows that $R_{ijkl} = 0$. Hence $x(M)$ locally is a Euclidean space. We choose local coordinates $u^1, u^2, \ldots, u^n$ such that the Blaschke metric is given by

$$
G = (du^1)^2 + (du^2)^2 + \cdots + (du^n)^2.
$$

Then $A_{ijk} = \text{const.}$ Without loss of generality we can assume that the center of $x(M)$ is at the origin and that $L_1 = -1$ (otherwise we may make a similarity transformation $\tilde{x} = cx$ such that $L_1 = -1$). We now choose the local equiaffine frame field $e_i = \frac{\partial}{\partial u^i}$ $(i = 1, \ldots, n)$ and $e_{n+1} = Y = x$. We have

$$
\begin{cases}
de_i = \sum \omega_i^j e_j + du^i x, \\
dx = \sum du^i e_i.
\end{cases}
\tag{3.33}
$$

Since the Levi–Civita connection of G is flat, we have $\omega_i^j = \sum A_{ijk} du^k$. Since the characterization in our theorem is local, we shall limit our considerations to a coordinate neighborhood. Let $p \in M$ be a fixed point with coordinates $(0, \ldots, 0)$, and $x \in M$ be an arbitrary point with coordinates $(v^1, v^2, \ldots, v^n)$. We draw a curve connecting both points:

$$
u^i(t) = v^i t, \quad 0 \leq t \leq 1.
$$

Along this curve the equations (3.33) become

$$
\begin{cases}
\dfrac{de_i}{dt} = \sum A_{ijk} v^k e_j + v^i x, \\
\dfrac{dx}{dt} = \sum v^i e_i.
\end{cases}
\tag{3.34}
$$

Define the matrices

$$
E := \begin{pmatrix} e_1 \\ \vdots \\ e_n \\ x \end{pmatrix}, \quad
B_k := \begin{pmatrix}
A_{11k} & \cdots & A_{1kk} & \cdots & A_{1nk} & 0 \\
\vdots & & \vdots & & \vdots & \vdots \\
A_{k1k} & \cdots & A_{kkk} & \cdots & A_{knk} & 1 \\
\vdots & & \vdots & & \vdots & \vdots \\
A_{n1k} & \cdots & A_{nkk} & \cdots & A_{nnk} & 0 \\
0 & \cdots & 1 & \cdots & 0 & 0
\end{pmatrix}. \tag{3.35}
$$

The equations (3.34) can be written in matrix form:

$$
\frac{dE}{dt} = \sum v^k B_k E. \tag{3.36}
$$

Since $R_{ijkl} = 0$ and $L_1 = -1$, the integrability condition (3.29) reads

$$
\sum A_{iml} A_{jmk} - \sum A_{imk} A_{jml} + \delta_{il}\delta_{jk} - \delta_{jl}\delta_{ik} = 0.
$$

This means

$$
B_i B_j = B_j B_i, \quad \forall i, j \in \{1, 2, \dots, n\}.
$$

It follows that there exists a constant matrix $C = (c_{AB})$ such that

$$
B_k = C \begin{pmatrix}
\lambda_1(k) & & & 0 \\
& \lambda_2(k) & & \\
& & \ddots & \\
0 & & & \lambda_{n+1}(k)
\end{pmatrix} C^{-1} \tag{3.37}
$$

where $\lambda_1(k), \dots, \lambda_n(k), \lambda_{n+1}(k)$ are constants. We can choose C such that $\det(C) = 1$. From the apolarity condition $\sum A_{iik} = 0$ we have

$$
\lambda_1(k) + \lambda_2(k) + \cdots + \lambda_{n+1}(k) = 0, \quad k = 1, 2, \dots, n. \tag{3.38}
$$

Define

$$
F = C^{-1}E, \quad F = \begin{pmatrix} f_1 \\ f_2 \\ \vdots \\ f_{n+1} \end{pmatrix}. \tag{3.39}
$$

From (3.36), (3.37), and (3.39) we get

$$
\frac{df_A}{dt} = \sum v^k \lambda_A(k) f_A, \quad A = 1, 2, \dots, n + 1.
$$

Hence

$$
f_A(t) = f_A(0) \exp\left\{ \sum v^k \lambda_A(k) t \right\}. \tag{3.40}
$$

By a unimodular affine transformation we may assume that

$$f_1(0) = (1, 0, \ldots, 0), \quad f_2(0) = (0, 1, 0, \ldots, 0), \ldots, f_{n+1}(0) = (0, \ldots, 0, 1).$$

Now (3.35), (3.39), and (3.40) imply that

$$x^A(t) = c_{n+1,A} \exp\left\{\sum v^k \lambda_A(k)t\right\}, \quad \text{for } A = 1, 2, \ldots, n + 1.$$

We evaluate the last expression for $t = 1$ and use (3.38). Then we obtain

$$x^1 x^2 \cdots x^{n+1} = c_{n+1,1} c_{n+1,2} \cdots c_{n+1,n+1} = \text{const.} \qquad \square$$

Corollary 3.9. *Let* $x : M \to A^{n+1}$ *be an affine hypersphere with normalized scalar curvature* $\chi = 0$. *Then* $x(M)$ *is contained either in a paraboloid or in an affine image of the hypersurface* $Q(c, n)$.

The proof is immediate.

Remark 3.10. In Section 3.5 below we will prove the following differential inequality for the Laplacian ΔJ on affine hyperspheres:

$$\Delta J \geq 2(n + 1)J\chi + \frac{2}{n(n - 1)} \sum (A_{ijk,l})^2.$$

If $\chi = \text{const.} > 0$, formula (2.125) gives $J = \text{const.}$, and the foregoing differential inequality then implies $J = 0$. Thus an affine hypersphere with $\chi = \text{const.} > 0$ must be part of a hyperellipsoid.

We would like to pose the following problem:

Problem. *Classify all affine hyperspheres with constant negative scalar curvature; obviously they must be hyperbolic.*

The hypersurfaces in Examples 3.4, 3.6, and 3.7 of Section 3.2.1 are examples of affine hyperspheres with $R = \text{const.} < 0$. Are there other examples of affine hyperspheres with $R = \text{const.} < 0$?

The following Theorem 3.11 first was proved by L. Vrancken, but his original proof was never published; instead a modified version of the proof appeared in the paper [387]; it uses the de Rham decomposition theorem.

Here we give another proof using only elementary tools.

Theorem 3.11. *Let* $x : M \to A^{n+1}$ *be an affine hypersphere with constant sectional curvature. Then* $x(M)$ *is contained either in a quadric or in an affine image of the hypersurface* $Q(c, n)$.

To prove Theorem 3.11 we need two lemmas.

Lemma 3.12. *Let g be a positive definite metric on $\mathbb{R}^n$ and let*

$$T : \mathbb{R}^n \times \mathbb{R}^n \times \mathbb{R}^n \to \mathbb{R}$$

be a symmetric multilinear mapping. On $S^{n-1} := \{v \in \mathbb{R}^n | g(v, v) = 1\}$ we define a function F by $F(v) = T(v, v, v)$. Let u be a vector at which the function F attains an extremal value and let $w \in S^{n-1}$ be such that $g(u, w) = 0$. Then $T(u, u, w) = 0$. Moreover, if F attains a relative maximum at u, then we have also that $T(u, u, u) - 2T(u, w, w) \geq 0$.

Proof. Assume that $w \in S^{n-1}$ and $g(u, w) = 0$. We define $f(t) = F(u \cos t + w \sin t)$. Since $f(t)$ attains an extremal value at $t = 0$, we have $f'(0) = 0$. If f attains a relative maximum at u, we also have $f''(0) \leq 0$. A direct computation gives Lemma 3.12. $\square$

Lemma 3.13. *Let $x : M \to A^{n+1}$ be a locally strongly convex affine hypersphere with constant sectional curvature C. We assume also that $x(M)$ is not an open part of a quadric. Then for any point $p \in M$, there is an orthonormal frame $\{e_1, \ldots, e_n\}$ of T_pM such that, at p,*

$$A_{1jk} = \begin{cases} 0, & \text{if } j \neq k, \\ -\sqrt{\frac{J}{n}}, & \text{if } j = k \neq 1, \\ (n-1)\sqrt{\frac{J}{n}}, & \text{if } j = k = 1, \end{cases} \quad (3.41)$$

where the Pick invariant J has constant value $J = C - L_1$.

Proof. Recall that the Fubini-Pick form $A : T_pM \times T_pM \times T_pM \to \mathbb{R}$ is a symmetric multilinear mapping. Define a function F on

$$S_p^{n-1} := \{v \in T_pM | G(v, v) = 1\}$$

by $F(v) = A(v, v, v)$. Let $e_1 \in S_p^{n-1}$ be a vector at which F attains an extremal value. Then, by Lemma 3.12,

$$A(e_1, e_1, w) = 0 \quad \text{for} \quad w \in S_p^{n-1} \quad \text{and} \quad G(e_1, w) = 0. \quad (3.42)$$

For fixed e_1, $A(e_1, \cdot, \cdot)$ is a symmetric bilinear function on T_pM. Denote by A_{e_1} the self-adjoint linear operator from T_pM into T_pM associated with $A(e_1, \cdot, \cdot)$, which means

$$G(A_{e_1}(v), w) = A(e_1, v, w) \quad \text{for all } v, w \in T_pM.$$

Recall from linear algebra that e_1 is an eigenvector of the operator A_{e_1}. As A_{e_1} is self-adjoint there exist n (not necessarily distinct) real eigenvalues $\mu_1, \mu_2, \ldots, \mu_n$ and a corresponding basis of T_pM of orthonormal eigenvectors $e_1, e_2, \ldots, e_n$:

$$G(e_i, e_j) = \delta_{ij},$$
$$A_{e_1}(e_j) = \mu_j e_j.$$

With respect to $\{e_1,\ldots,e_n\}$, we have $A_{1ij} = \mu_j \delta_{ij}$. Since $x(M)$ is an affine hypersphere with constant sectional curvature, from (2.123') of Proposition 2.16 and $J = C - L_1$ we have that

$$(\mu_k)^2 - \mu_k\mu_1 - J = 0, \quad k = 2,\ldots,n.$$

Hence

$$\mu_k = \frac{1}{2}\left(\mu_1 \pm \sqrt{(\mu_1)^2 + 4J}\right). \tag{3.43}$$

From (3.43) and the apolarity condition $0 = \sum G(A_{e_1}(e_j), e_j) = \sum \mu_j$ we conclude that μ_1 can take only finitely many values. If F attains a maximum at e_1, then $\mu_k \leq \mu_1$, therefore

$$\mu_2 = \cdots = \mu_n = \frac{1}{2}\left(\mu_1 - \sqrt{(\mu_1)^2 + 4J}\right).$$

The apolarity condition implies that

$$\mu_1 = (n-1)\sqrt{\frac{J}{n}} \quad \text{and} \quad \mu_2 = \cdots = \mu_n = -\sqrt{\frac{J}{n}}. \qquad \Box$$

Proof of Theorem 3.11. From the assumptions of the theorem we have $J = \text{const.}$ If $J = 0$ then $x(M)$ is contained in a quadric. Assume $J \neq 0$. First we show that we can choose a smooth local orthonormal frame field $\{e_1,\ldots,e_n\}$ such that (3.41) holds. Let $p \in M$. We choose a C^∞ orthonormal frame field $\{E_1,\ldots,E_n\}$ on a neighborhood of p such that at p the function $F(v) = A(v,v,v)$ defined on S_p^{n-1} attains its maximum at E_1. Then (3.41) holds at p. To find locally a differentiable unit vector field e_1 such that $F(v)$ attains its maximum for e_1 we apply Lagrange's multiplier method. Denote by A_{ijk} the components of the Fubini–Pick form relative to the frame field $\{E_1,\ldots,E_n\}$. Considering the function $f(x,y^1,\ldots,y^n,\lambda) := \sum A_{ijk}(x)y^i y^j y^k - \lambda((y^1)^2 + \cdots + (y^n)^2 - 1)$, we get a system of equations:

$$\begin{cases} \dfrac{\partial f}{\partial y^1} = 3\sum A_{1jk}(x)y^j y^k - 2y^1\lambda = 0, \\[2mm] \dfrac{\partial f}{\partial y^2} = 3\sum A_{2jk}(x)y^j y^k - 2y^2\lambda = 0, \\[2mm] \qquad\vdots \\[2mm] \dfrac{\partial f}{\partial y^n} = 3\sum A_{njk}(x)y^j y^k - 2y^n\lambda = 0, \\[2mm] (y^1)^2 + \cdots + (y^n)^2 - 1 = 0. \end{cases} \tag{3.44}$$

Define $F_i(x,y^1,\ldots,y^n,\lambda) := 3\sum A_{ijk}(x)y^j y^k - 2y^i\lambda$ for $1 \leq i \leq n$, and

$$F_{n+1}(x,y^1,\ldots,y^n) = (y^1)^2 + \cdots + (y^n)^2 - 1.$$

We take $e_1(p) := E_1(p)$, i.e. at p we have $(y^1,\ldots,y^n) = (1,0,\ldots,0)$. Since (3.41) holds at p, we have $\lambda(p) = \frac{3}{2}A_{111}$. Denote $y^{n+1} = \lambda$. A straightforward calculation gives

$$\det\left(\frac{\partial F_A}{\partial y^B}\right)(p) \neq 0.$$

From the implicit function theorem there are C^∞-functions $y^1(x), \ldots, y^n(x), \lambda(x)$ satisfying (3.44), and at p, $(y^1, \ldots, y^n) = (1, 0, \ldots, 0)$. Define the differentiable unit vector field $e_1 = y^1 E_1 + \cdots + y^n E_n$. Then, in a neighborhood of p, the function $A(v, v, v)$ attains its maximum at e_1. From the proof of Lemma 3.13 we find that, acting on the orthogonal complement $\{e_1\}^\perp$, $A_{e_1} = -\sqrt{\frac{J}{n}}\,\mathrm{id}$. Thus, we can choose C^∞-vector fields $\{e_2, \ldots, e_n\}$ such that $\{e_1, e_2, \ldots, e_n\}$ form an orthonormal frame field with respect to which we have

$$A_{1jk}(x) = \begin{cases} 0, & \text{if } j \neq k, \\ -\sqrt{\frac{J}{n}}, & \text{if } j = k \neq 1, \\ (n-1)\sqrt{\frac{J}{n}}, & \text{if } j = k = 1. \end{cases} \tag{3.45}$$

Denote by $\{\omega^i\}$ the dual frame field of $\{e_i\}$, and by $\tilde{\omega}^j_i$ the Levi–Civita connection forms of the Blaschke metric. Put

$$\tilde{\omega}^j_i = \sum \Gamma^j_{ik} \omega^k.$$

Then

$$\Gamma^j_{ik} = -\Gamma^i_{jk}.$$

Since

$$\sum A_{11m,i} \omega^i = dA_{11m} - 2 \sum A_{k1m} \tilde{\omega}^k_1 - \sum A_{11k} \tilde{\omega}^k_m$$
$$= (n+1)\sqrt{\frac{J}{n}}\, \tilde{\omega}^m_1,$$

we have

$$A_{11m,i} = (n+1)\sqrt{\frac{J}{n}}\, \Gamma^m_{1i}, \tag{3.46}$$

$$A_{111,i} = 0, \quad 1 \leq i, m \leq n. \tag{3.47}$$

From (3.46), (3.47), and the equality $A_{ijk,l} = A_{ijl,k}$ (see Lemma 3.6), we get

$$\Gamma^m_{11} = 0. \tag{3.48}$$

On the other hand, for $i, m \geq 2$, we have

$$\sum A_{1im,j} \omega^j = -\sum A_{kim} \tilde{\omega}^k_1 - \sum A_{1km} \tilde{\omega}^k_i - \sum A_{1ik} \tilde{\omega}^k_m$$
$$= -\sum A_{kim} \tilde{\omega}^k_1 - A_{1mm} \tilde{\omega}^m_i - A_{1ii} \tilde{\omega}^i_m$$
$$= -\sum A_{kim} \tilde{\omega}^k_1.$$

Hence

$$A_{1im,1} = -\sum A_{kim} \Gamma^k_{11} = 0. \tag{3.49}$$

Then (3.46) and (3.49) give

$$\tilde{\omega}^m_1 = 0.$$

It follows that $\chi = 0$. By Theorem 3.8, $x(M)$ must be contained in an affine image of the hypersurface $Q(c, n)$. $\qquad\square$

3.3 Affine hypersurfaces with parallel Fubini–Pick form

From Theorem 2.13 we know that a locally strongly convex hypersurface M in A^{n+1} is a hyperquadric if and only if it has vanishing Fubini–Pick form: $A \equiv 0$. Then, as a generalization of $A \equiv 0$, in [24] the authors considered affine hypersurfaces with parallel cubic form, namely *parallel Fubini–Pick form* $\tilde{\nabla}A = 0$, where $\tilde{\nabla}$ is the Levi-Civita connection of the affine metric. From Lemma 3.6 we know that the class of such affine hypersurfaces is a subclass of the class of affine hyperspheres; below we will sketch a classification of the subclass in the case of local strong convexity.

3.3.1 Implications from a parallel Fubini–Pick form

The condition $\tilde{\nabla}A = 0$ turns out to be significant; it has important implications. Particularly, by Lemma 3.6, a locally strongly convex affine hypersurface with parallel Fubini–Pick form is an affine hypersphere. Further implications of the condition $\tilde{\nabla}A = 0$ were proved in [84]:

Proposition 3.14. *Let M be a locally strongly convex affine hypersurface in A^{n+1} with $\tilde{\nabla}A = 0$. Then M is a locally homogeneous affine hypersphere.*

Proof. By Lemma 3.6, the condition $\tilde{\nabla}A = 0$ implies that M is an affine hypersphere: $\hat{B} = L_1 G$. Hence by the integrability condition we have

$$R_{ijkl} = \sum (A_{il}^m A_{mjk} - A_{ik}^m A_{mjl}) + L_1(G_{ik}G_{jl} - G_{il}G_{jk})$$

and thus $\tilde{\nabla}R = 0$. Let $p, q \in M$, and let $\{e_i\}$ be any orthonormal basis of T_pM. By parallel translation along geodesics through p we obtain a local frame $\{E_i\}$ on a normal neighborhood U_p of p. Since $\tilde{\nabla}A = 0$ and $\tilde{\nabla}R = 0$, the components $A_{ijk} = A(E_i, E_j, E_k)$ and $R_{ijkl} = G(R(E_i, E_j)E_l, E_k)$ are constant on U_p. If we translate $\{e_i\}$ parallel to q, we obtain an orthonormal basis $\{f_i\}$ of T_qM. Let $\ell : T_pM \to T_qM$ be the linear isometry such that $\ell(e_i) = f_i$, then ℓ preserves curvature and from Theorem 8.14 of [266], we know that there is an isometry $\mathfrak{T} : U \to M$ from a neighborhood $U \subset U_p$ around p such that $\mathfrak{T}(p) = q$ and $\mathfrak{T}_{*p} = \ell$. Let $F_i = \mathfrak{T}_*(E_i)$, then the frame $\{F_i\}$ is obtained from $\{f_i\}$ by parallel translation as above. Moreover, $A(F_i, F_j, F_k) = A_{ijk}$ and $G(R(F_i, F_j)F_l, F_k) = R_{ijkl}$ are the same constants. Therefore, $\mathfrak{T}$ preserves both G and A, so by the fundamental uniqueness theorem (Theorem 2.19), we again obtain that there is an equiaffine transformation T of A^{n+1} such that $T(x) = \mathfrak{T}(x)$ for all $x \in U$. Hence M is locally homogeneous. $\square$

Proposition 3.15. *Let M be a locally strongly convex affine hypersurface in A^{n+1} with $\tilde{\nabla}A = 0$ and $A \neq 0$. Then M is a hyperbolic affine hypersphere.*

Proof. By Lemma 3.6 we know that M is an affine hypersphere with constant affine mean curvature L_1. As in the proof of Lemma 3.13, we consider the function F on S_p^{n-1}, defined by $F(v) := A(v, v, v)$. Let $e_1 \in T_pM$ be a unit vector at which F attains a relative maximum $\mu_1 > 0$. Then the proofs of Lemmas 3.12 and 3.13 imply the existence of an orthonormal basis $\{e_1, \ldots, e_n\}$ of T_pM, which possesses the following properties:

$$G(e_i, e_j) = \delta_{ij}, \quad A_{1ij} = \mu_j \delta_{ij}, \quad 1 \le i, j \le n;$$
$$\mu_1 - 2\mu_i \ge 0, \quad i = 2, \ldots, n. \tag{3.50}$$

Noting that the function $f(t) = F(e_1 \cos t + e_i \sin t)$ for each $i \ge 2$ satisfies: $f'(0) = 0$, $f''(0) \le 0$, and if $f''(0) = 0$ then $f'''(0) = 0$. Thus, the computation

$$f'''(0) = 6 F(e_i) - 21 A(e_1, e_1, e_i) = 6F(e_i)$$

shows that

$$\text{if } \mu_1 = 2\mu_i, \text{ then } A(e_i, e_i, e_i) = 0, \text{ for } i \in \{2, 3, \ldots, n\}. \tag{3.51}$$

From the apolarity condition we have

$$\mu_1 + \mu_2 + \cdots + \mu_n = 0. \tag{3.52}$$

On the other hand, using $\tilde{\nabla}A = 0$ and the Ricci identity

$$A_{11i,1i} - A_{11i,i1} = 2 \sum A_{j1i} R_{j11i} + \sum A_{11j} R_{ji1i}, \quad i \ge 2,$$

we obtain

$$(\mu_1 - 2\mu_i)((\mu_i)^2 - \mu_1 \mu_i + L_1) = 0. \tag{3.53}$$

If $\mu_1 = 2\mu_i$ for all $i \in \{2, 3, \ldots, n\}$, then (3.52) implies that $\mu_1 = 0$ which is a contradiction. Therefore there is a number k, $1 \le k < n$ such that, after rearranging the ordering,

$$\mu_2 = \mu_3 = \cdots = \mu_k = \frac{1}{2}\mu_1, \text{ and } \mu_{k+1} < \frac{1}{2}\mu_1, \ldots, \mu_n < \frac{1}{2}\mu_1.$$

Moreover, if $i > k$, then (3.53) implies that

$$(\mu_i)^2 - \mu_1 \mu_i + L_1 = 0. \tag{3.54}$$

Subtracting (3.53) for $i, j > k$, we obtain

$$(\mu_i - \mu_j)(\mu_i + \mu_j - \mu_1) = 0.$$

But for $i, j > k$ one can check that $\mu_i + \mu_j - \mu_1 \ne 0$. Thus $\mu_i = \mu_j$ for $k < i, j \le n$. Setting $\mu_{k+1} = \cdots = \mu_n =: \lambda$ and using (3.52) and (3.54), we have

$$\lambda = -\frac{k+1}{2(n-k)}\mu_1 \text{ and } L_1 = -\frac{(k+1)(2n-k+1)}{4(n-k)^2}(\mu_1)^2 < 0. \qquad \square$$

3.3.2 Classification of affine hypersurfaces with parallel Fubini–Pick form

In this subsection we review results in several steps about the classification of locally strongly convex affine hypersurfaces with $\tilde{\nabla}A = 0$. First of all, the results of the last subsection show that the condition $\tilde{\nabla}A = 0$ implies that M is an affine hypersphere with constant affine scalar curvature. Thus Theorem 3.7 can be restated as follows.

Theorem 3.16. *Let* $x : M \to A^3$ *be a locally strongly convex affine surface with* $\tilde{\nabla}A = 0$ *and* $A \neq 0$. *Then, up to an affine transformation,* $x(M)$ *lies on the surface* $Q(c, 2)$.

For the study of the higher dimensional situation, from Section 3.1.3 we recall:
(i) the Calabi composition of two hyperbolic affine hyperspheres both having parallel Fubini–Pick form (with respect to the Levi–Civita connection) remains to be a
hyperbolic affine hypersphere with parallel Fubini-Pick form;
(ii) the Calabi composition of a hyperbolic affine hypersphere having parallel Fubini–
Pick form and a point also remains a hyperbolic affine hypersphere with parallel
Fubini–Pick form.

Thus, to achieve a complete classification of affine hypersurfaces with parallel Fubini–
Pick form, we need some reduction theorem about the Calabi compositions. We are
going to determine all lower dimensional affine hypersurfaces of this type as well as
those which can not be obtained by the Calabi composition. Fortunately, this can be
successfully done.

Since the hyperboloid has vanishing Fubini–Pick form, the Calabi compositions
of two hyperboloids or one hyperboloid with a point will produce another hyperbolic
affine hypersphere, which has parallel but nonvanishing Fubini–Pick form.

Next we present the classification of 3-dimensional affine hypersurfaces with parallel Fubini-Pick form, due to Dillen and Vrancken [79].

Theorem 3.17. *Let* M *be a 3-dimensional locally strongly convex hypersurface in* $\mathbb{R}^4$
with $\tilde{\nabla}A = 0$. *Then, up to an affine transformation, either* M *is an open part of a locally
strongly convex quadric (i.e.* $A = 0$*) or* M *is an open part of one of the following two
hypersurfaces:*
(i) $x_1 x_2 x_3 x_4 = 1$;
(ii) $(x_1^2 - x_2^2 - x_3^2)^3 x_4^2 = 1$.

Proof. If $A = 0$, then M is an open part of a locally strongly convex quadric. Hence from
now on, we will assume that $A \neq 0$, and thus M is a hyperbolic affine hypersphere with
affine mean curvature $L_1 < 0$. Fix any $p \in M$, and let $\{e_1, e_2, e_3\}$ be an orthonormal
basis of $T_p M$ as chosen in Proposition 3.15 with the properties

$$G(e_i, e_j) = \delta_{ij}, \quad A_{1ij} = \mu_j \delta_{ij}, \quad 1 \le i, j \le 3; \quad \mu_1 + \mu_2 + \mu_3 = 0, \tag{3.55}$$

$$\mu_1 - 2\mu_i \ge 0 \text{ and if } \mu_1 = 2\mu_i, \text{ then } A(e_i, e_i, e_i) = 0 \text{ for } i = 2, 3. \tag{3.56}$$

If $\mu_1 = 2\mu_2$, then $\mu_3 = -\frac{3}{2}\mu_1$. Let $u = -\frac{1}{2} e_1 + \frac{1}{2} \varepsilon \sqrt{3} e_3$, and choose $\varepsilon = \pm 1$ such
that $\varepsilon F(e_3) \ge 0$. Then we notice that

$$F(u) = A(u, u, u) = \frac{25}{16} \mu_1 + \frac{3}{8} \sqrt{3} \varepsilon F(e_3) > \mu_1,$$

which contradicts to the maximality of μ_1, showing that $\mu_1 > 2\mu_2$. Similarly, it holds $\mu_1 > 2\mu_3$. Hence by the proof of Proposition 3.15 we get

$$\mu_2 = \mu_3 = -\frac{1}{2}\mu_1 \text{ and } L_1 = -\frac{3}{4}(\mu_1)^2.$$

We can further choose e_2 as a unit vector for which the function F, restricted to $\{v \in T_pM \,|\, G(v,v) = 1\} \cap \{e_1\}^\perp$, attains its maximum $a := A(e_2, e_2, e_2) \geq 0$. It then follows that $A(e_2, e_2, e_3) = 0$. Hence

$$A(e_2, e_3, e_3) = -A(e_2, e_1, e_1) - A(e_2, e_2, e_2) = -a.$$

From the Ricci identity we have

$$A_{223,23} - A_{223,32} = 2\sum A_{j23}R_{j223} + \sum A_{22j}R_{j323}$$

and $\tilde{\nabla}A = 0$, and thus we obtain $3a(L_1 + 2a^2 - \frac{1}{4}(\mu_1)^2) = 0$. It follows that we have either $a = \sqrt{-2L_1/3}$ or $a = 0$ at the point $p \in M$. Accordingly, we have

$$\text{either } J = \frac{1}{6}\sum_{i,j,k}(A_{ijk})^2 = -L_1 > 0 \text{ or } J = \frac{1}{6}\sum_{i,j,k}(A_{ijk})^2 = -\frac{5}{9}L_1 > 0.$$

Since J is different for these two cases, it follows that the assertion holds at every point of M.

Case 1. If $a = \sqrt{-2L_1/3}$, then a direct computation shows that $R_{ijkl} = 0$ for all i, j, k, l. As $J = \frac{1}{6}\sum_{i,j,k}(A_{ijk})^2 = -L_1 > 0$, by Theorem 3.8, up to an affine transformation, M is an open part of the hypersurface $Q(1, 3)$.

Case 2. If $a = 0$, we extend the above basis $\{e_1, e_2, e_3\}$ at p, by parallel translation along geodesics (with respect to $\tilde{\nabla}$) through p to a normal neighbourhood around p. By the properties of parallel translation this gives a G-orthonormal basis $\{E_1, E_2, E_3\}$, with dual basis $\{\omega_1, \omega_2, \omega_3\}$, defined on a neighbourhood of p. Since $\tilde{\nabla}A = 0$, it also follows that A has the desired form at every point of a normal neighbourhood. Next, as usual denoting $\tilde{\nabla}E_i = \sum \tilde{\omega}_i^j E_j$, from $\tilde{\nabla}A = 0$ and

$$\sum A_{11i,j}\omega^j = dA_{11i} - 2\sum A_{j1i}\tilde{\omega}_1^j - \sum A_{11j}\tilde{\omega}_i^j$$

we get $2\mu_1\tilde{\omega}_1^i = 0$ for $i = 2, 3$. Thus $\tilde{\omega}_1^2 = \tilde{\omega}_1^3 = 0$. On the other hand, we have the calculation

$$R_{2323} = \sum_m A_{m23}A_{m23} - \sum_m A_{m22}A_{m33} + L_1 = -(\mu_1)^2 = \frac{4}{3}L_1.$$

Hence, when considered as a Riemannian manifold, (M, G) is locally isometric to $\mathbb{R} \times \mathbb{H}^2(4L_1/3)$, where $\mathbb{H}^2(4L_1/3)$ is the hyperbolic plane of constant negative curvature $4L_1/3$, and after identification, the local vector field E_1 is tangent to $\mathbb{R}$. Further, we will denote the immersion of M into $\mathbb{R}^4$ by f. Then, after applying a translation, we may assume that the affine normal satisfies the relation $Y = -L_1f$. So, using the standard parametrization of the hypersphere model of $\mathbb{H}^2(4L_1/3)$, we see that there exists local coordinates $\{u, v, w\}$ on M, such that $E_1 = f_w$, and f_u and $\left(\sinh\left(\sqrt{-4L_1/3}\,u\right)\right)^{-1}f_v$, together with f_w form a G-orthonormal basis. We may assume that $E_2 = f_u$ and $\sinh(\sqrt{-4L_1/3}\,u)E_3 = f_v$. Then a straightforward computation shows that

$$\tilde{\nabla}_{f_u} f_u = 0,$$

$$\tilde{\nabla}_{f_u} f_v = \tilde{\nabla}_{f_v} f_u = \sqrt{-4L_1/3}\,\coth\left(\sqrt{-4L_1/3}\,u\right)f_v,$$

$$\tilde{\nabla}_{f_v} f_v = -\sqrt{-4L_1/3}\,\sinh\left(\sqrt{-4L_1/3}\,u\right)\cosh\left(\sqrt{-4L_1/3}\,u\right)f_u.$$

Using the definition of A, we get the following system of differential equations, where, in order to simplify the equations, we have put $c = \sqrt{-L_1/3}$.

$$f_{ww} = 2cf_w + 3c^2 f, \tag{3.57}$$

$$f_{uw} = -cf_u, \tag{3.58}$$

$$f_{vw} = -cf_v, \tag{3.59}$$

$$f_{uu} = -cf_w + 3c^2 f, \tag{3.60}$$

$$f_{uv} = 2c\coth(2cu)f_v, \tag{3.61}$$

$$f_{vv} = -c\,\sinh^2(2cu)f_w - 2c\,\sinh(2cu)\cosh(2cu)f_u + 3c^2\,\sinh^2(2cu)f. \tag{3.62}$$

To solve the above equations, first we see from (3.57) that there exist vector valued functions $P_1(u,v)$ and $P_2(u,v)$ such that

$$f = P_1(u,v)e^{3cw} + P_2(u,v)e^{-cw}.$$

From (3.58) and (3.59) it then follows that the vector valued function P_1 is independent of u and v. Hence there exists a constant vector A_1 such that $P_1(u,v) = A_1$. Next, it follows from (3.60) that P_2 satisfies the following differential equation:

$$(P_2)_{uu} = 4c^2 P_2.$$

Hence we can write

$$P_2(u,v) = Q_1(v)\cosh(2cu) + Q_2(v)\sinh(2cu).$$

From (3.61), we then deduce that there exists a constant vector A_2 such that $Q_1(v) = A_2$. Finally, from (3.62), we get the following differential equation for Q_2:

$$(Q_2)_{vv} = -4c^2 Q_2.$$

This last formula implies that there exist constant vectors A_3 and A_4 such that

$$Q_2(v) = A_3\cos(2cv) + A_4\sin(2cv).$$

Since M is nondegenerate, M lies linearly full in $\mathbb{R}^4$. Hence A_1, A_2, A_3, A_4 are linearly independent vectors. Thus there exists an affine transformation such that

$$f = \left(\cosh(2cu)e^{-cw},\ \cos(2cv)\sinh(2cu)e^{-cw},\ \sin(2cv)\sinh(2cu)e^{-cw},\ e^{3cw}\right).$$

It follows that if $a = 0$ then up to an affine transformation, M locally lies on the hypersurface $(x_1^2 - x_2^2 - x_3^2)^3 x_4^2 = 1$. $\qquad\square$

Remark 3.18. It is easy to see that both hypersurfaces in Theorem 3.17 are Calabi compositions:

(i) $x_1 x_2 x_3 x_4 = 1$ can be regarded either as the Calabi composition of $x_1 x_2 = 1$ and $x_3 x_4 = 1$, or the Calabi composition of $x_1 x_2 x_3 = 1$ with a point;

(ii) $(x_1^2 - x_2^2 - x_3^2)^3 x_4^2 = 1$ is the Calabi composition of $x_1^2 - x_2^2 - x_3^2 = 1$ with a point.

Finally, before stating the complete classification of locally strongly convex affine hypersurfaces in $\mathbb{R}^{n+1}$ with $\tilde{\nabla} A = 0$ for all dimensions, we will further present briefly typical examples of hyperbolic affine hyperspheres with parallel Fubini-Pick form. For full details we refer to [133].

Example 3.8. $SL(m, \mathbb{R})/SO(m) \hookrightarrow \mathbb{R}^{n+1}$, $n + 1 = \frac{1}{2}m(m + 1)$, $m \geq 3$.

Let $GL(m, \mathbb{R})$ be the set of all $m \times m$ real matrices, which is identified with $\mathbb{R}^{m^2}$. We may regard $SL(m, \mathbb{R}) = \{X \in GL(m, \mathbb{R}) : \det X = 1\}$ as an $(m^2 - 1)$-dimensional hypersurface imbedded in $GL(m, \mathbb{R})$.

Let $s(m)$ be the set of real symmetric $m \times m$ matrices, which is identified with $\mathbb{R}^{n+1}$ with $n + 1 = \frac{1}{2}m(m + 1)$. We have a representation ϕ of $SL(m, \mathbb{R})$ on $s(m)$ by $\phi(A)X = AXA^t$ for $X \in s(m)$. It can be easily shown that

$$\det \phi(A) = 1 \quad \text{for each } A \in SL(m, \mathbb{R}),$$

i.e. for each $A \in SL(m, \mathbb{R})$, $\phi(A) : s(m) \to s(m)$ is an equiaffine transformation. In this way, we have a representation $\phi : SL(m, \mathbb{R}) \to SA(n + 1, \mathbb{R})$, where $SA(n + 1, \mathbb{R})$ is the group of unimodular affine transformations of $\mathbb{R}^{n+1}$.

Consider the hypersurface of $s(m)$ satisfying the equation $\det(X) = 1$; we take its connected component M that lies in the open interior of $s(m)$ consisting of all positive definite matrices. Then the mapping

$$f' : SL(m, \mathbb{R}) \to s(m) \text{ defined by } f'(A) := AA^t$$

is a submersion onto M, and the mapping f' is equivariant in the sense that

$$f'(AB) = \phi(A)f'(B) \quad \text{for all } A, B \in SL(m, \mathbb{R}).$$

M is the orbit of the $m \times m$ identity matrix e under the action ϕ. We have $f'(A) = f'(B)$ if and only if $B^{-1}A \in SO(m)$, where $SO(m)$ denotes the set of orthogonal $m \times m$ matrices with determinant 1. Hence, the isotropy group of $f' : SL(m, \mathbb{R}) \to s(m)$ is $SO(m)$, and M is diffeomorphic to $SL(m, \mathbb{R})/SO(m)$. It is known [140, vol. II, Chap. XI] that the latter is an irreducible and homogeneous symmetric space of noncompact type, and the involution at e is given by $\tau : A \mapsto (A^{-1})^t$. The map $f' : SL(m, \mathbb{R}) \to s(m)$ thus induces a *standard* imbedding hypersurface $f : SL(m, \mathbb{R})/SO(m) \to s(m)$. Let $\pi : SL(m, \mathbb{R}) \to M$ be the natural projection, then we have $f' = f \circ \pi$. According to Nomizu and Sasaki (see [260, pp. 106–113]), $f : SL(m, \mathbb{R})/SO(m) \to s(m)$ is an affine homogeneous hypersurface, i.e. for any two points of the image of f there exists an equiaffine transformation of $\mathbb{R}^{n+1} = s(m)$ preserving the image of f. As a consequence, every point of M has the same equiaffine properties. In the following, we will restrict our discussion to the point $x_0 = f(e) = I_m$.

To find $f_*(T_{x_0}M)$, we note that, at e, the tangent space $T_e\mathrm{SL}(m, \mathbb{R})$ is equal to

$$\mathfrak{sl}(m, \mathbb{R}) = \{a \in \mathrm{GL}(m, \mathbb{R}) : \operatorname{tr} a = 0\},$$

which can be identified with the Lie algebra of all left-invariant vector fields on $\mathrm{SL}(m, \mathbb{R})$. Taking the zero matrix as the origin, then, as $\operatorname{tr} e = m$, the position vector $e \in \mathrm{SL}(m, \mathbb{R})$ is transversal to $T_e\mathrm{SL}(m, \mathbb{R})$, and we may regard $\mathrm{SL}(m, \mathbb{R})$ as a centro-affine hypersurface of $\mathrm{GL}(m, \mathbb{R})$. It follows that

$$f : \mathrm{SL}(m, \mathbb{R})/\mathrm{SO}(m) \to s(m) = \mathbb{R}^{n+1}$$

can be considered as a centroaffine hypersurface with $\xi_x = f(x)$ as transversal vector for each $x \in \mathrm{SL}(m, \mathbb{R})/\mathrm{SO}(m)$.

Consider the Cartan decomposition of the Lie algebra

$$\mathfrak{sl}(m, \mathbb{R}) = \mathfrak{s}_0(m) \oplus \mathfrak{o}(m), \tag{3.63}$$

where $\mathfrak{s}_0(m) = \{X \in s(m) \mid \operatorname{tr} X = 0\}$ and $\mathfrak{o}(m)$ denotes the set of skew-symmetric $m \times m$-matrices. In fact, equation (3.63) gives the standard decomposition of the Lie algebra according to the involutive automorphism $\tau_*(X) = -X^t$ for the symmetric homogeneous space $M = \mathrm{SL}(m, \mathbb{R})/\mathrm{SO}(m)$. Now we can use equation (3.63) to represent any invariant structure on the space M (see [140, vol. II, Chap. XI]). In particular, $\mathfrak{s}_0(m)$ represents $T_{x_0}M$.

Assume $X \in \mathfrak{s}_0(m) = T_{x_0}M$. To find $f_*(X)$, let $a(s) = \exp(sX) \in \mathrm{SL}(m, \mathbb{R})$ and $\gamma(s) = \pi(a(s)) \in M$ with $\gamma'(0) = X$. Then $f(\gamma(s)) = a(s)a(s)^t$, and hence

$$f_*(X) = (f \circ \gamma)'(0) = \frac{d}{ds}\Big|_{s=0}[a(s)a(s)^t] = 2X. \tag{3.64}$$

We see that $f_*(T_{x_0}M)$ also consists of symmetric $m \times m$ matrices with trace 0. Moreover, the following computation

$$(f \circ \gamma)'(s) = f_*(\gamma'(s)) = f_*(a_sX) = \phi(a(s))f_*(X) = 2a(s)Xa(s)^t,$$

$$(f \circ \gamma)''(s) = \frac{d}{ds}2a(s)Xa(s)^t = 4a(s)X^2a(s)^t,$$

$$(f \circ \gamma)'''(s) = \frac{d}{ds}4a(s)X^2a(s)^t = 8a(s)X^3a(s)^t,$$

implies that

$$(f \circ \gamma)'(0) = 2X, \quad (f \circ \gamma)''(0) = 4X^2, \quad (f \circ \gamma)'''(0) = 8X^3. \tag{3.65}$$

On the other hand, by the Gauß structure equation,

$$(f \circ \gamma)''(s) = D_{\partial/\partial s}(f \circ \gamma)'(s) = D_{\partial/\partial s}f_*(\gamma'(s))$$
$$= f_*(\nabla_{\partial/\partial s}\gamma'(s)) + G(\gamma'(s), \gamma'(s))f(\gamma(s)).$$

Hence, at $s = 0$, as $f(\gamma(0)) = f(e) = I_m$, we get that

$$4X^2 = f_*(\nabla_X X) + G(X, X)I_m. \tag{3.66}$$

The fact $f_*(\nabla_X X) \in \mathfrak{s}_0(m)$ implies that $G(X, X) = \frac{4}{m} \operatorname{tr}(X^2)$. Polarization gives that

$$G(X, Y) = \frac{4}{m} \operatorname{tr}(XY), \tag{3.67}$$

which shows that the affine metric is the canonical positive definite metric on $SL(m, \mathbb{R})/SO(m)$ up to a constant multiple. Hence, by construction, $\gamma(s)$ is a geodesic with respect to the Levi–Civita connection $\tilde{\nabla}$ of the affine metric G, thus $\tilde{\nabla}_{\partial/\partial s}\gamma'(s) = 0$. Then (3.66) implies that the difference tensor K satisfies

$$f_*(K(X, X)) = 4X^2 - \frac{4}{m} \operatorname{tr}(X^2)I_m.$$

Thus, by (3.64), it holds that $K(X, X) = 2X^2 - \frac{2}{m} \operatorname{tr}(X^2)I_m$. Polarization then gives

$$K(X, Y) = XY + YX - \frac{2}{m} \operatorname{tr}(XY)I_m := K_X Y. \tag{3.68}$$

Denote by E_{jk} the $m \times m$ matrix which has a (j, k) entry 1 and all others 0. Put

$$\begin{cases} e_1 = \dfrac{1}{\sqrt{4(m-1)}}((m-1)E_{mm} - E_{11} - \cdots - E_{m-1,m-1}), \\[2ex] v_j = \sqrt{\dfrac{m}{8}}(E_{mj} + E_{jm}), \quad 1 \le j \le m-1; \\[2ex] w_j = \sqrt{\dfrac{m}{4j(j+1)}}\left(\sum_{r=1}^{j} E_{rr} - jE_{j+1,j+1}\right), \quad 1 \le j \le m-2; \\[2ex] w_{j\ell} = \sqrt{\dfrac{m}{8}}(E_{\ell j} + E_{j\ell}), \quad 1 \le j < \ell \le m-1. \end{cases} \tag{3.69}$$

Then it is easy to show that, with respect to the metric $G(X, Y) = \frac{4}{m} \operatorname{tr}(XY)$,

$$\{e_1; \ v_j \mid_{1 \le j \le m-1}; \ w_j \mid_{1 \le j \le m-2}; \ w_{j\ell} \mid_{1 \le j < \ell \le m-1}\}$$

forms an orthonormal basis of $SL(m, \mathbb{R})/SO(m)$ at I_m.

Using the formula (3.68) and $E_{jk}E_{pq} = \delta_{kp}E_{jq}$, we have

$$K(e_1, e_1) = \frac{1}{2(m-1)}(E_{11} + \cdots + E_{m-1,m-1} + (m-1)^2 E_{mm}) - \frac{1}{2}I_m,$$

$$K(v_j, v_j) = \frac{m}{4}(E_{jj} + E_{mm}) - \frac{1}{2}I_m, \quad 1 \le j \le m-1,$$

$$K(w_j, w_j) = \frac{m}{2j(j+1)}\left(\sum_{r=1}^{j} E_{rr} + j^2 E_{j+1,j+1}\right) - \frac{1}{2}I_m, \quad 1 \le j \le m-2,$$

$$K(w_{j\ell}, w_{j\ell}) = \frac{m}{4}(E_{jj} + E_{\ell\ell}) - \frac{1}{2}I_m, \quad 1 \le j < \ell \le m-1.$$

Then the identity

$$\sum_{j=1}^{m-2} \frac{m}{2j(j+1)}\left(j^2 a_{j+1} + \sum_{r=1}^{j} a_r\right) = \frac{m(m-2)}{2(m-1)} \sum_{j=1}^{m-1} a_j$$

and the above computations immediately give

$$K(e_1, e_1) + \sum_{j=1}^{m-1} K(v_j, v_j) + \sum_{j=1}^{m-2} K(w_j, w_j) + \sum_{1 \le j < \ell \le m-1} K(w_{j\ell}, w_{j\ell}) = 0.$$

It follows that $\mathrm{tr}_G(K_X) = 0$ for all $X \in T_{x_0}M$. Consequently, the position vector is proved to be the affine normal at x_0 and, by homogeneity of the hypersurface, we conclude that M is a locally strongly convex affine hypersphere.

To show that $\tilde{\nabla}K = 0$, we use

$$(f \circ \gamma)''(s) = f_*(K(\gamma'(s), \gamma'(s))) + G(\gamma'(s), \gamma'(s))f(\gamma(s))$$

to obtain

$$(f \circ \gamma)'''(s) = D_{\partial/\partial s}f_*(K(\gamma'(s), \gamma'(s))) + G(\gamma'(s), \gamma'(s))(f \circ \gamma)'(s)$$

$$= f_*(\nabla_{\partial/\partial s}(K(\gamma'(s), \gamma'(s))))$$

$$\quad + G(\gamma'(s), K(\gamma'(s), \gamma'(s)))f(\gamma(s)) + G(\gamma'(s), \gamma'(s))(f \circ \gamma)'(s)$$

$$= f_*\left((\tilde{\nabla}K)(\gamma'(s), \gamma'(s), \gamma'(s)) + K(\gamma'(s), K(\gamma'(s), \gamma'(s)))\right)$$

$$\quad + G(\gamma'(s), K(\gamma'(s), \gamma'(s)))f(\gamma(s)) + G(\gamma'(s), \gamma'(s))f_*(\gamma'(s)).$$

Hence, at $s = 0$, using (3.67) and (3.68), we obtain

$$8X^3 = 2(\tilde{\nabla}K)(X, X, X) + 2K(X, K(X, X)) + G(X, K(X, X))I_m + \frac{8}{m}\mathrm{tr}\,(X^2)X$$

$$= 2(\tilde{\nabla}K)(X, X, X) + G(X, K(X, X))I_m + \frac{8}{m}\mathrm{tr}\,(X^2)X$$

$$\quad + 2\left\{X\left(2X^2 - \frac{2}{m}\mathrm{tr}\,(X^2)I_m\right) + \left(2X^2 - \frac{2}{m}\mathrm{tr}\,(X^2)I_m\right)X - \frac{1}{2}G(X, K(X, X))I_m\right\}$$

$$= 2(\tilde{\nabla}K)(X, X, X) + 8X^3.$$

This proves that $(\tilde{\nabla}K)(X, X, X) = 0$ and by linearization we obtain $\tilde{\nabla}K = 0$.

We summarize the above discussions as follows.

Proposition 3.19. *The above standard imbedding*

$$f : \mathrm{SL}(m, \mathbb{R})/\mathrm{SO}(m) \to \mathbb{R}^{n+1} = s(m), \quad n + 1 = \frac{1}{2}m(m + 1) \qquad (3.70)$$

gives a locally strongly convex hyperbolic affine hypersphere with parallel Fubini–Pick form.

Example 3.9. $\mathrm{SL}(m, \mathbb{C})/\mathrm{SU}(m) \hookrightarrow \mathbb{R}^{n+1}$, $n + 1 = m^2$, $m \geq 3$.

Let $s(m)$ be the set of Hermitian $m \times m$ matrices, $\mathrm{SL}(m, \mathbb{C})$ be the set of complex $m \times m$ matrices of determinant 1, and

$$\mathrm{SU}(m) = \{A \in \mathrm{SL}(m, \mathbb{C}) \mid A\bar{A}^t = I_m\}$$

be the set of unitary $m \times m$ matrices with determinant 1. Let $\mathfrak{X}$ be the action of $\mathrm{SL}(m, \mathbb{C})$ on $s(m)$ defined as follows:

$$\mathfrak{X} : \mathrm{SL}(m, \mathbb{C}) \times s(m) \to s(m) \quad \text{s.t.} \quad (A, X) \mapsto \mathfrak{X}_A(X) = AX\bar{A}^t.$$

Consider the hypersurface of $s(m)$ satisfying the equation $\det(X) = 1$; we take the connected component M that lies in the open set of $s(m)$ consisting of all Hermitian positive definite matrices. Then the map $f' : \mathrm{SL}(m, \mathbb{C}) \to s(m)$, defined by $f'(A) := A\bar{A}^t$, is a submersion onto M, and it satisfies

$$f'(AB) = \mathfrak{X}_A(f'(B)).$$

Hence f' is equivariant. M is the orbit of I_m under the action $\mathfrak{X}$. The isotropy group is $\mathrm{SU}(m)$. Hence M is diffeomorphic to $\mathrm{SL}(m, \mathbb{C})/\mathrm{SU}(m)$. This is an irreducible, homogeneous, symmetric space of noncompact type, and the involution at the identity matrix I_m is given by $A \mapsto (\bar{A}^{-1})^t$.

Clearly, $f'(A) = f'(B)$ if and only if $B^{-1}A \in \mathrm{SU}(m)$ and therefore the mapping $f' : \mathrm{SL}(m, \mathbb{C}) \to s(m)$ induces a *standard* imbedding denoted by

$$f : \mathrm{SL}(m, \mathbb{C})/\mathrm{SU}(m) \to s(m).$$

Then, totally similar to the proof of Proposition 3.19, we have

Proposition 3.20. *The standard imbedding*

$$f : \mathrm{SL}(m, \mathbb{C})/\mathrm{SU}(m) \to \mathbb{R}^{n+1} = s(m), \quad n + 1 = m^2 \qquad (3.71)$$

gives a locally strongly convex hyperbolic affine hypersphere with parallel Fubini-Pick form.

Example 3.10. $SU^*(2m)/Sp(m) \hookrightarrow \mathbb{R}^{2m^2-m}$, $m \geq 3$.

Let $\mathbb{H}$ be the quaternion field over $\mathbb{R}$. Then the quaternionic general linear group $GL(m, \mathbb{H})$ has a well-known complex representation

$$U^*(2m) = \{A \in GL(2m, \mathbb{C}) \mid AJ = J\bar{A}\}, \quad \text{where} \quad J = \begin{pmatrix} 0 & I_m \\ -I_m & 0 \end{pmatrix}$$

and I_m is the $m \times m$ identity matrix.

The quaternionic analogues of the set of Hermitian $m \times m$ matrices and that of the complex special linear group $SL(m, \mathbb{C})$ are given by

$$S^*(m) = \{A \in U^*(2m) \mid \bar{A} = A^t\}$$

and

$$SU^*(2m) = \{A \in SL(2m, \mathbb{C}) \mid AJ = J\bar{A}\} = SL(2m, \mathbb{C}) \cap U^*(2m).$$

The compact Lie subgroup $Sp(m)$, which is also called the quaternion unitary group, is defined by

$$Sp(m) = \{A \in SU(2m) \mid AJ = J\bar{A}\} = SU^*(2m) \cap SU(2m).$$

Let ψ be the action of $SU^*(2m)$ on $S^*(m)$ as follows:

$$\psi : SU^*(2m) \times S^*(m) \rightarrow S^*(m) \text{ s.t. } (A, X) \mapsto \psi_A(X) = AX\bar{A}^t.$$

Consider the hypersurface of $S^*(m)$ satisfying the equation $\det(X) = 1$; we take the connected component M that lies in the open set of $S^*(m)$ consisting of all Hermitian positive definite matrices. Then the mapping

$$f' : SU^*(2m) \rightarrow S^*(m), \text{ defined by } f'(A) := A\bar{A}^t,$$

is a submersion onto M, and it satisfies $f'(AB) = \psi_A(f'(B))$; hence f' is equivariant. M is the orbit of I_{2m} under the action ψ. The isotropy group is $Sp(m)$. Hence M is diffeomorphic to $SU^*(2m)/Sp(m)$. This is an irreducible, homogeneous, symmetric space of noncompact type, and the involution at the identity matrix I_{2m} is given by $A \mapsto (\bar{A}^{-1})^t$.

Clearly, $f'(A) = f'(B)$ if and only if $B^{-1}A \in Sp(m)$ and therefore the mapping $f' : SU^*(2m) \rightarrow S^*(m)$ induces a *standard* imbedding denoted by

$$f : SU^*(2m)/Sp(m) \rightarrow S^*(m).$$

Then, totally similar to the proof of Proposition 3.19 again, we have the following:

Proposition 3.21. *The above standard imbedding*

$$f : SU^*(2m)/Sp(m) \rightarrow \mathbb{R}^{n+1} = S^*(m), \quad n + 1 = 2m^2 - m \tag{3.72}$$

gives a locally strongly convex hyperbolic affine hypersphere with parallel Fubini–Pick form.

Example 3.11. $E_{6(-26)}/F_4 \hookrightarrow \mathbb{R}^{27}$.

This last typical example is much more involved, and is related to the exceptional Lie group. For references about the exceptional noncompact Lie groups $E_{6(-26)}$ and compact Lie group F_4, among others, we refer to Baez [7] and Yokota [410].

Let $\mathbb{O}$ be the octonions, i.e. the division Cayley algebra over the field $\mathbb{R}$ of real numbers, which is an 8-dimensional $\mathbb{R}$-vector space with basis

$$\{e_0 = 1, e_1, \ldots, e_7\},$$

and define a multiplication between them, with e_0 being the unit, as in the following table:

$\cdot$	e_1	e_2	e_3	e_4	e_5	e_6	e_7
e_1	-1	e_3	$-e_2$	e_5	$-e_4$	$-e_7$	e_6
e_2	$-e_3$	-1	e_1	e_6	e_7	$-e_4$	$-e_5$
e_3	e_2	$-e_1$	-1	e_7	$-e_6$	e_5	$-e_4$
e_4	$-e_5$	$-e_6$	$-e_7$	-1	e_1	e_2	e_3
e_5	e_4	$-e_7$	e_6	$-e_1$	-1	$-e_3$	e_2
e_6	e_7	e_4	$-e_5$	$-e_2$	e_3	-1	$-e_1$
e_7	$-e_6$	e_5	e_4	$-e_3$	$-e_2$	e_1	-1

In $\mathbb{O}$, the conjugate $\bar{x}$, the real part $R(x)$, an inner product (x, y) and the length $|x|$ are defined respectively by

$$\overline{\sum_{j=0}^{7} a_j e_j} = a_0 e_0 - \sum_{j=1}^{7} a_j e_j, \quad R\left(\sum_{i=0}^{7} a_j e_j\right) = a_0,$$

$$\left(\sum_{j=0}^{7} a_j e_j, \sum_{j=0}^{7} b_j e_j\right) = \sum_{j=0}^{7} a_j b_j = R(x\bar{y}), \quad |x| = \sqrt{(x, x)} = \sqrt{x\bar{x}},$$

where $x = \sum_{j=0}^{7} a_j e_j$, $y = \sum_{j=0}^{7} b_j e_j$ and $a_j, b_j \in \mathbb{R}$. It can be easily seen that $\overline{xy} = \bar{y}\bar{x}$ for $x, y \in \mathbb{O}$.

Let $\mathcal{M}_3(\mathbb{O})$ be the vector space of all 3×3 matrices with entries in $\mathbb{O}$ and $\mathfrak{h}_3(\mathbb{O})$ be the subset of all Hermitian matrices with entries in $\mathbb{O}$:

$$\mathfrak{h}_3(\mathbb{O}) = \{X \in \mathcal{M}_3(\mathbb{O}) \mid X^* = X\},$$

where $X^* = \bar{X}^t$ denotes the conjugate transpose of X. Any element $X \in \mathfrak{h}_3(\mathbb{O})$ is of the form

$$X = X(\xi, \eta) = \begin{bmatrix} \xi_1 & \eta_3 & \bar{\eta}_2 \\ \bar{\eta}_3 & \xi_2 & \eta_1 \\ \eta_2 & \bar{\eta}_1 & \xi_3 \end{bmatrix}, \quad \xi_i \in \mathbb{R}, \quad \eta_i \in \mathbb{O}.$$

By identifying $X = X(\xi, \eta) \in \mathfrak{h}_3(\mathbb{O})$ with $(\xi_1, \xi_2, \xi_3, \eta_1, \eta_2, \eta_3) \in \mathbb{R}^{27}$, thus $\mathfrak{h}_3(\mathbb{O}) \cong \mathbb{R}^{27}$ is a 27-dimensional $\mathbb{R}$-vector space. In $\mathfrak{h}_3(\mathbb{O})$, the multiplication $X \circ Y$, called the Jordan multiplication, is defined by

$$X \circ Y = \frac{1}{2}(XY + YX).$$

$\mathfrak{h}_3(\mathbb{O})$ equipped with the product $\circ$ is a real Jordan algebra. In $\mathfrak{h}_3(\mathbb{O})$, we also define the trace $\mathrm{tr}(X)$ and a symmetric inner product (X, Y) respectively by

$$\mathrm{tr}(X) = \xi_1 + \xi_2 + \xi_3; \quad (X, Y) = \mathrm{tr}\,(X \circ Y).$$

Moreover, in $\mathfrak{h}_3(\mathbb{O})$ there is a symmetric cross product $X \times Y$, called the Freudenthal multiplication, defined by

$$X \times Y = \frac{1}{2}\left[2X \circ Y - \mathrm{tr}\,(X)Y - \mathrm{tr}\,(Y)X + (\mathrm{tr}\,(X)\mathrm{tr}\,(Y) - \mathrm{tr}\,(X \circ Y))I\right]$$

(where I is the 3×3 unit matrix), and a totally symmetric trilinear form (X, Y, Z) by

$$(X, Y, Z) = (X \times Y, Z) = (X, Y \times Z).$$

Despite noncommutativity and nonassociativity, the determinant of a matrix in $\mathfrak{h}_3(\mathbb{O})$ can be well-defined by

$$\det X = \frac{1}{3}(X, X, X).$$

For $X = X(\xi, \eta) \in \mathfrak{h}_3(\mathbb{O})$, noting that $X \circ X = X^2$, $X \circ X \circ X = X^3$ and

$$R(\eta_1(\eta_2\eta_3)) = R(\eta_2(\eta_3\eta_1)) = R(\eta_3(\eta_1\eta_2)) \ (= R(\eta_1\eta_2\eta_3)),$$

we have the calculation

$$\begin{aligned} \det X &= \frac{1}{3}\mathrm{tr}\,(X^3) - \frac{1}{2}\mathrm{tr}\,(X)\,\mathrm{tr}\,(X^2) + \frac{1}{6}(\mathrm{tr}\,(X))^3 \\ &= \xi_1\xi_2\xi_3 - \xi_1\eta_1\bar{\eta}_1 - \xi_2\eta_2\bar{\eta}_2 - \xi_3\eta_3\bar{\eta}_3 + 2R(\eta_1\eta_2\eta_3). \end{aligned} \tag{3.73}$$

This shows that the determinant is invariant under all automorphisms of $\mathfrak{h}_3(\mathbb{O})$. However, as stated in Baez [7, p. 182] and Yokota [410, Sect. 3.15], the determinant is invariant under an even bigger group of linear transformations, which is a 78-dimensional noncompact real form of the exceptional Lie group E_6. More precisely, the group of determinant-preserving linear transformations of $\mathfrak{h}_3(\mathbb{O})$ turns out to be a noncompact real form of E_6. This real form is sometimes called $E_{6(-26)}$, because its Killing form has signature -26. Hence we have

$$E_{6(-26)} = \{\alpha \in \mathrm{Iso}_{\mathbb{R}}(\mathfrak{h}_3(\mathbb{O}))\mid \det\,(\alpha X) = \det\,(X)\},$$

where $\mathrm{Iso}_{\mathbb{R}}(\mathfrak{h}_3(\mathbb{O}))$ denotes all $\mathbb{R}$-linear isomorphisms of $\mathfrak{h}_3(\mathbb{O})$.

Let F_4 denote the full automorphism group of the Jordan algebra $\mathfrak{h}_3(\mathbb{O})$:

$$F_4 = \{\alpha \in \mathrm{Iso}_{\mathbb{R}}(\mathfrak{h}_3(\mathbb{O}))\mid \alpha(X \circ Y) = \alpha X \circ \alpha Y\}.$$

Using Lemma 2.2.4 of Yokota [410], we conclude that

$$F_4 = \{\alpha \in \mathrm{Iso}_{\mathbb{R}}(\mathfrak{n}_3(\mathbb{O})) \mid \det(\alpha X) = \det(X), \; \alpha I = I\}$$
$$= \{\alpha \in \mathrm{Iso}_{\mathbb{R}}(\mathfrak{n}_3(\mathbb{O})) \mid \det(\alpha X) = \det(X), \; (\alpha X, \alpha Y) = (X, Y)\}$$
$$= \{\alpha \in \boldsymbol{E}_{6(-26)} \mid \alpha I = I\}.$$

Hence, we get the inclusion $F_4 \to E_{6(-26)}$ and that, within $E_{6(-26)}$, F_4 is the stabilizer of I in $\mathfrak{n}_3(\mathbb{O})$. Moreover, as a closed subgroup of the orthogonal group

$$O(27) = O(\mathfrak{n}_3(\mathbb{O})) = \{\alpha \in \mathrm{Iso}_{\mathbb{R}}(\mathfrak{n}_3(\mathbb{O})) \mid (\alpha X, \alpha Y) = (X, Y)\};$$

F_4 is a compact Lie group. This shows that F_4 is a compact subgroup of $E_{6(-26)}$. From Baez [7, p. 196], it is in fact maximal. It follows that the Killing form of the Lie algebra $\mathfrak{e}_{6(-26)}$ is negative definite on its 52-dimensional maximal compact Lie algebra $\mathfrak{f}_4$ and positive definite on the complementary 26-dimensional subspace, giving a signature of $26 - 52 = -26$.

Hence, using Yokota [410, Thm. 3.15.1], we have a noncompact homogeneous space $\tilde{M} = E_{6(-26)}/F_4 \simeq \mathbb{R}^{26}$. To obtain $T_I\tilde{M}$, we now consider the decomposition of the Lie algebra $\mathfrak{e}_6$ of $E_{6(-26)}$. From Baez [7, p. 191, Thm. 5], we have

$$\mathfrak{f}_4 = \mathfrak{der}(\mathbb{O}) \oplus \{X \in \mathcal{M}_3(\mathbb{O}) \mid X^* = -X, \mathrm{tr}(X) = 0\},$$

where $\mathfrak{der}(\mathbb{O})$ denotes the derivations of the octonions and it is the Lie algebra $\mathfrak{g}_2$ of the 14-dimensional automorphism group G_2 of the octonions. It follows that

$$T_I\tilde{M} \simeq \{X \in \mathcal{M}_3(\mathbb{O}) \mid X^* = X, \; \mathrm{tr}(X) = 0\}$$
$$= \{X \in \mathfrak{n}_3(\mathbb{O}) \mid \mathrm{tr}(X) = 0\}. \tag{3.74}$$

Then $E_{6(-26)}/F_4$ is locally isomorphic to (will be identified hereafter) the connected component A of I in $\{X \in \mathcal{M}_3(\mathbb{O}) \mid X^* = X, \det(X) = 1\}$, which can be naturally embedded into $\mathfrak{n}_3(\mathbb{O}) \simeq \mathbb{R}^{27}$. We denote this by $f : E_{6(-26)}/F_4 \hookrightarrow \mathbb{R}^{27}$ and call it the *standard imbedding*. Then the following is known (we refer to [133] for its full proof):

Proposition 3.22. *The above standard imbedding*

$$f : E_{6(-26)}/F_4 \hookrightarrow \mathbb{R}^{27} \tag{3.75}$$

gives a locally strongly convex hyperbolic affine hypersphere with parallel Fubini–Pick form.

Finally, after having shown the lower dimensional classification (Theorem 3.16 and Theorem 3.17) of locally strongly convex affine hypersurfaces in $\mathbb{R}^{n+1}$ with $\tilde{\nabla}A = 0$, and most importantly having introduced the above four typical examples, we can now state the complete classification theorem for all dimensions; for the proof in full detail see [133].

Classification Theorem 3.23. Let M be an n-dimensional ($n \geq 2$) locally strongly convex affine hypersurface in $\mathbb{R}^{n+1}$ with $\tilde{\nabla}A = 0$. Then M is either a quadric (i.e. $A = 0$) or one of the hyperbolic affine hyperspheres with $A \neq 0$ of the following types, where the imbeddings appear as standard imbeddings:

(i) $n = \frac{1}{2}m(m + 1) - 1, m \geq 3$, and M is affinely equivalent to the standard imbedding of $SL(m, \mathbb{R})/SO(m) \hookrightarrow \mathbb{R}^{n+1}$;

(ii) $n = m^2 - 1$, $m \geq 3$, and M is affinely equivalent to the standard imbedding of $SL(m, \mathbb{C})/SU(m) \hookrightarrow \mathbb{R}^{m^2}$;

(iii) $n = 2m^2 - m - 1$, $m \geq 3$, and M is affinely equivalent to the standard imbedding of $SU^*(2m)/Sp(m) \hookrightarrow \mathbb{R}^{n+1}$;

(iv) $n = 26$, and M is affinely equivalent to the imbedding $E_{6(-26)}/F_4 \hookrightarrow \mathbb{R}^{27}$;

(v) M is obtained as the Calabi composition of a lower dimensional hyperbolic affine hypersphere with parallel cubic form and a point;

(vi) M is obtained as the Calabi composition of two lower dimensional hyperbolic affine hyperspheres with parallel cubic form.

Related to the Classification Theorem 3.23, we recall that a locally strongly convex hypersurface M in A^{n+1} is called λ-*isotropic* if it satisfies

$$G(K(X,X), K(X,X)) = \lambda(p)[G(X,X)]^2$$

for some function λ, and all $X \in T_pM$ for any $p \in M$. For a locally strongly convex affine hypersphere of A^{n+1}, Birembaux and Djorić [17] proved that affine isotropic hyperspheres are constant isotropic, i.e. λ is constant, and that the condition of constant isotropy implies that $\tilde{\nabla}A = 0$. Accordingly, the following type of a restricted version of Theorem 3.23 holds (cf. [17]).

Theorem 3.24. *Let M be an n-dimensional locally strongly convex constant isotropic affine hypersphere in $\mathbb{R}^{n+1}$. Then M is either a quadric, or one of the following four cases occurs, where the imbeddings appear as standard imbeddings:*

(i) $n = 5$, *and M is affinely equivalent to the imbedding of $SL(3, \mathbb{R})/SO(3)$ into $\mathbb{R}^6$;*

(ii) $n = 8$, *and M is affinely equivalent to the imbedding of $SL(3, \mathbb{C})/SU(3)$ into $\mathbb{R}^9$;*

(iii) $n = 14$, *and M is affinely equivalent to the imbedding of $SU^*(6)/Sp(3)$ into $\mathbb{R}^{15}$;*

(iv) $n = 26$ *and M is affinely equivalent to the imbedding of $E_{6(-26)}/F_4$ into $\mathbb{R}^{27}$.*

3.4 Affine completeness, Euclidean completeness, and Calabi completeness

From this section on we will focus on complete affine hyperspheres. In affine differential geometry there are several notions of completeness (cf. [236]). In the theory of hypersurfaces we study three notions of completeness:

(1) *affine completeness*, that is, the completeness of the Blaschke metric G;
(2) *Euclidean completeness*, that is, the completeness of the Riemannian metric on M induced from a Euclidean metric on A^{n+1};
(3) *Calabi completeness*, that is, the completeness of the Calabi metric $\mathcal{H}$ for a graph hypersurface in A^{n+1}.

First we point out that the notion of *Euclidean completeness* is independent of the choice of the Euclidean metric on A^{n+1}. In fact, let $(\,,\,)$ and $\ll\,,\,\gg$ be two Euclidean inner products on V, where the vector space V is associated to A^{n+1}. Let $\{\eta_1,\ldots,\eta_{n+1}\}$ and $\{\bar{\eta}_1,\ldots,\bar{\eta}_{n+1}\}$ be orthonormal bases relative to $(\,,\,)$ and $\ll\,,\,\gg$, respectively, related by

$$\eta_A = \sum C_A^B \bar{\eta}_B, \quad A = 1, 2, \ldots, n + 1.$$

Then the two induced metrics ds^2 and $d\bar{s}^2$ on M satisfy

$$ds^2 = (dx^1, \ldots, dx^{n+1}) \begin{pmatrix} dx^1 \\ \vdots \\ dx^{n+1} \end{pmatrix},$$

$$d\bar{s}^2 = \left[C \begin{pmatrix} dx^1 \\ dx^2 \\ \vdots \\ dx^{n+1} \end{pmatrix} \right]^t C \begin{pmatrix} dx^1 \\ dx^2 \\ \vdots \\ dx^{n+1} \end{pmatrix} = (dx^1, \ldots, dx^{n+1}) C^t C \begin{pmatrix} dx^1 \\ dx^2 \\ \vdots \\ dx^{n+1} \end{pmatrix},$$

where $C = (C_A^B)$, and C^t is the transposed matrix of C. Let μ and λ denote the largest and the smallest eigenvalues of $C^t C$, respectively. Then

$$\lambda\, ds^2 \le d\bar{s}^2 \le \mu\, ds^2.$$

This means that the two notions of Euclidean completeness are equivalent.

3.4.1 Euclidean completeness and affine completeness

For affine hypersurfaces, the affine completeness and the Euclidean completeness are not equivalent. There are many such examples, see e.g. the following example of R. Schneider.

Example 3.12 (cf. [312]). Consider the function

$$f(x^1, x^2) = \frac{1}{2}\left(\frac{1}{x^1} + (x^2)^2\right)$$

and its graph $M := \{(x^1, x^2, f(x^1, x^2)) \mid 0 < x^1 < \infty, -\infty < x^2 < \infty\}$. It is easy to check that M is locally strongly convex and Euclidean complete. On the other hand, the Blaschke metric of M is given by

$$G_{11} = (x^1)^{-9/4}, \quad G_{12} = 0, \quad G_{22} = (x^1)^{3/4}.$$

Consider the curve

$$x^1(t) = t, \quad x^2(t) = 0, \quad 1 \le t < \infty.$$

Its affine arc length is

$$l = \int_1^\infty \sqrt{G_{11}(t)}dt = \int_1^\infty t^{-9/8}dt < \infty.$$

This shows that (M, G) is not complete.

Nevertheless, S. Y. Cheng and S. T. Yau proved that for an *affine hypersphere* the Euclidean completeness implies the affine completeness (cf. [57]). In the following we shall give a generalization of the result of Cheng and Yau.

Let M be a noncompact, Euclidean complete, locally strongly convex hypersurface in A^{n+1}. From the theorem of Hadamard-Sacksteder-Wu (cf. [407]) M is the graph of a strongly convex function $x^{n+1} = f(x^1, \ldots, x^n)$ defined on a convex domain $\Omega \subset A^n$. Hence M is globally strongly convex. Moreover, we may assume that the hyperplane $x^{n+1} = 0$ is the tangent hyperplane of M at some point $\bar{x} \in M$, and that $\bar{x}$ has the coordinates $(0, \ldots, 0)$. [In fact, if we define

$$\tilde{f}(x^1, \ldots, x^n) = f(x^1, \ldots, x^n) - f(\bar{x}^1, \ldots, \bar{x}^n) - \sum \frac{\partial f}{\partial x^i}(\bar{x}^1, \ldots, \bar{x}^n)(x^i - \bar{x}^i),$$

$$\tilde{x}_i = x^i - \bar{x}^i, \quad 1 \le i \le n,$$

for any $(x^1, \ldots, x^n) \in \Omega$, then the graph of $\tilde{f}(\tilde{x})$ has the required properties. Since the above transformation is affine, our claim is proved]. With respect to this coordinate system we have $f \ge 0$ and for any number $C > 0$ the level set

$$M_C = \{x \in M \mid x^{n+1} = f(x^1, \ldots, x^n) \le C\}$$

is compact. Assume that there are constants $N > 0, p \ge 0$ such that

$$(\lambda_1^2 + \lambda_2^2 + \cdots + \lambda_n^2)^{1/2} \le N[\ln(2 + f)]^p, \tag{3.76}$$

where $\lambda_1, \ldots, \lambda_n$ are the affine principal curvatures of M. First of all, we derive an estimate for $|\Delta f|$. Consider the function

$$\varphi = (C - f)\frac{|\Delta f|}{(2 + f)[\ln(2 + f)]^p} \tag{3.77}$$

defined on M_C. Obviously, φ attains its supremum at some interior point x^* of M_C. Without loss of generality we may assume that $|\Delta f| \neq 0$ at x^*; then $\operatorname{grad}\varphi = 0$ at x^*. Choose a local orthonormal frame field $\{e_1, \dots, e_n\}$ of the Blaschke metric G on M such that, at x^*, $f_{,1} = \|\operatorname{grad} f\|$, $f_{,i} = 0$ $(2 \leq i \leq n)$, where $f_{,i} = e_i(f)$. Then, at x^*,

$$\frac{-f_{,1}|\Delta f|}{(2+f)[\ln(2+f)]^p} - (C-f)\frac{f_{,1}|\Delta f|}{(2+f)^2[\ln(2+f)]^p}$$
$$- p(C-f)\frac{f_{,1}|\Delta f|}{(2+f)^2[\ln(2+f)]^{p+1}} + (C-f)\frac{|\Delta f|_{,1}}{(2+f)[\ln(2+f)]^p} = 0.$$

It follows that

$$\frac{f_{,1}|\Delta f|}{2+f} \leq |\Delta f|_{,1}.$$

Taking the $(n+1)$-st component of the identity

$$(\Delta x)_{,i} = nY_{,i} = -n\sum B_{ij}x_j$$

we get

$$(\Delta f)_{,i} = -n\sum B_{ij}f_{,j},$$
$$|\Delta f|_{,1} \leq n(\lambda_1^2 + \cdots + \lambda_n^2)^{1/2} \cdot f_{,1}.$$

Hence, from the assumptions, at x^*,

$$\frac{|\Delta f| f_{,1}}{2+f} \leq nN[\ln(2+f)]^p f_{,1}.$$

In the case $x^* \neq \bar{x}$ we have $f_{,1} = \|\operatorname{grad} f\| > 0$. It follows that

$$\frac{|\Delta f|}{(2+f)[\ln(2+f)]^p} \leq nN \quad \text{and} \quad \varphi \leq (C-f)n \cdot N \leq CnN. \tag{3.78}$$

In fact, (3.78) holds at any point where φ attains its supremum.
In the case $x^* = \bar{x}$ we have

$$\varphi \leq C\frac{|\Delta f|}{(2+f)[\ln(2+f)]^p}\bigg|_{\bar{x}} = \frac{C}{2(\ln 2)^p}|\Delta f|(\bar{x}).$$

Let $N' = \max\left\{nN, \frac{1}{2(\ln 2)^p}|\Delta f|(\bar{x})\right\}$. Then at any point of M_C

$$\frac{|\Delta f|}{(2+f)[\ln(2+f)]^p} \leq \frac{CN'}{C-f}.$$

Let $C \to \infty$; then

$$\frac{|\Delta f|}{(2+f)[\ln(2+f)]^p} \leq N'. \tag{3.79}$$

Next, we derive a gradient estimate for f.

Consider the function

$$\psi := \exp\left\{\frac{-m}{C-f}\right\} \frac{\|\operatorname{grad} f\|^2}{(2+f)^2[\ln(2+f)]^p} \tag{3.80}$$

defined on M_C, where m is a positive constant to be determined later. Clearly, ψ attains its supremum at some interior point x^* of M_C. We can assume that $\|\operatorname{grad} f\| > 0$ at x^*. Choose a local orthonormal frame field $\{e_1, \ldots, e_n\}$ of the Blaschke metric on M such that, at $x^*, f_{,1} = \|\operatorname{grad} f\| > 0, f_{,i} = 0 \ (i \geq 2)$. Then, at x^*,

$$\psi_{,i} = 0,$$
$$\sum \psi_{,ii} \leq 0.$$

We calculate both expressions explicitly:

$$\frac{-mf_{,i}}{(C-f)^2} \sum (f_{,j})^2 - \left[\frac{2}{2+f} + \frac{p}{(2+f)\ln(2+f)}\right] f_{,i} \sum (f_{,j})^2 \tag{3.81}$$
$$+ 2 \sum f_{,j} f_{,ji} = 0,$$

$$-2\left[\frac{m}{(C-f)^2} + \frac{2}{2+f} + \frac{p}{(2+f)\ln(2+f)}\right] f_{,11}(f_{,1})^2 \tag{3.82}$$
$$+ \left[\frac{-2m}{(C-f)^3} + \frac{2}{(2+f)^2} + \frac{p(1+\ln(2+f))}{(2+f)^2[\ln(2+f)]^2}\right] (f_{,1})^4 + 2 \sum (f_{,ij})^2$$
$$- \left[\frac{m}{(C-f)^2} + \frac{2}{2+f} + \frac{p}{(2+f)\ln(2+f)}\right] (\Delta f)(f_{,1})^2 + 2 \sum f_{,j} f_{,jii} \leq 0.$$

Inserting (3.81) into (3.82) and noting that $\frac{2}{(2+f)^2} + \frac{p(1+\ln(2+f))}{(2+f)^2[\ln(2+f)]^2} > 0$ we get

$$-\left[\frac{m}{(C-f)^2} + \frac{2}{2+f} + \frac{p}{(2+f)\ln(2+f)}\right]^2 \cdot (f_{,1})^4 - \frac{2m}{(C-f)^3}(f_{,1})^4 \tag{3.83}$$
$$- \left[\frac{m}{(C-f)^2} + \frac{2}{2+f} + \frac{p}{(2+f)\ln(2+f)}\right] (\Delta f)(f_{,1})^2$$
$$+ 2 \sum (f_{,ij})^2 + 2 \sum f_{,j} f_{,jii} \leq 0.$$

Now let us compute the terms $f_{,ij}$ and $f_{,jii}$. An application of the Ricci identity shows that

$$\sum f_{,j} f_{,jii} = \sum f_{,j}(\Delta f)_{,j} + \sum R_{ij} f_{,i} f_{,j}.$$

We use Corollary 2.17 (ii) and again the Weingarten structure equation

$$(\Delta f)_{,j} = -n \sum B_{ij} f_{,i},$$
$$R_{ij} = \sum A_{mli} A_{mlj} + \frac{n-2}{2} B_{ij} + \frac{n}{2} L_1 \delta_{ij}$$

to obtain

$$\sum f_{,j} f_{,jii} = -nB_{11}(f_{,1})^2 + \sum (A_{ml1})^2 (f_{,1})^2 + \frac{n-2}{2} B_{11}(f_{,1})^2 + \frac{n}{2} L_1 (f_{,1})^2. \tag{3.84}$$

Taking the $(n + 1)$-st component of $x_{,ij} = \sum A_{ijk}x_{,k} + \frac{1}{n}\Delta x\delta_{ij}$, we have

$$f_{,ij} = A_{ij1}f_{,1} + \frac{1}{n}\Delta f\delta_{ij},$$

$$\sum(f_{,ij})^2 = \sum\left(A_{ij1}f_{,1} + \frac{1}{n}\Delta f\delta_{ij}\right)^2 = \sum(A_{ij1})^2(f_{,1})^2 + \frac{1}{n}(\Delta f)^2. \tag{3.85}$$

Combination of (3.84) and (3.85) gives

$$\sum f_{,i}f_{,jii} = \sum(f_{,ij})^2 + \frac{1}{2}nL_1(f_{,1})^2 - \frac{n+2}{2}B_{11}(f_{,1})^2 - \frac{1}{n}(\Delta f)^2. \tag{3.86}$$

Applying the inequality of Cauchy–Schwarz we obtain

$$\sum(f_{,ij})^2 \geq (f_{,11})^2 + \sum_{i>1}(f_{,ii})^2$$

$$\geq (f_{,11})^2 + \frac{1}{n-1}\left(\sum_{i>1}f_{,ii}\right)^2$$

$$= \frac{n}{n-1}(f_{,11})^2 + \frac{1}{n-1}(\Delta f)^2 - \frac{2}{n-1}f_{,11}\cdot\Delta f \tag{3.87}$$

$$\geq \left(\frac{n}{n-1} - \delta\right)(f_{,11})^2 - \frac{1-\delta(n-1)}{\delta(n-1)^2}(\Delta f)^2$$

for any $\delta > 0$. Inserting (3.86) and (3.87) into (3.83), from (3.81) we conclude that the following inequality holds:

$$\left(\frac{1}{n-1} - \delta\right)\left[\frac{m}{(C-f)^2} + \frac{2}{2+f} + \frac{p}{(2+f)\ln(2+f)}\right]^2 \cdot (f_{,1})^4$$

$$-\frac{2m}{(C-f)^3}(f_{,1})^4 - \left[\frac{m}{(C-f)^2} + \frac{2}{2+f} + \frac{p}{(2+f)\ln(2+f)}\right](\Delta f)(f_{,1})^2$$

$$+ nL_1(f_{,1})^2 - (n+2)B_{11}(f_{,1})^2 - \left[\frac{4-4(n-1)\delta}{\delta(n-1)^2} + \frac{2}{n}\right](\Delta f)^2$$

$$\leq 0.$$

We choose the following values for δ and m, namely: $\delta < \frac{1}{2(n-1)}$, $m = 4(n-1)C$. We use the following inequality to simplify the foregoing one:

$$\left(\frac{1}{n-1} - \delta\right)\left[\frac{m}{(C-f)^2} + \frac{2}{2+f} + \frac{p}{(2+f)\ln(2+f)}\right]^2 \cdot (f_{,1})^4 - \frac{2m}{(C-f)^3}(f_{,1})^4$$

$$\geq \left(\frac{1}{2(n-1)} - \delta\right)\left[\frac{m}{(C-f)^2} + \frac{2}{2+f} + \frac{p}{(2+f)\ln(2+f)}\right]^2 \cdot (f_{,1})^4.$$

We use the abbreviations

$$g := \frac{m}{(C-f)^2} + \frac{2}{2+f} + \frac{p}{(2+f)\ln(2+f)},$$

$$a := \frac{1}{2(n-1)} - \delta > 0,$$

$$b := \frac{4-4(n-1)\delta}{\delta(n-1)^2} + \frac{2}{n} > 0$$

and get the following form of the inequality:

$$ag^2(f_{,1})^4 - (g|\Delta f| - nL_1 + (n+2)B_{11}) \cdot (f_{,1})^2 - b(\Delta f)^2 \le 0.$$

The left hand side is a quadratic expression in $(f_{,1})^2$. If one considers its zeroes, it follows that

$$(f_{,1})^2 \le \frac{1}{ag^2}[g|\Delta f| - nL_1 + (n+2)B_{11} + g\sqrt{ab}|\Delta f|].$$

To further estimate this expression we use the assumptions, (3.79) and the definition of g; we have the inequalities

$$g \ge \frac{2}{2+f},$$

$$|\Delta f| \le N'(2+f)[\ln(2+f)]^p,$$

$$|-nL_1 + (n+2)B_{11}| \le [n+(n+2)](\lambda_1^2 + \cdots + \lambda_n^2)^{1/2}$$

$$< 2(n+1)N[\ln(2+f)]^p,$$

and insert these inequalities to get an upper bound for $(f_{,1})^2$:

$$(f_{,1})^2 \le \left(\frac{N'}{2a} + \frac{\sqrt{ab}}{2a}N' + \frac{n+1}{2a}N \right)(2+f)^2[\ln(2+f)]^p.$$

From (3.80) we thus get, with our special choice of δ and m:

$$\psi \le \frac{N'}{2a} + \frac{\sqrt{ab}}{2a}N' + \frac{n+1}{2a}N$$

which holds at x^*, where ψ attains its supremum. Hence, at any point of M_C, we have

$$\frac{\|\text{grad} f\|^2}{(2+f)^2[\ln(2+f)]^p} \le \left(\frac{N'}{2a} + \frac{\sqrt{ab}}{2a}N' + \frac{n+1}{2a}N \right)\exp\left\{ \frac{4(n-1)C}{C-f} \right\}.$$

Let $C \to \infty$, then

$$\frac{\|\text{grad} f\|}{(2+f)[\ln(2+f)]^{p/2}} \le e^{2(n-1)}\sqrt{\frac{N'}{2a} + \frac{\sqrt{ab}}{2a}N' + \frac{n+1}{2a}N} := Q, \tag{3.88}$$

where Q is a constant. Using the gradient estimate (3.88) we can prove:

Theorem 3.25. *Let M be a locally strongly convex, Euclidean complete hypersurface in A^{n+1}. Choose a coordinate system $(x^1, \ldots, x^{n+1})$ in A^{n+1} such that M is represented as a graph by a positive, strongly convex function $x^{n+1} = f(x^1, \ldots, x^n)$, defined on a convex domain $\Omega \subset A^n$, where the hyperplane $x^{n+1} = 0$ is the tangent hyperplane of M at some point $\bar{x}$, and $\bar{x}$ has coordinates $(0, \ldots, 0)$. If there is a constant $N > 0$ such that the affine principal curvatures of the Blaschke hypersurface satisfy the following inequality on M*

$$(\lambda_1^2 + \cdots + \lambda_n^2)^{1/2} \le N[\ln(2+f)]^2,$$

then M is also affine complete.

Proof. For any unit speed geodesic starting from $\bar{x}$

$$\sigma : [0, L) \to M,$$

by (3.88) we have

$$\frac{df(\sigma(s))}{ds} \le \|\mathrm{grad}\, f\| \le Q(2 + f) \ln(2 + f), \quad 0 \le s < L.$$

It follows that

$$L \ge \frac{1}{Q} \int_0^{x^{n+1}(\sigma(L))} \frac{df}{(2 + f) \ln(2 + f)}. \tag{3.89}$$

Since

$$\int_0^{\infty} \frac{df}{(2 + f) \ln(2 + f)} = \infty$$

and $f : \Omega \to \mathbb{R}$ is proper (i.e. the inverse image of any compact set is compact), equation (3.89) implies the affine completeness of M. □

As direct consequences of Theorem 3.25 we have the following two results:

Theorem 3.26. *Let M be a locally strongly convex, Euclidean complete hypersurface in A^{n+1}. If there is a constant $C > 0$ such that the affine principal curvatures satisfy the following inequality on M:*

$$(\lambda_1^2 + \cdots + \lambda_n^2)^{1/2} \le C,$$

then M is also affine complete.

Theorem 3.27. *Every Euclidean complete affine hypersphere is affine complete.*

3.4.2 The equivalence between Calabi metrics and Euclidean metrics

Let $\Omega \subset \mathbb{R}^n$ be a convex domain such that the origin $O \in \Omega$. In this subsection, we compare the Calabi metric $\mathcal{H}$ on Ω, defined by a smooth and strongly convex function f, with the standard Euclidean metric on Ω under the condition that the Pick invariant $J^{(C)}$ is bounded from above. This comparison is based on Lemma 3.29 below, which gives an estimate for eigenvalues of the Hessian matrix (f_{ij}). A similar estimate was first proved in [176], but the lemma given here is slightly stronger.

For $p \in \Omega$, denote by $D_a(p)$ and $B_a(p)$ the balls of radius a that are centered at p with respect to the Euclidean metric and Calabi metric, respectively. Similarly, denote by d_E and d_f the distance functions with respect to the Euclidean metric and Calabi metric, respectively.

First we prove the following lemma.

Lemma 3.28. *Let $\Gamma : x = x(t)$, $t \in [0, t_0]$, be a curve starting from the origin $x(0) = O$ to $p_0 = x(t_0)$ in Ω. Assume that, along Γ, $J^{(C)} \leq N^2$ for some positive constant N. Then, along Γ, there hold*

$$\left| \frac{d \ln T}{ds} \right| \leq 2n^2 N, \qquad \left| \frac{d \ln \det(f_{ij})}{ds} \right| \leq (n + 2)N, \tag{3.90}$$

where $J^{(C)}$ is defined as in (2.160), $T = \mathrm{tr}\,(f_{ij}) = \sum f_{ii}$ and s denotes the arc length parameter of Γ with respect to the Calabi metric $\mathcal{H}$.

Proof. For any $p \in \Gamma$, by an orthogonal coordinate transformation, we can choose a coordinate system $(x_1, \ldots, x_n)$ such that $f_{ij}(p) = \lambda_i \delta_{ij}$. Denote by $\lambda_{\max}$ (resp. $\lambda_{\min}$) the maximal (resp. minimal) eigenvalue of (f_{ij}). Suppose that, at p,

$$\frac{\partial}{\partial t} = \sum a_i \frac{\partial}{\partial x_i}.$$

Then we have, at p,

$$\frac{1}{T^2}\left(\frac{\partial T}{\partial t}\right)^2 = \frac{1}{T^2}\left(\sum_{i,j} f_{iij}a_j\right)^2 \leq \frac{n}{T^2} \sum_j \left(\sum_i f_{iij}\right)^2 a_j^2$$

$$\leq \frac{n}{T^2}\left(\sum_j \frac{1}{\lambda_j}\left(\sum_i f_{iij}\right)^2\right)\left(\sum \lambda_j a_j^2\right)$$

$$\leq n^2 \left(\sum f^{im}f^{jn}f^{kl}f_{ijk}f_{mnl}\right)\left(\sum f_{ij}a_i a_j\right)$$

$$\leq 4n^4 J^{(C)} \sum f_{ij}a_i a_j = 4n^4 J^{(C)} \cdot \left(\frac{ds}{dt}\right)^2.$$

Here the third "$\leq$" in the above inequalities is derived by using the following computation:

$$\sum_{i,j,k} f^{im}f^{jn}f^{kl}f_{ijk}f_{mnl} = \sum_{i,j,k} f_{ijk}f_{mnl} \frac{1}{\lambda_l}\delta_l^k \frac{1}{\lambda_i}\delta_i^m \frac{1}{\lambda_j}\delta_j^n$$

$$\geq \sum_{i,k}(f_{iik})^2 \frac{1}{\lambda_k \lambda_{\max}^2} \geq \frac{1}{n}\frac{\sum_k(\sum_i f_{iik})^2 \frac{1}{\lambda_k}}{(\sum_i \lambda_i)^2}.$$

Therefore, we have the following differential inequalities:

$$\left| \frac{d \ln T}{ds} \right| \leq 2n^2 \sqrt{J^{(C)}} \leq 2n^2 N, \qquad \left| \frac{d \ln \det(f_{ij})}{ds} \right| \leq (n + 2)N.$$

This completes the proof of Lemma 3.28. □

Lemma 3.29. *Let f be a smooth, strongly convex function defined on Ω. Assume that there exists a constant N such that*

$$J^{(C)} \leq N^2 \quad \text{in } \Omega, \tag{3.91}$$

and the Hessian matrix (f_{ij}) satisfies $f_{ij}(0) = \delta_{ij}$. Let $\Gamma : x = x(s), s \in [0, a]$, be a curve lying in Ω, starting from $x(0) = O$ with arc length parameter with respect to the Calabi metric $\mathcal{H} = \sum f_{ij}dx_idx_j$. As before let $\lambda_{\min}$ and $\lambda_{\max}$ be the minimal and maximal eigenvalues of (f_{ij}) along Γ. Then there exists a constant $c_1 > 0$ such that

(i) $\exp\{-c_1 a\} \leq \lambda_{\min} \leq \lambda_{\max} \leq \exp\{c_1 a\}$;

(ii) $\Gamma \subset D_{a\exp\left\{\frac{1}{2}c_1 a\right\}}(O)$.

Proof. (i) With the initial values given by $f_{ij}(0) = \delta_{ij}$ and integrating (3.90) with respect to s, we have

$$n\exp\left\{-2n^2 Na\right\} \leq T(q) \leq n\exp\left\{2n^2 Na\right\},$$

$$\exp\left\{-(n + 2)Na\right\} \leq \det(f_{ij})(q) \leq \exp\left\{(n + 2)Na\right\}$$

for any $q \in B_a(O)$. Then we have the following bounds for the eigenvalues of (f_{ij}):

$$\frac{\exp\{-(n + 2)Na\}}{(n\exp\{2n^2 Na\})^{n-1}} \leq \lambda_{\min} \leq \lambda_{\max} \leq n\exp\{2n^2 Na\}.$$

It follows that we have $c_1 > 0$ such that (i) holds.

By definition of the Calabi metric, the Euclidean length of Γ is less than $a\exp\{\frac{1}{2}c_1 a\}$. Hence we have (ii). $\qquad\square$

Corollary 3.30. *Let f be given as in Lemma 3.29.*
(i)   *Assume that $D_r(O) \subset \Omega$; then there exists a constant a_1 depending only on N and r such that $B_{a_1}(O) \subset D_r(O)$.*
(ii) *Assume that $B_a(O) \subset \Omega$; then there exists a constant r_1 depending only on N and a such that $D_{r_1}(O) \subset B_a(O)$.*

Proof. (i) Let v be any nonzero vector at O and $\Gamma_v(s)$ be a geodesic ray with respect to $\mathcal{H}$ initiating from O in the direction of v with arc length parameter. Then by (ii) of Lemma 3.29 we have

$$d_E(O, \Gamma_v(s)) \leq s\exp\left\{\frac{1}{2}c_1 s\right\}.$$

Hence $\Gamma_v(s) \subset D_r(O)$ if $s\exp\{\frac{1}{2}c_1 s\} < r$. By taking

$$a_1 = \min\left\{1, \frac{1}{2}r\exp\{-\frac{1}{2}c_1\}\right\},$$

we immediately get the assertion (i).

To show (ii), we note that by (i) of Lemma 3.29 the eigenvalues of (f_{ij}) are uniformly bounded. Let $\tilde{\Gamma}_v(t) = tv, 0 \leq t \leq t_0$, be a ray with $|v|_E = 1$, so that t is the arc length parameter with respect to the Euclidean metric. Then, if $\tilde{\Gamma}_v([0, t_0]) \subset B_a(O)$, the length of $\tilde{\Gamma}_v([0, t_0])$ with respect to $\mathcal{H} = \mathcal{H}_f$ is less than $t_0 \exp(\frac{1}{2}c_1 a)$. Hence we have the estimate:

$$d_f(O, \tilde{\Gamma}_v(t)) \le t \exp\left\{\frac{1}{2}c_1 a\right\}.$$

If we take $r_1 = \frac{1}{2}a \exp\{-\frac{1}{2}c_1 a\}$, then assertion (ii) follows. □

We now establish the following convergence theorem.

Theorem 3.31. *Let $\Omega \subset \mathbb{R}^n$ be a convex domain such that the origin $O \in \Omega$, and let $\{f_k\}_{k=1}^{\infty}$ be a sequence of smooth and strongly convex functions defined on Ω. Assume that $\{f_k\}$ are already normalized such that, for each k and all $p \in \Omega$*

$$f_k(p) \ge f_k(0) = 0, \quad \partial_{ij}^2 f_k(0) = \delta_{ij}.$$

Assume that $J_{f_k}^{(C)} \le N^2$ holds for some positive constant N, then we have:
(i)   there exists a constant $a > 0$ such that $B_{a, f_k}(O) \subset \Omega$;
(ii) there exists a subsequence of $\{f_k\}$ that locally C^2-converges to a strongly convex function f_∞ in $B_{a, f_\infty}(O)$;
(iii) moreover, if $(\Omega, \mathcal{H}_{f_k})$ is complete; $(\Omega, \mathcal{H}_{f_\infty})$ is complete.

Here $B_{a, f_k}(O)$ denotes the ball with center O and radius a with respect to the Calabi metric $\mathcal{H}_{f_k}$.

Proof. Let $D_r(O) \subset \Omega$, then by (i) of Corollary 3.30 we have a constant a_1 such that

$$B_{a_1, f_k}(O) \subset D_r(O) \subset \Omega.$$

Now by (ii) of Corollary 3.30, there exists a constant r_1 such that

$$D_{r_1}(O) \subset B_{a_1, f_k}(O) \subset \Omega.$$

From $J_{f_k}^{(C)} \le N^2$ we see that f_k C^2-converges (passing to a subsequence) to a strongly convex function f_∞ in $D_{r_1/2}(O)$. Take a constant a such that

$$B_{a, f_k}(O) \subset D_{r_1/2}(O).$$

Then we get both assertions (i) and (ii).

After proving that f_k C^2-converges to a strongly convex function f_∞ in $D_{r_1/2}(O)$ and $B_{a, f_\infty}(O) \subset D_{r_1/2}(O)$, we now prove (iii): For any point $q \in \partial B_{a, f_\infty}(O)$, we normalize f_k by the following transformation:

$$\tilde{x} = A_k(x - x(q)), \quad \tilde{f}_k(\tilde{x}) = f_k(x) - \operatorname{grad} f_k(q) \cdot (x - x(q)) - f_k(q),$$

where we choose matrices A_k such that $\partial_{ij}^2 \tilde{f}_k(0) = \delta_{ij}$.

Considering $B_{a_1\tilde{f}_k}(q)$, we find $B_{a\tilde{f}_k}(q)$ and $D_{r_1/2}(q)$ as above. Then from assertion (ii) we obtain the convergence of $\tilde{f}_k$ in $B_{a\tilde{f}_\infty}(q)$. It follows that

$$\lim_{k\to\infty} A_k = A_\infty, \quad \lim_{k\to\infty} \text{grad} f_k(q) = \text{grad} f_\infty(q), \quad \lim_{k\to\infty} f_k(q) = f_\infty(q), \qquad (3.92)$$

where A_∞ is a matrix such that $\partial^2_{ij}\tilde{f}_\infty(0) = \delta_{ij}$.

Since $\partial B_{a f_\infty}(0)$ is compact, we can do this for every $q \in \partial B_{a f_\infty}(0)$. Returning to the coordinates x and f_k, we conclude that f_k C^2-converges to a strongly convex function f_∞ in $B_{2a f_\infty}(0)$. Note that a is a uniform constant: repeating this procedure we finally get (iii). □

The following theorem provides an equivalence between the Calabi metric and the Euclidean metric in a convex domain Ω under the condition that $J^{(C)}$ is bounded in the sense that the Calabi metric is complete if and only if the Euclidean metric is complete.

Theorem 3.32. *Let $\Omega \subset \mathbb{R}^n$ be a convex domain. Let f be a smooth and strongly convex function on Ω, and assume that its graph hypersurface satisfies $J^{(C)} \leq N^2$, where N is a positive constant. Then $(\Omega, \mathcal{H})$ is complete if and only if the graph of f in $\mathbb{R}^{n+1}$ is Euclidean complete.*

Proof. For the "only if" part, without loss of generality, we assume that the origin $O \in \Omega$ and that for all $p \in \Omega$

$$f(p) \geq f(0), \quad f_{ij}(0) = \delta_{ij}.$$

We will prove that $f|_{\partial\Omega} =+ \infty$. Namely, for any Euclidean unit vector v and the ray from O to $\partial\Omega$ along the direction v, given by

$$\ell_v : [0, t) \to \Omega, \quad \ell_v(s) = sv, \quad \lim_{s\to t} \ell_v(s) \in \partial\Omega,$$

we show that $f(\ell_v(s)) \to \infty$ as $s \to t$.

On $[0, t)$, we choose a sequence of points

$$0 = s_0 < s_1 < \cdots < s_k < \cdots$$

such that $d_f(\ell_v(s_i), \ell_v(s_{i+1})) = 1$. From the completeness of $(\Omega, \mathcal{H})$, we can assume that this is an infinite sequence. Set $f_i = f(\ell_v(s_i))$, then we have the following:

Claim. *There is a constant $\delta > 0$ such that, for each i, $f_{i+1} - f_i \geq \delta$ holds.*

Proof of the Claim. First we show this for $i = 0$. By Corollary 3.30, we conclude that $B_1(0)$ contains a Euclidean ball $D_{a_1}(0)$. Since the eigenvalues of (f_{ij}) are bounded from below and $\text{grad} f(0) = 0$, we conclude that there exists a positive constant δ such that $f(x) - f(0) \geq \delta$ for any $x \in B_1(0) - D_{a_1}(0)$. In particular, $f_1 - f_0 \geq \delta$.

Next, we consider an arbitrary i. Let $p = \ell_v(s_i)$ and $q = \ell_v(s_{i+1})$. Without loss of generality we assume that $f_{ij}(p) = \delta_{ij}$. We normalize f to get a new function $\tilde{f}$ such that

p is the minimal point of $\tilde{f}$:

$$\tilde{f}(x) = f(x) - f(p) - \operatorname{grad} f(p) \cdot (x - p). \tag{3.93}$$

We notice that the claim is invariant under the transformation (3.93), then by the same argument as for the case $i = 0$, we have $\tilde{f}(q) - \tilde{f}(p) \geq \delta$. Then we have

$$f(q) - f(p) = \tilde{f}(q) - \tilde{f}(p) + \operatorname{grad} f(p) \cdot (q - p) \geq \tilde{f}(q) - \tilde{f}(p) \geq \delta.$$

Here the first inequality holds due to the convexity of f. This completes the proof of the claim. □

From the above claim, we have shown that for each k there holds $f_k \geq k\delta$, and therefore f_k goes to ∞ as $k \to \infty$. From this we obtain the assertion of the "only if" part.

For the "if" part, we assume also that $f|_\Omega \geq f(0) = 0$. Since the graph of f is Euclidean complete, for any constant $C > 0$, the section

$$S_f(0, C) := \{x \in \Omega \mid f(x) \leq C\} \tag{3.94}$$

is compact. Consider the function

$$F = \exp\left\{-\frac{2C}{C - f}\right\} \frac{\|\operatorname{grad} f\|_{\mathcal{H}}^2}{(1 + f)^2},$$

defined in $S_f(0, C)$. We assume that F attains its maximum at some interior point p^*. We may assume $\|\operatorname{grad} f\|_{\mathcal{H}}(p^*) > 0$. Choose an orthonormal frame field $\{e_i\}$ of the Calabi metric $\mathcal{H}$ around p^* such that $e_i(p^*) = \frac{\partial}{\partial x^i}$ and

$$f_{ij}(p^*) = \delta_{ij}, \quad \|\operatorname{grad} f\|_{\mathcal{H}}(p^*) = f_{,1}(p^*), \quad f_{,i}(p^*) = 0, \; i \geq 2.$$

Then $F_{,i}(p^*) = 0$ implies that, at p^*,

$$\left(-\frac{2f_{,1}}{1 + f} - \frac{2Cf_{,1}}{(C - f)^2}\right)(f_{,1})^2 + 2f_{,1}f_{,11} = 0.$$

Using $f_{,1}(p^*) > 0$ and $\frac{2C}{(C-f)^2} > 0$, we conclude that

$$\frac{(f_{,1})^2}{(1 + f)^2} \leq \frac{f_{,11}}{1 + f}. \tag{3.95}$$

We write $\nabla_{e_1} e_1 = \sum_i \Gamma_{11}^i e_i$; then we have

$$f_{,11} = f_{11} - \Gamma_{11}^1 f_{,1} = 1 - \frac{1}{2} f_{111} f_{,1} \leq 1 + n\sqrt{J} f_{,1} \leq 1 + nN f_{,1}.$$

Inserting this into (3.95) and applying the inequality of Cauchy–Schwarz, we have

$$\frac{(f_{,1})^2}{(1 + f)^2} \leq 2 + n^2 N^2.$$

Then $F(p^*) \leq e^{-2}(2 + n^2 N^2)$. This implies that $F \leq c_1$ since F attains its maximum at p^*. Hence in $S_f(0, \frac{C}{2})$,

$$\|\text{grad } \ln(1 + f)\|_{\mathcal{H}} \leq c_2, \tag{3.96}$$

for some constant c_2 that is independent of C. Letting $C \to \infty$, we have shown that (3.96) holds everywhere.

Take a point $p_1 \in \partial S_f(0, C)$ such that $d_f(0, p_1) = d_f(0, \partial S_f(0, C))$. Let γ be the minimal geodesic from 0 to p_1. From (3.96) we have

$$c_2 \geq \|\text{grad } \ln(1 + f)\|_{\mathcal{H}} \geq \left| \frac{d \ln(1 + f)}{ds} \right|_{\gamma},$$

where s denotes the arc length parameter with respect to the metric $\mathcal{H}$. Applying this and a direct integration we obtain

$$d_f(0, p_1) = \int_{\gamma} ds \geq (c_2)^{-1} \ln(1 + f)|_{p_1} = (c_2)^{-1} \ln(1 + C).$$

This implies that $d_f(0, \partial S_f(0, C)) \to+ \infty$ as $C \to \infty$, and therefore $(\Omega, \mathcal{H})$ is complete. □

3.4.3 Remarks

We add some remarks on the relations between the three notions of affine completeness, Euclidean completeness, and Calabi completeness:

(i) The notions of *Calabi completeness* and *Euclidean completeness* on M are generally not equivalent. For example, the global graph over $\mathbb{R}^n$ in $\mathbb{R}^{n+1}$, given by

$$f(x) = \exp\{x^1\} + \sum_{i=2}^{n} (x^i)^2,$$

is Euclidean complete, but not Calabi complete.

(ii) Analogously, the notions of *Calabi completeness* and *affine completeness* on M are generally not equivalent. For example, the one-sheeted hyperboloid

$$M := \{(x^1, x^2, \sqrt{1 + (x^1)^2 + (x^2)^2}) \mid (x^1, x^2) \in \mathbb{R}^2\}$$

is Euclidean complete and affine complete, but not Calabi complete.

3.5 Affine complete elliptic affine hyperspheres

The aim of this section is twofold:

(i) The main results of this section is Theorem 3.35, called the theorem of Blaschke and Deicke: "Any closed affine hypersphere is an ellipsoid". This result was one of the first global affine results, in dimension $n = 2$ it is due to Blaschke (see [20, §77] and [22, p. 32 and §III.5.5]).

(ii) The proof given here is a typical application of E. Hopf's maximum principle; this method is one of the most important methods in global differential geometry.

First of all, we derive an estimate for the Laplacian of the Pick invariant of an affine hypersphere. We have (cf. formula (3.32))

$$\frac{n(n-1)}{2}\Delta J = \sum (A_{ijk,l})^2 + \sum (R_{ij})^2 + \sum (R_{ijkl})^2 - (n+1)RL_1.$$

To get a differential inequality we need the following lemma, which is due to E. Calabi.

Lemma 3.33. *Let M be an n-dimensional Riemannian C^∞-manifold. With respect to any orthonormal frame field, the Riemannian curvature tensor R_{ijkl}, the Ricci tensor R_{ij}, and the scalar curvature R satisfy*

$$\sum (R_{ij})^2 \geq \frac{1}{n}R^2, \tag{3.97}$$

$$\sum (R_{ijkl})^2 \geq \frac{2}{n-1} \sum (R_{ij})^2. \tag{3.98}$$

Proof. Define

$$R'_{ij} = R_{ij} - \frac{1}{n}R\delta_{ij}.$$

Then the trace of the (0,2)-tensor R' vanishes:

$$\sum R'_{ii} = 0,$$

and

$$0 \leq \sum (R'_{ij})^2 = \sum (R_{ij})^2 - \frac{1}{n}R^2.$$

This proves (3.97).

Before proving (3.98) we observe that for dimension $n = 2$, both terms of this inequality are identical, so that we can limit our consideration to $n \geq 3$. Consider the conformal curvature tensor of H. Weyl:

$$W_{ijkl} = R_{ijkl} - \frac{1}{n-2}(\delta_{ik}R_{jl} + \delta_{jl}R_{ik} - \delta_{il}R_{jk} - \delta_{jk}R_{il}) + \frac{R}{(n-1)(n-2)}(\delta_{ik}\delta_{jl} - \delta_{il}\delta_{jk}).$$

The conformal curvature tensor is traceless, which means that the contraction of any two of the four indices of W_{ijkl} gives the zero tensor. Define

$$R'_{ijkl} := R_{ijkl} - W_{ijkl}.$$

Then from the definition of W we get

$$\sum (R'_{ijkl})^2 = \frac{4}{n-2} \sum (R_{ij})^2 - \frac{2R^2}{(n-1)(n-2)}.$$

On the other hand, from the definition of the (0,4)-tensor R' we get

$$\begin{aligned}
\sum (R_{ijkl})^2 &= \sum (R'_{ijkl})^2 + \sum (W_{ijkl})^2 + 2 \sum W_{ijkl} R'_{ijkl} \\
&= \sum (R'_{ijkl})^2 + \sum (W_{ijkl})^2 \\
&\geq \sum (R'_{ijkl})^2 \\
&= \frac{4}{n-2} \sum (R_{ij})^2 - \frac{2R^2}{(n-1)(n-2)} \\
&\geq \frac{2}{n-1} \sum (R_{ij})^2.
\end{aligned}$$

This completes the proof of the lemma. □

Inserting (3.97) and (3.98) into (3.32) we conclude that

$$\frac{n(n-1)}{2} \Delta J \geq \sum (A_{ijk,l})^2 + \frac{n+1}{n(n-1)} R^2 - (n+1)RL_1. \tag{3.99}$$

Combining this with the affine theorema egregium

$$R = n(n-1)(J + L_1),$$

we find that

$$\frac{n(n-1)}{2} \Delta J \geq \sum (A_{ijk,l})^2 + n(n-1)(n+1)J(J + L_1). \tag{3.100}$$

Remark 3.34. The above calculation is due to E. Calabi. For $n = 2$, (3.100) was obtained by W. Blaschke [20, §76]. For higher dimensions it was obtained by E. Calabi ([39], for parabolic affine hyperspheres) and R. Schneider [313]. U. Simon calculated ΔJ for general hypersurfaces and applied his formula to get some new characterizations of ellipsoids (cf. [327, 329]).

Using (3.100) we prove the following theorem of Blaschke ($n = 2$) and Deicke ($n \geq 2$):

Theorem 3.35. *Let $x : M \to A^{n+1}$ be a compact affine hypersphere without boundary. Then $x(M)$ is an ellipsoid.*

Proof. First of all, we assert that for any compact convex hypersurface M without boundary there is at least one point $x_0 \in M$ at which $L_1 > 0$. To see this, we consider the function

$$f = \langle U, a \rangle,$$

where U is the affine conormal vector field on M, $a \in A^{n+1}$ is any constant vector, and $\langle \, , \, \rangle : V^* \times V \to \mathbb{R}$ is the standard scalar product from Section 1.1.1. Then, by Corollary 2.11, the Laplacian of f satisfies

$$\Delta f = -nL_1 f.$$

Therefore

$$\frac{1}{2}\Delta f^2 = \|\operatorname{grad} f\|^2 + f\Delta f = \sum (f_i)^2 - nL_1 f^2.$$

Integrating over M and using Stokes' formula, we get

$$\int_M \left(\sum (f_i)^2 \right) dV - \int_M nL_1 f^2 dV = 0.$$

If L_1 is nonpositive everywhere, the formula above implies $f = $ const. Since a is arbitrary, it follows that U is a constant covector. This is impossible, and our assertion is proved. Since an affine hypersphere satisfies $L_1 = $ const., we have $L_1 > 0$. It follows immediately from (3.100) that $\Delta J \geq 0$, i.e. J is a subharmonic function on a compact manifold without boundary. From the maximum principle we get that $J = $ const. From (3.100) we conclude that $J = 0$ and thus from Theorem 2.13, $x(M)$ is an ellipsoid. □

Corollary 3.36. *Let M be an affine complete elliptic affine hypersphere. Then M is an ellipsoid.*

Proof. From Corollary 2.17 (ii) we have

$$R_{ij} \geq (n-1)L_1 \delta_{ij}, \quad L_1 = \text{const.} > 0.$$

Since M is affine complete, Myers' theorem (see [140, vol. II, p. 88]) tells us that M is compact. Then the corollary follows from Theorem 3.35. □

From the proof of Theorem 3.35 one obtains the following.

Theorem 3.37. *There does not exist a locally strongly convex, compact hypersurface without boundary and with nonpositive affine mean curvature.*

3.6 A differential inequality on a complete Riemannian manifold

In Section 3.5 we gave an exemplary proof of Theorem 3.35 with the maximum principle of E. Hopf. It was necessary to prove Hopf's theorem in the case of functions that are not necessarily differentiable, but for which one verifies a certain weak elliptic differential inequality. Calabi [39] gave a precise definition of the inequality in question, showing that, if applied to twice differentiable functions, it reduces to the elementary notion of that inequality, and that in the extended sense it provides a more general condition under which E. Hopf's theorem is satisfied.

Definition 3.38. Let u be an upper semicontinuous function on an open domain D, and let v be any function on D, and let f be a differentiable function defined on $\mathbb{R}^2$. We say that u is a *weak solution* of the differential inequality

$$\Delta u + f(u, \|\operatorname{grad} u\|) \geq v$$

on D, if for each point $x_0 \in D$ and any $\varepsilon > 0$, there exists a neighborhood $N_{x_0,\varepsilon}$ of x_0 in D and a C^2-function $u_{x_0,\varepsilon}$ defined on $N_{x_0,\varepsilon}$ such that
(i)   the difference function $u - u_{x_0,\varepsilon}$ satisfies

$$u(x) - u_{x_0,\varepsilon}(x) \geq u(x_0) - u_{x_0,\varepsilon}(x_0) = 0, \quad \text{and}$$

(ii) $u_{x_0,\varepsilon}$ satisfies the inequality

$$\Delta u_{x_0,\varepsilon} + f(u_{x_0,\varepsilon}, \|\operatorname{grad} u_{x_0,\varepsilon}\|) \geq v(x_0) - \varepsilon$$

at the point x_0 in the strict sense.

In a modified terminology one sometimes says: *"u satisfies the given differential inequality in a weak sense"*.
Similarly, for a lower semicontinuous function u, we can define a weak solution of

$$\Delta u + f(u, \|\operatorname{grad} u\|) \leq v.$$

Let M be a complete Riemannian manifold with Ricci curvature bounded from below by a constant $-K, K > 0$, and let p be a fixed point on M. Denote by $r(p, x)$ the geodesic distance function from p, and by $C(p)$ the cut locus of the point p. In $M - C(p)$, r^2 is differentiable and satisfies (cf. Appendix B)

$$\|\operatorname{grad} r^2\| = 2r, \tag{3.101}$$

$$\Delta(r^2) \leq 2[1 + (n-1)(1 + \sqrt{K}r)]. \tag{3.102}$$

For any positive C^2- function ϕ on M we consider the function

$$F = \frac{1}{(a^2 - r^2)^\alpha \phi},$$

defined in a geodesic ball $B_a(p) = \{x \in M \mid r(p, x) \le a\}$, where α is a positive number.

The function F attains its minimum at some point q in the interior of $B_a(p)$. First we assume that r^2 is C^∞-differentiable in a neighborhood of q. Then, at the point q, we have

$$\mathrm{grad}\, F = 0,$$

$$\Delta F \ge 0.$$

Hence

$$\frac{\mathrm{grad}\, \phi}{\phi} - \frac{\alpha}{a^2 - r^2}\, \mathrm{grad}\, r^2 = 0, \tag{3.103}$$

$$\frac{\|\mathrm{grad}\, \phi\|^2}{\phi^2} - \frac{\Delta\phi}{\phi} + \frac{\alpha\Delta r^2}{a^2 - r^2} + \frac{\alpha\|\mathrm{grad}\, r^2\|^2}{(a^2 - r^2)^2} \ge 0. \tag{3.104}$$

We insert (3.103) into (3.104) and get

$$\frac{\alpha\Delta r^2}{a^2 - r^2} + \frac{\alpha(\alpha + 1)\|\mathrm{grad}\, r^2\|^2}{(a^2 - r^2)^2} \ge \frac{\Delta\phi}{\phi}, \tag{3.105}$$

which, together with $a \ge r$ and the formulas (3.101) and (3.102), gives at q

$$4\alpha(\alpha + 1)a^2 + 2\alpha[1 + (n - 1)(1 + a\sqrt{K})](a^2 - r^2) \ge (a^2 - r^2)^2\phi^{-1}\Delta\phi.$$

Thus

$$[(a^2 - r^2)^\alpha\phi]^{2/\alpha}\phi^{-1-(2/\alpha)}\Delta\phi \le 4\alpha(\alpha + 1)a^2 + 2\alpha[1 + (n - 1)(1 + \sqrt{K}a)]a^2. \tag{3.106}$$

Obviously, (3.106) holds at any point where F attains a local minimum.

Secondly, if $q \in C(p)$, we choose a point p' on a minimal geodesic joining p and q. Denote by $r(p', x)$ the geodesic distance function from p'. Clearly, $q \notin C(p')$, that is, $r(p', x)$ is differentiable at q. We consider the function

$$\tilde{F} := \frac{1}{[a^2 - (r(p, p') + r(p', x))^2]^\alpha\phi}.$$

It is easy to check that $\tilde{F}$ also attains its minimum at q. Thus we can apply the argument used for F to $\tilde{F}$. Considering the limit $p' \to p$, we find that (3.106) holds at points where the function r^2 is not differentiable. Now we consider a continuous positive function ψ on M. Suppose that ψ is a weak solution of the differential inequality

$$\Delta\psi \ge f(\psi),$$

where $f(\psi)$ is a function with the property: for some positive constants α, b and N, the inequality $\psi > N$ implies $f(\psi) \ge b\psi^{1+(2/\alpha)}$. We define the function

$$F := \frac{1}{(a^2 - r^2)^\alpha\psi}$$

in $B_a(p)$. F attains its minimum at some interior point, say q, of $B_a(p)$. By Definition 3.38 there exist a neighborhood $N_{q,\varepsilon}$ of q and a C^2-function $\psi_{q,\varepsilon}$ on $N_{q,\varepsilon}$ such that

$$\psi(x) - \psi_{q,\varepsilon}(x) \geq \psi(q) - \psi_{q,\varepsilon}(q) = 0, \quad \text{for all } x \in N_{q,\varepsilon},$$

and $\psi_{q,\varepsilon}$ satisfies the differential inequality at q in the strict sense

$$\Delta \psi_{q,\varepsilon} \geq f(\psi_{q,\varepsilon}) - \varepsilon.$$

If $\psi(q) > N$, we have, at q,

$$\Delta \psi_{q,\varepsilon} \geq f(\psi_{q,\varepsilon}) - \varepsilon = f(\psi(q)) - \varepsilon \geq b\psi(q)^{1+(2/\alpha)} - \varepsilon. \tag{3.107}$$

It is easy to check that the function $\frac{1}{(a^2-r^2)^\alpha \psi_{q,\varepsilon}}$ also attains its local minimum at q. Thus $\psi_{q,\varepsilon}$ satisfies the inequality (3.106). Using (3.107) we conclude that, at q,

$$b[(a^2 - r^2)^\alpha \psi(q)]^{2/\alpha} - \varepsilon(a^2 - r^2)^2 \psi^{-1}(q)$$
$$\leq 4\alpha(\alpha + 1)a^2 + 2\alpha[1 + (n - 1)(1 + \sqrt{K}a)]a^2.$$

For $\varepsilon \to 0$, we obtain

$$b[(a^2 - r^2)^\alpha \psi(q)]^{2/\alpha} \leq p(a);$$

where $p(a)$ denotes the following polynomial in a:

$$p(a) = 4\alpha(\alpha + 1)a^2 + 2\alpha[1 + (n - 1)(1 + \sqrt{K}a)]a^2.$$

Since $[(a^2 - r^2)^\alpha \psi]^{2/\alpha}$ attains its maximum at q, for any $x \in B_a(p)$ we have

$$b[(a^2 - r^2(x))^\alpha \psi(x)]^{2/\alpha} \leq p(a),$$

that is

$$\psi(x) \leq \left[\frac{p(a)}{b(a^2 - r^2)^2}\right]^{\alpha/2}.$$

If $\psi(q) \leq N$ we have $(a^2 - r^2)^\alpha \psi \leq a^{2\alpha} \cdot N$. Therefore for any x we have

$$\psi(x) \leq \max\left\{\left[\frac{p(a)}{b(a^2 - r^2)^2}\right]^{\alpha/2}, \frac{a^{2\alpha}N}{(a^2 - r^2)^\alpha}\right\}.$$

Let $a \to \infty$, then $\psi(x) \leq N$. We have proved

Theorem 3.39. *Let M be a complete Riemannian manifold with Ricci curvature bounded from below by a negative constant and let ψ be a continuous positive function on M. Assume that ψ is a weak solution of the differential inequality*

$$\Delta \psi \geq f(\psi),$$

where $f(\psi)$ is a function with the property: for some positive constants α, b and N, the inequality $\psi > N$ implies $f(\psi) \geq b\psi^{1+(2/\alpha)}$. Then, on M, we have

$$\psi \leq N.$$

Corollary 3.40. *Let M be a complete Riemannian manifold with Ricci curvature bounded from below. Assume that ϕ is a nonnegative continuous function and a weak solution of the differential inequality*

$$\Delta\phi \geq b_0\phi^k - b_1\phi^{k-1} - \cdots - b_{k-1}\phi - b_k,$$

where $k > 1$ is an integer, and $b_0 > 0$, $b_1 \geq 0, \ldots, b_k \geq 0$. Let N be the largest root of the polynomial equation

$$b_0\phi^k - b_1\phi^{k-1} - \cdots - b_{k-1}\phi - b_k = 0.$$

Then

$$\phi(x) \leq N, \quad \text{for all } x \in M.$$

Proof. For any $\varepsilon > 0$, put $\psi := \phi + \varepsilon$. Then $\psi > 0$ and ψ is a weak solution of

$$\Delta\psi \geq b_0(\psi - \varepsilon)^k - b_1(\psi - \varepsilon)^{k-1} - \cdots - b_{k-1}(\psi - \varepsilon) - b_k.$$

Let $N(b, \varepsilon)$ be the largest root of the polynomial equation

$$b_0(\psi - \varepsilon)^k - b\psi^k - b_1(\psi - \varepsilon)^{k-1} - \cdots - b_k = 0,$$

where $0 < b < b_0$. Then, if $\psi > N(b, \varepsilon)$, ψ is a weak solution of

$$\Delta\psi \geq b\psi^k.$$

Thus from Theorem 3.39 we obtain the inequality

$$\psi \leq N(b, \varepsilon).$$

Let $b \to 0$, then $N(b, \varepsilon) \to N + \varepsilon$, which is the largest root of the equation

$$b_0(u - \varepsilon)^k - b_1(u - \varepsilon)^{k-1} - \cdots - b_k = 0.$$

Therefore,

$$\phi \leq N. \qquad \square$$

As an interesting application of Corollary 3.40, we prove the following result due to Li, Sheng and Simon (cf. [179]).

Theorem 3.41. *Let M be a locally strongly convex, affine complete hypersphere in A^{n+1}. If M possesses nonnegative scalar curvature, i.e. $R \geq 0$, then it is either a quadratic hypersurface or a hyperbolic hypersphere of type $Q(c, n)$.*

Proof. We consider two cases according to the sign of L_1:
(i) if $L_1 > 0$, according to Corollary 3.36, M is an ellipsoid;

(ii) if $L_1 \leq 0$, by using (3.100), $L_1 = $ const. and the relation

$$R = n(n-1)(J + L_1),$$

we have the differential inequality for R

$$\Delta R \geq \frac{2(n+1)}{n(n-1)} R^2. \tag{3.108}$$

Recall that for an affine hypersphere the Ricci curvature is bounded from below:

$$R_{ij} \geq (n-1)L_1 G_{ij}.$$

Therefore, by (3.108) and the completeness of (M, G), it is immediate from Corollary 3.40 that if $R \geq 0$ on M, then it should be the case $R \equiv 0$. It follows that we have two subcases:
(a) the relations $R = 0$ and $L_1 = 0$ imply that $J = 0$ on M, thus the hypersurface is an elliptic paraboloid;
(b) the relations $R = 0$ and $J = -L_1 > 0$ imply that the hypersurface is equiaffinely equivalent to $Q(c, n)$ for an appropriate constant c.

This completes the proof of Theorem 3.41. □

3.7 Estimates of the Ricci curvatures of affine complete affine hyperspheres of parabolic or hyperbolic type

The purpose of this section is to prove that the Ricci curvature of an affine complete, parabolic or hyperbolic affine hypersphere is negative semidefinite. As a consequence of this result, an affine complete parabolic affine hypersphere must be an elliptic paraboloid.

The proof is based on [41] and results of Cheng and Yau [57].

Following Calabi [41], we consider the function

$$\psi(x) = \sup_{\xi \in T_x M} \left\{ \sum A^i_{jk} A^j_{il} \xi^k \xi^l \mid \sum G_{kl} \xi^k \xi^l = 1 \right\}, \quad x \in M. \tag{3.109}$$

Clearly, $\psi(x)$ is a continuous function on M.

Proposition 3.42. ψ satisfies the weak differential inequality

$$\Delta \psi \geq (n+1) \left(\frac{\psi^2}{n-1} + L_1 \psi \right).$$

Proof. Let $x_0 \in M$ and let $\xi(x_0)$ be a unit tangent vector at x_0 that maximizes $\sum A^i_{jk} A^j_{il} \xi^k \xi^l$. Extend $\xi(x_0)$ to a differentiable vector field η in a neighborhood N of

x_0 by translating $\xi(x_0)$ by Levi–Civita parallelism along the geodesics from x_0. Consider the differentiable function

$$\bar{\psi}(x) = \sum A^i_{jk} A^j_{il} \eta^k \eta^l,$$

where η is a unit vector field; we have $\bar{\psi} \le \psi$ everywhere in N with equality holding at x_0. According to Definition 3.38, it is sufficient to prove that at x_0

$$\Delta\bar{\psi} \ge (n+1)\left(\frac{\psi^2}{n-1} + L_1\psi\right) = (n+1)\left(\frac{\bar{\psi}^2}{n-1} + L_1\bar{\psi}\right).$$

Note that η is generated by parallel translation along every geodesic from x_0; thus we have

$$\operatorname{grad}\eta(x_0) = 0, \quad \Delta\eta(x_0) = 0.$$

As M is a hyperbolic or parabolic affine hypersphere, from the integrability conditions (3.28), (3.29), and the Ricci identity (cf. A.3.1.8) we obtain that

$$\Delta A^j_{ik} = \sum A^{jl}_t A^t_{lm} A^m_{ik} + \sum A^l_{im} A^{tj}_l A^m_{tk} + \sum A^l_{km} A^m_{lt} A^{tj}_i$$
$$- 2\sum A^{jl}_m A^m_{ti} A^t_{kl} + (n+1)L_1 A^j_{ik}. \tag{3.110}$$

Hence we have, at x_0,

$$\frac{1}{2}\Delta\bar{\psi}(x_0) = \left\{\sum\left(\sum A^i_{jk}\Delta A^j_{il} + \sum A^i_{jk,m}A^j_{il,m}\right)\eta^k\eta^l\right\}\Big|_{x_0}$$

$$\ge \left\{\sum A^i_{jk}\Delta A^j_{il}\eta^k\eta^l\right\}\Big|_{x_0}$$

$$= \left\{\sum A^i_{jk}\eta^k\left(\sum A^j_{im}A^{mi'}_{j'}A^{j'}_{i'l} + \sum A^j_{lm}A^{mi'}_{j'}A^{j'}_{i'i}\right.\right.$$

$$\left.\left. + \sum A_{i'lm}A^{i'm}_{j'}A^{i'j}_i - 2\sum A^{ji'}_{j'}A^j_{im}A^m_{i'l} + (n+1)L_1 A^j_{il}\right)\eta^l\right\}\Big|_{x_0}$$

$$= \left\{\sum_m\left[\left(\sum\eta^k A^i_{jk}A^j_{im}\right)\left(\sum\eta^l A^{i'}_{j'l}A^{j'm}_{i'}\right)\right] + (n+1)L_1\bar{\psi}\right.$$

$$\left. + \sum\left[\sum\eta^k(A^i_{jk}A^{j'}_{i'i} - A^i_{i'k}A^{j'}_{ji})\sum\eta^l(A^j_{ml}A^{mi'}_{j'} - A^{i'}_{ml}A^{mj}_{j'})\right]\right\}\Big|_{x_0}$$

$$\ge \left\{\bar{\psi}^2 + (n+1)L_1\bar{\psi} + 2\frac{\bar{\psi}^2}{n-1}\right\}\Big|_{x_0}$$

$$= (n+1)\left(\frac{\bar{\psi}^2}{n-1} + L_1\bar{\psi}\right)\Big|_{x_0}.$$

The last inequality is a consequence of an elementary quadratic inequality of norms in tensor spaces. □

Theorem 3.43. *Let M be an affine complete hyperbolic or parabolic affine hypersphere in A^{n+1}. Then the Ricci curvature of M is negative semidefinite.*

Proof. It follows from Proposition 3.42 and Corollary 3.40 that

$$\psi(x) \leq -(n-1)L_1, \quad \text{for all } x \in M.$$

Then the theorem follows from the formula

$$R_{ij} = \sum A_{iml}A_{jml} + (n-1)L_1\delta_{ij}. \qquad \Box$$

Corollary 3.44. *For a complete affine hypersphere of hyperbolic or parabolic type,*

$$J \leq -L_1.$$

This has the following consequence due to E. Calabi (cf. [41]).

Theorem 3.45. *Every affine complete, parabolic affine hypersphere is an elliptic paraboloid.*

Proof. This follows immediately from $0 \leq J \leq -L_1 = 0$. $\qquad \Box$

Combining Theorems 3.27 and 3.45 gives a new proof of the following well-known theorem of Jörgens–Calabi–Pogorelov (due to Jörgens for $n = 2$ [138], Calabi for $n = 3, 4, 5$ [40], and Pogorelov for general n [284]).

Theorem 3.46. *Let $f(x^1,\ldots,x^n)$ be a strongly convex differentiable function defined for all $x^1,\ldots,x^n$. If f satisfies*

$$\det\left(\frac{\partial^2 f}{\partial x^i \partial x^j}\right) = 1,$$

then f must be a quadratic polynomial.

Proof. Consider the graph of the function f:

$$M = \{(x^1,\ldots,x^n,x^{n+1}) \mid x^{n+1} = f(x^1,\ldots,x^n), (x^1,\ldots,x^n) \in A^n\}.$$

Then M is a Euclidean complete, parabolic, affine hypersphere. By Theorem 3.27, M is affine complete. The assertion now follows from Theorem 3.45. $\qquad \Box$

Using Theorem 3.43 we can prove the following result (cf. [184]).

Theorem 3.47. *Let $x : M \to A^3$ be an affine complete hyperbolic affine sphere. If there is a constant $\varepsilon > 0$ such that $J \geq \varepsilon$ on M, then $x(M)$ must be an affine image of the surface $Q(c, 2)$.*

Proof. Without loss of generality we may assume that M is simply connected (otherwise we can pass to the universal covering surface of M). We choose isothermal parameters u, v on M such that the Blaschke metric is given by $G = E((du)^2 + (dv)^2)$.

Then

$$\Delta = E^{-1}\left(\frac{\partial^2}{\partial u^2} + \frac{\partial^2}{\partial v^2}\right),$$

$$\chi = \Delta \ln E^{-1/2}.$$

We introduce a new metric

$$\tilde{G} = J^{1/3}G = J^{1/3}E((du)^2 + (dv)^2).$$

It is easy to see from $J \geq \varepsilon$ and the completeness of (M, G) that $(M, \tilde{G})$ is complete. The normalized scalar curvature $\tilde{\chi}$ of $(M, \tilde{G})$ is

$$\tilde{\chi} = J^{-1/3}E^{-1}\left(\frac{\partial^2}{\partial u^2} + \frac{\partial^2}{\partial v^2}\right)\ln(J^{-1/6}E^{-1/2}) = J^{-1/3}(-\Delta \ln J^{1/6} + \chi) = 0.$$

The last equality above is due to the formula (3.26). Hence $(M, \tilde{G})$ is a Euclidean space, therefore (M, G) is conformally equivalent to the complex plane $\mathbb{C}$. By Theorem 3.43 we know that $\chi \leq 0$, i.e.

$$\left(\frac{\partial^2}{\partial u^2} + \frac{\partial^2}{\partial v^2}\right)\ln J^{1/6} \leq 0,$$

which means that $\ln J^{1/6}$ is a superharmonic function on $\mathbb{C}$. From $J \geq \varepsilon$ we conclude now that $J = \text{const.} > 0$. Application of Theorem 3.7 then gives the assertion. □

3.8 Qualitative classification of complete hyperbolic affine hyperspheres

In Sections 3.5–3.7 we proved that every affine complete elliptic affine hypersphere is an ellipsoid (cf. Corollary 3.36), and that every affine complete parabolic affine hypersphere is an elliptic paraboloid (cf. Theorem 3.45). But an affine complete hyperbolic affine hypersphere need not be a hyperboloid. It is easy to verify that the affine hyperspheres in Examples 3.4–3.7 of Section 3.2.1 are Euclidean complete; therefore they are affine complete (Theorem 3.27). E. Calabi has made the following conjecture about hyperbolic affine hyperspheres:

(i) *Every affine complete n-dimensional hyperbolic affine hypersphere is asymptotic to the boundary of a convex cone with its vertex at the center of the hypersphere.*

(ii) *Every pointed, nondegenerate convex cone determines a hyperbolic affine hypersphere, asymptotic to the boundary of the cone and uniquely determined by the value of its affine mean curvature.*

This section is devoted to the study of Calabi's conjecture above.

3.8.1 Euclidean complete hyperbolic affine hyperspheres

In this subsection we shall prove the first part of the Calabi conjecture under the assumption of Euclidean completeness, mainly following S. Y. Cheng and S. T. Yau [57].

Let M be a hyperbolic affine hypersphere with affine mean curvature $L_1 < 0$ in A^{n+1}. Define a Euclidean inner product on V. We assume that M is Euclidean complete. Theorem 3.27 states that M is also affine complete, and M cannot be compact (see Theorem 3.37). We can conclude from the theorem of Hadamard–Sacksteder–Wu (cf. [407]) that M is the graph of a strongly convex function $x^{n+1} = f(x^1, \ldots, x^n)$ defined on a convex domain $\Omega \subset A^n$. Hence M is globally strongly convex. By a unimodular affine transformation we may assume that the origin O is at the center of M, that the hyperplane $x^{n+1} = 0$ is parallel to the tangent hyperplane of M at some point $x_0 \in M$, and that x_0 has the coordinates $(0, \ldots, 0, 1)$. Then f attains its minimum 1 at $O = (0, \ldots, 0) \in \Omega$. Moreover, f must tend to infinity at points of $\partial\Omega$. We have the following:

Proposition 3.48. $\Omega = A^n$.

Proof. If $\partial\Omega \neq \emptyset$, we may assume that $\bar{x} = (\bar{x}^1, \bar{x}^2, \ldots, \bar{x}^n) \in \partial\Omega$ and that, for $0 \le t < 1$, we have $t\bar{x} = (t\bar{x}^1, \ldots, t\bar{x}^n) \in \Omega$. Then, for $t \to 1, f(t\bar{x}) \to \infty$. The line segment joining the origin and the point $(t\bar{x}^1, \ldots, t\bar{x}^n, f(t\bar{x}))$ must intersect the graph of f at more than one point when $t \to 1$. Since the affine normal vector is a positive multiple of the position vector with the origin at the center, this is in contradiction to the fact that the center must lie on the convex side of the hyperbolic affine hypersphere. Hence $\partial\Omega = \emptyset$ and the proposition is proved. □

Define the set

$$D := \{(\partial_1 f(x), \ldots, \partial_n f(x)) \mid x = (x^1, \ldots, x^n) \in \Omega\}. \qquad (3.111)$$

Then D is the Legendre transformation domain of Ω relative to f. The Legendre transformation function u of f (see Definition 2.27) is

$$u(\xi) = u(\xi_1, \xi_2, \ldots, \xi_n) = \sum x^i \frac{\partial f}{\partial x^i} - f(x),$$

$$\xi_i = \frac{\partial f}{\partial x^i}, \quad x \in \Omega. \qquad (3.112)$$

According to (3.13), the Blaschke metric G_{ij} of M is

$$G_{ij} = \rho \frac{\partial^2 u}{\partial \xi_i \partial \xi_j},$$

where $\rho = [\det(\frac{\partial^2 f}{\partial x^i \partial x^j})]^{-1/(n+2)}$. Since $(\frac{\partial^2 f}{\partial x^i \partial x^j})$ and (G_{ij}) are positive definite, $(\frac{\partial^2 u}{\partial \xi_i \partial \xi_j})$ is also positive definite, and therefore u is strongly convex. On the other hand, since M is a hyperbolic affine hypersphere, (3.16′) gives that

$$\left[\det\left(\frac{\partial^2 u}{\partial\xi_i\partial\xi_j}\right)\right]^{1/(n+2)} = \frac{1}{L_1 u}.$$

We conclude from $L_1 < 0$ that

$$u(\xi) < 0, \quad \text{for all } \xi = (\xi_1, \xi_2, \ldots, \xi_n) \in D. \tag{3.113}$$

From Lemma 2.28, the gradient map

$$F : \Omega = A^n \to D,$$

$$x = (x^1, \ldots, x^n) \mapsto \left(\frac{\partial f}{\partial x^1}(x), \ldots, \frac{\partial f}{\partial x^n}(x)\right) = (\xi_1, \ldots, \xi_n)$$

is a diffeomorphism. Thus D is an open set in A^n. Consider the set

$$C := Cl\{t(x^1, x^2, \ldots, x^n, f(x)) \mid t > 0, x = (x^1, \ldots, x^n) \in A^n\}, \tag{3.114}$$

where Cl denotes the closure of the given set. Using the standard Euclidean inner product $(\cdot, \cdot) : V \times V \to \mathbb{R}$ we define the polar cone C^* of C by

$$C^* := \{\xi \in V \mid (\xi, x) \le 0, \forall\, x \in C\}. \tag{3.115}$$

As usual we use the same notation for a point and its position vector with respect to the origin. Then we can regard C^* as a convex cone in A^{n+1}.

Let θ be a Euclidean normal vector of M at a point p. We call θ an outer normal vector of M at p if θ points to the convex side of M.

Lemma 3.49. *The set E of all Euclidean outer normal vectors of M is equal to the interior C^{*0} of C^*, i.e.*

$$E = C^{*0} = \{\xi \in A^{n+1} \mid (\xi, x) < 0, \forall\, x \in C, x \ne 0\}. \tag{3.116}$$

Proof. Let $\bar{x} \in A^n$ and let θ be a Euclidean outer normal vector of M at the point $(\bar{x}, f(\bar{x}))$. We can write

$$\theta = s(\partial_1 f(\bar{x}), \ldots, \partial_n f(\bar{x}), -1)$$

for some nonzero real number s. The inner product of θ with $(\bar{x}, f(\bar{x}))$ is equal to

$$s\left(\sum \bar{x}^i \partial_i f(\bar{x}) - f(\bar{x})\right) = su(\bar{x}).$$

Since the position vector $(\bar{x}, f(\bar{x}))$ points towards the concave side of M and θ points to the convex side, it follows from $u < 0$ that $s > 0$. On the other hand, for any point $p = t(x, f(x))$, where $t > 0$ and $x = (x^1, \ldots, x^n) \in A^n$, the inner product of θ with p is equal to

$$(\theta, p) = st\left(\sum x^i \frac{\partial f}{\partial x^i}(\bar{x}) - f(x)\right).$$

Since the function $\sum x^i \cdot \frac{\partial f}{\partial x^i}(\bar{x}) - f(x)$ attains its maximum $u(\bar{x})$ at $\bar{x}$, and since $u(\bar{x}) < 0$, we have $(\theta, p) \le stu(\bar{x}) < 0$. Then it is easy to conclude that, for any $p \in C$, we have $(\theta, p) \le 0$ and thus $E \subset C^*$. Since the Legendre transformation domain D is an open set in A^n, it follows that E is an open set in A^{n+1}. Therefore $E \subset C^{*0}$.

Conversely, assume that $\eta = (\eta^1, \ldots, \eta^{n+1}) \in C^{*0}$, then $(\eta, x) < 0$, for all $x \in C$ with $x \ne 0$, and for any positive number d, the set

$$\{x \in C \mid (\eta, x) \ge -d\}$$

is bounded. Hence, for a large positive number d, the set

$$M_d = \{x \in M \mid (\eta, x) \ge -d\}$$

is compact and has nonempty interior. Therefore, there exists an interior point $\overset{*}{x}$ of M_d such that the function (η, x) defined on M_d attains its supremum at $\overset{*}{x}$, that is, η is an outer normal vector of M at $\overset{*}{x}$. $\qquad \square$

Now we can prove the following:

Proposition 3.50. *The set D from (3.111) is a bounded convex domain.*

Proof. We identify D with the set

$$\{(\partial_1 f(x), \ldots, \partial_n f(x), -1) \mid x \in A^n\}.$$

From Lemma 3.49 we have

$$D = C^{*0} \cap \{\xi_{n+1} = -1\}.$$

Hence D is convex as it is the intersection of two convex sets.

Next we prove that D is bounded. We choose a small positive number r such that the circular cone

$$C(r) = \{t(\bar{x}^1, \ldots, \bar{x}^{n+1}) \mid t \ge 0, (\bar{x}^1)^2 + \cdots + (\bar{x}^n)^2 \le r, \ \bar{x}^{n+1} = 1\}$$

is contained in C. Then $C^* \subset C(r)^*$. Obviously, the set

$$C(r)^{*0} \cap \{\xi_{n+1} = -1\}$$

is bounded. It follows that D is bounded. $\qquad \square$

The following result is a step to the proof of the first part of the Calabi conjecture. Its proof is due to Cheng and Yau [57].

Theorem 3.51. *Let M be an n-dimensional Euclidean complete hyperbolic affine hypersphere in A^{n+1}. Then M is asymptotic to the boundary of the convex cone, which is defined as the convex hull of the center and M.*

Proof. Adopt the notation from above. Let C be the convex cone defined by (3.114). Let g be the function defining the boundary ∂C of the cone C. Then clearly $f \geq g$ holds. It is sufficient to prove that, for all $\varepsilon > 0$, there is a compact set W in A^n such that, for $x \notin W$, $0 \leq f(x) - g(x) \leq \varepsilon$.

Set

$$D_\varepsilon := \{(\xi_1, \xi_2, \ldots, \xi_n) \in D \mid u(\xi_1, \ldots, \xi_n) < -\varepsilon\}.$$

From (3.13) and (3.16), the Blaschke metric induced by the affine hypersphere on D_ε is

$$G = (L_1 u)^{-1} \sum \frac{\partial^2 u}{\partial \xi_i \partial \xi_j} d\xi_i d\xi_j.$$

We use the relation between f and its Legendre transformation function

$$f(x^1, \ldots, x^n) = f\left(\frac{\partial u}{\partial \xi_1}(\xi), \ldots, \frac{\partial u}{\partial \xi_n}(\xi)\right) = \sum \xi_i \frac{\partial u}{\partial \xi_i} - u(\xi)$$

and apply the gradient estimate (3.88) (taking $p = 0$ in (3.88)):

$$L_1 u \sum u^{ij} \frac{\partial f}{\partial \xi_i} \frac{\partial f}{\partial \xi_j} \leq \alpha^2 \left(\sum \xi_i \frac{\partial u}{\partial \xi_i} - u + 2\right)^2. \tag{3.117}$$

Here α is a positive constant depending only on n and $-L_1$, and (u^{ij}) denotes the inverse matrix of $\left(\frac{\partial^2 u}{\partial \xi_i \partial \xi_j}\right)$. (3.9) implies $\frac{\partial f}{\partial \xi_i} = \sum \xi_j \frac{\partial^2 u}{\partial \xi_j \partial \xi_i}$; we insert this into (3.117) and verify that

$$(L_1 u) \sum \frac{\partial^2 u}{\partial \xi_i \partial \xi_j} \xi_i \xi_j \leq \alpha^2 \left(\sum \xi_i \frac{\partial u}{\partial \xi_i} - u + 2\right)^2. \tag{3.118}$$

Now we assert that the closure $\bar{D}_\varepsilon$ of D_ε is contained in the interior of D. Otherwise we have $\partial D_\varepsilon \cap \partial D \neq \phi$. Assume $\bar{\xi} = (\bar{\xi}_1, \bar{\xi}_2, \ldots, \bar{\xi}_n) \in \partial D_\varepsilon \cap \partial D$ and define

$$\sigma(t) = \left\{t(\bar{\xi}_1, \bar{\xi}_2, \ldots, \bar{\xi}_n) \mid 0 \leq t \leq 1\right\}$$

to be the line segment joining the origin O and the point $\bar{\xi}$. Since on D_ε,

$$L_1 u \geq -\varepsilon L_1 > 0,$$

the affine arc length of $\sigma(t)$ is

$$l = \int_0^1 \left(\frac{1}{L_1 u} \sum \frac{\partial^2 u}{\partial \xi_i \partial \xi_j} \bar{\xi}_i \bar{\xi}_j\right)^{1/2} dt$$

$$= \left(\int_0^{1/2} + \int_{1/2}^1\right) \left(\frac{1}{L_1 u} \sum \frac{\partial^2 u}{\partial \xi_i \partial \xi_j} \bar{\xi}_i \bar{\xi}_j\right)^{1/2} dt.$$

The first integral of the last line above is clearly finite. We compute the second one. It follows from (3.118) that

$$
\int_{1/2}^{1} \left[\frac{1}{L_1 u t^2} \sum \frac{\partial^2 u}{\partial \xi_i \partial \xi_j} (t\bar{\xi}_i)(t\bar{\xi}_j) \right]^{1/2} dt
$$

$$
\leq \int_{1/2}^{1} \frac{\alpha}{L_1 u t} \left(\sum t\bar{\xi}_i \frac{\partial u}{\partial \xi_j} - u + 2 \right) dt
$$

$$
\leq -\frac{\alpha}{L_1 \varepsilon} \int_{1/2}^{1} du(\sigma(t)) - \int_{1/2}^{1} \frac{2\alpha}{L_1} \left(1 + \frac{2}{\varepsilon} \right) dt.
$$

This integral is also finite. Therefore the arc length of $\sigma(t)$ is finite. This contradicts the facts that the line segment tends to ∂D and the metric on D is complete. Hence we arrive at $\bar{D}_\varepsilon \subset D$.

Since D is bounded, $\bar{D}_\varepsilon$ is compact. Recalling F from (3.7) we see that $W = F^{-1}(\bar{D}_\varepsilon)$ is a compact set in A^n. For $\xi \notin D_\varepsilon$ we have $-\varepsilon \leq u(\xi) < 0$. We obtain that for $\bar{x} \notin W$,

$$
-\varepsilon \leq \sum \bar{x}^i \frac{\partial f}{\partial x^i}(\bar{x}) - f(\bar{x}) < 0,
$$

or

$$
0 < f(\bar{x}) - \sum \bar{x}^i \partial_i f(\bar{x}) \leq \varepsilon. \tag{3.119}
$$

On the other hand, since $v = (\partial_1 f(\bar{x}), \ldots, \partial_n f(\bar{x}), -1)$ is a Euclidean outer normal vector of M at the point $(\bar{x}^1, \ldots, \bar{x}^n, f(\bar{x}))$, it follows from Lemma 3.49 that, for the point $p = (\bar{x}^1, \ldots, \bar{x}^n, g(\bar{x}))$ of ∂C, $(v, p) \leq 0$, that is,

$$
\sum \bar{x}^i \frac{\partial f}{\partial x^i}(\bar{x}) - g(\bar{x}) \leq 0.
$$

We conclude from (3.119) and the last inequality above that

$$
0 \leq f(\bar{x}) - g(\bar{x}) \leq \varepsilon.
$$

This completes the proof of Theorem 3.51. □

3.8.2 Affine complete hyperbolic affine hyperspheres

Theorem 3.51 answers the first part of the Calabi conjecture only for Euclidean complete, hyperbolic affine hyperspheres. Let M be a locally strongly convex, affine complete, hyperbolic affine hypersphere: *is M Euclidean complete?* This question was finally answered in the affirmative by A.-M. Li (cf. [169, 170]). In this subsection we will explain A.-M. Li's solution of this question closely following [170].

Let M be a connected manifold without boundary and let $x : M \to A^{n+1}$ be a locally strongly convex immersion. Assume that x is an affine complete, hyperbolic affine hypersphere. Then (see Theorem 3.43 and its Corollary 3.44)

$$J \le -L_1.$$

Let p_0 be a point on M. We choose an affine coordinate system $(x^1, \ldots, x^{n+1})$ in A^{n+1} such that the origin coincides with the center of $x(M)$ and the hyperplane $x^{n+1} = 0$ is parallel to the tangent hyperplane of $x(M)$ at $x(p_0)$, and $x(p_0)$ has the coordinates $(0, \ldots, 0, 1)$. We consider the canonical Euclidean metric,

$$(x, y) = x^1 y^1 + \cdots + x^{n+1} y^{n+1},$$

for $x = (x^1, \ldots, x^{n+1})$, $y = (y^1, \ldots, y^{n+1})$. Now there are two metrics on M: the metric g induced from the Euclidean metric of the ambient space, and the Blaschke metric G. By assumption, (M, G) is a complete Riemannian manifold. We identify the hyperplane $x^{n+1} = 0$ with A^n and denote by j the canonical projection, i.e. the map

$$j : A^{n+1} \to A^n,$$
$$(x^1, \ldots, x^{n+1}) \mapsto (x^1, \ldots, x^n).$$

Obviously, there is an open set $N, p_0 \in N \subset M$, such that $x|_N$ is an imbedding of N in A^{n+1} and $j \circ x|_N$ is a diffeomorphism of N onto $j \circ x(N)$. Then $x(N)$ is the graph of a strongly convex function $x^{n+1} = f(x^1, \ldots, x^n)$ defined on a domain $\Omega \subset A^n$, and $\partial_i f(0, \ldots, 0) = 0$ for $1 \le i \le n$. We cut $x(M)$ with a hyperplane given by $x^{n+1} = h$, where $h > 1$. Denote by M_h the connected component of the set $\{p \in M | x^{n+1}(p) < h\}$ containing p_0. We denote by $\bar{M}_h$ the closure of M_h in (M, g), and by $\overline{x(M_h)}$ the closure of $x(M_h)$ in the ambient Euclidean space.

For h near to 1, one can easily verify that M_h has the following properties:
(i) $x|_{M_h}$ is an imbedding, and $x(M_h)$ is given as the graph of a strongly convex function $x^{n+1} = f(x^1, \ldots, x^n)$, defined on a bounded convex domain $\Omega_h \subset A^n$, and

$$\partial_i f(0, \ldots, 0) = 0, \quad \text{for } 1 \le i \le n,$$
$$f(x^1, \ldots, x^n) \to h, \quad \text{when } (x^1, \ldots, x^n) \to \partial \Omega_h;$$

(ii) $x(\bar{M}_h) = \overline{x(M_h)}$.

On the one hand, since f is convex, f satisfies the inequality

$$f(x) - f(0) \le \sum x^i \partial_i f.$$

Here and later we simply write $f(x^1, \ldots, x^n)$ as $f(x)$; the last inequality is equivalent to $u \ge -1$ for the Legendre transform function. On the other hand, since $x(M)$ is a hyperbolic affine hypersphere, u satisfies (3.16′); hence, being negative at $(0, \ldots, 0)$, the function u remains negative for all $(x^1, \ldots, x^n) \in \Omega_h$. So we have the pinching relation

$$-1 \le u < 0.$$

Restricting the considerations to M_h, the affine conormal vector field U can be identified with

$$[\det(\partial_j \partial_i f)]^{-1/(n+2)}(-\partial_1 f, \ldots, -\partial_n f, 1).$$

Let $b := (0, \ldots, 0, 1)$; then

$$\langle U, b \rangle = \left[\det \left(\frac{\partial^2 f}{\partial x^i \partial x^j} \right) \right]^{-1/(n+2)} = \left[\det \left(\frac{\partial^2 u}{\partial \xi_i \partial \xi_j} \right) \right]^{1/(n+2)}$$

$$= \frac{1}{L_1 u} \geq \frac{1}{|L_1|} > 0.$$

By continuity we have $\langle U, b \rangle \geq \frac{1}{|L_1|} > 0$ at any point $p \in \bar{M}_h$. This means that the line $x(p) + tb$, $-\infty < t < +\infty$, and $x(\bar{M}_h)$ intersect transversally at $x(p)$. Hence there is an open set N', $p \in N'$, such that $j \circ x|_{N'}$ is a diffeomorphism of N' onto $j \circ x(N')$. Since f is strongly convex and $\partial_i f(0, \ldots, 0) = 0$, for any point $p \in \partial M_h$, by continuity, we can prove that $|\mathrm{grad}\, x^{n+1}(p)| \neq 0$. Thus ∂M_h is a regular submanifold of M of dimension $n - 1$.

Lemma 3.52. *If for some number $d > 1$, the set M_d has the above property* (i), *then $j \circ x|_{\bar{M}_d}$ is injective.*

Proof. First we show that $\sum (\partial_i f)^2$ is bounded from above in Ω_d. Let $(\dot{x}^1, \ldots, \dot{x}^n)$ be an arbitrary point in Ω_d, let $\dot{x} = (\dot{x}^1, \ldots, \dot{x}^n, f(\dot{x}^1, \ldots, \dot{x}^n))$, and let θ be normal at $\dot{x}$, where $\theta := (\partial_1 f(\dot{x}), \ldots, \partial_n f(\dot{x}), -1)$. Then $(\theta, \dot{x}) = u(\dot{x}) < 0$. As the graph is convex we have $(\theta, x) < 0$ for any point $(x^1, \ldots, x^n) \in \Omega_d$ and $x = (x^1, \ldots, x^n, f(x^1, \ldots, x^n))$. We choose a positive number l such that $\{(x^1, \ldots, x^n) \in A^n | (x^1)^2 + \cdots + (x^n)^2 < l\} \subset \Omega_d$. Consider the circular cone

$$\mathcal{G} = \{t(x^1, \ldots, x^{n+1}) \mid t > 0,\ x^{n+1} = d,\ (x^1)^2 + \cdots + (x^n)^2 < l\}$$

and its polar cone

$$\mathcal{G}^* = \{y \in A^{n+1} \mid (y, x) \leq 0,\ \text{for all } x \in \mathcal{G}\}.$$

For any $x \in \mathcal{G}$ we have $(\theta, x) < 0$, which means that

$$\theta \in \{x^{n+1} = -1\} \cap \mathcal{G}^* = \left\{ x \in A^{n+1} | x^{n+1} = -1,\ (x^1)^2 + \cdots + (x^n)^2 \leq \frac{d}{l} \right\}.$$

Since $(\dot{x}^1, \ldots, \dot{x}^n)$ is an arbitrary point in Ω_d, we have in Ω_d

$$\sum (\partial_i f)^2 \leq \frac{d}{l}. \qquad (3.120)$$

If there are $p, q \in \bar{M}_d$, $p \neq q$, such that $j \circ x(p) = j \circ x(q)$. Denote by $\tilde{r}(p, q)$ the geodesic distance from p to q with respect to the induced Riemann metric g, we have $\tilde{r}(p, q) > 0$. We can choose sequences $\{p_m\}, \{q_m\}$ in M_d such that $\lim p_m = p$, $\lim q_m = q$ and $\tilde{r}(p_m, q_m) > \frac{1}{2}\tilde{r}(p, q) > 0$.

Let $\alpha_m = j \circ x(p_m)$, $\beta_m = j \circ x(q_m)$. Then we have $\alpha_m, \beta_m \in \Omega_d$ and $\lim \alpha_m = \lim \beta_m$. Hence $\alpha_m - \beta_m \to 0$. An application of (3.120) shows that $\lim \tilde{r}(p_m, q_m) = 0$ and we get a contradiction. □

As a consequence of Lemma 3.52, $x|_{\bar{M}_d}$ is an imbedding.

Lemma 3.53. *If, for some number $d > 1$, M_d has the above properties* (i) *and* (ii), *then there exists a positive number δ such that $M_{d+\delta}$ also has the properties* (i) *and* (ii).

Proof. For any positive number $\varepsilon > 0$ let

$$N(p, \varepsilon) = \{q \in M | \tilde{r}(p, q) < \varepsilon\},$$
$$N(\bar{M}_d, \varepsilon) = \bigcup_{p \in \bar{M}_d} N(p, \varepsilon).$$

First, we prove that there exists a small positive number $\tilde{\varepsilon}$ such that $j \circ x|_{N(\bar{M}_d, \tilde{\varepsilon})}$ is injective. If not, then for any natural number m there are two points p_m, q_m in $N(\bar{M}_d, \frac{1}{m})$, $p_m \neq q_m$, such that $j \circ x(p_m) = j \circ x(q_m)$. Then there are points $\tilde{p}_m, \tilde{q}_m \in \bar{M}_d$ such that $p_m \in N(\tilde{p}_m, \frac{1}{m})$ and $q_m \in N(\tilde{q}_m, \frac{1}{m})$. For $m \to \infty$ we get sequences $\{p_m\}, \{\tilde{p}_m\}, \{q_m\}, \{\tilde{q}_m\}$. It is easy to see from (i) and (ii) that $\bar{M}_d$ is compact. Then we can choose subsequences $\{\tilde{p}_{m_k}\}, \{\tilde{q}_{m_k}\}$ such that $\{\tilde{p}_{m_k}\}$ converges to a point $p \in \bar{M}_d$ and $\{\tilde{q}_{m_k}\}$ converges to a point $q \in \bar{M}_d$. Obviously, $\lim_{k \to \infty} p_{m_k} = p$, $\lim_{k \to \infty} q_{m_k} = q$. By continuity we have the relation $j \circ x(p) = j \circ x(q)$. Since $j \circ x|_{\bar{M}_d}$ is injective we have $p = q$. We already proved that the vector $b = (0, \dots, 0, 1)$ is transversal to $x(\bar{M}_d)$ at $x(p)$; then there is a positive number ε' such that $j \circ x|_{N(p, \varepsilon')}$ is injective. But for large $k \in \mathbb{N}$ we have $p_{m_k}, q_{m_k} \in N(p, \varepsilon')$ and $j \circ x(p_{m_k}) = j \circ x(q_{m_k})$. This is a contradiction. The conclusion is proved. Then $x(N(\bar{M}_d, \tilde{\varepsilon}))$ is the graph of a strongly convex function $x^{n+1} = f(x^1, \dots, x^n)$ defined on an open set $j \circ x(N(\bar{M}_d, \tilde{\varepsilon})) \supset \bar{\Omega}_d$. Since f is strongly convex, $\partial_i f(0, \dots, 0) = 0$, and $f = d$ on $\partial \Omega_d$, we can find a positive number δ such that $M_{d+\delta}$ has the properties (i) and (ii). □

To finally prove that the affine completeness of the hyperbolic affine hypersphere implies its Euclidean completeness, we proceed as follows. Define

$$S := \{h > 1 | M_h \text{ has the properties (i) and (ii)}\}.$$

One verifies that if $d \in S$ then $(1, d) \subset S$. Let $C := \sup S$. We show that $j \circ x|_{M_C}$ is injective. Assume that there are points $p, q \in M_C, p \neq q$, such that $j \circ x(p) = j \circ x(q)$. Then we can draw curves $\tau_1 : [0, 1] \to M_C$ and $\tau_2 : [0, 1] \to M_C$ such that $\tau_1(0) = p_0$, $\tau_1(1) = p$, $\tau_2(0) = p_0$, $\tau_2(1) = q$ and $x^{n+1}(\tau_1(t)) < C$, $x^{n+1}(\tau_2(t)) < C$ for any $t \in [0, 1]$. We can choose $h < C$ such that $x^{n+1}(\tau_1(t)) < h$, $x^{n+1}(\tau_2(t)) < h$ for any $t \in [0, 1]$, which means that $p, q \in M_h$. But M_h has the properties (i) and (ii). By definition of C, we get a contradiction. This proves that $x(M_C)$ is the graph of a strongly convex function $x^{n+1} = f(x^1, \dots, x^n)$ defined on a domain $\Omega_C = j \circ x(M_C)$. Since $\Omega_C = \bigcup_{h < C} \Omega_h$ and

$\Omega_{h_1} \subset \Omega_{h_2}$ for $h_1 < h_2$, Ω_C is a convex set. If $C = \infty$, it is clear that $x(M)$ is a graph and M is Euclidean complete. In the following we assume that $C < \infty$. In this case it is easy to verify that Ω_C is bounded and M_C has the property (i). By continuity $x(\bar{M}_C) \subset \overline{x(M_C)}$. We notice that $x(\bar{M}_C) \neq \overline{x(M_C)}$, otherwise we may enlarge C again by Lemma 3.53. Hence there exists a point $\bar{x} \in \overline{x(M_C)}$, but $\bar{x} \notin x(\bar{M}_C)$.

Now we derive a gradient estimate for the function $\langle U, b \rangle$ on M_C. Denote by $r(p_0, p)$ the geodesic distance function from p_0 with respect to the Blaschke metric. For any positive number a let $B_a(p_0) = \{p \in M \mid r(p_0, p) \leq a\}$. Consider the function

$$\varphi := (a^2 - r^2)(C - f)^{1/2} \frac{\|\mathrm{grad}\,\langle U, b \rangle\|_G}{\langle U, b \rangle}$$

defined on $B_a(p_0) \cap M_C$, where $\| \ \|_G$ denotes the norm with respect to the Blaschke metric, $f(p) := f(x^1(p), \ldots, x^n(p))$, $\langle U, b \rangle(p) := \langle U(x(p)), b \rangle$ for $p \in M_C$. Obviously, φ attains its supremum at some interior point p^*. As in Section 3.6, we may assume that r^2 is a C^2-function in a neighborhood of p^*, and $\|\mathrm{grad}\,\langle U, b \rangle\|_G > 0$ at p^*. Then at p^*, grad $\varphi = 0$. Choose a local orthonormal frame field $\{p, e_1, \ldots, e_n\}$ on M_C with respect to the Blaschke metric such that, at p^*,

$$\langle U, b \rangle_{,1} = \|\mathrm{grad}\,\langle U, b \rangle\|_G > 0,$$
$$\langle U, b \rangle_{,i} = 0, \quad \text{for } 2 \leq i \leq n.$$

Then by differentiation we obtain at p^*

$$\frac{1}{2}(a^2 - r^2)(C - f)^{-1/2} \frac{f_{,1}\langle U, b \rangle_{,1}}{\langle U, b \rangle} + (a^2 - r^2)(C - f)^{1/2} \left(\frac{\langle U, b \rangle_{,1}}{\langle U, b \rangle} \right)^2$$
$$= (a^2 - r^2)(C - f)^{1/2} \frac{\langle U, b \rangle_{,11}}{\langle U, b \rangle} - (r^2)_{,1}(C - f)^{1/2} \frac{\langle U, b \rangle_{,1}}{\langle U, b \rangle}.$$

Denote the inverse matrix of $(f_{ij}) = (\partial_i \partial_j f)$ by (f^{ij}). Since

$$\frac{-f_{,1}\langle U, b \rangle_{,1}}{\langle U, b \rangle} = -\frac{\sum f_{,i}\langle U, b \rangle_{,i}}{\langle U, b \rangle} = \sum \frac{f_{,i} u_{,i}}{u}$$
$$= \frac{1}{u} \sum [\det(f_{kl})]^{1/(n+2)} f^{ij} \partial_i f \partial_j u$$
$$= \frac{1}{u} [\det(f_{kl})]^{1/(n+2)} \sum x^i \partial_i f$$
$$= [\det(f_{kl})]^{1/(n+2)} \cdot \frac{u + f}{u} \leq 0$$

and

$$|(r^2)_{,1}| \leq 2a, \quad \langle U, b \rangle_{,11} = -A_{111}\langle U, b \rangle_{,1} - L_1\langle U, b \rangle,$$
$$|A_{111}| \leq \sqrt{n(n-1)J} \leq \sqrt{-n(n-1)L_1},$$

we have

$$(a^2 - r^2)(C - f)^{1/2} \left(\frac{\|\text{grad } \langle U, b \rangle\|_G}{\langle U, b \rangle} \right)^2$$

$$\leq (2a + a^2 \sqrt{-n(n-1)L_1})C^{1/2} \frac{\|\text{grad } \langle U, b \rangle\|_G}{\langle U, b \rangle} + a^2 C^{1/2}|L_1|.$$

Multiplying both sides by $(a^2 - r^2)(C - f)^{1/2}$ we get

$$\varphi^2 \leq (2a + a^2 \sqrt{-n(n-1)L_1}) \, C^{1/2} \, \varphi + a^4 C|L_1|.$$

It follows that

$$\varphi \leq (2a + a^2 \sqrt{-n(n-1)L_1})C^{1/2} + a^2 C^{1/2} \sqrt{-L_1}$$

holds at p^*, where φ attains its supremum. Hence, at any interior point of $M_C \cap B_a(p)$, we have

$$\frac{\|\text{grad } \langle U, b \rangle\|_G}{\langle U, b \rangle} \leq \frac{1}{a^2 - r^2}(C - f)^{-1/2}(2a + a^2 \sqrt{-n(n-1)L_1} + a^2 \sqrt{-L_1})C^{1/2}.$$

Let $a \to \infty$, then

$$\frac{\|\text{grad } \langle U, b \rangle\|_G}{\langle U, b \rangle} \leq m(C - f)^{-1/2} \tag{3.121}$$

where

$$m := (\sqrt{-n(n-1)L_1} + \sqrt{-L_1})C^{1/2}.$$

Now we use the gradient estimate (3.121) to get a contradiction in the case $C < \infty$. Without loss of generality we can assume that $\bar{x}$, defined above, has the coordinates $(0, \ldots, 0, \bar{x}^n, \bar{x}^{n+1})$, where $\bar{x}^n > 0$. Draw a line segment joining $(0, \ldots, 0)$ and $(0, \ldots, 0, \bar{x}^n)$ on A^n :

$$x^i(t) = 0, \quad 1 \leq i \leq n - 1,$$

$$x^n(t) = \bar{x}^n \cdot t, \quad 0 \leq t < 1.$$

Since M is affine complete, the affine arc length of the curve

$$\left\{ (x(t), f(x(t))) \,\Big|\, \frac{1}{2} \leq t \leq T < 1 \right\},$$

where $x(t) := (x^1(t), \ldots, x^n(t))$, must tend to infinity when $T \to 1$. Now we compute the affine arc length of this curve. First, we have

$$\frac{\|\text{grad } u\|_G^2}{(L_1 u)^2} = \frac{1}{(L_1 u)^2}[\det(f_{kl})]^{1/(n+2)} \sum f^{ij}\partial_i u \, \partial_j u$$

$$= \frac{1}{(L_1 u)^2}[\det(f_{kl})]^{1/(n+2)} \sum f_{ij}x^i x^j$$

$$= [\det(f_{kl})]^{-1/(n+2)} \sum f_{ij}x^i x^j.$$

The affine arc length of the curve $\{(x(t), f(x(t))) \mid \frac{1}{2} \le t < 1\}$ is

$$l = \int_{1/2}^{1} \left\{ [\det(f_{kl})]^{-1/(n+2)} \sum f_{ij} \bar{x}^i \bar{x}^j \right\}^{1/2} dt$$

$$= \int_{1/2}^{1} \frac{1}{t} \left\{ [\det(f_{kl})]^{-1/(n+2)} \sum f_{ij} x^i x^j \right\}^{1/2} dt$$

$$= \int_{1/2}^{1} \frac{\|\operatorname{grad} u\|_G}{L_1 u t} \, dt$$

$$\le \frac{2}{|L_1|} \int_{1/2}^{1} \frac{\|\operatorname{grad} u\|_G}{|u|} \, dt.$$

From (3.121) and $\langle U, b \rangle = \frac{1}{L_1 u}$ we get

$$\frac{\|\operatorname{grad} u\|_G}{|u|} \le m(C - f)^{-1/2}.$$

Since

$$\frac{df}{dt} = \frac{\partial f}{\partial x^n}(0, \ldots, 0, \bar{x}^n t) \bar{x}^n$$

$$\ge \frac{\partial f}{\partial x^n}\left(0, \ldots, 0, \frac{1}{2}\bar{x}^n\right) \bar{x}^n$$

we have

$$l \le \frac{2m}{|L_1| \bar{x}^n \cdot \frac{\partial f}{\partial x^n}\left(0, \ldots, 0, \frac{1}{2}\bar{x}^n\right)} \int_{f(0,\ldots,0,\frac{1}{2}\bar{x}^n)}^{C} \frac{df}{(C - f)^{1/2}}$$

$$= \frac{4m}{|L_1| \bar{x}^n \cdot \frac{\partial f}{\partial x^n}\left(0, \ldots, 0, \frac{1}{2}\bar{x}^n\right)} \left[C - f\left(0, \ldots, 0, \frac{1}{2}\bar{x}^n\right) \right]^{1/2}$$

$$< \infty,$$

which contradicts the affine completeness of M. Thus we can state the following theorem due to A.-M. Li (cf. [170]).

Theorem 3.54. *Every locally strongly convex, affine complete, hyperbolic affine hypersphere is Euclidean complete.*

Remark 3.55. (i) From Theorems 3.54, 3.45, and Corollary 3.36 one concludes that every affine complete affine hypersphere is Euclidean complete. This, together with Theorem 3.27, shows that for affine hyperspheres the Euclidean completeness and the affine completeness are equivalent.

(ii) Theorem 3.54 implies that an affine complete hyperbolic affine hypersphere can be globally represented as a graph of a positive convex function:

$$x^{n+1} = f(x^1, \ldots, x^n).$$

On the other hand, by Corollary 3.44 we have $J \leq -L_1$. Then, similar to the proof of (3.96), we can derive the gradient estimate of $\ln(1 + f)$ with respect to the Blaschke metric, by which we can prove that the hyperbolic affine hypersphere is asymptotic to the boundary of a convex cone with vertex at the center.

Theorem 3.54 and Theorem 3.51 together give a complete solution of the first part of the Calabi conjecture:

Theorem 3.56. *Every locally strongly convex, affine complete, hyperbolic affine hypersphere is asymptotic to the boundary of a convex cone with vertex at the center.*

3.8.3 Proof of the second part of the Calabi conjecture

In this subsection we shall prove the second part of the Calabi conjecture, mainly following T. Sasaki (cf. [295]).

Theorem 3.57. *Given a nondegenerate convex cone C in A^{n+1} and a negative constant L_1, there is a unique, Euclidean complete hyperbolic affine hypersphere M with affine mean curvature L_1, asymptotic to the boundary of C.*

Proof. The proof is divided into five steps.

Step 1. Without loss of generality, we can assume that the vertex of C is at the origin O. Denote by C^* the polar cone of C. Choose a unimodular affine coordinate system $(\xi_1, \xi_2, \ldots, \xi_{n+1})$ in $\mathbb{R}^{n+1}$ and denote by $(x^1, x^2, \ldots, x^{n+1})$ the dual coordinates in A^{n+1}. The intersection of the interior C^{*0} of C^* and the hyperplane $\xi_{n+1} = -1$ is a bounded convex open set:

$$D := C^{*0} \cap \{\xi_{n+1} = -1\}.$$

For simplicity, we write $(\xi_1, \xi_2, \ldots, \xi_n)$ as coordinates of a point in D.

From Theorem 6 in [56, p. 67] there exists a unique continuous convex function u on $\bar{D}$ satisfying

$$\begin{cases} \det\left(\dfrac{\partial^2 u}{\partial \xi_i \partial \xi_j}\right) = (L_1 u)^{-n-2} & \text{in } D, \\ u = 0 & \text{on } \partial D. \end{cases} \tag{3.122}$$

Using the function u, we define a set Ω in A^n by

$$\Omega := \left\{ (x^1, \ldots, x^n) \big| \; x^i = \frac{\partial u}{\partial \xi_i}(\xi), \; \xi \in D \right\}$$

and $f = f(x^1, \ldots, x^n)$ on Ω as the Legendre transformation function:

$$f(x^1, \ldots, x^n) = \sum \xi_i \frac{\partial u}{\partial \xi_i}(\xi) - u(\xi).$$

It is easy to see from the convexity of u that f is also convex. Consider the graph of the convex function f

$$M := \left\{ (x^1, \ldots, x^n, f(x^1, \ldots, x^n)) \mid (x^1, \ldots, x^n) \in \Omega \right\}.$$

Since the Legendre transformation function u of f satisfies the differential equation $\det \left(\frac{\partial^2 u}{\partial \xi_i \partial \xi_j} \right) = (L_1 u)^{-n-2}$, it follows from Theorem 3.5 that M is a hyperbolic affine hypersphere with center at the origin and with affine mean curvature L_1.

Step 2. The hypersphere M constructed above depends on the bounded convex domain D, while D depends on the choice of the coordinate system in $\mathbb{R}^{n+1}$. Now assume that $(\eta_1, \eta_2, \ldots, \eta_{n+1})$ is another unimodular affine coordinate system in $\mathbb{R}^{n+1}$ such that $\hat{D} := C^{*0} \cap \{\eta_{n+1} = -1\}$ is also a bounded convex domain. We denote by $(y^1, \ldots, y^{n+1})$ the dual coordinate system of $(\eta_1, \eta_2, \ldots, \eta_{n+1})$. As in Step 1, we can construct a hyperbolic affine hypersphere $\hat{M}$ with center at the origin and with affine mean curvature L_1. We are going to investigate the relation between $\hat{M}$ and M. The new coordinates $(\eta_1, \eta_2, \ldots, \eta_{n+1})$ can be expressed in terms of the old ones $(\xi_1, \xi_2, \ldots, \xi_{n+1})$ as follows:

$$\begin{pmatrix} \eta_1 \\ \vdots \\ \eta_{n+1} \end{pmatrix} = \begin{pmatrix} t_1^1 & \cdots & t_1^{n+1} \\ \vdots & & \vdots \\ t_{n+1}^1 & \cdots & t_{n+1}^{n+1} \end{pmatrix} \begin{pmatrix} \xi_1 \\ \vdots \\ \xi_{n+1} \end{pmatrix}.$$

Then the dual coordinates satisfy

$$\begin{pmatrix} y^1 \\ \vdots \\ y^{n+1} \end{pmatrix} = \begin{pmatrix} T_1^1 & \cdots & T_{n+1}^1 \\ \vdots & & \vdots \\ T_1^{n+1} & \cdots & T_{n+1}^{n+1} \end{pmatrix} \begin{pmatrix} x^1 \\ \vdots \\ x^{n+1} \end{pmatrix},$$

where (T_B^A) is the transpose of the inverse matrix of (t_B^A). Clearly, (t_B^A) and (T_B^A) are unimodular.

We need the following lemma.

Lemma 3.58. (i) *The coordinates $(\xi_1, \xi_2, \ldots, \xi_n, -1)$ of the points in D and the coordinates $(\eta_1, \eta_2, \ldots, \eta_n, -1)$ of the points in $\hat{D}$ satisfy the relation*

$$\xi_i = \left(\sum T_i^j \eta_j - T_i^{n+1} \right) \left(T_{n+1}^{n+1} - \sum T_{n+1}^j \eta_j \right)^{-1}.$$

(ii) *Let $u(\xi)$ be the convex solution of the equation (3.122) and $v(\eta)$ be the convex solution of the corresponding equation*

$$\begin{cases} \det \left(\dfrac{\partial^2 v}{\partial \eta_i \partial \eta_j} \right) = (L_1 v)^{-n-2} & \text{in } \hat{D}, \\ v = 0 & \text{on } \partial \hat{D}. \end{cases}$$

Then

$$u(\xi) = v(\eta)\left(T_{n+1}^{n+1} - \sum T_{n+1}^j \eta_j\right)^{-1}.$$

Proof. The proof of (i) is straightforward.

To prove (ii), we can use (i) to verify that

$$\frac{\partial}{\partial \eta_i}\left[u\left(T_{n+1}^{n+1} - \sum T_{n+1}^j \eta_j\right)\right] = -T_{n+1}^i u + \sum \frac{\partial u}{\partial \xi_j}(T_j^i + \xi_j T_{n+1}^i),$$

$$\frac{\partial^2}{\partial \eta_i \partial \eta_j}\left[u\left(T_{n+1}^{n+1} - \sum T_{n+1}^l \eta_l\right)\right]$$

$$= \left(T_{n+1}^{n+1} - \sum T_{n+1}^l \eta_l\right)^{-1} \sum \frac{\partial^2 u}{\partial \xi_k \partial \xi_l}(T_k^i + \xi_k T_{n+1}^i)(T_l^j + \xi_l T_{n+1}^j).$$

Analogously, (i) gives

$$\det(T_k^i + \xi_k T_{n+1}^i) = \begin{vmatrix} T_1^1 & \cdots & T_1^n & \xi_1 \\ \vdots & & \vdots & \vdots \\ T_n^1 & \cdots & T_n^n & \xi_n \\ -T_{n+1}^1 & \cdots & -T_{n+1}^n & 1 \end{vmatrix} = \left(T_{n+1}^{n+1} - \sum T_{n+1}^j \eta_j\right)^{-1}.$$

Then

$$\det\left(\frac{\partial^2}{\partial \eta_i \partial \eta_j}\left[u\left(T_{n+1}^{n+1} - \sum T_{n+1}^l \eta_l\right)\right]\right) = \left[L_1 u\left(T_{n+1}^{n+1} - \sum T_{n+1}^l \eta_l\right)\right]^{-n-2}.$$

From the convexity of u and that $u|_{\partial D} = 0$, we see that $u(T_{n+1}^{n+1} - \sum T_{n+1}^l \eta_l)$ is a convex solution of equation (3.122).

Then (ii) follows from the uniqueness of the solution of (3.122). □

Now, we continue the proof of Theorem 3.57.

We denote the Legendre transformation function of v by

$$g(y^1, y^2, \ldots, y^n) = \sum \eta_k \frac{\partial v}{\partial \eta_k} - v(\eta),$$

where

$$y^k = \frac{\partial v}{\partial \eta_k}(\eta), \quad \eta \in \hat{D}.$$

The graph

$$\hat{M} = \left\{(y^1, y^2, \ldots, y^n, g(y^1, \ldots, y^n)) \,\Big|\, y^k = \frac{\partial v}{\partial \eta_k}(\eta), \eta \in \hat{D}\right\}$$

of g is clearly a hyperbolic affine hypersphere with its center at the origin and affine mean curvature L_1. From Lemma 3.58 the coordinates of the points in $\hat{M}$ satisfy

$$y^i = \frac{\partial v}{\partial \eta_i} = \sum \frac{\partial u}{\partial \xi_k} \frac{\partial \xi^k}{\partial \eta_i} \left(T^{n+1}_{n+1} - \sum T^l_{n+1}\eta_l\right) - T^i_{n+1}u$$

$$= \sum T^i_k x^k + T^i_{n+1}f, \quad 1 \le i \le n,$$

$$y^{n+1} = g = \sum y^i \eta_i - v(\eta)$$

$$= \sum (T^i_k x^k + T^i_{n+1}f)\eta_i - u\left(T^{n+1}_{n+1} - \sum T^l_{n+1}\eta_l\right)$$

$$= \sum T^{n+1}_k x^k + T^{n+1}_{n+1}f.$$

This shows that $\hat{M}$ and M are equiaffinely equivalent.

Step 3. In this step we prove that M is Euclidean complete. We need the following lemma (cf. [57]):

Lemma 3.59. *Let W be a bounded domain in $\mathbb{R}^n$. Let φ be a function defined on $W \times \mathbb{R}$ such that $\varphi(x, u)$ is increasing in the variable u for $u \le 0$. Let u and v be two strongly convex C^2-solutions of the equation $\det(u_{ij}) = \varphi(x, u)$ such that $\lim_{x \to \partial W} (u(x) - v(x)) \le 0$ uniformly. Then $u \le v$ on W.*

Proof. Let $a^{ij}(t)$ be the cofactor components of the matrix $(v_{ij} + t(u_{ij} - v_{ij}))$. Then we have the following equations:

$$(u - v)\left(\frac{\varphi(x, u) - \varphi(x, v)}{u - v}\right) = \det(u_{ij}) - \det(v_{ij})$$

$$= \int_0^1 \frac{d}{dt} \det(v_{ij} + t(u_{ij} - v_{ij})) dt$$

$$= \sum \left(\int_0^1 a^{ij}(t) dt\right)(u - v)_{ij}.$$

By the strict convexity of u and v, one knows that

$$\sum \left(\int_0^1 a^{ij}(t) dt\right) \frac{\partial^2}{\partial x^i \partial x^j}$$

is an elliptic operator. The conclusion of the lemma then follows from the maximum principle. □

We continue to prove the third step and show that M is Euclidean complete. It is sufficient to prove that $\Omega = A^n$. We consider the function

$$\Psi(x) := \sup_{\xi \in \bar{D}} \left\{-u(\xi) + \sum x^i \xi_i\right\}, \quad x \in A^n.$$

Clearly, $\Psi(x) = f(x)$ for $x \in \Omega$. From the definition of $\Psi(x)$, one can see that for any $\bar{x} \in A^n$, there is a point $\bar{\xi} \in \bar{D}$ such that

$$\Psi(\bar{x}) = -u(\bar{\xi}) + \sum \bar{x}^i \bar{\xi}_i.$$

If $\bar{\xi}$ is in the interior of D, we consider the function

$$\Phi(\xi) = -u(\xi) + \sum \bar{x}^i \xi_i$$

defined on a neighborhood of $\bar{\xi}$. Obviously, $\Phi(\xi)$ attains its maximum at $\bar{\xi}$. Hence we have

$$\frac{\partial \Phi}{\partial \xi_i}(\bar{\xi}) = -\frac{\partial u}{\partial \xi_i}(\bar{\xi}) + \bar{x}^i = 0.$$

This implies that $\bar{x}^i = \frac{\partial u}{\partial \xi_i}(\bar{\xi})$ and $\bar{x} \in \Omega$. We assert that, for any $x \in A^n$, there exists an interior point ξ of D such that $\Psi(x) = -u(\xi) + \sum x^i \xi_i$. If not, we assume that there exists a point $\bar{x}$ such that $\Psi(\bar{x}) = -u(\bar{\xi}) + \sum \bar{x}^i \bar{\xi}_i$ for some $\bar{\xi} \in \partial D$ and $\Psi(\bar{x}) > -u(\xi) + \sum \bar{x}^i \xi_i$ for every $\xi \in D$. Since D is convex, we can find a closed simplex W contained in $\bar{D}$ such that $\partial W \cap \partial D = \{\bar{\xi}\}$. Let u_W be the convex solution of the boundary value problem

$$\begin{cases} \det\left(\dfrac{\partial^2 u_W}{\partial \xi_i \partial \xi_j}\right) = (L_1 u_W)^{-n-2} \text{ in } W, \\ u_W = 0 \text{ on } \partial W. \end{cases}$$

Obviously, the function u satisfies the equation $\det\left(\frac{\partial^2 u}{\partial \xi_i \partial \xi_j}\right) = (L_1 u)^{-n-2}$ in the interior of W, and $u|_{\partial W} \leq 0$. It follows from Lemma 3.59 that $u_W \geq u$ on W. Hence it follows from $u(\bar{\xi}) = u_W(\bar{\xi})$ that

$$-u_W(\bar{\xi}) + \sum \bar{x}^i \bar{\xi}_i = -u(\bar{\xi}) + \sum \bar{x}^i \bar{\xi}_i$$
$$= \Psi(\bar{x}) > -u(\xi) + \sum \bar{x}^i \xi_i$$
$$\geq -u_W(\xi) + \sum \bar{x}^i \xi_i$$

for every interior point ξ of D. This means that the function

$$\Phi_W(\xi) = -u_W(\xi) + \sum \bar{x}^i \xi_i$$

attains its maximum at a boundary point of W. But this is a contradiction, because we shall prove in Step 4 below that the Legendre transformation domain Ω_W of the interior of W is the entire space A^n. From this we can conclude that $\Omega = A^n$.

Step 4. The proof of $\Omega_W = A^n$. In Section 3.2.1 we have seen that the hypersurface, denoted by $Q(c, n)$, defined by

$$x^1 x^2 \cdots x^{n+1} = c, \quad x^i > 0, \quad 1 \leq i \leq n+1, \quad c = \text{const.}$$

is a hyperbolic affine hypersphere with $-(n+1)^{-(n+1)/(n+2)} c^{-2/(n+2)}$ as its affine mean curvature. By a suitable choice of the constant c, one can assume that the affine mean curvature of $Q(c, n)$ is L_1. Clearly, $Q(c, n)$ is asymptotic to the boundary of the convex cone

$$C_Q := \left\{ x \in A^{n+1} \mid x^1 \geq 0, x^2 \geq 0, \ldots, x^{n+1} \geq 0 \right\}.$$

The polar cone of C_Q is (for simplicity, we identify A^{n+1} with $(A^{n+1})^*$ through the Euclidean inner product):

$$C_Q^* = \left\{ x \in A^{n+1} \mid x^1 \leq 0, x^2 \leq 0, \ldots, x^{n+1} \leq 0 \right\}.$$

Taking a rotation of the coordinate system, we can assume that the y^{n+1}-axis of the new coordinate system is parallel to the vector $(1, 1, \ldots, 1)$, and regard $Q(c, n)$ as the graph of some positive convex function $q(y^1, y^2, \ldots, y^n)$ defined on the entire hyperplane $y^{n+1} = 0$. From Lemma 3.49 the domain of the Legendre transformation relative to the function $q(y^1, \ldots, y^n)$ in the hyperplane $y^{n+1} = 0$ is $C_Q^{*0} \cap \{y^{n+1} = -1\}$. It is a standard n-simplex. Make successive coordinate transformations

$$\begin{cases} \bar{y}^i = \sum a_j^i y^j, \\ \bar{y}^{n+1} = \dfrac{1}{a} y^{n+1}, \quad a = \det(a_j^i) \neq 0, \end{cases}$$

and

$$\begin{cases} \tilde{y}^i = \bar{y}^i, \\ \tilde{y}^{n+1} = \bar{y}^{n+1} + \sum b_j \bar{y}^j. \end{cases}$$

Assume that relative to the new coordinates $(\tilde{y}^1, \tilde{y}^2, \ldots, \tilde{y}^{n+1})$, $Q(c, n)$ is represented by $\tilde{y}^{n+1} = \tilde{q}(\tilde{y}^1, \ldots, \tilde{y}^n)$. Choosing (a_j^i) and b_i suitably, we can assume that the domain of the Legendre transformation relative to $\tilde{y}^{n+1} = \tilde{q}(\tilde{y}^1, \ldots, \tilde{y}^n)$ in the hyperplane $\tilde{y}^{n+1} = 0$ is W. Since $\tilde{q}(\tilde{y}^1, \ldots, \tilde{y}^n)$ is also the Legendre transformation function of u_W, we conclude that $\Omega_W = A^n$.

Step 5. Step 3 and Theorem 3.51 imply that M is a Euclidean complete hyperbolic affine hypersphere that is asymptotic to the boundary of the convex cone

$$\tilde{C} = Cl \left\{ t(x^1, \ldots, x^n, f(x^1, \ldots, x^n)) \mid t > 0, \ (x^1, \ldots, x^n) \in A^n \right\}.$$

The polar cone of $\tilde{C}$ is

$$\tilde{C}^* = \left\{ \xi \in (A^{n+1})^* \mid (\xi, p) \leq 0, \text{ for all } p \in \tilde{C} \right\}.$$

Put

$$\tilde{D} := \tilde{C}^{*0} \cap \{\xi_{n+1} = -1\}.$$

It follows from the proof of Proposition 3.50 that $\tilde{D}$ is the Legendre transformation domain of A^n relative to the function f. Hence $\tilde{D} = D$ and therefore $\tilde{C}$ coincides with the closure of $C : \tilde{C} = Cl(C)$. This shows that M is asymptotic to the boundary of C.

Now we prove the uniqueness. By a unimodular affine transformation we may assume that $(0, \ldots, 0, -1) \in C^{*0}$. Let M' be another Euclidean complete hyperbolic affine hypersphere asymptotic to the boundary of C and with affine mean curvature L_1. Let p be the center of M'. From Theorem 3.51, M' is asymptotic to the boundary of the convex cone

$$C' = Cl\left\{t(x-p) \mid t > 0, x = (x^1, \ldots, x^{n+1}) \in M'\right\}.$$

Hence $C' = Cl(C)$, and the center of M' is at the origin. Since $(0, \ldots, 0, -1) \in C^{*0}$, by Lemma 3.49 the hyperplane $x^{n+1} = 0$ is parallel to the tangent hyperplane of M' at some point $x_0 \in M'$. Since M' is Euclidean complete, it is the graph of a strongly convex function defined on a convex domain of the hyperplane $x^{n+1} = 0$. Without loss of generality we can assume that, with respect to the coordinate system $(x^1, \ldots, x^{n+1})$, M' can be represented by a strongly convex function

$$x^{n+1} = f'(x^1, \ldots, x^n)$$

defined on a convex domain $\Omega' \subset A^n = \{x^{n+1} = 0\}$. The Legendre transformation domain of Ω' is

$$D' = C^{*0} \cap \{\xi_{n+1} = -1\} = D.$$

By Theorem 3.51 the Legendre transformation function u' satisfies $u'|_{\partial D} = 0$. Hence u' and u satisfy the same equation (3.122), therefore $u = u'$. It follows that $f' = f$, hence M' and M coincide.

This finally finishes the proof of Theorem 3.57. □

Then Sections 3.8.2 and 3.8.3 together give a proof of the Calabi conjecture on hyperbolic affine hyperspheres.

3.9 Complete hyperbolic affine 2-spheres

C. P. Wang [395] found a method to construct all examples of complete hyperbolic affine 2-spheres which admit the action of a discrete subgroup of the equiaffine group with compact quotient. In this section we shall describe his method.

First of all let us express the structure equations of an affine surface in terms of holomorphic parameters (see [44, 395]). Let $x : M \to A^3$ be a locally strongly convex surface. The affine structure of A^3 induces on M an orientation and a conformal structure, induced by suitably oriented relative normalizations (see Section 1.10 and Remark 2.1). In particular, the conformal class contains the (positive definite) second fundamental form of x

$$II = \sum h_{ij}\omega^i\omega^j \tag{3.123}$$

induced from an appropriate Euclidean structure additionally introduced on A^3.

M can be naturally regarded as a Riemann surface. Choose isothermal parameters u, v with respect to (3.123) and let $z = u + \sqrt{-1}v$. The Blaschke metric can be written as

$$G = 2F(z, \bar{z})|dz|^2, \tag{3.124}$$

where $dz = du + \sqrt{-1}dv$. Assume that the components of the cubic Fubini–Pick form and the affine Weingarten form (see Lemma 2.5) with respect to the local coordinate system (u, v) are A_{ijk} and B_{ij} respectively. Define α and β by

$$\alpha := \frac{1}{2}(A_{111} + \sqrt{-1}A_{222}), \tag{3.125}$$

$$\beta := \frac{1}{2}\left(\frac{B_{11} - B_{22}}{2} - \sqrt{-1}B_{12}\right). \tag{3.126}$$

Then the Fubini–Pick form and the Weingarten form can be expressed by

$$A = \alpha(dz)^3 + \bar{\alpha}(d\bar{z})^3, \tag{3.127}$$

$$\hat{B} = \beta(dz)^2 + 2L_1 F dz d\bar{z} + \bar{\beta}(d\bar{z})^2. \tag{3.128}$$

We use the Cauchy–Riemann operators

$$\frac{\partial}{\partial z} = \frac{1}{2}\left(\frac{\partial}{\partial u} - \sqrt{-1}\frac{\partial}{\partial v}\right),$$

$$\frac{\partial}{\partial \bar{z}} = \frac{1}{2}\left(\frac{\partial}{\partial u} + \sqrt{-1}\frac{\partial}{\partial v}\right).$$

From the theory of complex manifolds, every complex tensor bundle on M is reduced to a direct sum of bigraded complex line bundles $E_{r,s}$, where r and s are integers. Locally, $E_{r,s}$ is generated by $dz^r \otimes d\bar{z}^s$; here dz^r and $d\bar{z}^s$, for r or s negative, denote the contravariant tensors $\left(\frac{\partial}{\partial z}\right)^{-r}$ or $\left(\frac{\partial}{\partial \bar{z}}\right)^{-s}$, respectively. The tensor products are regarded as commutative unless otherwise specified. Thus the metric coefficient F is the fibre coordinate of a cross section in $E_{1,1}$. The tangent bundle of M tensored with $\mathbb{C}$ splits into the direct sum of $E_{-1,0}$ and $E_{0,-1}$, generated locally by $\frac{\partial}{\partial z}$ and $\frac{\partial}{\partial \bar{z}}$, respectively.

3.9.1 A splitting of the Levi–Civita operator

The Levi–Civita operator ∇ of covariant derivation on smooth sections in $E_{r,s}$ splits into

$$\nabla = \nabla' + \nabla'',$$

where ∇' is of bidegree $(1, 0)$ and ∇'' of bidegree $(0, 1)$. They satisfy the following:
(i) ∇, ∇' and ∇'' are all linear derivation operators, i.e. for any complex number c and arbitrary smooth sections f, f_1, f_2 in $E_{r,s}$ and g in $E_{p,q}$, each of them, say ∇, satisfies
   (a) $\nabla(cg) = c\nabla g$;
   (b) $\nabla(f_1 + f_2) = \nabla f_1 + \nabla f_2$;
   (c) $\nabla(f \otimes g) = (\nabla f) \otimes g + f \otimes \nabla g$;
(ii) if $f = f(z, \bar{z})$ and $g = g(z, \bar{z})$ are the local coefficients of smooth sections in $E_{r,0}$ and in $E_{0,s}$, respectively, then the local coefficients of $\nabla''f$ in $E_{r,1}$ and of $\nabla'g$ in $E_{1,s}$ are $\frac{\partial f(z,\bar{z})}{\partial \bar{z}}$ and $\frac{\partial g(z,\bar{z})}{\partial z}$;
(iii) for scalar f, $\nabla'f = \frac{\partial f}{\partial z}$, $\nabla''f = \frac{\partial f}{\partial \bar{z}}$;
(iv) for the metric coefficient F we have $\nabla'F = \nabla''F = 0$.

From the properties above one can derive the formulas for the general case, where f is a smooth section in $E_{r,s}$:

$$\nabla' f = F^r \frac{\partial}{\partial z}(F^{-r}f) = \frac{\partial f}{\partial z} - rf \frac{\partial \ln F}{\partial z}, \qquad (3.129)$$

$$\nabla'' f = F^s \frac{\partial}{\partial \bar{z}}(F^{-s}f) = \frac{\partial f}{\partial \bar{z}} - sf \frac{\partial \ln F}{\partial \bar{z}}. \qquad (3.130)$$

Computing the second covariant derivatives, one obtains the Ricci identity

$$[\nabla', \nabla''] f = \nabla' \nabla'' f - \nabla'' \nabla' f = (s - r)F\chi f, \qquad (3.131)$$

where

$$\chi = -F^{-1} \frac{\partial^2 \ln F}{\partial z \partial \bar{z}} \qquad (3.132)$$

is the intrinsic Gauß curvature of the Blaschke metric. The Laplace operator of the Blaschke metric is

$$\Delta = \frac{2}{F} \frac{\partial^2}{\partial z \partial \bar{z}}. \qquad (3.133)$$

Since u and v are isothermal parameters relative to (3.123), we have

$$\begin{aligned} \mathrm{Det}(x_z, x_{\bar{z}}, x_{zz}) &= \mathrm{Det}(x_z, x_{\bar{z}}, x_{\bar{z}\bar{z}}) = 0, \\ \mathrm{Det}(x_z, x_{\bar{z}}, x_{z\bar{z}}) &= \sqrt{-1}\, F^2, \end{aligned} \qquad (3.134)$$

where $x_z = \frac{\partial x}{\partial z}$, $x_{\bar{z}} = \frac{\partial x}{\partial \bar{z}}$, etc. The affine normal vector field Y and the conormal vector field U of $x(M)$ satisfy the relations

$$Y = F^{-1} x_{z\bar{z}}, \qquad (3.135)$$

$$U = -\sqrt{-1}\, F^{-1}[x_z, x_{\bar{z}}], \qquad (3.136)$$

where the brackets denote the complex "cross" product on the complexification $TA^3 \otimes \mathbb{C}$ of TA^3 (see also Section 1.1.4). $\{x_z, x_{\bar{z}}, Y\}$ is a complex frame field on $x(M)$, and $\{U, U_z, U_{\bar{z}}\}$ is a complex frame field on the immersed surface $U : M \to (A^3)^*$. These two frames satisfy the following relations:

$$\mathrm{Det}(x_z, x_{\bar{z}}, Y) = \mathrm{Det}(U, U_z, U_{\bar{z}}) = \sqrt{-1}F, \qquad (3.137)$$

$$\begin{pmatrix} U \\ U_z \\ U_{\bar{z}} \end{pmatrix} \cdot (x_z, x_{\bar{z}}, Y) = \begin{pmatrix} 0 & 0 & 1 \\ 0 & -F & 0 \\ -F & 0 & 0 \end{pmatrix}, \qquad (3.138)$$

where "$\cdot$": $((A^3)^* \otimes \mathbb{C}) \times (A^3 \otimes \mathbb{C}) \to \mathbb{C}$ is the complex inner product. The structure equations

$$x_{,ij} = \sum A_{ij}^k x_k + G_{ij} Y,$$

$$Y_i = -\sum B_i^k x_k,$$

$$U_{,ij} = -\sum A_{ij}^k U_k - B_{ij} U$$

can be written in the form

$$\begin{cases} \nabla'(x_z, x_{\bar{z}}, Y) = (F^{-1}\alpha x_{\bar{z}}, FY, -L_1 x_z - F^{-1}\beta x_{\bar{z}}), \\ \nabla''(x_z, x_{\bar{z}}, Y) = (FY, F^{-1}\bar{\alpha} x_z, -L_1 x_{\bar{z}} - F^{-1}\bar{\beta} x_z), \end{cases} \tag{3.139}$$

$$\begin{cases} \nabla'(U, U_z, U_{\bar{z}}) = (U_\xi, -\beta U - F^{-1}\alpha U_{\bar{z}}, -L_1 FU), \\ \nabla''(U, U_z, U_{\bar{z}}) = (U_{\bar{z}}, -L_1 FU, -\bar{\beta} U - F^{-1}\bar{\alpha} U_z). \end{cases} \tag{3.140}$$

From (3.139) and (3.140) we obtain

$$\alpha = -\sqrt{-1}\,\mathrm{Det}(Y, x_z, x_{zz}) = \sqrt{-1}\,\mathrm{Det}(U, U_z, U_{zz}), \tag{3.141}$$

$$\beta = \sqrt{-1}\,\mathrm{Det}(Y, x_z, Y_z) = \sqrt{-1}\,F^{-1}\mathrm{Det}(U_z, U_{\bar{z}}, U_{zz}). \tag{3.142}$$

Define locally the forms

$$\alpha^{\#} := \alpha\,(dz)^3, \quad \beta^{\#} := \beta\,(dz)^2.$$

Obviously, the cubic form $\alpha^{\#}$ and the quadratic form $\beta^{\#}$ are independent of the choice of the complex parameters; therefore they are globally defined forms on M. We will call $\alpha^{\#}$ the *Pick form* for M. It is easy to see that the zeros of $\alpha^{\#}$ are the zeros of the Pick invariant, and the zeros of $\beta^{\#}$ are the umbilic points on M. The Codazzi equations from Proposition 2.16 are expressible as follows:

$$\nabla''\alpha = \frac{\partial\alpha}{\partial\bar{z}} = -F\beta,$$

$$\nabla'\bar{\alpha} = \frac{\partial\bar{\alpha}}{\partial z} = -F\bar{\beta}, \tag{3.143}$$

$$\nabla''\beta = \frac{\partial\beta}{\partial\bar{z}} = F\frac{\partial L_1}{\partial z} + F^{-1}\alpha\bar{\beta},$$

$$\nabla'\bar{\beta} = \frac{\partial\bar{\beta}}{\partial z} = F\frac{\partial L_1}{\partial\bar{z}} + F^{-1}\bar{\alpha}\beta. \tag{3.144}$$

The Gauß integrability condition (theorema egregium) is given by

$$\chi = F^{-3}\alpha\bar{\alpha} + L_1. \tag{3.145}$$

Now we consider affine 2-spheres. We follow C. P. Wang's approach. Let M be a noncompact, simply connected surface. Let $x : M \to A^3$ be a locally strongly convex hyperbolic affine 2-sphere. Since M is simply connected, we can introduce global isothermal coordinates $z = u + \sqrt{-1}v$ such that $G = 2F|dz|^2$. Denote $2F = e^{2w}$, where w is a real valued function. In this case the integrability conditions are reduced to

$$\alpha_{\bar{z}} = 0, \tag{3.146}$$

$$w_{z\bar{z}} + 2|\alpha|^2 e^{-4w} + \frac{1}{4}L_1 e^{2w} = 0, \tag{3.147}$$

where $L_1 = \mathrm{const}$. The first equation shows that $\alpha^{\#}$ is a holomorphic cubic form on M.

Using the preceding notations, we can state the fundamental theorem for affine 2-sphere as follows:

Theorem 3.60. (i) *Let* $x : M \to A^3$ *be an affine 2-sphere with Blaschke metric G. Then G defines a Riemann structure on M such that G is a Hermitian metric on M and the Pick form* $\alpha^{\#}$ *is a holomorphic cubic form on M. Moreover, G,* $\alpha^{\#}$ *and the constant mean curvature* L_1 *are related by the Gauß equation* (3.147).

(ii) *Let M be a simply connected Riemann surface. Let G be a Hermitian metric on M and* $\alpha^{\#}$ *a holomorphic cubic form on M. If there is a constant* L_1 *such that* (3.147) *holds, then* $(G, \alpha^{\#})$ *determines an affine 2-sphere* $x : M \to A^3$ *with G as its Blaschke metric,* $\alpha^{\#}$ *its Pick form and* L_1 *its mean curvature. Moreover, two such pairs* $(G, \alpha^{\#})$ *and* $(\tilde{G}, \tilde{\alpha}^{\#})$ *determine the same affine sphere (modulo equiaffine transformations) if and only if there is a diffeomorphism* $\sigma : M \to M$ *such that*

$$G = \sigma^* \tilde{G}, \quad \alpha^{\#} = \sigma^* \tilde{\alpha}^{\#}.$$

Let $(G, \alpha^{\#}, L_1)$ with $L_1 < 0$ be a triple that satisfies (3.147). Then the associated triple $(|L_1|G, |L_1|\alpha^{\#}, -1)$ also satisfies (3.147), which defines an affine sphere. One can easily show that

Proposition 3.61. *For* $L_1 < 0$, *the hyperbolic affine 2-sphere determined by* $(G, \alpha^{\#}, L_1)$ *is equiaffinely equivalent to a dilatation of that determined by* $(|L_1|G, |L_1|\alpha^{\#}, -1)$.

Thus, in order to give examples of complete hyperbolic affine 2-spheres, we need only to find examples of triples $(G, \alpha^{\#}, -1)$ satisfying the Gauß equation (3.147). This equation is very similar to that of constant mean curvature surfaces in pseudo-Riemann space forms, which was obtained by B. Palmer in [273]. We apply his method in our case.

Let M_g be a compact Riemann surface with genus $g \geq 1$. The Riemann metric G induces a complex structure on M. If the genus of M equals 1, then M is holomorphic to $\mathbb{C}/\Lambda$, where Λ is a lattice in $\mathbb{C}$. In this case there exists a flat metric on $\mathbb{C}/\Lambda$, and by pull back we get a flat Riemannian metric on M, which is conformal to the given metric G. If the genus g of M satisfies $g \geq 2$, the Riemann uniformization theorem guarantees the existence of a conformal covering $\Pi : D \to M$; the covering transformation group Γ is a discrete subgroup of the biholomorphic transformation group $\text{Aut}(D)$. Thus we have $M = D/\Gamma$. Consider $\tilde{G}$ to be the metric on D, invariant under $\text{Aut}(D)$; this metric is also Γ-invariant. Restrict $\tilde{G}$ to $M = D/\Gamma$; this restricted metric, say G_0, is conformal to G and has constant negative curvature -1.

Let $H^3(TM_g)$ denote the complex linear space of all holomorphic cubic forms on M_g. The Riemann-Roch formula (see [122, pp. 36–37]) gives the formula $\dim_{\mathbb{C}} H^3(TM_1) = 1$, and $\dim_{\mathbb{C}} H^3(TM_g) = 5g - 5$ for $g \geq 2$. We can define an invariant norm for $H^3(TM_g)$ by

$$\|\alpha^{\#}\|^2 = |\alpha|^2 e^{-3\phi}, \tag{3.148}$$

where $\alpha^{\#} = \alpha(dz)^3$, $G_0 = e^{\phi}|dz|^2$, and z denotes a local complex coordinate on M_g.

Proposition 3.62. *Let G be a Hermitian metric on M_g and $\alpha^\# \in H^3(TM_g)$. We write $G = e^\psi G_0$. Then $(G, \alpha^\#, -1)$ satisfies (3.147) if and only if the function $\psi : M_G \to \mathbb{R}$ satisfies*

$$\Delta\psi + 16\|\alpha^\#\|^2 e^{-2\psi} - 2e^\psi = 2k_0, \tag{3.149}$$

where $\Delta = 4e^{-\phi} \frac{\partial^2}{\partial z \partial \bar{z}}$ is the Laplace-Beltrami operator on M_g with respect to G_0.

Proof. Since $G = e^\psi G_0 = e^{\psi+\phi}|dz|^2$, we have $2w = \psi + \phi$. From the fact that G_0 is of constant curvature k_0, we get $k_0 = -\frac{1}{2}\Delta\phi = -2e^{-\phi}\phi_{z\bar{z}}$. Hence w is a solution of (3.147) if and only if ψ satisfies

$$
\begin{aligned}
\Delta\psi &= 4e^{-\phi}\psi_{z\bar{z}} \\
&= 4e^{-\phi}(2w_{z\bar{z}} - \phi_{z\bar{z}}) \\
&= 2e^{-\phi}(-8e^{-4w}|\alpha|^2 + e^{2w}) + 2k_0 \\
&= -16e^{-2\psi}\|\alpha^\#\|^2 + 2e^\psi + 2k_0.
\end{aligned}
$$

This completes the proof of the assertion. □

The following theorem is the key point in constructing complete hyperbolic affine 2-spheres.

Theorem 3.63. *Given a nonzero element $\alpha^\# \in H^3(TM_g)$, there exists a unique Hermitian metric G such that $(G, \alpha^\#, -1)$ satisfies the Gauß equation (3.149).*

Before giving the proof of Theorem 3.63, we first explain how to use it to construct complete hyperbolic affine 2-spheres.

Let $\sigma : M \to M_g$ be the conformal universal covering. Given a nonzero element $\alpha^\# \in H^3(TM_g)$, by Theorem 3.63 we get a unique triple $(G, \alpha^\#, -1)$ satisfying (3.149) on M_g. Then by Proposition 3.62 it is clear that $(\sigma^*G, \sigma^*\alpha^\#, -1)$ satisfies (3.147) on M. Since M is simply connected and $\sigma^*\alpha^\#$ is a holomorphic cubic form on M, by Theorem 3.60 we know that $(\sigma^*G, \sigma^*\alpha^\#)$ determines an affine 2-sphere $x : M \to A^3$ with σ^*G as its Blaschke metric. This metric is complete because $\sigma : (M, \sigma^*G) \to (M_g, G)$ is a local isometry.

Now let $\pi_1(M_g)$ be the covering transformation group of $\sigma : M \to M_g$. For any $\tau \in \pi_1(M_g)$ we have $\tau^*\sigma^*G = \sigma^*G$ and $\tau^*\sigma^*\alpha^\# = \sigma^*\alpha^\#$. Since the Blaschke metric σ^*G and the Pick form $\sigma^*\alpha^\#$ determine completely the affine sphere x up to equiaffine transformations, the mapping $\tau : x(p) \to x(\tau(p))$, for all $p \in M$, must be a restriction of an equiaffine transformation in A^3. Thus $\pi_1(M_g)$ can be regarded as a discrete subgroup of the equiaffine group acting on x. In summary, we have

Theorem 3.64. *Let M_g be a compact Riemann surface with genus $g \geq 2$. Then, any nonvanishing holomorphic cubic form $\alpha^\#$ on M_g determines, in the way described above,* *a complete hyperbolic affine sphere* $x : M \to A^3$ *which admits the action of a discrete subgroup of the equiaffine group.*

Remark 3.65. It is easy to see that the given construction yields all complete hyperbolic affine 2-spheres which admit the action of a discrete subgroup of the equiaffine group in A^3 with compact quotient.

Now we use a method similar to that of B. Palmer in [273] to give an outlined proof of Theorem 3.63.

Proof of Theorem 3.63. By Proposition 3.62 we need only to show that, given a nonzero element $\alpha^{\#} \in H^3(TM_g)$, the equation

$$\Delta\psi + 16\|\alpha^{\#}\|^2 e^{-2\psi} - 2e^{\psi} = 2k_0 \tag{3.150}$$

has a unique solution $\psi : M_g \to \mathbb{R}$. We divide the proof into four steps.

Step 1. Define the Hilbert space $W^{1,2}$ of all real functions ψ on M_g such that ψ and grad ψ are square integrable, equipped with the inner product

$$\langle \psi, \eta \rangle := \int G_0(\operatorname{grad}\psi, \operatorname{grad}\eta) + \int \psi\eta, \quad \|\psi\|^2_{W^{1,2}} = \langle\psi,\psi\rangle, \tag{3.151}$$

where grad is the gradient operator on M_g with respect to G_0. Let E be the subspace of $W^{1,2}$ defined by

$$E = \left\{ \psi \in W^{1,2} \ \middle| \ \int \psi = 0 \right\}. \tag{3.152}$$

In this section we make the convention that all integrals are over M_g with respect to G_0. From Rayleigh's theorem [48, p. 16] we know that the norm $\| \ \|_{W^{1,2}}$ and the norm $\| \ \|_E$, defined by $\|\psi\|^2_E = \int G_0(\operatorname{grad}\psi, \operatorname{grad}\psi)$, are equivalent in E. We introduce implicitly the operators $\mathbb{L}$ and $\mathbb{N}$ by

$$\langle \mathbb{L}\psi, \eta \rangle = \int G_0(\operatorname{grad}\psi, \operatorname{grad}\eta), \tag{3.153}$$

$$\langle \mathbb{N}\psi, \eta \rangle = \int (2e^{\psi} - 16\|\alpha^{\#}\|^2 e^{-2\psi})\eta. \tag{3.154}$$

Then (3.150) can be written as an operator equation

$$\mathbb{L}\psi + \mathbb{N}\psi = -2k_0. \tag{3.155}$$

By an inequality of Moser (see [6, p. 63]) there exist constants c and c' such that, for any $\psi \in W^{1,2}$, we have

$$\int e^{\psi} \le c \exp\{c'\|\psi\|^2_{W^{1,2}}\}. \tag{3.156}$$

Now we can show that $\mathbb{L}$ is a continuous self-adjoint map of $W^{1,2}$ to itself, $\mathbb{N}$ is a continuous, Frechet differentiable map of $W^{1,2}$ to itself, and the map $w \to D\mathbb{N}(w)$ is a continuous map of $W^{1,2}$ into the space of bounded linear functionals.

Step 2. Decompose $W^{1,2}$ as

$$W^{1,2} = \ker \mathbb{L} + E$$

and denote by $\mathbb{P}$ the orthogonal projection of $W^{1,2}$ onto E. We can write any solution ψ of (3.155) as $\psi = \eta + m$, where $\eta = \mathbb{P}\psi$ and $m \in \ker \mathbb{L}$. Thus (3.155) splits into the following two equations:

$$\mathbb{L}\eta + \mathbb{P}\mathbb{N}(\eta + m) = \mathbb{P}f, \tag{3.157}$$

$$e^m \int e^\eta - 8e^{-2m} \int \|\alpha^\#\|^2 e^{-2\eta} = 4\pi(g - 1), \tag{3.158}$$

where $f \in W^{1,2}$ is implicitly defined by $\langle f, \eta \rangle = -2 \int k_0 \eta$. We write (3.158) as

$$p(y) = qy^3 + 4\pi(1 - g)y^2 - r = 0, \tag{3.159}$$

where $y = e^m$, $r = r(\eta) = 8 \int \|\alpha^\#\|^2 e^{-2\eta} > 0$ and $q = q(\eta) = \int e^\eta > 0$. It is easy to see that all real roots of (3.159) are positive. Since $p'(y) = 3qy^2 + 8\pi(1 - g)y$ has only one positive root, we conclude that (3.159) has only one positive root. Therefore we can write $m = m(\eta)$. Thus the existence and uniqueness of the solution ψ of (3.155) are equivalent to that of the solution η of (3.157) on E with $m = m(\eta)$.

Step 3. Define $\mathbb{F} : E \to E$ by

$$\mathbb{F}(\eta) = \mathbb{L}\eta + \mathbb{P}\mathbb{N}(m(\eta) + \eta). \tag{3.160}$$

We can show that, for any $\psi \in E$,

$$\langle D\mathbb{F}(\eta)\psi, \psi \rangle = \int G_0(\text{grad}\,\psi, \text{grad}\,\psi) \tag{3.161}$$
$$+ m' \int (2e^{\eta+m} + 32\|\alpha^\#\|^2 e^{-2(m+\eta)})\psi$$
$$+ \int (2e^{\eta+m} + 32\|\alpha^\#\|^2 e^{-2(\eta+m)})\psi^2,$$

where $m' = \frac{d}{dt} m(\eta + t\psi)_{t=0}$. But (3.158) implies

$$- m' \int (2e^{m+\eta} + 32\|\alpha^\#\|^2 e^{-2(m+\eta)}) = \int (2e^{m+\eta} + 32\|\alpha^\#\|^2 e^{-2(m+\eta)})\psi. \tag{3.162}$$

So the sum of the last two terms in (3.161) is nonnegative. We have

$$\langle D\mathbb{F}(\eta)\psi, \psi \rangle \geq \int G_0(\text{grad}\,\psi, \text{grad}\,\psi) = \|\psi\|_E^2. \tag{3.163}$$

The Lax–Milgram lemma [101, p. 83, Thm. 5.8] implies that

$$D\mathbb{F}(\eta) : E \to E$$

is invertible and $\|(D\mathbb{F}(\eta))^{-1}\| \leq 1$ for all $\eta \in E$.

Step 4. Now $\mathbb{F} : E \to E$ is a Frechet differentiable local homeomorphism with derivative $D\mathbb{F}$ satisfying the relation $\|(D\mathbb{F})^{-1}\| \leq 1$. From Hadamard's theorem $\mathbb{F}$ is a global homeomorphism. Thus there exists a unique solution η for (3.157). We get a solution $\psi = \eta + m(\eta)$ for (3.155), which is a weak solution for (3.149). We write (3.157) in the form

$$\Delta\psi = \Psi.$$

Since (3.156) implies that $\Psi \in L^p$ for all $p \geq 1$, we obtain a unique classical solution which equals ψ almost everywhere. Thus we complete the proof of Theorem 3.63. $\square$

4 Rigidity and uniqueness theorems

Chapter 3 presents several local and global uniqueness results, in particular a qualitative global classification of locally strongly convex affine hyperspheres. These results are based on developments in the theory of partial differential equations.

In Euclidean hypersurface theory most of the global uniqueness results are based on three meanwhile classical methods:
(i)   integral formulas;
(ii) the index method;
(iii) the maximum principle.

The Seminar Lecture Notes [127] introduce the methods and present Euclidean uniqueness results. There are also many applications of these methods in affine differential geometry; the monographs [20] and [311] contain sections with global results. The previous chapter, in particular, gives examples of the application of the maximum principle for elliptic partial differential equations and its extension (e.g. in Sections 3.5–3.6).

The aim of Chapter 4 is twofold: we would like to give examples of important global affine uniqueness results, and at the same time we would like to give an introduction to the methods (i) and (ii) mentioned above: integral formulas in Section 4.1 and the index method in Section 4.2.

4.1 Integral formulas for affine hypersurfaces and their applications

Integral formulas are one of the most powerful tools for proving global rigidity and uniqueness theorems in Riemannian geometry. K. P. Grotemeyer [107] and later U. Simon [323, 324] and C. C. Hsiung and J. K. Shahin [126] generalized the well-known Minkowski integral formulas to affine (relative) hypersurfaces and obtained many characterizations of ellipsoids by the higher affine mean curvatures. With integral formulas as the main tool, R. Schneider (cf. [312, 313]) studied the global solutions of the equation

$$\Delta f = -nL_1 f$$

on an ovaloid and proved that if two ovaloids in A^{n+1} have the same Blaschke metric and the same affine mean curvature function, then they differ by a unimodular affine transformation. He also answered affirmatively the conjecture of W. Blaschke: an ovaloid in A^3 with constant scalar curvature is an ellipsoid. This result was generalized later by M. Kozlowski and U. Simon [145] to Einstein hypersurfaces in A^{n+1}. Moreover there are interesting results on compact hypersurfaces with boundary based on integral formulas of Bochner–Lichnerowicz type (cf. [145, 167, 321, 322, 330]). In this section we collect some important results in affine differential geometry obtained by application of integral formulas.

4.1.1 Minkowski-type integral formulas for affine hypersurfaces

In this section we derive Minkowski-type integral formulas for affine hypersurfaces using a method of S. S. Chern (cf. [62]).

Let $x : M \to A^{n+1}$ be a compact, locally strongly convex hypersurface (with or without boundary). Following S. S. Chern we introduce the differential forms

$$A_r = \mathrm{Det}(x, Y, \underbrace{dY, \ldots, dY}_{r}, dx, \ldots, dx), \tag{4.1}$$

$$D_r = \mathrm{Det}(Y, \underbrace{dY, \ldots, dY}_{r}, dx, \ldots, dx). \tag{4.2}$$

Each of these expressions is a determinant of order $n + 1$, whose columns are the components of the respective vectors or vector-valued differential forms, with the convention that in the expansion of the determinant the multiplication of differential forms is in the sense of exterior multiplication. The subscript r denotes the number of entries dY in these determinants.

Choose a local equiaffine frame field $\{x; e_1, \ldots, e_n, e_{n+1}\}$ on M such that

$$G_{ij} = \delta_{ij}.$$

Since dY and dx are linear combinations of $e_1, \ldots, e_n$, only, exterior differentiation of (4.1) gives

$$dA_r = -\Lambda D_{r+1} - D_r, \quad \text{for } r = 0, 1, \ldots, n - 1, \tag{4.3}$$

where $\Lambda = \langle U, -x \rangle$ is the affine support function relative to the origin (cf. Section 2.3.2). Integrating both sides of the equation (4.3) over M, and applying Stokes' formula, we get

$$\int_M (\Lambda D_{r+1} + D_r) = - \int_{\partial M} A_r, \quad \text{for } r = 0, 1, \ldots, n - 1. \tag{4.4}$$

We insert

$$dx = \sum \omega^i e_i$$

and

$$dY = - \sum B_j^i \omega^j e_i$$

into equation (4.2) and get

$$D_r = (-1)^{n+r} n! L_r dV, \tag{4.5}$$

where

$$dV = \omega^1 \wedge \omega^2 \wedge \cdots \wedge \omega^n$$

denotes the Riemannian volume element of the Blaschke metric on M (see Appendix A.3.2.1) in terms of the equiaffine frame (and its dual). Inserting (4.5) into (4.4) we obtain the following integral formula

$$n! \int_M (L_r - \Lambda L_{r+1})dV = (-1)^{n+r+1} \int_{\partial M} A_r, \quad r = 0, 1, \ldots, n - 1. \tag{4.6}$$

If M is a compact, locally strongly convex hypersurface without boundary, we have the following Minkowski-type integral formulas for affine hypersurfaces:

$$\int_M (L_r - \Lambda L_{r+1})dV = 0, \quad \text{for } r = 0, 1, \ldots, n - 1. \tag{4.7}$$

4.1.2 Characterization of ellipsoids

We begin with the following:

Definition 4.1. A connected compact locally strongly convex hypersurface without boundary in A^{n+1} is called an *ovaloid*.

Theorem 4.2. *Let M be an ovaloid in A^{n+1}. If the affine mean curvature of M is constant then M is an ellipsoid.*

Proof. It follows from (4.7) that

$$\int_M (L_1 - \Lambda L_2)dV = 0,$$

$$\int_M (1 - \Lambda L_1)dV = 0.$$

Since $L_1 = $ const., we can deduce from both equations that

$$\int_M \Lambda [(L_1)^2 - L_2]dV = 0. \tag{4.8}$$

Choose the origin in the interior of M so that $\Lambda > 0$ on M. Using Newton's inequality $(L_1)^2 \geq L_2$, from (4.8) we obtain that $(L_1)^2 = L_2$. Hence on M:

$$\lambda_1 = \lambda_2 = \cdots = \lambda_n.$$

This shows that M is an affine hypersphere. Consequently, it is an ellipsoid (cf. Theorem 3.35). □

To generalize Theorem 4.2 to higher affine mean curvatures L_r, $r \geq 2$, we follow ideas of A. Ros and A.-M. Li (see [171]) and prove the following:

Lemma 4.3. *Let M be an ovaloid in A^{n+1}. Then there exists at least one point $x_0 \in M$ such that at x_0,*

$$\lambda_1 > 0,\ \lambda_2 > 0, \ldots, \lambda_n > 0.$$

Proof. We choose the origin in the interior of M such that $\Lambda > 0$. Since M is compact, the affine support function $\Lambda = \langle U, -x \rangle$ must attain its maximum at some point, say x_0. At the point x_0

$$\Lambda_{,i} = \langle U_{,i}, -x \rangle = 0, \quad \text{for } i = 1, 2, \ldots, n,$$

and $(\Lambda_{,ij})$ is negative semi-definite. Then formula (2.73) reads at x_0:

$$\Lambda_{,ij} = -\Lambda B_{ij} + G_{ij}.$$

It follows that (B_{ij}) is positive definite at x_0. Therefore at x_0

$$\lambda_1 > 0, \lambda_2 > 0, \ldots, \lambda_n > 0. \qquad \square$$

Lemma 4.4. *Let M be an ovaloid in A^{n+1}. If $L_n \neq 0$ everywhere on M then $\lambda_i > 0$ on M for $i = 1, \ldots, n$.*

Proof. Assume that there is a point x_1 on M such that some affine principal curvature, say λ_1, is negative at x_1. From Lemma 4.3 there is a point x_0 such that λ_1 is positive at x_0. Since λ_1 is a continuous function on M, there must be a point x_2 such that $\lambda_1(x_2) = 0$. This contradicts the condition $L_n \neq 0$ on M. $\qquad \square$

Let $S_r : \mathbb{R}^n \to \mathbb{R}$ $(1 \leq r \leq n)$ be the *normalized r-th elementary symmetric function* in the variables $x^1, \ldots, x^n$. We denote by C_r the *connected component* of the open set $\{x \in \mathbb{R}^n \mid S_r(x) > 0\}$ containing the vector $a = (1, \ldots, 1)$. Gårding showed $C_r \subset C_k$ if $r \geq k$. For $x \in C_r$ the Gårding inequality gives us (cf. [96])

$$[S_k(x)]^{(k-1)/k} \leq S_{k-1}(x), \quad k = 1, \ldots, r. \tag{4.9}$$

Theorem 4.5. *Let M be an ovaloid in A^{n+1}. If for some fixed r, $1 \leq r \leq n$, $L_r = const.$ on M then M must be an ellipsoid.*

Proof. From Lemma 4.3 and the condition $L_r = const.$ we know that $L_r > 0$ on M. Set $(L_r)^{1/r} = C > 0$. Then for $k < r$, $L_k > 0$ everywhere on M from Gårding's result, and

$$L_1 \geq (L_2)^{1/2} \geq \cdots \geq (L_{r-1})^{1/(r-1)} \geq (L_r)^{1/r} = C.$$

Choose the origin in the interior of M so that $\Lambda > 0$. The Minkowski formulas give

$$\int_M L_r \Lambda \, dV = \int_M L_{r-1} \, dV \geq C^{r-1} \cdot \int_M dV.$$

On the other hand we have

$$\int_M L_r \Lambda \, dV = C^{r-1} \int_M \Lambda \cdot C \cdot dV \le C^{r-1} \int_M \Lambda \cdot L_1 dV = C^{r-1} \cdot \int_M dV.$$

It follows that $L_1 = C$ on M. Now Theorem 4.2 gives the assertion. $\qquad\square$

Theorem 4.5 has a long history. Under stronger assumptions it was first stated in [353], and extended in [126, 323, 324]. The above proof was given by A.-M. Li [171].

Theorem 4.6. *Let $r \in \mathbb{N}$, $2 \le r \le n$, be fixed and let $x : M \to A^{n+1}$ be an ovaloid with $L_n > 0$ on M. If there exists $r - 1$ nonnegative constants $C_1, C_2, \ldots, C_{r-1}$ such that on M*

$$L_r = \sum_{j=1}^{r-1} C_j L_j,$$

then M is an ellipsoid.

Proof. As $L_n > 0$, all principal curvatures and all curvature functions are positive on M (Lemma 4.4), thus, recalling that $L_0 = 1$,

$$L_j L_{r-1} - L_r L_{j-1} \ge 0, \quad \text{for } j < r. \tag{4.10}$$

Equality holds if and only if $\lambda_1 = \lambda_2 = \cdots = \lambda_n$; see [112, p. 52, formula (2.22.1)] . The assumptions imply

$$1 = \sum_{j=1}^{r-1} C_j \frac{L_j}{L_r} \ge \sum_{j=1}^{r-1} C_j \frac{L_{j-1}}{L_{r-1}} \quad \text{or} \quad L_{r-1} - \sum_{j=1}^{r-1} C_j L_{j-1} \ge 0,$$

where equality holds as above. We use the Minkowski integral formulas in (4.7):

$$0 \le \int \Big(L_{r-1} - \sum_{j=1}^{r-1} C_j L_{j-1} \Big) dV = \int \Lambda \Big(L_r - \sum_{j=1}^{r-1} C_j L_j \Big) dV = 0;$$

as a consequence

$$L_{r-1} - \sum_{j=1}^{r-1} C_j L_{j-1} = 0,$$

which gives equality in the inequality (4.10) and thus implies $\lambda_1 = \cdots = \lambda_n$; that means: the ovaloid is an affine hypersphere. Then Theorem 3.35 gives the assertion. $\qquad\square$

Remark 4.7. Theorem 4.2 is due to W. Blaschke for $n = 2$.

If M is an ovaloid in A^{n+1} and $L_n > 0$ on M, we set

$$R_1 = \frac{1}{\lambda_1}, \quad R_2 = \frac{1}{\lambda_2}, \ldots, R_n = \frac{1}{\lambda_n},$$

$$Q_r = \frac{1}{\binom{n}{r}} \sum_{i_1 < \cdots < i_r} R_{i_1} R_{i_2} \cdots R_{i_r}, \quad \text{for } r = 1, 2, \ldots, n,$$

$$Q_0 := 1,$$

and consider the volume element dm of the Weingarten form, which is the metric of the spherical indicatrix (see Section 4.6 in [341]):

$$dm = \omega_{n+1}^1 \wedge \cdots \wedge \omega_{n+1}^n.$$

Then the integral formulas (4.7) can be rewritten in the form

$$\int_M (Q_{r+1} - \Lambda Q_r) dm = 0, \quad \text{for } r = 0, 1, \ldots, n-1.$$

If $L_n \neq 0$ one can replace L_r by Q_r in the theorems above.

4.1.3 Some further characterizations of ellipsoids

W. Blaschke made the following conjecture (cf. [20]):

Conjecture 4.1. *An ovaloid in A^3 with constant scalar curvature is an ellipsoid.*

In [312] R. Schneider proved that if an ovaloid M and an ellipsoid are equiaffinely isometric, that is, they have the same Blaschke metric, then M is an ellipsoid. This result solves the above conjecture. Later, M. Kozlowski and U. Simon gave the following generalization (cf. [145]):

Theorem 4.8. *Let $x : M \to A^{n+1}$ be an ovaloid. If M is an Einstein space relative to the Blaschke metric for $n \geq 3$ and if the scalar curvature R is constant for $n = 2$, then $x(M)$ is an ellipsoid.*

To prove this theorem we need a result of Lichnerowicz on the spectrum of a closed Riemannian manifold (M, g); for an introduction see [14, Chap. III], or [48]. The proof of Theorem 4.8 follows from Lemma 4.10 later.

Let (M, g) be a closed Riemannian C^∞-manifold. The *spectrum* $Sp(M)$ is a set defined by

$$Sp(M) = \{\lambda \in \mathbb{R}, \text{ such that there exists } f \in C^2(M), f \neq 0, \text{ satisfying}$$
$$\Delta f = -\lambda f, \text{ where the Laplacian is defined with respect to the}$$
$$\text{Riemannian metric } g\}.$$

λ is called an *eigenvalue* of the Laplacian and f an *eigenfunction*. If $\lambda \in Sp(M)$, then $\lambda \geq 0$, because of Green's theorem:

$$\lambda \int_M f^2 dV = -\int_M f \Delta f dV = \int_M \|\text{grad} f\|^2 dV \geq 0.$$

Here the gradient, the norm and the volume element dV are defined with respect to the Riemannian metric. The eigenvalues of the Laplacian operator form an infinite sequence $0 = \lambda_0 < \lambda_1 < \lambda_2 < \cdots$, going to $+\infty$; for each eigenvalue λ_i, the set of the corresponding eigenfunctions forms a vector space of finite dimension; see [14, 48] or [344].

The following inequality is well known (cf. [14, p. 186], [48, p. 16]):

$$\int_M \|\operatorname{grad} f\|^2 dV \geq \lambda_1 \int_M f^2 dV \qquad (4.11)$$

for all $f \in C^\infty(M)$ such that $\int_M f dV = 0$.

Lemma 4.9 (Lichnerowicz; see [14, p. 179]). *Let (M, g) be an n-dimensional compact Riemannian manifold without boundary. If there exists a constant $K > 0$ such that its Ricci tensor satisfies*

$$\sum R_{ij} X^i X^j \geq (n-1) K g(X, X)$$

for any tangent vector X on M, then the first eigenvalue λ_1 of the Laplace operator Δ satisfies $\lambda_1 \geq nK$.

Proof. Let $f \in C^\infty(M)$ be an eigenfunction corresponding to λ_1. In terms of an orthonormal frame field, we have

$$\left(\sum (f_{,i})^2\right)_{,j} = 2 \sum f_{,i} f_{,ij},$$

$$\Delta \sum (f_{,i})^2 = 2 \sum (f_{,ij})^2 + 2 \sum f_{,i} f_{,ijj}.$$

We get the so-called Bochner–Lichnerowicz formula:

$$\Delta \sum (f_{,i})^2 = 2 \sum (f_{,ij})^2 + 2 \sum R_{ij} f_{,i} f_{,j} + 2 \sum (\Delta f)_{,i} f_{,i}.$$

Thus, since f is an eigenfunction, we have

$$\Delta \sum (f_{,i})^2 \geq \frac{2}{n} (\Delta f)^2 + 2(n-1) K \|\operatorname{grad} f\|^2 - 2\lambda_1 \|\operatorname{grad} f\|^2$$

$$= \frac{2}{n} \lambda_1^2 f^2 + (2(n-1) K - 2\lambda_1) \|\operatorname{grad} f\|^2.$$

We use Green's formula and the fact f is a first eigenfunction; then

$$0 \geq \frac{2}{n} \cdot (\lambda_1)^2 \int_M f^2 dV + 2((n-1) K - \lambda_1) \int_M \|\operatorname{grad} f\|^2 dV$$

$$\geq \left[\frac{2}{n} (\lambda_1)^2 + 2((n-1) K - \lambda_1) \cdot \lambda_1 \right] \cdot \int_M f^2 dV.$$

It follows that $\lambda_1 \geq nK$. □

Lemma 4.10. *Let M be an ovaloid in A^{n+1}, and let a be a fixed vector in V. Then*

$$\int_M \langle U, a \rangle dV = 0.$$

Proof. In analogy to (4.1) one considers the $(n-1)$-form

$$\bar{\omega} := \text{Det}(a, x, dx, \ldots, dx)$$

and proves

$$d\bar{\omega} = n! \langle U, a \rangle dV.$$

Application of Stokes' theorem gives the assertion. □

Proof of Theorem 4.8. Let us consider the function $f = \langle U, a \rangle$ on M. From the affine theorema egregium (2.125), $\chi = \frac{R}{n(n-1)} = J + L_1$, we have

$$n \int_M \chi f^2 dV \geq \int_M n L_1 f^2 dV = - \int_M f \Delta f dV = \int_M \|\text{grad} f\|^2 dV$$

$$\geq \lambda_1 \int_M f^2 dV.$$

Hence we have $n\chi \geq \lambda_1 > 0$. On the other hand, by Lemma 4.9 and the assumptions of the theorem we have $\lambda_1 \geq n\chi$. Hence $\lambda_1 = n\chi$. It follows that $Jf^2 = 0$ everywhere. Since the set of zeros of f has measure zero, we conclude that $J \equiv 0$ on M and thus $x(M)$ is an ellipsoid. □

U. Simon and M. Kozlowski made the following conjecture:

Conjecture 4.2. *Let $x : M \to A^{n+1}$ be an ovaloid. If the scalar curvature of M is constant, then $x(M)$ is an ellipsoid.*

Up to now this conjecture is open for $n \geq 3$.

We have seen that for an ovaloid in A^3 each one of the conditions: (1) $L_1 = $ const., (2) $R = $ const., implies that the ovaloid is an ellipsoid. This suggests that we investigate the third problem: *is every ovaloid with $J = $ const. an ellipsoid?* The following theorem of Švec, which was also obtained by Li and Wang independently, answers this question. The following proof was given by Li in [186].

Theorem 4.11. *Let $x : M \to A^3$ be an ovaloid. If $J = $ const. on M then $x(M)$ is an ellipsoid.*

Proof. Choose an equiaffine frame field $\{x; e_1, e_2, e_3\}$ on $x(M)$ such that $G_{ij} = \delta_{ij}$. Then

$$A_{ijk,l} - A_{ijl,k} = \frac{1}{2}(-\delta_{jl}B_{ik} - \delta_{il}B_{jk} + \delta_{ik}B_{jl} + \delta_{jk}B_{il}).$$

Application of the Ricci identity A.3.1.8 gives

$$\Delta J = 6\chi J + \sum_{i,j,k,l} (A_{ijk,l})^2 - 2 \sum A_{ijk}B_{ij,k}. \tag{4.12}$$

Since
$$\sum A_{ijk}B_{ij,k} = \sum (A_{ijk}B_{ij})_{,k} - \sum B_{ij}A_{ijk,k},$$

Corollary 2.17 gives
$$\sum A_{ijk,k} = L_1\delta_{ij} - B_{ij},$$

and one verifies the following integral formula (see [327] and [329] for $n \geq 2$)
$$\int_M \{6\chi J + \sum (A_{ijk,l})^2 - (\lambda_1 - \lambda_2)^2\}dV = 0. \tag{4.13}$$

If $J = 0$, the theorem is proved. Assume that $J = $ const. > 0. Let $x \in M$ be an arbitrary fixed point. We may choose e_1 and e_2 such that $A_{111} = 0$ at x (cf. Section 3.2.2). Since
$$J = 2((A_{111})^2 + (A_{222})^2) = \text{const.},$$

we have
$$A_{111}A_{111,1} + A_{222}A_{222,1} = 0,$$
$$A_{111}A_{111,2} + A_{222}A_{222,2} = 0.$$

Hence at x
$$A_{222,1} = 0,$$
$$A_{222,2} = 0,$$

and the apolarity condition together with the Codazzi equations for A implies
$$A_{111,1} = -A_{221,1} = -A_{121,2} - \frac{1}{2}(B_{11} - B_{22}) = -\frac{1}{2}(B_{11} - B_{22}),$$
$$A_{111,2} = A_{112,1} + B_{12} = -A_{222,1} + B_{12} = B_{12}.$$

It follows that at x
$$\sum (A_{ijk,l})^2 = 4((A_{111,1})^2 + (A_{111,2})^2 + (A_{222,1})^2 + (A_{222,2})^2) \tag{4.14}$$
$$= (\lambda_1 - \lambda_2)^2.$$

This last relation is independent of the choice of the special coordinate system at x, thus the relation holds everywhere on M.

Inserting (4.14) into (4.13), we obtain
$$6J \int_M \chi dV = 0.$$

On the other hand, the Gauß–Bonnet formula states that the integral $\int_M \chi dV = 4\pi$. We conclude that $J = 0$ on M, and thus $x(M)$ is an ellipsoid. $\qquad\square$

So far, the corresponding problem for $n > 2$ is still unsolved. We state:

Problem 4.3. Let $x : M \to A^{n+1}$ $(n \geq 3)$ be an ovaloid. If $J = $ const. on M, is $x(M)$ an ellipsoid?

4.1.4 Global solutions of a differential equation of Schrödinger type

The following theorem was proved by W. Blaschke [20] for $n = 2$ and was extended by R. Schneider [312] to higher dimensions.

Theorem 4.12. *Let* $x : M \to A^{n+1}$ *be an ovaloid and f be a* C^∞-*function on M. If f satisfies the differential equation*

$$\Delta f = -nL_1 f,$$

then there exists a constant vector $a \in A^{n+1}$ *such that*

$$f = \langle U, a \rangle.$$

Proof. Choose a local equiaffine frame field $\{x; e_1, \ldots, e_n, e_{n+1}\}$ on M such that $G_{ij} = \delta_{ij}$. The differential df satisfies

$$df = \sum f_i \omega^i.$$

We construct a vector field

$$a = -\sum f_i e_i + f e_{n+1} \tag{4.15}$$

on M, then

$$\langle U, a \rangle = f.$$

In the following we prove that a is a constant vector field. From (4.15), together with the structure equations (2.5) and (2.6), we get

$$da = -\sum df_i e_i - \sum f_i \omega_i^j e_j - \sum f_i \omega_i^{n+1} e_{n+1} + df e_{n+1} + f \sum \omega_{n+1}^i e_i$$
$$= \sum \left(-f_{,ji} - \sum f_m A_{mji} - f B_{ji} \right) \omega^i e_j.$$

This implies that $da \in T_x M$. Put

$$da = \sum T_{ji} \omega^i e_j,$$

where

$$T_{ji} := -f_{,ji} - \sum f_m A_{mji} - f B_{ji}.$$

Obviously, T is symmetric, and from the assumptions it satisfies

$$\sum T_{ii} = -(\Delta f + nL_1 f) = 0.$$

We are going to prove $T_{ji} = 0$. Put

$$\theta := \mathrm{Det}\left(x, a, da, \overbrace{dx, \ldots, dx}^{n-2} \right).$$

Then

$$d\theta = \mathrm{Det}(x, da, da, dx, \ldots, dx) - \mathrm{Det}(a, da, dx, \ldots, dx).$$

Since da and dx are tangent to M, we have

$$\text{Det}(x, da, da, dx, \ldots, dx) = \langle U, x \rangle \text{Det}(e_{n+1}, da, da, dx, \ldots, dx),$$
$$\text{Det}(a, da, dx, \ldots, dx) = \langle U, a \rangle \text{Det}(e_{n+1}, da, dx, \ldots, dx).$$

Inserting $dx = \sum \omega^i e_i$ and $da = \sum T_{ji}\omega^i e_j$ into these relations and integrating over M we obtain

$$\int_M \langle U, x \rangle S_2 \, dV = \int_M \langle U, a \rangle S_1 \, dV,$$

where S_1 and S_2 are the first and the second elementary symmetric functions of the eigenvalues of (T_{ji}), respectively. (The beginner can verify this considering the matrix T to be diagonalized). It follows from $S_1 = \frac{1}{n} \sum T_{ii} = 0$ that

$$\int_M \langle U, x \rangle S_2 \, dV = 0. \tag{4.16}$$

Choose the origin in the interior of $x(M)$, then $\Lambda = \langle U, -x \rangle > 0$. We conclude from (4.16) and Newton's inequality $S_2 \leq (S_1)^2 = 0$ that also $S_2 = 0$; but $S_1 = 0, S_2 = 0$ give $T_{ji} = 0$. This implies that a is a constant field. □

4.1.5 Rigidity theorems for ovaloids

Using Theorem 4.12, R. Schneider [312] obtained the following rigidity theorem:

Theorem 4.13. *Let* $x, \bar{x} : M \to A^{n+1}$ *be two ovaloids. If, at corresponding points,* $\bar{G} = G$ *and* $\bar{L}_1 = L_1$, *then there is a unimodular affine transformation* σ *in* A^{n+1} *such that*

$$\bar{x} = \sigma \circ x.$$

Proof. Let $\{\eta_1, \eta_2, \ldots, \eta_{n+1}\}$ be a fixed unimodular affine frame in A^{n+1}. Then, for the hypersurface x, the conormal U satisfies Corollary 2.11, thus

$$\Delta \langle U, \eta_A \rangle = -nL_1 \langle U, \eta_A \rangle, \quad \text{for } A = 1, 2, \ldots, n+1.$$

There is a corresponding equation for the conormal $\bar{U}$ of $\bar{x}$. It follows from the conditions $\bar{G} = G$ and $\bar{L}_1 = L_1$ that

$$\Delta \langle \bar{U} - U, \eta_A \rangle = -nL_1 \langle \bar{U} - U, \eta_A \rangle.$$

From Theorem 4.12 there exist constant vectors $\xi_1, \xi_2, \ldots, \xi_{n+1}$ such that

$$\langle \bar{U}, \eta_A \rangle = \langle U, \xi_A \rangle, \quad \text{for } A = 1, 2, \ldots, n+1.$$

We claim that $\xi_1, \xi_2, \ldots, \xi_{n+1}$ are linearly independent. In fact, if there are constants $a_1, a_2, \ldots, a_{n+1}, (a_1, \ldots, a_{n+1}) \neq (0, \ldots, 0)$, such that

$$\sum a_A \xi_A = 0,$$

then

$$\langle \bar{U}, \sum a_A \eta_A \rangle = \langle U, \sum a_A \xi_A \rangle = 0.$$

This means that $\sum a_A \eta_A$ is tangent to $\bar{x}(M)$ at every point of the ovaloid $\bar{x}(M)$. Hence

$$\sum a_A \eta_A = 0.$$

This contradicts the fact that $\eta_1, \eta_2, \ldots, \eta_{n+1}$ are linearly independent. As $\{\xi_A\}$ and $\{\eta_A\}$ are bases of the vector space V, there exists a regular, linear transformation $T : V \to V$ such that

$$\xi_A = \sum T_A^B \eta_B.$$

Then

$$\langle \bar{U}, \xi_A \rangle = \sum T_A^B \langle \bar{U}, \eta_B \rangle = \sum T_A^B \langle U, \xi_B \rangle. \tag{4.17}$$

Let $T^* : V^* \to V^*$ be its dual linear transformation. Then (4.17) implies

$$\bar{U} = T^* U. \tag{4.18}$$

Define a hypersurface in A^{n+1} by

$$\tilde{x} = |\det(T_A^B)|^{-1/(n+1)} T\bar{x}, \tag{4.19}$$

then the affine normal vector field $\tilde{Y}$ on $\tilde{x}(M)$ is

$$\tilde{Y} = |\det(T_A^B)|^{-1/(n+1)} T\bar{Y}.$$

Since

$$\langle \bar{U}, \bar{Y} \rangle = 1 = \langle \tilde{U}, \tilde{Y} \rangle = |\det(T_A^B)|^{-1/(n+1)} \langle \tilde{U}, T\bar{Y} \rangle$$
$$= |\det(T_A^B)|^{-1/(n+1)} \langle T^* \tilde{U}, \bar{Y} \rangle$$

and

$$\langle \bar{U}, \bar{x}_i \rangle = 0 = \langle \tilde{U}, \tilde{x}_i \rangle = |\det(T_A^B)|^{-1/(n+1)} \langle T^* \tilde{U}, \bar{x}_i \rangle,$$

we conclude

$$\bar{U} = |\det(T_A^B)|^{-1/(n+1)} T^* \tilde{U}.$$

From (4.18), we have

$$\tilde{U} = |\det(T_A^B)|^{1/(n+1)} U.$$

Since $\tilde{G}_{ij} = \bar{G}_{ij} = G_{ij}$, we get

$$\langle \tilde{U}_i, \tilde{x}_j \rangle = \langle U_i, x_j \rangle.$$

This, together with the equations $\langle \tilde{U}, \tilde{x}_j \rangle = 0 = \langle U, x_j \rangle$ and the linear dependence of $\tilde{U}$ and U, implies that

$$x_j = |\det(T_A^B)|^{1/(n+1)} \tilde{x}_j$$

and so

$$x = |\det(T_A^B)|^{1/(n+1)} \tilde{x} + b, \tag{4.20}$$

where b is a fixed vector in A^{n+1}. Comparing (4.19) with (4.20), we obtain

$$x = T\bar{x} + b.$$

Since this affine transformation leaves the Blaschke metric invariant, we have

$$\det(T^B_A) = 1. \qquad \qquad \square$$

The well-known uniqueness theorem of Minkowski in the theory of hypersurfaces in Euclidean space has the following affine explanation, which was pointed out by W. Blaschke [20, p. 207] for $n = 2$, and by others in higher dimensions; see the commentary [333, Sect. III.5.3], [334, 336].

Theorem 4.14. *Let $x, \bar{x} : M \to A^{n+1}$ be two ovaloids. If, at corresponding points, $U = \bar{U}$, then there exists a parallel translation σ such that $\bar{x} = \sigma \circ x$.*

Proof. Introduce an arbitrary Euclidean inner product with the same volume element as the given one on A^{n+1}. As the conormals coincide, the tangent planes are parallel at corresponding points, thus the Euclidean normals $\mu, \bar{\mu}$, which must be parallel to the equiaffine conormals, coincide: $\mu = \bar{\mu}$. An easy calculation gives the relation between conormal and Euclidean normal for a hypersurface (see e.g. [341, §6.2.4]), thus

$$K^{-1/(n+2)}\mu = U = \bar{U} = \bar{K}^{-1/(n+2)}\bar{\mu},$$

where K and $\bar{K}$ denote the Euclidean Gauß–Kronecker curvatures of $x(M)$ and $\bar{x}(M)$, respectively. Hence $K = \bar{K}$. The Euclidean uniqueness theorem of Minkowski (see [62]) gives the assertion. $\qquad \square$

By considering the connection ∇^* induced by the conormal immersion, U. Simon obtained the following (cf. [282, 334]):

Theorem 4.15. *Let $x, \bar{x} : M \to A^{n+1}$ be two ovaloids. If, at corresponding points, the conormal connections $\nabla^*, \bar{\nabla}^*$ coincide: $\nabla^* = \bar{\nabla}^*$, then x and $\bar{x}$ differ by an affine transformation.*

Proof. Let $u^1, u^2, \ldots, u^n$ be local coordinates on M. The connection ∇^* and the affine Weingarten form $\hat{B}$ of x (with components Γ_{ij}^{*k} and B_{ij}) satisfy the system of differential equations in Proposition 2.10

$$\begin{cases} dU = \sum du^i U_i, \\ dU_i = \sum \Gamma_{ij}^{*k} du^j U_k - \sum B_{ij} du^j U, \quad i = 1, \ldots, n. \end{cases} \qquad (4.21)$$

Let R_{ij}^* be the Ricci tensor of ∇^*. Then $R_{ij}^* = (n-1)B_{ij}$ (see [341, Sect. 4.8.3]). The condition $\bar{\nabla}^* = \nabla^*$ implies that $\bar{R}_{ij}^* = R_{ij}^*$ and therefore $\bar{B}_{ij} = B_{ij}$ (quantities with a bar correspond in an obvious way to $\bar{x}$). Suppose that $p_0 \in M$ is a fixed point. Applying an affine transformation σ to $\bar{x}(M)$ and denoting $\sigma\bar{x}$ by $x^\#$, we can assume that

$$U(p_0) = U^\#(p_0), \quad U_i(p_0) = U_i^\#(p_0), \quad \text{for } i = 1, 2, \ldots, n,$$

where $U^\#$ is the conormal vector field on $x^\#(M)$. It is easy to verify that $\bar{\Gamma}_{ij}^{*k} = \Gamma_{ij}^{*k\#}$ and $\bar{B}_{ij} = B_{ij}^\#$.

Now let p be any point on M and $r(t) = (u^1(t), \ldots, u^n(t))$ any curve connecting p_0 with p. Along the curve $r(t)$, the system (4.21) of p.d.e.'s becomes a system of ordinary differential equations:

$$\begin{cases} \dfrac{dU}{dt} = \sum \dfrac{du^i}{dt} U_i, \\[2mm] \dfrac{dU_i}{dt} = \sum \Gamma_{ij}^{*k} \dfrac{du^j}{dt} U_k - \sum B_{ij} \dfrac{du^j}{dt} U, \quad \text{for } i = 1, \ldots, n. \end{cases}$$

Since $\Gamma_{ij}^{*k} = \bar{\Gamma}_{ij}^{*k} = \Gamma_{ij}^{*k\#}$ and $B_{ij} = \bar{B}_{ij} = B_{ij}^\#$, it follows from the uniqueness theorem on the solution of a system of ordinary differential equations that it holds $U = U^\#$ on $r(t)$. Because the point p is arbitrary and M is simply connected, we have $U = U^\#$ on M. Then the theorem follows from Theorem 4.14. □

It is an obvious question whether one can prove a similar result for the induced connection ∇. We will investigate this problem in Section 4.2.4 below.

4.1.6 Hypersurfaces with boundary

In this subsection we investigate compact hypersurfaces $x : M \to A^{n+1}$ with boundary ∂M and study the question: how does the geometry at the boundary determine the geometry of the hypersurface?

Exemplary, we study two such types of problems; in the first case (i) we are interested in a certain type of results; in the second case (ii) our emphasis is on the method.

(i)   For locally strongly convex hypersurfaces the vanishing of the Pick invariant J locally characterizes quadrics. *Assume now that, at the boundary of a compact hypersurface x, the Pick invariant vanishes. Under which conditions for x is the hypersurface a quadric itself?*

(ii) In [312] Schneider gave a partial solution to the following problem:

Consider a compact affine hypersphere with simply connected boundary, assume the boundary to have the following two properties:

(a) *the boundary $x(\partial M)$ lies in a hyperplane;*

(b) *there is a constant vector field $b \in V$ such that, at each point of the boundary ∂M, the tangent plane is parallel to b; analytically this means $\langle U, b \rangle = 0$ at ∂M. For obvious geometric reasons the boundary is called a shadow submanifold, if b represents the direction of parallel light.*

An affine hypersphere with the boundary properties (a) and (b) is half an ellipsoid.

The final solution of Schneider's problem was given by A. Schwenk. She reformulated the boundary conditions in terms of two Dirichlet problems, which finally gave a solution (see [321, 330]).

We will now examine the first type (i) of results.

R. Schneider proved the following theorem (cf. [312]):

Theorem 4.16. *Let* $x : M \to A^{n+1}$ *be a compact affine hypersphere with boundary. If the Pick invariant J vanishes on the boundary ∂M, then $x(M)$ lies on a quadric.*

Proof. There are two cases, corresponding to the sign of the mean curvature.

(i) $L_1 > 0$ or $L_1 = 0$. From (3.100) we have

$$\frac{1}{2}n(n-1)\Delta J \ge \sum (A_{ijk,l})^2 + n(n-1)(n+1)J(J + L_1) \ge 0.$$

This means that J is a subharmonic function on M. Thus it attains its maximum at a point on ∂M. It follows immediately that $J = 0$ on M and $x(M)$ lies on a quadric.

(ii) $L_1 < 0$. Choose the center of $x(M)$ as origin. Then $Y = -L_1 x$. Choose a local equiaffine frame field $\{x; e_1, \ldots, e_n, e_{n+1}\}$ on $x(M)$ such that $G_{ij} = \delta_{ij}$. Let $f := \langle B^*, x \rangle$, where $B^* \in V^*$ is a fixed covector. Then

$$\begin{cases} f_i = \langle B^*, e_i \rangle, \\ f_{,ij} = \sum A_{ijl} f_l - L_1 \delta_{ij} f, \\ \Delta f = -n L_1 f. \end{cases} \tag{4.22}$$

Consider the vector field

$$a = \sum f_i e_i - fY$$

on $x(M)$. Then $\langle U, a \rangle = -f$. We now prove that a is a constant vector field. Covariant differentiation gives

$$da = \sum T_{ij} \omega^i e_j,$$

where

$$T_{ij} = f_{,ij} + \sum A_{ijl} f_l + L_1 \delta_{ij} f = 2 \sum A_{ijl} f_l.$$

Here the last equality is due to (4.22). Moreover, we have $\sum T_{ii} = 0$ on M and $T_{ij} = 0$ on ∂M. As in the proof of Theorem 4.12 we can prove an integral formula

$$\int_M \langle U, x \rangle S_2 dV - \int_M \langle U, a \rangle S_1 dV = \int_{\partial M} \theta,$$

where S_1, S_2 and θ are defined as in the proof of Theorem 4.12. As $da = \sum T_{ij} \omega^i e_j = 0$ on ∂M, the 1-form θ vanishes on ∂M; thus

$$\int_M \langle U, x \rangle S_2 dV = \int_M \langle U, a \rangle S_1 dV.$$

As in (4.16), it follows from $S_1 = 0$ and $\langle U, x \rangle = -\frac{1}{L_1} > 0$ on M that $S_2 = 0$ on M. Again we get $T_{ij} = 0$, so $a : M \rightarrow V$ is a constant vector field; therefore the tensor norm of $T_{ij} := 2 \sum A_{ijk} f_k$ vanishes: $0 = \| \sum A_{ijk} f_k \|$. This holds for any covector B^* and any function $f := \langle B^*, x \rangle$. Thus $A_{ijk} = 0$; this means that $x(M)$ lies on a quadric. $\square$

In order to generalize Theorem 4.2 to hypersurfaces with boundary, we use an integral formula of A. Schwenk and U. Simon [322], which we will now derive.

Let $a \in V$ be a fixed vector and let M be a compact, locally strongly convex hypersurface with C^∞-boundary in A^{n+1}. Consider the function $f := \langle U, a \rangle$. Then the structure equations in Proposition 2.10 give

$$
\begin{cases}
f_i := f_{,i} = \langle U_{,i}, a \rangle, \\
f_{,ij} = -\sum A_{ijl} f_l - B_{ij} f, \\
\Delta f = -n L_1 f.
\end{cases}
\tag{4.23}
$$

From the Bochner–Lichnerowicz formula (see the proof of Lemma 4.9), we have

$$
\frac{1}{2} \Delta \sum (f_i)^2 = \sum (f_{,ij})^2 + \sum f_i (\Delta f)_i + \sum R_{ij} f_i f_j.
$$

The Ricci tensor and B satisfy Corollary 2.17, thus we have

$$
R_{ij} = \sum A_{iml} A_{mlj} + \frac{1}{2}(n-2) B_{ij} + \frac{n}{2} L_1 \delta_{ij},
$$

$$
\sum B_{ij,i} = \sum A_{ijm} B_{mi} + n(L_1)_j.
$$

We insert both of the above relations and the following identity

$$
\sum B_{ij} f_i f_j = \sum (B_{ij} f_i f)_j - \sum f B_{ij,j} f_i - \sum f f_{,ij} B_{ij}
$$

$$
= \sum (B_{ij} f_i f)_j + \sum (B_{ij})^2 f^2 - n \sum f_i (L_1)_i f
$$

into the Bochner-Lichnerowicz formula; using $L_1 = $ const. and (4.23), we obtain

$$
\frac{n}{2} \Delta \sum (f_i)^2 - \frac{n(n-2)}{2} \sum (B_{ij} f_i f)_j + \frac{1}{4} n^2 L_1 \Delta (f^2)
$$

$$
= [n \sum (f_{,ij})^2 - (\Delta f)^2] + n \sum A_{iml} A_{mlj} f_i f_j + \frac{n-2}{2} S f^2, \tag{4.24}
$$

where

$$
S := \sum_{i<j} (\lambda_i - \lambda_j)^2 = n \left(\sum (B_{ij})^2 - n(L_1)^2 \right).
$$

The function S vanishes exactly at umbilics.

Integrating both sides of equation (4.24) and applying Stokes' formula (Theorem A.36), we obtain an integral formula for hypersurfaces with constant equiaffine mean curvature:

$$\int_{\partial M} \left[n \sum f_i f_{,ij} v_j - \frac{n(n-2)}{2} \sum f f_i B_{ij} v_j + \frac{1}{2} n^2 L_1 \sum f f_i v_i \right] dO$$

$$= \int_M \left\{ [n \sum (f_{,ij})^2 - (\Delta f)^2] + n \sum A_{iml} A_{mlj} f_i f_j + \frac{1}{2}(n-2) S f^2 \right\} dV, \quad (4.25)$$

where v_i denotes the i-th component of the exterior unit normal vector to ∂M relative to the Blaschke metric on M and to the induced orientation on ∂M, and dO denotes the induced volume element on ∂M.

It is not difficult to check that the integral formula (4.25) still holds if one replaces f by the affine support function $\Lambda(x) = \langle U, b - x \rangle$; see [322].

Using (4.25), A. Schwenk and U. Simon proved that a compact, locally strongly convex hypersurface M must lie on a quadric if L_1 = const. on M, and if on the boundary ∂M it holds $J = 0$ and $\lambda_1 = \cdots = \lambda_n$. We note that, when $n \geq 3$, the condition $\lambda_1 = \cdots = \lambda_n$ on ∂M is not needed (cf. [167]).

Theorem 4.17. *Let $x : M \to A^{n+1} (n \geq 3)$ be a compact, locally strongly convex hypersurface with C^∞-boundary. If*

(i) L_1 = const. *on M,*

(ii) $J = 0$ *on ∂M,*

then $x(M)$ lies on a quadric.

To prove this theorem we need the following two lemmas.

Lemma 4.18. *Let $x : M \to A^{n+1}$ be a compact, locally strongly convex hypersurface with C^∞-boundary. If L_1 = const. on M and the affine normals on ∂M intersect at one point, or are mutually parallel, then $x(M)$ is an affine hypersphere.*

Proof. There are two cases.

Case 1. The affine normals on ∂M are mutually parallel. Assume that they are parallel to a fixed vector $b \in V$. Let $f := \langle U, b \rangle$ and $b = \sum b_i e_i + b_{n+1} Y$. We have

$$f_j = \langle U_j, b \rangle = -b_j, \quad \text{for } j = 1, 2, \ldots, n.$$

It follows from the fact that Y is parallel to b on ∂M that $b_j = 0$ and thus $f_j = 0$ on ∂M.

Case 2. The affine normals on ∂M intersect at one point. Choose the intersection point as the origin and consider the function $\Lambda = \langle U, -x \rangle$. Then $\Lambda_{,i} = \langle U_{,i}, -x \rangle$. Since x is parallel to Y on ∂M, we conclude that $\Lambda_{,i} = 0$ on ∂M.

In both cases the boundary integrals in (4.25) vanish. Since all integrands on the right-hand side of (4.25) are nonnegative, we have

$$S f^2 = 0,$$

or

$$S\Lambda^2 = 0.$$

Since $x(M)$ is locally strongly convex, the set of zero points of f or Λ, respectively, has measure zero. Thus S is almost everywhere zero. It follows from a continuity argument that $S = 0$ on M; consequently, $x(M)$ is an affine hypersphere. □

Lemma 4.19. *Let $x : M \to A^{n+1}(n \geq 3)$ be a compact locally strongly convex hypersurface with C^∞-boundary. If $J = 0$ on ∂M, then the affine normals on ∂M intersect at one point or are mutually parallel.*

Proof. Choose a local equiaffine frame field $\{x; e_1, \ldots, e_n, e_{n+1}\}$ on M such that $G_{ij} = \delta_{ij}$ and, when we restrict them to $\partial M, e_1, \ldots, e_{n-1}$ are tangent to ∂M. We agree on the index ranges:

$$1 \leq \alpha, \beta, \gamma, \cdots \leq n-1, \quad 1 \leq i, j, \cdots \leq n.$$

Since $J = 0$ on ∂M, we have $A_{ijk} = 0$ and $A_{ijk,\alpha} = 0$ on ∂M. Then from the integrability conditions we have

$$\frac{1}{2}(\delta_{i\alpha}B_{j\beta} - \delta_{j\beta}B_{i\alpha} + \delta_{j\alpha}B_{i\beta} - \delta_{i\beta}B_{j\alpha}) = A_{ij\alpha,\beta} - A_{ij\beta,\alpha} = 0.$$

It follows that

$$B_{\alpha\beta} = \frac{1}{n-1}\left(\sum B_{\gamma\gamma}\right)\delta_{\alpha\beta} := \lambda\delta_{\alpha\beta},$$
$$B_{n\alpha} = 0.$$

Since on ∂M

$$B_{\alpha\beta,\gamma} - B_{\alpha\gamma,\beta} = \sum A_{\alpha\beta i}B_{i\gamma} - \sum A_{\alpha\gamma i}B_{i\beta} = 0,$$

we obtain $\lambda_{,\alpha} = 0$, i.e. $\lambda = $ const. Hence along ∂M we have $dY = -\lambda\,dx$. □

Proof of Theorem 4.17. A combination of the Lemmas 4.18, 4.19 and Theorem 4.16 gives the proof of Theorem 4.17. □

Theorem 4.20. *Let $x : M \to A^{n+1}$ $(n \geq 3)$ be a compact, connected, locally strongly convex hypersurface with C^∞-boundary. If $L_1 \neq 0$, $\frac{L_2}{L_1} = $ const. $\neq 0$ on M and $J = 0$ on ∂M, then $x(M)$ lies on a quadric.*

Proof. By Lemma 4.19 and $L_2 \neq 0$ the affine normals on ∂M intersect at one point. We choose this point as the origin of A^{n+1}. From Minkowski's integral formulas (4.6) we get

$$n! \int_M (1 - \Lambda L_1)dV = (-1)^{n+1} \int_{\partial M} A_0,$$

$$n! \int_M (L_1 - \Lambda L_2)dV = (-1)^{n+2} \int_{\partial M} A_1.$$

Since x is parallel to Y on ∂M, we have $A_0 = A_1 = 0$ on ∂M. Set $\frac{L_2}{L_1} =: c$, then

$$\int_M (L_1 - c)dV = -\int_M \Lambda (cL_1 - L_2)dV = 0. \tag{4.26}$$

Since $L_1 \neq 0$, L_1 does not change sign. Using the inequality $(L_1)^2 \geq L_2$, we get $L_1 \geq c$, and the integral formula (4.26) gives $L_1 = c$. This is possible only if $\lambda_1 = \cdots = \lambda_n$. Then Theorem 4.20 follows directly from Theorem 4.16. $\qquad\square$

We switch now to the second case (ii) which we briefly described in the beginning of Section 4.1.6. Let $x : M \to A^{n+1}$ be a locally strongly convex hypersurface, and let $0 \neq b \in V$ be a fixed vector. Then the set $\{q \in M | \langle U(q), b \rangle = 0\}$ is the *shadow boundary* with respect to the parallel light ray in the direction of b. As already mentioned, the following theorem was conjectured by R. Schneider [312], and was solved by A. Schwenk ([321]; also [330]).

Theorem 4.21. *Let $x : M \to A^{n+1}$ be a compact affine hypersphere with C^∞-boundary $x(\partial M)$. If $x(\partial M)$ is a shadow boundary with respect to parallel light and if $x(\partial M)$ lies on a hyperplane Π, then $x(M)$ is a "half" ellipsoid, i.e. the center of the ellipsoid lies in Π.*

Proof. First of all, we assume that M is an elliptic or hyperbolic affine hypersphere. Choose the center of M as origin in A^{n+1}. Let $b \neq 0$ be a fixed vector in the direction of parallel light and F a fixed covector in V^* such that the hyperplane Π is described by $\langle F, x \rangle = $ const. $=: C$. We define two functions f and q on M by

$$f = \langle U, b \rangle,$$
$$q = \langle F, x \rangle.$$

Then the Gauß structure equations (2.61) and Proposition 2.10 together with the apolarity condition (2.46) and $Y = -L_1 x$ imply

$$\begin{cases} \Delta f = -nL_1 f & \text{in } M, \\ f = 0 & \text{on } \partial M, \end{cases} \tag{4.27}$$

$$\begin{cases} \Delta q = -nL_1 q & \text{in } M, \\ q = C & \text{on } \partial M. \end{cases} \tag{4.28}$$

Choose a local equiaffine frame field $\{x; e_1, \ldots, e_{n-1}, e_n, e_{n+1}\}$ on M such that $G_{ij} = \delta_{ij}$, and, when restricted to ∂M, the vectors $e_1, \ldots, e_{n-1}$ are tangent to ∂M and e_n is the exterior unit normal vector field of ∂M with respect to the Blaschke metric on M. According to the preceding choice of the frame we agree on the following notational convention:

If we consider a function (or an invariant, respectively) on the boundary ∂M and if we use components then the indices $\alpha, \beta, \ldots \in \{1, \ldots, n-1\}$ indicate that we restrict it to the boundary.

Decompose b:

$$b = \sum_{\alpha=1}^{n-1} a_\alpha e_\alpha + a_n e_n + a_{n+1} e_{n+1}.$$

The condition $\langle U, b \rangle|_{\partial M} = 0$ implies that

$$a_{n+1}|_{\partial M} = 0 \text{ and } f_\alpha|_{\partial M} = \langle U_\alpha, b \rangle|_{\partial M} = 0 \text{ for } 1 \le \alpha \le n-1.$$

It follows that along ∂M

$$a_\alpha = 0 \text{ for } 1 \le \alpha \le n-1 \text{ and } b = a_n e_n,$$

and $a_n \ne 0$ everywhere on ∂M.

From (4.27), (4.28), and Green's formula (A.34) we have

$$0 = \int_M (f \Delta q - q \Delta f) dV = -C \int_{\partial M} f_n dO.$$

Since $f_n = \langle U_n, b \rangle = -a_n \ne 0$ on ∂M we get $C = 0$. This means that the center of M lies on Π. Differentiation of $b = a_n e_n$ along ∂M gives

$$\Gamma^\alpha_{n\beta} = 0 \text{ on } \partial M, \text{ for } 1 \le \alpha, \beta \le n-1,$$

while differentiation of $q = \langle F, x \rangle = 0$ along ∂M gives, on ∂M,

$$q_\alpha = \langle F, e_\alpha \rangle = 0, \quad 1 \le \alpha \le n-1. \tag{4.29}$$

Differentiating (4.29) again and using $\langle F, x \rangle|_{\partial M} = 0$, we get on ∂M

$$\Gamma^n_{\alpha\beta} = 0, \text{ for } 1 \le \alpha, \beta \le n-1.$$

It follows from (2.43), namely $\sum A_{ijk} \omega^k = \frac{1}{2}(\omega_i^j + \omega_j^i)$, that on ∂M

$$A_{\alpha\beta n} = 0, \text{ for } 1 \le \alpha, \beta \le n-1.$$

This last relation and the apolarity condition (2.46) imply

$$A_{nnn} = 0 \text{ on } \partial M,$$

which together with (4.27) and (4.22) shows that all boundary-integrals in (4.25) vanish.

Furthermore, since $n \sum (f_{,ij})^2 - (\Delta f)^2 \ge 0$ (Newton's inequality), $S = 0$ and $\sum_{m,l} A_{iml} A_{mlj}$ is semi-positive definite, it follows from (4.25) that

$$f_{ij} + L_1 f G_{ij} = 0 \text{ and } \sum A^r_{ij} f_r = 0$$

on M. From these relations one easily verifies on M:

$$G(\text{grad } J, \text{grad } f) = 2 f J L_1.$$

On ∂M the function f vanishes and the vector field $\text{grad } f$ is parallel to the normal vector field v. This gives $G(\text{grad } J, v) = 0$ on ∂M, thus

$$\int_M \Delta J dV = \int_{\partial M} G(\text{grad } J, v) dO = 0.$$

Again we use the differential inequality (3.100) for ΔJ and obey $L_1 > 0$ which is a consequence of (4.27) and Green's formula. This implies

$$\Delta J \geq 2(n + 1)J^2 \geq 0,$$

and therefore $J \equiv 0$ on M from the last integral. Hence $x(M)$ lies on a quadric with positive affine mean curvature, i.e. $x(M)$ lies on an ellipsoid. □

Remark 4.22. If we assume that $x(\partial M)$ lies in a closed half-space determined by a hyperplane containing the center of the affine sphere, instead of the assumption that $x(\partial M)$ lies on a hyperplane, the assertion of Theorem 4.21 still holds. In fact, assume that the hyperplane is given by $q = \langle F, x \rangle = 0$, then $q|_{\partial M} \geq 0$ (or ≤ 0), and

$$0 = \int_M (f\Delta q - q\Delta f)dV = -\int_{\partial M} qf_n dO.$$

From $f_n \neq 0$ on ∂M it follows that $q|_{\partial M} = 0$; therefore $x(\partial M)$ lies on a hyperplane. Application of the Schneider–Schwenk theorem gives the assertion. This result is a little more general than the Schneider–Schwenk theorem.

4.2 The index method

So far we applied different methods to prove global rigidity and uniqueness results in Chapters 3 and 4. First we applied methods in relation with special types of p.d.e.'s, in particular gradient estimates and the maximum principle; then, in Section 4.1, we derived and applied integral formulas.

We are going to present now another method which is one of the most important global tools for the investigation of surfaces: the index method. This method was first introduced by Cohn–Vossen in the proof of a rigidity theorem for ovaloids in Euclidean space of dimension 3. Cohn–Vossen's proof was originally restricted to the analytic case. Later, the assumption of analyticity could be dropped by using results about partial differential equations. The index method was applied in affine differential geometry first by H. F. Münzner (cf. [225, 227, 228]) and later by U. Simon [336, 343]. In this section we introduce the index method (see also [127]) and apply it to study affine Weingarten surfaces.

4.2.1 Fields of line elements and nets

Let M be a 2-dimensional C^∞-manifold. Below we first state:

Definition 4.23. A one-dimensional (*differentiable*) distribution θ on M is given if we assign to each point p of M a one-dimensional subspace S_p of T_pM. θ is called a C^k-distribution if for every $p \in M$ there exists a neighborhood $N(p)$ of p and a C^k-vector field Z on $N(p)$ such that Z_q spans S_q for all $q \in N(p)$. A C^0-distribution is also called a continuous distribution.

There exist compact C^∞-manifolds without boundary which do not admit C^0-distributions, due to topological obstructions.

By the theory of ordinary differential equations, every one-dimensional distribution on M determines locally a unique family of integral curves on M.

Let N be an open subset of M. Then N is a two-dimensional open C^∞-submanifold of M.

Definition 4.24. Let N be an open subset of M. A one-dimensional distribution θ on N is called a *field of line elements on M*. The points $p \in M - N$ are called the *singularities* of the field of line elements. When θ is a C^k-distribution on N, we call it a C^k-*field of line elements* on M. Let p be a singularity of the field θ of line elements. If there exists a neighborhood of p in which p is the only singularity, then p is called an *isolated singularity* of θ.

On a compact surface without boundary, if a field of line elements has only isolated singularities, then the number of singularities is finite.

In the following we define the index of an isolated singularity of a continuous field of line elements. Let M be a two-dimensional, oriented Riemann manifold. Let θ be a continuous field of line elements on M, and let p be an isolated singularity of θ. Let $c : [0, 1] \to M$ be a simple closed curve such that p is the only singularity in the interior of c and there are no singularities on c. Then θ induces a field $\theta(c(t))$ of line elements on c. Choose one of the two possible directions along $\theta(c(0))$ at $c(0)$. This determines a direction at $c(t)$ for every t, $0 \le t \le 1$, by continuity. We denote this oriented field of line elements by $\overrightarrow{\theta(c(t))}$. We wish to measure the "total change" of this field going once around $c(t)$. In order to do this, we assume that c is small enough to be contained in a chart of M with a fixed local coordinate system (u, v). Within such a region there is defined a vector field without singularities, e.g., $\frac{\partial}{\partial u}$. This determines a direction at each point, which will be denoted by $\overrightarrow{W}$. We denote by $\{\overrightarrow{\theta(c(t))}, \overrightarrow{W(c(t))}\}$ the oriented angle between $\overrightarrow{\theta(c(t))}$ and $\overrightarrow{W(c(t))}$. Let $\delta_c(\{\overrightarrow{\theta(c(t))}, \overrightarrow{W(c(t))}\})$ be the total change of this angle while going around c once in the positive direction.

Definition 4.25. The *index j* of an isolated singularity p of a field θ of line elements is defined by

$$j = \frac{1}{2\pi} \delta_c \left(\left\{ \overrightarrow{\theta(c(t))}, \overrightarrow{W(c(t))} \right\} \right).$$

(It is easy to see that $j = \frac{n}{2}$ where n is an integer.)

Theorem 4.26. *The index j does not depend on the choice of $\overrightarrow{W}$ or of c.*

Proof. (i) Let V be another field of line elements without singularity. Then

$$\delta_c(\{\vec{\theta}, \vec{W}\}) = \delta_c(\{\vec{\theta}, \vec{V}\}) + \delta_c(\{\vec{V}, \vec{W}\}).$$

However, by choosing c small enough, we can make $\delta_c(\{\vec{V}, \vec{W}\})$ arbitrarily small since both are continuous fields without singularities. Hence, since $\delta_c(\{\vec{V}, \vec{W}\})$ is an integer multiple of π, necessarily one has $\delta_c(\{\vec{V}, \vec{W}\}) = 0$.

(ii) Since $\frac{1}{\pi}\delta_c$ depends continuously on c, while $2j$ is an integer, it is clear that j does not depend on c. $\square$

Theorem 4.27. *The index j does not depend on the Riemannian metric used.*

Proof. Let (g_{ij}) and (h_{ij}) be two positive definite Riemannian metrics on M. Then

$$f_{ij}(t) = (1 - t)g_{ij} + th_{ij}, \ 0 \le t \le 1$$

is also a positive definite Riemannian metric for every t. But the angles change continuously with t and $2j$ is an integer. Hence, j does not depend on the metric. $\square$

We state the following theorem without proof. It is the key of the index method.

Theorem 4.28 (Poincaré index formula)**.** *Let M be a closed orientable, 2-dimensional manifold of genus g. Let θ be a given field of line elements on M with at most finitely many singularities. Then the sum of the indices j of the singularities of the field is $2 - 2g$ (see* [125, p. 113, 2.2, Thm. II])*.*

Definition 4.29. Let N be an open subset of M. A *net* $\mathcal{N}$ on N is given if we assign to each point p of N two 1-dimensional linear subspaces $I_1(p)$ and $I_2(p)$ of T_pM, such that $I_1(p) \ne I_2(p)$ for every $p \in N$. $\mathcal{N}$ is called a C^k-net if for each $p \in N$ there exist a neighborhood $D(p)$ and two C^k-fields Z_1 and Z_2 on $D(p)$, such that $Z_1(q)$ spans $I_1(q)$ and $Z_2(q)$ spans $I_2(q)$ for all $q \in D(p)$. The points in $M - N$ are called the *singularities* of $\mathcal{N}$.

Locally, a net can be generated by two fields of line elements. In general, this is not true globally. If there exist two distributions θ_1 and θ_2 on N such that $\theta_1(p) = I_1(p)$ and $\theta_2(p) = I_2(p)$ for every $p \in N$, we say that the net $\mathcal{N}$ is *decomposable*.

Theorem 4.30. *Let M be a closed, 2-dimensional Riemannian manifold of genus 0. Let L be a covariant, symmetric, quadratic, continuous tensor field on M satisfying the conditions:*
(i) *for every $p \in M$, $\det(L_{ij}(p)) \le 0$, where L_{ij} denote the local components of L;*
(ii) *if $\det(L_{ij}(p)) = 0$, then $L_{ij}(p) = 0$ for all i and j.*

Then the equation

$$L_{11}(du)^2 + 2L_{12}dudv + L_{22}(dv)^2 = 0 \qquad (4.30)$$

determines a decomposable continuous net $\mathcal{N}$ *with singularities*

$$Q = \{p \in M \mid L_{ij}(p) = 0\}.$$

Proof. Topologically M is diffeomorphic to a Euclidean sphere S^2. Assume p is not in Q, and let p be bijectively related to a pole $P \in S^2$. By stereographic projection from P and applying the diffeomorphism, we obtain a coordinate system (u, v) on $M - \{p\}$. Multiplying both sides of (4.30) by L_{11}, we get

$$(L_{11}du + L_{12}dv)^2 + (L_{11}L_{22} - (L_{12})^2)(dv)^2 = 0.$$

This equation can be decomposed into two equations

$$L_{11}du + (L_{12} + \sqrt{(L_{12})^2 - L_{11}L_{22}})dv = 0 \qquad (\text{I})$$

and

$$L_{11}du + (L_{12} - \sqrt{(L_{12})^2 - L_{11}L_{22}})dv = 0. \qquad (\text{II})$$

They determine two families of integral curves on $M - (\{p\} \cup Q)$ unless the coefficients of (I) or (II) vanish identically. Similarly, we can obtain

$$(L_{12} - \sqrt{(L_{12})^2 - L_{11}L_{22}})du + L_{22}dv = 0 \qquad (\text{I})'$$

and

$$(L_{12} + \sqrt{(L_{12})^2 - L_{11}L_{22}})du + L_{22}dv = 0. \qquad (\text{II})'$$

It is easy to see that the coefficients of the equations

$$L_{11}du + (L_{12} + \sqrt{(L_{12})^2 - L_{11}L_{22}})dv = 0,$$

$$(L_{12} - \sqrt{(L_{12})^2 - L_{11}L_{22}})du + L_{22}dv = 0$$

are not all zero on $M - (Q \cup \{p\})$ and that the equations (I) and (I') are equivalent. Thus they determine a distribution θ_1 on $M - (Q \cup \{p\})$ such that $\theta_1(q)$ is tangential to the integral curve through q for every $q \in M - (Q \cup \{p\})$. Similarly, the equations

$$L_{11}du + (L_{12} - \sqrt{(L_{12})^2 - L_{11}L_{22}})dv = 0,$$

$$(L_{12} + \sqrt{(L_{12})^2 - L_{11}L_{22}})du + L_{22}dv = 0$$

determine another distribution θ_2 on $M - (Q \cup \{p\})$. Obviously, θ_1 and θ_2 are independent of the choice of the local coordinate system, and $\theta_1(q) \neq \theta_2(q)$ for every $q \in M - (Q \cup \{p\})$. We can choose another local coordinate system containing the point p to find two different linear subspaces at p. Therefore we obtain actually two distributions θ_1 and θ_2 on $M - Q$, and hence a decomposable net on M. $\square$

Remark and Definition 4.31. Let q be an isolated singularity of a continuous net. Then there exists a local chart U such that the net is decomposable on $U\backslash\{q\}$, and that the two distributions θ_1, θ_2 defined in this way on $U\backslash\{q\}$ are different at each point. A simple argument shows that, at q, the indices of both distributions coincide: $j_1 = j_2$. Thus we can define $j := j_1 = j_2$ to be the *index* of the net at the isolated singularity q.

Theorem 4.32. Let $\mathbb{N}$, θ_1 and θ_2 be as in Theorem 4.30 and its proof. Assume that the point p is an isolated singularity of the net $\mathbb{N}$. Let j_1 and j_2 denote the indices of θ_1 and θ_2, respectively. Then

$$j_1 = j_2 = j = -\frac{1}{2}\frac{1}{2\pi}\delta_c[\arg(L_{11} - \sqrt{-1}L_{12})].$$

Proof. Consider the equation

$$L_{11}(du)^2 + 2L_{12}dudv + [(1 - t)L_{22} - tL_{11}](dv)^2 = 0, \tag{4.31}$$

for $0 \le t \le 1$. Since, when $(L_{11}, L_{12}, L_{22}) \ne (0, 0, 0)$,

$$(1 - t)L_{11}L_{22} - (1 - t)(L_{12})^2 - t(L_{12})^2 - t(L_{11})^2$$
$$= (1 - t)(L_{11}L_{22} - (L_{12})^2) - t((L_{11})^2 + (L_{12})^2) < 0,$$

it follows from Theorem 4.30 that, for every t, (4.31) determines a decomposable continuous net $\mathbb{N}(t)$ with an isolated singularity at p. The index $j(t)$ of p changes continuously with t, and $2j(t)$ is an integer, thus j does not depend on t and $j(t) = j$ for $0 \le t \le 1$. Taking $t = 1$, we get

$$L_{11}(du)^2 + 2L_{12}dudv - L_{11}(dv)^2 = 0.$$

This equation determines two distributions θ_1 and θ_2. Let c be a simple closed curve contained in a local coordinate chart such that p is the only singularity in the interior of c and there are no singularities on c. We orient θ_1 along c and denote by τ the angle between θ_1 and $\frac{\partial}{\partial u}$. Then

$$du : dv = \cos\tau : \sin\tau.$$

Thus

$$L_{11}\cos 2\tau + L_{12}\sin 2\tau = 0. \tag{4.32}$$

One can see that there are no points on the curve c such that L_{11} and L_{12} vanish simultaneously, otherwise it follows from the conditions (i) and (ii) in Theorem 4.30 that $L_{11} = L_{12} = L_{22} = 0$, which contradicts the fact that there are no singularities on c. We put

$$\alpha := \arg(L_{11} - \sqrt{-1}L_{12}).$$

From (4.32) we get

$$\cos(\alpha + 2\tau) = 0.$$

So we have

$$\tau = -\frac{\alpha}{2} + \frac{k\pi + \frac{\pi}{2}}{2}$$

and

$$\delta_c(\tau) = -\frac{1}{2}\delta_c(\alpha).$$

Consequently.

$$j = -\frac{1}{2}\frac{1}{2\pi}\delta_c\left[\arg(L_{11} - \sqrt{-1}L_{12})\right]. \qquad \square$$

4.2.2 Vekua's system of partial differential equations

Let D be a bounded open subdomain of $\mathbb{R}^2$. Consider the following system of partial differential equations of the first order in Hilbert normal form

$$\begin{cases} \dfrac{\partial f}{\partial u} - \dfrac{\partial g}{\partial v} = a(u,v)f + b(u,v)g, \\[2mm] \dfrac{\partial f}{\partial v} + \dfrac{\partial g}{\partial u} = c(u,v)f + d(u,v)g \end{cases} \qquad (4.33)$$

on D, where $a(u,v), b(u,v), c(u,v), d(u,v)$ are four given C^∞-functions on D. A solution (f,g) of the system (4.33) will be called a *completely regular solution* if $\frac{\partial f}{\partial u}, \frac{\partial f}{\partial v}, \frac{\partial g}{\partial u}$, and $\frac{\partial g}{\partial v}$ are continuous on D. The system (4.33) generalizes the well known Cauchy–Riemann p.d.e.'s from complex function theory. This suggests that we introduce a complex notation:

$$F := f + \sqrt{-1}g, \quad z := u + \sqrt{-1}v.$$

Now we can write the system (4.33) in the complex form

$$\frac{\partial F}{\partial \bar{z}} = PF + Q\bar{F}, \qquad (4.34)$$

where

$$\frac{\partial}{\partial \bar{z}} := \frac{1}{2}\left(\frac{\partial}{\partial u} + \sqrt{-1}\frac{\partial}{\partial v}\right),$$

$$P := \frac{1}{4}\left[(a + d) + \sqrt{-1}(c - b)\right],$$

$$Q := \frac{1}{4}\left[(a - d) + \sqrt{-1}(c + b)\right].$$

F is called a *completely regular solution* of the system (4.34), if (f,g) is a completely regular solution of the system (4.33). We need the following two results about the solutions of the equation (4.34), due to T. Carleman. Their proofs can be found in [375, p. 146, Thm 3.5] and [401, p. 192, Thm. 5.3.5].

Theorem 4.33. *Let F be a completely regular solution of equation (4.34). If F ≠ 0, then the zeros of F are isolated and of finite order.*

Let F be a completely regular solution of (4.34). We write (4.34) in the form

$$\frac{\partial F}{\partial \bar{z}} = P_0 F, \tag{4.35}$$

where

$$P_0 = P + Q\frac{\bar{F}}{F}.$$

P_0 is bounded in D and continuous at every point of D except, perhaps, at some discrete points. Thus P_0 is integrable on D. Set

$$W(z) := -\frac{1}{\pi} \iint_D \frac{P_0(\xi)}{\xi - z} d\xi. \tag{4.36}$$

Obviously, $W(z)$ is a continuous function on D. Moreover, W solves $\frac{\partial}{\partial \bar{z}} W = P_0$. Hence we have $\frac{\partial}{\partial \bar{z}}(e^{-W}F) = 0$, which means that $\varphi := e^{-W}F$ is a holomorphic function. Thus, the function F can be written in the form

$$F(z) = \varphi(z)e^{W(z)}. \tag{4.37}$$

Next we consider a system of equations with more than two unknown functions. The system of equations

$$\frac{\partial F}{\partial \bar{z}} = PF + Q\bar{F}, \tag{4.38}$$

where

$$F = \begin{pmatrix} f_1 \\ f_2 \\ \vdots \\ f_n \end{pmatrix} \quad \text{and } f_i = f_i(z),$$

is called a *homogeneous Pascali system* of partial differential equations of first order.

Theorem 4.34. *Every completely regular solution F ≠ 0 of Pascali's system (4.38) has only isolated zeros of finite order.*

4.2.3 Affine Weingarten surfaces

Weingarten surfaces generalize surfaces with constant curvature functions. A locally strongly convex surface whose affine principal curvatures satisfy a functional equation is called an *affine Weingarten surface*. The results of this subsection are taken from [164]. They generalize results of A. Švec [360].

In Theorems 4.35 and 4.39 below we will prove a series of global and local results about Weingarten surfaces. In the examples following these results we discuss special cases of our results which demonstrate the very general character of the results in Theorems 4.35 and 4.39.

Let $x : M \to A^3$ be a locally strongly convex surface. We choose isothermal parameters (u, v) on $x(M)$ such that the Blaschke metric is given by

$$(G_{ij}) = \begin{pmatrix} E & 0 \\ 0 & E \end{pmatrix}.$$

Relative to the local coordinate system (u, v) we consider the system (2.124′):

$$\frac{\partial B_{11}}{\partial v} - \frac{\partial B_{12}}{\partial u} = L_1 \frac{\partial E}{\partial v} + \frac{1}{E} A_{112}(B_{22} - B_{11}) + \frac{2}{E} A_{111} B_{12}, \tag{4.39}$$

$$\frac{\partial B_{12}}{\partial v} - \frac{\partial B_{22}}{\partial u} = -L_1 \frac{\partial E}{\partial u} + \frac{1}{E} A_{122}(B_{22} - B_{11}) - \frac{2}{E} A_{222} B_{12}. \tag{4.40}$$

Since

$$EL_1 = \frac{1}{2}(B_{11} + B_{22}),$$

we have

$$L_1 \frac{\partial E}{\partial v} = -E \frac{\partial L_1}{\partial v} + \frac{\partial}{\partial v}\left(\frac{B_{11} + B_{22}}{2}\right), \tag{4.41}$$

$$L_1 \frac{\partial E}{\partial u} = -E \frac{\partial L_1}{\partial u} + \frac{\partial}{\partial u}\left(\frac{B_{11} + B_{22}}{2}\right). \tag{4.42}$$

Inserting (4.41) and (4.42) into (4.39) and (4.40), we get

$$\frac{\partial}{\partial v}\left(\frac{B_{11} - B_{12}}{2}\right) - \frac{\partial B_{12}}{\partial u} = -E \frac{\partial L_1}{\partial v} + \frac{1}{E} A_{112}(B_{22} - B_{11}) + \frac{2}{E} A_{111} B_{12}, \tag{4.43}$$

$$\frac{\partial}{\partial u}\left(\frac{B_{11} - B_{22}}{2}\right) + \frac{\partial B_{12}}{\partial v} = E \frac{\partial L_1}{\partial u} + \frac{1}{E} A_{122}(B_{22} - B_{11}) - \frac{2}{E} A_{222} B_{12}. \tag{4.44}$$

Theorem 4.35. Let $x : M \to A^3$ be an ovaloid and let $F : \mathbb{R}^2 \to \mathbb{R}$ be a C^∞-function satisfying the following conditions:
(i) there exists a C^∞-function $\alpha : \mathbb{R}^2 \to \mathbb{R}$ such that

$$\frac{\partial F}{\partial y_2}(y_1, y_2) - \frac{\partial F}{\partial y_1}(y_1, y_2) = \alpha(y_1, y_2)(y_2 - y_1);$$

(ii) the equations

$$F(y) = 0 \text{ and } F'(y) = 0$$

have no common root, where $F(y) = F(y, y)$ and

$$F'(y) = \frac{\partial F}{\partial y}(y).$$

If the affine principal curvatures satisfy $F(\lambda_1, \lambda_2) = 0$ on M, then $x(M)$ is an ellipsoid.

Proof. Since $\lambda_1 = L_1 + \sqrt{(L_1)^2 - L_2}$ and $\lambda_2 = L_1 - \sqrt{(L_1)^2 - L_2}$, at a nonumbilic we have

$$\frac{\partial F}{\partial \lambda_1} \frac{\partial \lambda_1}{\partial u} + \frac{\partial F}{\partial \lambda_2} \frac{\partial \lambda_2}{\partial u} = 0,$$

$$2\frac{\partial F}{\partial \lambda_1} \frac{\partial L_1}{\partial u} + \left(\frac{\partial F}{\partial \lambda_2} - \frac{\partial F}{\partial \lambda_1} \right) \frac{\partial \lambda_2}{\partial u} = 0,$$

i.e.

$$2\frac{\partial F}{\partial \lambda_1} \frac{\partial L_1}{\partial u} + \alpha(\lambda_1, \lambda_2)(\lambda_2 - \lambda_1)\frac{\partial \lambda_2}{\partial u} = 0. \tag{4.45}$$

On the other hand,

$$\frac{\partial L_2}{\partial u} = \frac{\partial \lambda_1}{\partial u}\lambda_2 + \frac{\partial \lambda_2}{\partial u}\lambda_1 = 2\lambda_2\frac{\partial L_1}{\partial u} + (\lambda_1 - \lambda_2)\frac{\partial \lambda_2}{\partial u}.$$

Hence

$$(\lambda_2 - \lambda_1)\frac{\partial \lambda_2}{\partial u} = 2\lambda_2\frac{\partial L_1}{\partial u} - \frac{\partial L_2}{\partial u} \tag{4.46}$$

$$= 2\lambda_2\frac{\partial L_1}{\partial u} - \left[2\frac{B_{22}}{E}\frac{\partial L_1}{\partial u} + \frac{B_{11} - B_{22}}{E}\frac{\partial}{\partial u}\left(\frac{B_{22}}{E} \right) - 2\frac{B_{12}}{E}\frac{\partial}{\partial u}\left(\frac{B_{12}}{E} \right) \right].$$

Inserting (4.46) into (4.45), one verifies that

$$\left(2\frac{\partial F}{\partial \lambda_1} + 2\alpha\lambda_2 - 2\alpha\frac{B_{22}}{E} \right) \frac{\partial L_1}{\partial u} = a\frac{B_{11} - B_{22}}{2} + bB_{12} \tag{4.47}$$

where a and b are C^∞-functions on M. (4.47) can be written in the form

$$\left(2\frac{\partial F}{\partial \lambda_1} + 2\alpha\lambda_2 - 2\alpha\frac{B_{22}}{E} \right) \frac{\partial L_1}{\partial u} \equiv 0 \mod \left(\frac{B_{11} - B_{22}}{2}, B_{12} \right). \tag{4.48}$$

Similarly we have

$$\left(2\frac{\partial F}{\partial \lambda_1} + 2\alpha\lambda_2 - 2\alpha\frac{B_{22}}{E} \right) \frac{\partial L_1}{\partial v} \equiv 0 \mod \left(\frac{B_{11} - B_{22}}{2}, B_{12} \right). \tag{4.49}$$

(4.48) and (4.49) hold at nonumbilics. In the following we show that they hold also at umbilics. Let x_0 be an umbilic. If there is a neighborhood $D(x_0)$ of x_0 in M consisting of umbilics, then $L_1 = $ const. on $D(x_0)$ and (4.48), (4.49) obviously hold. If every neighborhood of x_0 contains a nonumbilic, we can take a sequence $x_k \to x_0$, where x_k is nonumbilic and at which (4.48) and (4.49) hold. Our claim follows by continuity.

We consider the net of affine curvature lines

$$-B_{12}(du)^2 + (B_{11} - B_{22})dudv + B_{12}(dv)^2 = 0.$$

Let $p \in M$ be an umbilic. Put $\lambda_0 = \lambda_1 = \lambda_2$, then $F(\lambda_0) = 0$. Since the equations $F(\lambda) = 0, F'(\lambda) = 0$ have no common root we have $F'(\lambda_0) \neq 0$, i.e.

$$\frac{\partial F}{\partial \lambda_1}(\lambda_0, \lambda_0) + \frac{\partial F}{\partial \lambda_2}(\lambda_0, \lambda_0) \neq 0,$$

hence $\frac{\partial F}{\partial \lambda_1}(\lambda_0, \lambda_0) \neq 0$. It follows that there is a neighborhood $D(p)$ of p such that

$$2\frac{\partial F}{\partial \lambda_1} + 2\alpha\lambda_2 - 2\alpha\frac{B_{22}}{E} \neq 0 \text{ in } D(p). \tag{4.50}$$

Inserting (4.48), (4.49), and (4.50) into (4.43) and (4.44), and introducing the functions

$$g := \frac{1}{2}(B_{11} - B_{22}), \; f := B_{12},$$

we obtain

$$\begin{cases} \dfrac{\partial g}{\partial v} - \dfrac{\partial f}{\partial u} \equiv 0 \mod(f,g), \\[2mm] \dfrac{\partial g}{\partial u} + \dfrac{\partial f}{\partial v} \equiv 0 \mod(f,g). \end{cases}$$

Set

$$z := u + \sqrt{-1}v, \text{ and } F := f + \sqrt{-1}g.$$

From Theorem 4.33 we conclude that either $F \equiv 0$ on $D(p)$, or the zeros of F are isolated and of finite order. If $F \equiv 0$ on $D(p)$, we will prove that M is totally umbilical (see below). Suppose that $F \not\equiv 0$, and p is an isolated zero. We compute the index of p. We write $F(z)$ in the form

$$F(z) = \varphi(z)e^{w(z)},$$

where $\varphi(z)$ is an analytic function. Let z_0 correspond to p, then z_0 is an isolated zero of $\varphi(z)$. Suppose that z_0 is of n-th order. We take a simple closed curve c such that z_0 is the only zero of $\varphi(z)$ in the interior of c and there are no zeros on c. Then we have

$$\delta_c(\arg F) = \delta_c(\arg \varphi) + \delta_c(\operatorname{Im} w(z)).$$

Since $w(z)$ is continuous, it follows from the argument principle that

$$\delta_c(\arg F) = 2\pi n.$$

From Theorem 4.32,

$$j = -\frac{1}{2}\frac{1}{2\pi}\delta_c\left\{\arg\left[-B_{12} - \sqrt{-1}\frac{B_{11} - B_{22}}{2}\right]\right\}$$

$$= -\frac{1}{2}\frac{1}{2\pi}\delta_c(\arg F) = -\frac{n}{2} < 0.$$

Since the genus of M is zero, it follows from the Poincaré index formula that

$$2 = \sum j < 0.$$

From the contradiction we obtain $F \equiv 0$. From this and the definition of

$$F := f + \sqrt{-1}g = B_{12} + \sqrt{-1}\cdot\frac{1}{2}(B_{11} - B_{22})$$

we get $B_{12} = 0$ and $B_{11} = B_{22} = L_1$. As G has isothermal form, this gives $\hat{B} = L_1 G$; that means M is totally umbilical and $x(M)$ is an affine sphere. It follows from Theorem 3.35 that $x(M)$ is an ellipsoid. $\qquad\square$

Theorem 4.36. *Let $x : M \to A^3$ be a connected, locally strongly convex surface and let $F : \mathbb{R}^2 \to \mathbb{R}$ be a C^∞-function satisfying the conditions* (i) *and* (ii) *in Theorem 4.35. If $F(\lambda_1, \lambda_2) = 0$ on M and if there is an arc $\Gamma \subset M$ such that $\lambda_1 = \lambda_2$ on Γ, then $x(M)$ lies on an affine sphere.*

Proof. Let $p_0 \in \Gamma$ and $p_1 \in M - \Gamma$. We draw a curve $\gamma : [0,1] \to M$ such that $\gamma(0) = p_0$ and $\gamma(1) = p_1$. Since the zeros of any nontrivial solution of the elliptic system (4.39) are isolated, the condition $\lambda_1 = \lambda_2$ on Γ implies that there is a neighborhood $D(p_0)$ of p_0 such that $\lambda_1 = \lambda_2$ on $D(p_0)$. Then there is a $t > 0$ such that $\lambda_1 = \lambda_2$ on $\gamma([0,t))$. Let

$$t^* = \sup\{t \mid 0 < t \le 1, \lambda_1 = \lambda_2 \text{ on } \gamma([0,t))\}.$$

By continuity we have $\lambda_1 = \lambda_2$ at t^*. Assume that $\lambda_1 \neq \lambda_2$ at p_1, then $t^* < 1$. As before, there is a neighborhood D^* of $\gamma(t^*)$ such that $\lambda_1 = \lambda_2$ on D^*, which contradicts the definition of t^*. Hence $\lambda_1 = \lambda_2$ at p_1. $\qquad\square$

Theorem 4.37. *Let $x : M \to A^3$ be an ovaloid and let $f(y^1, y^2)$ be a polynomial in y^1, y^2 satisfying*

$$\frac{\partial f}{\partial y^1}(y, y) = \frac{\partial f}{\partial y^2}(y, y). \tag{4.51}$$

Define $f(y) := f(y, y)$. Let $\{p_1, p_2, \ldots, p_s\}$ be the set of real roots of $f'(y) = 0$. If

$$f(\lambda_1, \lambda_2) = \text{const.} = c \quad \text{on } M,$$

where $c \in \mathbb{R} \setminus \{f(p_1), f(p_2), \ldots, f(p_s)\}$, then $x(M)$ is an ellipsoid.

Theorem 4.38. *Let $x : M \to A^3$ be a connected, locally strongly convex surface. Let $f(y^1, y^2)$ and $\{p_1, p_2, \ldots, p_s\}$ be defined as in Theorem 4.37. If on M*

$$f(\lambda_1, \lambda_2) = \text{const.} = c,$$

where $c \in \mathbb{R} \setminus \{f(p_1), f(p_2), \ldots, f(p_s)\}$, and if there is an arc $\Gamma \subset M$ such that $\lambda_1 = \lambda_2$ on Γ then $\lambda_1 = \lambda_2$ on M.

Proof of Theorems 4.37 and 4.38. Consider the polynomial function in y^2

$$F(y^2) = \frac{\partial f}{\partial y^2}(y^1, y^2) - \frac{\partial f}{\partial y^1}(y^1, y^2).$$

Then (4.51) gives $F(y^1) = 0$, i.e. y^1 is a root of $F(y^2) = 0$. Hence there is a polynomial $\alpha(y^1, y^2)$ such that

$$\frac{\partial f}{\partial y^2}(y^1, y^2) - \frac{\partial f}{\partial y^1}(y^1, y^2) = \alpha(y^1, y^2)(y^2 - y^1).$$

Since $c \in \mathbb{R}\backslash\{f(p_1), f(p_2), \ldots, f(p_s)\}$, the equations

$$f(y) - c = 0 \text{ and } f'(y) = 0$$

have no common real root. Application of Theorems 4.35 and 4.36 gives the assertions. □

In the following we consider the class of surfaces for which the Pick invariant J and the affine principal curvatures satisfy a relation of the type $J = F(\lambda_1, \lambda_2)$.

Theorem 4.39. *Let $x : M \rightarrow A^3$ be a connected, locally strongly convex C^∞-surface and let $F : \mathbb{R}^2 \rightarrow \mathbb{R}$ be defined as in Theorem 4.35. If $J = F(\lambda_1, \lambda_2)$ on M, and if there is an arc $\Gamma \subset M$ such that $\lambda_1 = \lambda_2$ and $J = 0$ on Γ then $x(M)$ lies on a quadric.*

Proof. From $J = F(\lambda_1, \lambda_2)$ we have

$$
\begin{aligned}
\frac{\partial F}{\partial \lambda_1}\frac{\partial \lambda_1}{\partial u} + \frac{\partial F}{\partial \lambda_2}\frac{\partial \lambda_2}{\partial u} &= \frac{\partial J}{\partial u}, \\
2\frac{\partial F}{\partial \lambda_1}\frac{\partial L_1}{\partial u} + \alpha \cdot (\lambda_2 - \lambda_1)\frac{\partial \lambda_2}{\partial u} &= \frac{\partial J}{\partial u}.
\end{aligned}
\tag{4.52}
$$

Since

$$J = \frac{2}{E^3}\left((A_{111})^2 + (A_{222})^2\right),$$

we have

$$\frac{\partial J}{\partial u} \equiv 0 \mod(A_{111}, A_{222}). \tag{4.53}$$

From (4.52), (4.53), and (4.46) we obtain

$$\left(2\frac{\partial F}{\partial \lambda_1} + 2\alpha\lambda_2 - 2\alpha\frac{B_{22}}{E}\right)\frac{\partial L_1}{\partial u} \equiv 0 \mod\left(\frac{B_{11} - B_{22}}{2}, B_{12}, A_{111}, A_{222}\right). \tag{4.54}$$

Similarly, we have

$$\left(2\frac{\partial F}{\partial \lambda_1} + 2\alpha\lambda_2 - 2\alpha\frac{B_{22}}{E}\right)\frac{\partial L_1}{\partial v} \equiv 0 \mod\left(\frac{B_{11} - B_{22}}{2}, B_{12}, A_{111}, A_{222}\right). \tag{4.55}$$

(4.54) and (4.55) hold at nonumbilics. An analogous argument as in the proof of Theorem 4.35 shows that both relations hold also at umbilics. Let $p \in M$ be a point with $\lambda_1 = \lambda_2 =: \lambda_0$ and $A_{111} = A_{222} = 0$. Then $F(\lambda_0) = 0$. It follows that $F'(\lambda_0) \neq 0$. Hence there is an open set D with $p \in D$ such that

$$2\frac{\partial F}{\partial \lambda_1} + 2\alpha\lambda_2 - 2\alpha\frac{B_{22}}{E} \neq 0 \text{ in } D.$$

Denote

$$
\begin{aligned}
g &:= \tfrac{1}{2}(B_{11} - B_{22}), & f &:= B_{12}, \\
e &:= A_{111}, & h &:= A_{222}.
\end{aligned}
$$

We get

$$\frac{\partial g}{\partial v} - \frac{\partial f}{\partial u} \equiv 0 \mod(f, g, e, h),$$

$$\frac{\partial g}{\partial u} + \frac{\partial f}{\partial v} \equiv 0 \mod(f, g, e, h).$$

On the other hand, from the integrability conditions $(2.119')$ in Proposition 2.16 we have

$$A_{111,1} - A_{222,2} = -\frac{1}{2}E(B_{22} - B_{11}),$$

$$A_{111,2} - A_{222,1} = B_{12}E.$$

We obtain the following system

$$\begin{cases} \dfrac{\partial g}{\partial v} - \dfrac{\partial f}{\partial u} \equiv 0 \mod(f, g, e, h), \\[2mm] \dfrac{\partial g}{\partial u} + \dfrac{\partial f}{\partial v} \equiv 0 \mod(f, g, e, h), \\[2mm] \dfrac{\partial e}{\partial u} - \dfrac{\partial h}{\partial v} \equiv 0 \mod(f, g, e, h), \\[2mm] \dfrac{\partial e}{\partial v} + \dfrac{\partial h}{\partial u} \equiv 0 \mod(f, g, e, h). \end{cases} \qquad (4.56)$$

Define the complex functions

$$W_1 := f + \sqrt{-1}g, \quad W_2 := e + \sqrt{-1}h,$$

$$W = \begin{pmatrix} W_1 \\ W_2 \end{pmatrix}.$$

Then the system (4.56) reads

$$\frac{\partial W}{\partial \bar{z}} = PW + Q\bar{W}, \qquad (4.57)$$

where P, Q are 2×2 matrices. This system is homogeneous and in Pascali normal form. By Theorem 4.34 the zeros of any nontrivial solution of (4.57) are isolated and of finite order. An analogous argument as in the proof of Theorem 4.36 now proves the assertion. $\qquad \square$

From Theorems 4.35–4.39 we can obtain many uniqueness theorems by choosing special functions. We give some examples. In the following we assume that $x : M \to A^3$ is an ovaloid.

Example 4.1. Let $f(y^1, y^2) = (y^1)^k + (y^2)^k$. Then

$$f(y) := f(y, y) = 2y^k, \quad f'(y) = 2ky^{k-1}.$$

The only root of $f'(y) = 0$ is $y = 0$. Hence if

$$(\lambda_1)^k + (\lambda_2)^k = \text{const.} > 0 \text{ on } M \text{ for } k \neq 0$$

then $\lambda_1 = \lambda_2$ on M and $x(M)$ is an ellipsoid.

Example 4.2. Let $f(y^1, y^2) = (y^1)^{k+1} + (y^2)^{k+1} - c((y^1)^k + (y^2)^k)$ where $c \neq 0$ is a constant. Then

$$f(y) = 2y^{k+1} - 2cy^k, \quad f'(y) = 2(k+1)y^k - 2cky^{k-1}.$$

Hence if

$$\frac{(\lambda_1)^{k+1} + (\lambda_2)^{k+1}}{(\lambda_1)^k + (\lambda_2)^k} = \text{const.} \neq 0, \quad (\lambda_1)^k + (\lambda_2)^k \neq 0, \text{ on } M,$$

then $\lambda_1 = \lambda_2$ on M and $x(M)$ is an ellipsoid.

Example 4.3. Let $f(y^1, y^2) = (L_1)^m (L_2)^n$, where $L_1 = \frac{1}{2}(y^1 + y^2)$, $L_2 = y^1 y^2$. Then

$$f(y) = y^{2n+m}, \quad f'(y) = (2n+m)y^{2n+m-1}.$$

Hence if

$$(L_1)^m (L_2)^n = \text{const.} \neq 0, \quad m + 2n \neq 0 \text{ on } M,$$

then $x(M)$ is an ellipsoid.

4.2.4 An affine analogue of the Cohn–Vossen theorem

Theorem 4.13 states that two hyperovaloids are equiaffinely equivalent if, at corresponding points, the Blaschke metrics and the equiaffine mean curvatures coincide. At a first glance, this theorem seems to be a certain analogue of the famous Euclidean Cohn–Vossen theorem, stating that two ovaloids in Euclidean 3-space E^3 are equivalent modulo the group of motions, if their Euclidean first fundamental forms coincide. A relative simple argument (see [334]) shows that the following is an adequate generalization of Cohn–Vossen's theorem:

Assume that for two ovaloids in Euclidean 3-space the Levi–Civita connections of the first fundamental forms are equal at corresponding points, then both ovaloids are homothetic.

This result suggests to investigate the analogous affine situation and to consider the connection induced by the affine normal; a similar result for the affine conormal connections was already proved in Theorem 4.15.

A more detailed comparison of such results, from the relative point of view, is contained in [336].

As we are going to use the index method, the results are restricted to dimension $n = 2$.

Theorem 4.40. *Let M be a closed, 2-dimensional differentiable manifold of genus zero and let $g, g^\#$ be two Riemannian metrics on M with the same Riemannian volume form $\omega(g) = \omega(g^\#)$. Let $\tilde{\nabla}$ be an affine connection without torsion on M. If g and $g^\#$ satisfy Codazzi equations with respect to $\tilde{\nabla}$, then $g = g^\#$ on M.*

Proof. Define $g' := g - g^\#$. Let (v^1, v^2) be any local coordinate system on M and denote differentiation by $\tilde{\nabla}_j = \tilde{\nabla}_{\partial_j}$, $\partial_j = \frac{\partial}{\partial v^j}$ in the local calculus. If g and $g^\#$ satisfy Codazzi equations with respect to $\tilde{\nabla}$, i.e.

$$\tilde{\nabla}_k g_{ij} = \tilde{\nabla}_j g_{ik}, \quad \tilde{\nabla}_k g^\#_{ij} = \tilde{\nabla}_j g^\#_{ik},$$

then g' satisfies

$$\partial_j g'_{ik} - \partial_k g'_{ij} = \sum \tilde{\Gamma}^l_{ij} g'_{lk} - \sum \tilde{\Gamma}^l_{ik} g'_{lj}, \tag{4.58}$$

where $\tilde{\Gamma}^k_{ij}$ denote the Christoffel symbols with respect to $\tilde{\nabla}$. Define the symmetric (2,0)-tensor D by

$$D^{ik} := g^{ik} + g^{\#ik},$$

where g^{ik} and $g^{\#ik}$ are the components of g^{-1} and $g^{\#-1}$, resp.; D is positive definite. We introduce isothermal coordinates (u^1, u^2) for D [127, pp. 137–139]. As the volume forms coincide, we get

$$\text{tr}_D g' = \sum D^{ik} g'_{ik} = 0, \tag{4.59}$$

and thus we have $g'_{11} = -g'_{22}$ in isothermal coordinates on D. (4.59) implies that $\det(g'_{ij}) \leq 0$. Now (4.58) is a linear elliptic system in g'_{11} and g'_{12}:

$$\partial_1 g'_{12} - \partial_2 g'_{11} = (\tilde{\Gamma}^1_{21} + \tilde{\Gamma}^2_{11})(-g'_{11}) + (\tilde{\Gamma}^1_{11} - \tilde{\Gamma}^2_{12})g'_{12},$$

$$\partial_2 g'_{12} + \partial_1 g'_{11} = (\tilde{\Gamma}^1_{22} + \tilde{\Gamma}^2_{21})g'_{11} - (\tilde{\Gamma}^1_{21} - \tilde{\Gamma}^2_{22})g'_{12}.$$

Now we use the index method. From Section 4.2.2 we know that either the solutions of this system are identically zero, or the points $p \in M$ with

$$g'_{11}(p) = 0 = g'_{12}(p)$$

are isolated and the net defined globally by the equation $g'(v, w) = 0$ has negative index in p. But the index sum on M must be positive from the Poincaré index formula in Theorem 4.28, so the only global solution is $g' = 0$, that means $g = g^\#$. □

It would be of interest to solve the following open problem similar to Theorem 4.40:

Problem 4.4. *Let M be defined as in Theorem 4.40, and let $\nabla, \nabla^\#$ be two affine connections without torsion. Assume that a Riemannian metric g satisfies Codazzi equations with respect to both connections:*

$$\nabla_k g_{ij} = \nabla_j g_{ik} \text{ and } \nabla^\#_k g_{ij} = \nabla^\#_j g_{ik}.$$

Under which additional conditions do the connections $\nabla, \nabla^\#$ coincide?

A sufficient condition for solving Problem 4.4 is that ∇ and $\nabla^\#$ are Levi–Civita connections of Riemannian metrics g and $g^\#$, respectively, and that the Riemannian volume forms coincide: $\omega(g) = \omega(g^\#)$; see [343].

As a consequence of Theorem 4.40 we obtain the following result.

Theorem 4.41. *Let $x, x^\# : M \to A^3$ be ovaloids. Assume that the induced connections coincide on M:*

$$\nabla = \nabla^\#.$$

Then x and $x^\#$ are affinely equivalent.

Proof. The assumptions imply that the volume forms of the Blaschke metrics coincide; this is a simple consequence of the equations:

$$\nabla + A = \nabla = \nabla^\# = \nabla^\# + A^\#.$$

Contraction and the apolarity of A and $A^\#$ give the following relation for the Christoffel symbols:

$$\frac{1}{2}\partial_i \ln \det G = \sum \Gamma^j_{ij} = \sum \Gamma^{\#j}_{ij} = \frac{1}{2}\partial_i \ln \det G^\#,$$

so $\omega(G^\#) = \alpha\omega(G)$ for a positive constant $\alpha \in \mathbb{R}^+$.

Define $x^{\#\#} : M \to A^3$ by $x^{\#\#} = \alpha^{(n+2)/2n(n+1)}x$, then

$$\omega(G^\#) = \omega(G^{\#\#}), \nabla^\# = \nabla^{\#\#}.$$

The assertion follows now from Theorem 4.40. □

Remark 4.42. Nomizu and Opozda [243] independently gave a proof of Theorem 4.41, under the additional assumption that $L_2 > 0$ on M, using an equiaffine Herglotz integral formula.

5 Variational problems and affine maximal surfaces

In Euclidean surface theory, the study of the first variation of the area functional led to an Euler–Lagrange equation of second order; this equation is equivalent to the vanishing of the Euclidean mean curvature. For vanishing mean curvature, the second variation is positive, thus this class of surfaces was called minimal surfaces, and it has been studied in detail from the last century through the present.

When Blaschke (see [20, § 68]) studied the first variation of the equiaffine area integral, he discovered a series of important facts. In particular, he found that the Euler–Lagrange equation is of fourth order and nonlinear, but this equation is equivalent to the vanishing of the affine mean curvature. This analogy leads to the notion "affine minimal surfaces" for the corresponding class — without studying before the second variation of the area integral.

It took about 60 years until Calabi [42] took up this study. He found the situation to be more complicated than expected, and proved that, for locally strongly convex hypersurfaces, the second variation is negative, if
(i)   the dimension n is two, or
(ii) the hypersurface is a graph for $n > 2$.

Thus, he proposed to call locally strongly convex hypersurfaces with vanishing affine mean curvature "affine maximal". The expression for the second variation is so complicated that he could give no estimate of the sign for $n > 2$ in the general case. For indefinite Blaschke metric there are examples showing that both signs for the second variation may occur; for such examples see [376].

All improper affine hyperspheres have vanishing affine mean curvature, so there are many examples with $L_1 \equiv 0$; the simplest is the elliptic paraboloid. We follow Calabi's proposal and name the class of hypersurfaces with $L_1 \equiv 0$ *affine maximal (hyper)surfaces*, as we restrict our study to locally strongly convex hypersurfaces.

Some of the important tools for local and global investigations in the theory of Euclidean minimal surfaces are:
(a) the Schwarz–Weierstraß representation;
(b) the Gauß map and the application of methods from complex function theory.

For both (a) and (b), there exist affine analogues.

Affine Weierstraß formulas are due to results of Calabi [44], Li [166, 167], and Terng [364], while the role of the affine conormal was discovered much earlier (see [20, § 70]; see also [183]); its importance for the global study of affine maximal surfaces was shown by Calabi [44–47] and Li [166, 167]. The global investigations concern the "affine Bernstein problem", an analogue of the Euclidean Bernstein problem. There are two conjectures due to Calabi [42] and Chern [60]; we present the solutions for the two-dimensional case in Sections 5.4, 5.5, and 5.6, respectively.

We begin this chapter in Section 5.1 with a problem posed by Voss in 1986: the calculation of the variational formulas for higher affine mean curvature functions. These formulas are due to Li [165]; see also Voss [378].

5.1 Variational formulas for higher affine mean curvatures

Let $x : M \to A^{n+1}$ be a compact, locally strongly convex hypersurface without boundary. In Section 2.2.3 we defined the affine curvature functions L_r; now we compute the first variation of $\int_M L_r dV$.

Choose a local equiaffine frame field $\{x; e_1, \ldots, e_{n+1}\}$ on $x(M)$. Then we have

$$dx = \sum \omega^A e_A,$$

$$de_A = \sum \omega_A^B e_B,$$

$$d\omega^A = \sum \omega^B \wedge \omega_B^A,$$

$$d\omega_A^B = \sum \omega_A^C \wedge \omega_C^B,$$

$$\sum \omega_A^A = 0.$$

We restrict to $x(M)$ and denote by $\bar{\omega}^A, \bar{\omega}_A^B$ the restrictions of ω^A, ω_A^B. Then

$$\bar{\omega}^{n+1} = 0,$$

$$\bar{\omega}_i^{n+1} = \sum h_{ij} \bar{\omega}^j, \qquad h_{ij} = h_{ji},$$

$$\bar{\omega}_{n+1}^i = -\sum l^{ik} \bar{\omega}_k^{n+1}, \quad l^{ik} = l^{ki},$$

$$\bar{\omega}_{n+1}^{n+1} = -\frac{1}{n+2} d_M \ln H, \quad H := \det(h_{ij}),$$

where d_M denotes the exterior differential operator on $x(M)$.

We define a differential form $\bar{\Omega}_r$ by

$$\bar{\Omega}_r = \frac{1}{n!} \sum \delta_{i_1 i_2 \cdots i_n}^{1 \, 2 \, \cdots \, n} \bar{\omega}_{n+1}^{i_1} \wedge \cdots \wedge \bar{\omega}_{n+1}^{i_r} \wedge \bar{\omega}^{i_{r+1}} \wedge \cdots \wedge \bar{\omega}^{i_n}. \tag{5.1}$$

A direct calculation gives

$$H^{(r+1)/(n+2)} \bar{\Omega}_r = (-1)^r H^{1/(n+2)} L_r \bar{\omega}^1 \wedge \cdots \wedge \bar{\omega}^n \tag{5.2}$$

$$= (-1)^r L_r dV,$$

where

$$dV = H^{1/(n+2)} \bar{\omega}^1 \wedge \cdots \wedge \bar{\omega}^n$$

is the volume element with respect to the Blaschke metric (compare the expression below (4.5) and note that there $G_{ij} = \delta_{ij}$).

Now we consider a one-parameter variation of $x(M)$. Denote by I the interval $[-\frac{1}{2}, \frac{1}{2}]$. Let $f : M \times I \to A^{n+1}$ be a smooth map such that its restriction to $M \times \{t\}, t \in I$, is a locally strongly convex immersion and such that $f(p, 0) = x(p)$ for every $p \in M$. We choose a local unimodular affine frame field $\{f(p, t); e_1(p, t), \ldots, e_{n+1}(p, t)\}$ on $M \times I$ such that, for each $t \in I$, $e_i(p, t)$ are tangent to $f(M \times \{t\})$, $e_{n+1}(p, t)$ is parallel to the affine normal to $f(M \times \{t\})$ at $f(p, t)$, and

$$h_{ij}(p, 0) = \delta_{ij}.$$

Then

$$e_{n+1}(p, 0) = Y, \quad \bar{\omega}_{n+1}^{n+1}(p, 0) = 0.$$

We introduce a differential form Ω_r:

$$\Omega_r = \frac{1}{n!} \sum \delta_{i_1 i_2 \cdots i_n}^{12 \cdots n} \omega_{n+1}^{i_1} \wedge \cdots \wedge \omega_{n+1}^{i_r} \wedge \omega^{i_{r+1}} \wedge \cdots \wedge \omega^{i_n}. \tag{5.3}$$

Taking the exterior differentiation of Ω_r and using the structure equations, we get

$$d\Omega_r = (r + 1)\omega_{n+1}^{n+1} \wedge \Omega_r + (n - r)\omega^{n+1} \wedge \Omega_{r+1}.$$

Hence

$$d(H^{(r+1)/(n+2)}\Omega_r) = \frac{r + 1}{n + 2} H^{(r+1)/(n+2)} d\ln H \wedge \Omega_r$$
$$+ H^{(r+1)/(n+2)}[(r + 1)\omega_{n+1}^{n+1} \wedge \Omega_r + (n - r)\omega^{n+1} \wedge \Omega_{r+1}]. \tag{5.4}$$

Pulling back under f and splitting off the terms in dt, we can write

$$\omega^i = \bar{\omega}^i + a^i dt, \quad 1 \leq i \leq n - 1,$$

$$\omega^{n+1} = a\, dt;$$

$$\omega_{n+1}^i = \bar{\omega}_{n+1}^i + a_{n+1}^i dt, \quad 1 \leq i \leq n - 1,$$

$$\omega_{n+1}^{n+1} = \bar{\omega}_{n+1}^{n+1} + a_{n+1}^{n+1} dt,$$

where $\bar{\omega}^i, \bar{\omega}_{n+1}^i, \bar{\omega}_{n+1}^{n+1}$ depend on p and t but don't contain dt. The coefficients a^i, a, a_{n+1}^i, a_{n+1}^{n+1} are functions of p and t. The functions $a^i(p, 0)$ and $a(p, 0)$ are the tangential components and the normal component of the variational vector field, respectively. Since $h_{ij}(p, 0) = \delta_{ij}$, we have

$$H(p, 0) = 1, \quad \omega_{n+1}^{n+1}(p, 0) = a_{n+1}^{n+1} dt.$$

The operator d on $M \times I$ can be decomposed as

$$d = d_M + dt \frac{\partial}{\partial t}.$$

We can write

$$H^{(r+1)/(n+2)}\Omega_r = H^{(r+1)/(n+2)}\bar{\Omega}_r + dt \wedge \varphi,$$

where

$$\varphi := H^{\frac{r+1}{n+2}} \left\{ \frac{r}{n!} \sum \delta^{12\cdots n}_{i_1 i_2 \cdots i_n} a^{i_1}_{n+1} \bar{\omega}^{i_2}_{n+1} \wedge \cdots \wedge \bar{\omega}^{i_r}_{n+1} \wedge \bar{\omega}^{i_{r+1}} \wedge \cdots \wedge \bar{\omega}^{i_n} \right.$$
$$\left. + (-1)^r \frac{n-r}{n!} \sum \delta^{12\cdots n}_{i_1 i_2 \cdots i_n} \bar{\omega}^{i_1}_{n+1} \wedge \cdots \wedge \bar{\omega}^{i_r}_{n+1} \wedge a^{i_{r+1}} \bar{\omega}^{i_{r+2}} \wedge \cdots \wedge \bar{\omega}^{i_n} \right\}.$$

Hence

$$d(H^{(r+1)/(n+2)}\Omega_r) = d_M(H^{(r+1)/(n+2)}\bar{\Omega}_r) - dt \wedge d_M\varphi + dt \wedge \frac{\partial(H^{(r+1)/(n+2)}\bar{\Omega}_r)}{\partial t}. \tag{5.5}$$

Comparing (5.4) with (5.5) and equating the terms in dt we obtain

$$\frac{\partial(H^{(r+1)/(n+2)}\bar{\Omega}_r)}{\partial t}\bigg|_{t=0} = d_M\varphi + \frac{r+1}{n+2}\frac{\partial \ln H}{\partial t}\bigg|_{t=0}\bar{\Omega}_r + (r+1)a^{n+1}_{n+1}(p,0)\bar{\Omega}_r + (n-r)\phi\,\bar{\Omega}_{r+1},$$

where we denote $\phi = a(p,0)$. Integrating over M we get

$$\frac{\partial}{\partial t}\int_M L_r\,dV\bigg|_{t=0} = (r+1)\int_M \left(a^{n+1}_{n+1} + \frac{1}{n+2}\frac{\partial \ln H}{\partial t} \right)\bigg|_{t=0} L_r\,dV - (n-r)\int_M L_{r+1}\phi\,dV. \tag{5.6}$$

Now we calculate $(a^{n+1}_{n+1} + \frac{1}{n+2}\frac{\partial \ln H}{\partial t})|_{t=0}$ (see finally (5.10)). Without loss of generality we can assume that the variation is given by

$$f(p,t) = x(p) + t\phi\,e_{n+1} + t\sum \psi^i e_i,$$

where $\{e_1, \ldots, e_n, e_{n+1}\}$ is a local equiaffine frame field on $x(M)$ such that

$$h_{ij} = \delta_{ij}, \quad e_{n+1} = Y. \tag{5.7}$$

We have

$$d_M f(p,t)$$
$$= d_M x + t d_M \phi\,e_{n+1} - t\phi \sum l^{ij}\bar{\omega}^{n+1}_i e_j + t\sum d_M\psi^i e_i + t\sum \psi^i \bar{\omega}^j_i e_j + t\sum \psi^i \bar{\omega}^{n+1}_i e_{n+1}$$
$$= \sum \left(e_i + t\phi_i e_{n+1} - t\phi \sum l^{ij}e_j + t\sum \psi^j_{,i}e_j + t\psi^i e_{n+1} + t\sum \psi^k A_{kij}e_j \right)\bar{\omega}^i,$$

where ϕ_i and $\psi^j_{,i}$ are defined via the equations

$$d_M\phi = \sum \phi_i \bar{\omega}^i,$$
$$d_M\psi^j + \frac{1}{2}\sum \psi^k(\bar{\omega}^j_k - \bar{\omega}^k_j) = \sum \psi^j_{,i}\bar{\omega}^i.$$

We choose the equiaffine frame field $\{e_A(p,t)\}$ over $M \times I$ such that

$$e_i(p,t) = e_i + t\phi_i e_{n+1} - t\phi \sum l^{ij}e_j + t\sum \psi^j_{,i}e_j + t\psi^i e_{n+1} + t\sum \psi^k A_{kij}e_j, \tag{5.8}$$

and $e_{n+1}(p,t)$ is parallel to the affine normal to $f(M \times \{t\})$. Then

$$\bar{\omega}^j(p,t) = \bar{\omega}^j(p,0) = \bar{\omega}^j.$$

Since

$$de_i(p,t) = d_M e_i(p,t) + dt \frac{\partial e_i(p,t)}{\partial t}$$
$$= \sum (\bar\omega_i^j + a_i^j dt) e_j(p,t) + (\bar\omega_i^{n+1} + a_i^{n+1} dt) e_{n+1}(p,t),$$

from (5.8) we get

$$a_i^j = -\phi l^{ij} + \psi_{,i}^j + \sum \psi^k A_{kij}.$$

Hence

$$a_{n+1}^{n+1} = -\sum a_i^i = n\phi L_1 - \sum \psi_{,i}^i.$$

The relations

$$d_M e_i(p,t) \equiv d_M e_i + t d_M \phi_i e_{n+1} - t\phi \sum l^{ik} \bar\omega_k^{n+1} e_{n+1} + t\sum \psi_{,i}^k \bar\omega_k^{n+1} e_{n+1}$$
$$+ t\sum \psi^k A_{kij} \bar\omega_j^{n+1} e_{n+1} + t d_M \psi^i e_{n+1} \quad \mod (e_1, e_2, \dots, e_n)$$

and

$$\mathrm{Det}(e_1(p,t), \dots, e_n(p,t), d_M e_i(p,t)) = \sum h_{ij}(p,t) \bar\omega^j$$

imply that

$$\sum \frac{\partial h_{ij}(p,t)}{\partial t}\Big|_{t=0} \bar\omega^j = \sum_{l=1}^n \mathrm{Det}\Big(e_1, \dots, \frac{\partial e_l(p,t)}{\partial t}\Big|_{t=0}, \dots, e_n, d_M e_i\Big)$$

$$+ \mathrm{Det}\Big(e_1, \dots, e_n, \frac{\partial d_M e_i(p,t)}{\partial t}\Big|_{t=0}\Big)$$

$$= (-n\phi L_1 + \sum \psi_{,k}^k) \sum \delta_{ij} \bar\omega^j + \sum \phi_{,ij} \bar\omega^j - \sum \phi_k A_{kij} \bar\omega^j$$

$$+ \sum \psi_{,j}^i \bar\omega^j - \phi \sum l^{ij} \bar\omega^j + \sum \psi_{,i}^j \bar\omega^j,$$

i.e.

$$\frac{\partial h_{ij}(p,t)}{\partial t}\Big|_{t=0} = -nL_1 \phi \delta_{ij} + \Big(\sum \psi_{,k}^k\Big) \delta_{ij} + \psi_{,j}^i + \psi_{,i}^j - \phi l^{ij} + \phi_{,ij} - \sum \phi_k A_{kij}. \tag{5.9}$$

From (5.7) and (5.9) it follows that

$$\frac{1}{n+2} \frac{\partial \ln H}{\partial t}\Big|_{t=0} = -\frac{n(n+1)}{n+2} L_1 \phi + \sum \psi_{,k}^k + \frac{1}{n+2} \Delta^{(B)} \phi,$$

where, as before, $\Delta^{(B)}$ denotes the Laplacian with respect to the Blaschke metric; thus we finally get the announced relation:

$$\Big(a_{n+1}^{n+1} + \frac{1}{n+2} \frac{\partial \ln H}{\partial t}\Big)\Big|_{t=0} = \frac{n}{n+2} L_1 \phi + \frac{1}{n+2} \Delta^{(B)} \phi. \tag{5.10}$$

Inserting (5.10) into (5.6), we obtain

$$\frac{\partial}{\partial t} \int_M L_r dV\Big|_{t=0} = \int_M \Big[\frac{r+1}{n+2} L_r \Delta^{(B)} \phi + \frac{n(r+1)}{n+2} L_1 L_r \phi - (n-r) L_{r+1} \phi\Big] dV, \tag{5.11}$$

and using Green's formula

$$\frac{\partial}{\partial t} \int_M L_r dV \Big|_{t=0} = \int_M \left[\frac{r+1}{n+2} \Delta^{(B)} L_r + \frac{n(r+1)}{n+2} L_1 L_r - (n-r) L_{r+1} \right] \phi \, dV. \tag{5.12}$$

When $r = 0$, (5.12) gives the well-known formula (see [20, 60])

$$\frac{\partial}{\partial t} \int_M dV \Big|_{t=0} = -\frac{n(n+1)}{n+2} \int_M L_1 \phi \, dV. \tag{5.13}$$

Remark 5.1. Let M be a compact manifold with C^∞-boundary ∂M. Then:
(i) the variational formulas (5.12) for $r \geq 1$ hold if $a^i = a^i_{n+1} = 0$ and $\phi = \phi_i = 0$ on ∂M;
(ii) the variational formula (5.13) holds if $a^i = 0$ and $\phi = \phi_i = 0$ on ∂M.

Using the variational formula (5.11), we can derive again the Minkowski integral formula in Section 4.1.1. Let $x : M \to A^{n+1}$ be a closed convex hypersurface. Consider the following variation of $x(M)$:

$$f(p, t) = (1 + t)x(p).$$

A direct calculation gives

$$\frac{\partial}{\partial t} \int_M L_r dV \Big|_{t=0} = \left[(n-r) - \frac{n(r+1)}{n+2} \right] \int_M L_r dV. \tag{5.14}$$

From (5.11) we get

$$\frac{\partial}{\partial t} \int_M L_r dV \Big|_{t=0} = -\int_M \left[\frac{r+1}{n+2} L_r \Delta^{(B)} \Lambda + \frac{n(r+1)}{n+2} L_1 L_r \Lambda - (n-r) L_{r+1} \Lambda \right] dV, \tag{5.15}$$

where Λ is the affine support function from Section 2.3.2. Comparing (5.14) with (5.15) and using $\Delta^{(B)} \Lambda = -nL_1 \Lambda + n$, for an ovaloid we obtain the affine Minkowski formulas (see Section 4.1.1):

$$\int_M (L_r - \Lambda L_{r+1}) dV = 0, \quad r = 0, 1, \ldots, n-1.$$

As an application of the variational formula we have the following theorem.

Theorem 5.2. Let $x : M \to A^{n+1}$ be an ovaloid with $L_n > 0$ on $x(M)$. If n is even and $x(M)$ is a solution of the variational problem $\delta \int_M L_{n/2} dV = 0$, then $x(M)$ is an ellipsoid.

Proof. From (5.12) it follows that

$$\int_M \left[\frac{1}{2} \Delta^{(B)} L_{n/2} + \frac{n}{2} (L_1 L_{n/2} - L_{n/2 +1}) \right] dV = 0.$$

Hence

$$\int_M (L_1 L_{n/2} - L_{n/2+1}) dV = 0.$$

Application of the inequality $L_1 L_{n/2} \geq L_{n/2+1}$ [112, p. 52] gives equality in each point of M, which is possible if and only if each point is umbilical. Theorem 3.35 of Blaschke and Deicke then gives the assertion. □

5.2 Affine maximal hypersurfaces

5.2.1 Definitions and fundamental results

Let $x : M \rightarrow A^{n+1}$ be a locally strongly convex hypersurface, where the parameter manifold M may be open, or compact with (possibly empty) smooth boundary ∂M.

Definition 5.3. A differentiable map $f : M \times I \rightarrow A^{n+1}$, where I is an open interval $(-\varepsilon, \varepsilon), \varepsilon > 0$, is called an *allowable interior deformation* of x, if it satisfies the following properties:
(i) for each $t \in I$ the map $x_t : M \rightarrow A^{n+1}$ defined by $x_t(p) = f(p, t)$ is a locally strongly convex hypersurface such that $x_0 = x$;
(ii) there exists a compact subdomain $\Sigma' \subset M$ (the closure of a connected, open subset of M) with smooth boundary $\partial \Sigma'$, where $\partial \Sigma'$ may contain, meet, or be disjoint from ∂M such that, for each $p \in M \backslash \Sigma'$ and all $t \in I, f(t, p) = x(p)$;
(iii) for each $p \in \partial \Sigma'$ and for all $t \in I, f(t, p) = x(p)$ and the tangent hyperplane $dx_t(p)$ coincides with $dx(p)$.

A locally strongly convex hypersurface $x^* : M \rightarrow A^{n+1}$ is said to be *interior-homotopic to x*, if there exists an allowable interior deformation

$$f : M \times I \rightarrow A^{n+1}$$

with $I = (-\varepsilon, 1 + \varepsilon)$ such that $x_0 = x$, $x_1 = x^*$.

In what follows, when we study variations of the affinely invariant volume of $x(M)$ under interior deformations, we may replace M, without loss of generality, by the compact subdomain $\Sigma' \subset M$, or else assume from the beginning that M is compact with smooth boundary.

Definition 5.4. Let $x : M \rightarrow A^{n+1}$ be a locally strongly convex hypersurface. If $L_1 = 0$ on M then $x(M)$ is called an *affine maximal hypersurface*.

From (5.13) and the remarks of the previous section it is easy to see that the affine maximal hypersurfaces are extremals of the interior variation of the affine invariant volume. Historically the hypersurfaces with $L_1 = 0$ were called "affine minimal hypersurfaces". Calabi (see [42]) suggested to call locally strongly convex hypersurfaces with $L_1 = 0$ "*affine maximal hypersurfaces*" because of the following result (in fact, Calabi's result is a little more general than Theorem 5.5 below; see [42]).

Theorem 5.5. *Let $x, x^\# : \Omega \to A^{n+1}$ be two graphs defined on a compact domain $\Omega \subset A^n$ by strongly convex functions $f, f^\#$, namely*

$$x^{n+1} = f(x^1, \dots, x^n) \text{ and } x^{n+1} = f^\#(x^1, \dots, x^n),$$

respectively. Assume that on $\partial\Omega : f = f^\#$ and $\frac{\partial f}{\partial x^i} = \frac{\partial f^\#}{\partial x^i}, 1 \le i \le n$. Denote by $L_1, L_1^\#$ the affine mean curvatures of x and $x^\#$, and by $dV, dV^\#$ their equiaffinely invariant volume elements, respectively. If $L_1 = 0$ on Ω, then

$$\int_\Omega dV \ge \int_\Omega dV^\#,$$

and equality holds if and only if $f = f^\#$ on Ω.

Proof. Choose an allowable interior deformation defined by the linear inter-polation

$$x_t : f_t(x) = (1 - t)f(x) + tf^\#(x)$$
$$= f(x) + t(f^\#(x) - f(x)), \quad x \in \Omega, \ 0 \le t \le 1.$$

Then $f_t(x)$ is locally strongly convex in Ω. Let $\mu_1(x), \dots, \mu_n(x)$, for each $x \in \Omega$, denote the eigenvalues of the matrix

$$C_j^i := \sum f^{ik}(f_{kj}^\# - f_{kj}),$$

where

$$f_{ij} = \partial_i \partial_j f, \ f_{ij}^\# = \partial_i \partial_j f^\#, \text{ and } (f^{ij}) := (f_{ij})^{-1}.$$

Then the eigenvalues of the matrix

$$\left(\sum_{k=1}^n f^{ik} \partial_k \partial_j f_t \right),$$

for each t, are positive, namely they are given by

$$1 + t\mu_i(x), \ 1 \le i \le n.$$

The volume element dV_t of x_t is given by

$$dV_t = [\det(\partial_i \partial_j f_t)]^{1/(n+2)} dx^1 \wedge \cdots \wedge dx^n$$
$$= \prod_{i=1}^n [1 + t\mu_i(x)]^{1/(n+2)} dV.$$

By the well-known inequality of the geometric and arithmetic mean (applied to the $(n + 2)$ positive numbers $1, 1, 1 + t\mu_1, \ldots, 1 + t\mu_n$) we get

$$dV_t \le \frac{1}{n + 2}\Big[2 + \sum(1 + t\mu_i(x))\Big]dV$$

$$= \Big[1 + \frac{t}{n + 2}\sum\mu_i(x)\Big]dV;$$

equality holds if and only if either $t = 0$ or $\mu_1(x) = \cdots = \mu_n(x) = 0$. Since

$$\frac{\partial(dV_t(x))}{\partial t}\Big|_{t=0} = \frac{1}{n + 2}\sum\mu_i(x)\,dV,$$

we get

$$dV_t \le dV + t\frac{\partial(dV_t)}{\partial t}\Big|_{t=0}.$$

Thus

$$\int_\Omega dV_t \le \int_\Omega dV + t\frac{\partial}{\partial t}\int_\Omega dV_t\Big|_{t=0}, \quad 0 \le t \le 1.$$

From formula (5.13) and the condition $L_1 = 0$ we have $\frac{\partial}{\partial t}\int_\Omega dV_t\big|_{t=0} = 0$, therefore, taking $t = 1$ in the preceding inequality, we have

$$\int_\Omega dV^{\#} \le \int_\Omega dV.$$

Equality $\int_\Omega dV^{\#} = \int_\Omega dV$ implies that $\mu_1(x) = \cdots = \mu_n(x) = 0$ for all $x \in \Omega$. This means that $(f_{ij} - f_{ij}^{\#})$ is identically zero. But since $f = f^{\#}$ and $f_i = f_i^{\#}$ on $\partial\Omega$, we then have $f = f^{\#}$ on Ω. $\qquad\Box$

For any affine maximal surface in A^3, Calabi further proved that the second variation of the equiaffine invariant volume under all interior deformations is negative definite (see [42, Thm. 1.3]).

The following theorem is an immediate consequence of Theorem 3.37.

Theorem 5.6. *There are no compact affine maximal hypersurfaces without boundary.*

Now we derive the differential equation of an affine maximal hypersurface. Let M be a graph hypersurface

$$M := \{(x^1, \ldots, x^{n+1}) \mid x^{n+1} = f(x), x = (x^1, \ldots, x^n) \in \Omega\},$$

where f is a strongly convex smooth function defined on a domain $\Omega \subset \mathbb{R}^n$. Then, from the equation (2.145) and by denoting

$$\rho := [\det(\partial_i\partial_j f)]^{-1/(n+2)},$$

we have $\varDelta^{(B)}\rho = -nL_1\rho$. Hence, we have the following theorem.

Theorem 5.7. *A locally strongly convex hypersurface in A^{n+1}, given as graph of a strongly convex smooth function f, is affine maximal if and only if f satisfies the p.d.e.*

$$\Delta^{(B)}\rho = 0. \tag{5.16}$$

Obviously, any parabolic affine hypersphere is an affine maximal hypersurface. In particular, the elliptic paraboloid

$$x^{n+1} = \frac{1}{2}[(x^1)^2 + \cdots + (x^n)^2], \quad (x^1, \ldots, x^n) \in A^n$$

is an affine complete affine maximal hypersurface.

5.2.2 An affine analogue of the Weierstraß representation

Again, we consider a Euclidean inner product on V with normed determinant form Det. For simplicity, we study maximal surfaces in $\mathbb{R}^3$. Moreover, when without causing confusion, we will omit the superscript "B" which has been used to mark invariants of the Blaschke geometry since Section 2.7.

Let $x : M \to \mathbb{R}^3$ be a locally strongly convex surface. Choose isothermal parameters u, v on M with respect to the Blaschke metric, let

$$e_1 = \partial_u x = x_u, \quad e_2 = \partial_v x = x_v,$$

and denote $\frac{\partial U}{\partial u} =: U_u$, $\frac{\partial U}{\partial v} =: U_v$. Then $G_{11} = G_{22} > 0$, $G_{12} = G_{21} = 0$. We use the canonical inner product on $\mathbb{R}^3$ and have

$$\begin{array}{lll}
\langle U, x_u \rangle = 0, & \langle U_u, x_u \rangle = -G_{11}, & \langle U_v, x_u \rangle = 0, \\
\langle U, x_v \rangle = 0, & \langle U_u, x_v \rangle = 0, & \langle U_v, x_v \rangle = -G_{22}, \\
\langle U, Y \rangle = 1, & \langle U_u, Y \rangle = 0, & \langle U_v, Y \rangle = 0.
\end{array} \tag{5.17}$$

We use the cross product construction from Section 1.1.4; then

$$\begin{aligned}
x_u &= \lambda[U, U_v], \\
x_v &= \mu[U, U_u], \\
\mathrm{Det}(x_u, x_v, Y) \cdot \mathrm{Det}(U_u, U_v, U) &= (G_{11})^2,
\end{aligned} \tag{5.18}$$

where λ, μ are differentiable functions. Since

$$\mathrm{Det}(x_u, x_v, Y) = |\det(h_{ij})|^{1/4} = G_{11},$$

we have

$$\mathrm{Det}(U_u, U_v, U) = G_{11} > 0.$$

From (5.17) and (5.18),

$$-G_{11} = \langle U_u, x_u \rangle = -\lambda \, \mathrm{Det}(U_u, U_v, U) = -\lambda \, G_{11}.$$

Hence $\lambda = 1$. Similarly, we have $\mu = -1$. Consequently,

$$\begin{cases} x_u = [U, U_v], \\ x_v = -[U, U_u]. \end{cases} \tag{5.19}$$

Thus we obtain the following analogue of the Weierstraß representation, due to E. Calabi [44], A.-M. Li [166], and C.-L. Terng [364]:

$$x = \int [U, U_v] du - [U, U_u] dv. \tag{5.20}$$

5.2.3 Construction of affine maximal surfaces

If $x(M)$ is an affine maximal surface, then

$$\Delta^{(B)} U = 0,$$

where the Laplacian in the above given coordinate system reads

$$\Delta^{(B)} = \frac{1}{\text{Det}(U_u, U_v, U)} \left(\frac{\partial^2}{\partial u^2} + \frac{\partial^2}{\partial v^2} \right).$$

It follows that the components $U^1(u, v), U^2(u, v)$ and $U^3(u, v)$ of U are harmonic functions.

Conversely, given a triple of functions

$$U = (U^1(u, v), U^2(u, v), U^3(u, v))$$

defined on a simply connected domain $\Omega \subset \mathbb{R}^2$ satisfying the following two conditions:
(i) U^1, U^2, U^3 are harmonic with respect to the canonical metric of $\mathbb{R}^2$;
(ii) $\text{Det}(U_u, U_v, U) > 0$ in Ω,
we can construct an affine maximal surface $x : \Omega \to A^3$ as follows:

$$x = \int_{(u_0, v_0)}^{(u, v)} [U, U_v] du - [U, U_u] dv, \tag{5.21}$$

where $(u_0, v_0), (u, v) \in \Omega$. The surface is well defined because the integrability conditions are satisfied:

$$[U, U_v]_v + [U, U_u]_u = [U, U_{uu} + U_{vv}] = 0.$$

Now let us prove that the surface, defined by (5.21), is a locally strongly convex affine maximal surface. From (5.21) we have

$$x_u = [U, U_v], \quad x_v = -[U, U_u], \tag{5.22}$$
$$[x_u, x_v] = \text{Det}(U_u, U_v, U) \cdot U. \tag{5.23}$$

Let

$$e_3 = \frac{U}{\mathrm{Det}(U_u, U_v, U) \cdot (U, U)},$$

then

$$\mathrm{Det}(x_u, x_v, e_3) = 1,$$

i.e. $\{x; x_u, x_v, e_3\}$ is a unimodular affine frame field, and the structure equations of Gauß read

$$x_{ij} = \sum \Gamma_{ij}^k x_k + h_{ij} e_3, \quad 1 \le i, j \le 2,$$

where, from (5.22) and (5.23), we have

$$h_{11} = h_{22} = [\mathrm{Det}(U_u, U_v, U)]^2, \quad h_{21} = h_{12} = 0.$$

Hence $x(M)$ is locally strongly convex and the Blaschke metric satisfies

$$G_{ij} = [\det(h_{ij})]^{-1/4} h_{ij},$$

i.e.

$$G_{11} = G_{22} = \mathrm{Det}(U_u, U_v, U), \quad G_{12} = G_{21} = 0.$$

The conormal vector is

$$[\det(h_{ij})]^{-1/4} [x_u, x_v] = U.$$

Finally, since $U^1(u, v), U^2(u, v)$ and $U^3(u, v)$ are harmonic functions, $x(\Omega)$ is an affine maximal surface. $\qquad\square$

In the following we give some examples of affine maximal surfaces.

Example 5.1. Consider $\Omega := \mathbb{R}^2$ and let us define $U : \mathbb{R}^2 \to \mathbb{R}^3$ by $U := (1, u, v)$; we get

$$x = \left(\frac{1}{2}(u^2 + v^2), -u, -v \right),$$

which is an elliptic paraboloid.

Example 5.2. Consider $\Omega := \{(u, v) \in \mathbb{R}^2 | u > 0\}$ and let us define $U : \Omega \to \mathbb{R}^3$ by $U := (1, u^2 - v^2, v)$, then

$$\mathrm{Det}(U_u, U_v, U) = 2u.$$

The construction above gives

$$x = \left(\frac{1}{3}u^3 + uv^2, -u, -2uv \right), \quad u > 0.$$

Example 5.3. Let $\Omega := \{(u, v) \in \mathbb{R}^2 \mid v < 0 < u\}$ and define $U := (u, v, 2uv)$, then

$$\mathrm{Det}(U, U_u, U_v) = -2uv.$$

The integration of (5.21) gives

$$x = \left(-\frac{2}{3}v^3, -\frac{2}{3}u^3, \frac{1}{2}(u^2 + v^2) \right), \quad v < 0 < u.$$

5.2.4 Construction of improper (parabolic) affine spheres

Since improper affine spheres are affine maximal surfaces, we can use (5.21) to give a construction of 2-dimensional improper affine spheres. For an improper affine sphere the affine normal Y is a constant vector field, therefore $x(\Omega)$ is a parabolic affine sphere if and only if the image of $U : \Omega \to \mathbb{R}^3$ lies in a plane, i.e. there are constants a, b, c, and d such that

$$aU^1(u, v) + bU^2(u, v) + cU^3(u, v) + d = 0, \quad \text{for } (u, v) \in \Omega.$$

Particularly, taking $U^1(u, v) = 1$, and taking $U^2(u, v), U^3(u, v)$ to be harmonic functions, satisfying

$$\det \begin{vmatrix} \dfrac{\partial U^2(u, v)}{\partial u} & \dfrac{\partial U^3(u, v)}{\partial u} \\ \dfrac{\partial U^2(u, v)}{\partial v} & \dfrac{\partial U^3(u, v)}{\partial v} \end{vmatrix} > 0, \quad (u, v) \in \Omega \subset \mathbb{R}^2,$$

we get an improper affine sphere.

As a further application of the Weierstraß representation we can derive the following generalization of the classical Liouville theorem.

Theorem 5.8. *Let $p(u, v), q(u, v), r(u, v)$ be harmonic functions defined on $\mathbb{R}^2$. If there is a constant $\delta > 0$ such that*

$$F := \det \begin{vmatrix} p & q & r \\ \dfrac{\partial p}{\partial u} & \dfrac{\partial q}{\partial u} & \dfrac{\partial r}{\partial u} \\ \dfrac{\partial p}{\partial v} & \dfrac{\partial q}{\partial v} & \dfrac{\partial r}{\partial v} \end{vmatrix} \geq \delta, \quad \text{for } (u, v) \in \mathbb{R}^2,$$

then $p(u, v), q(u, v), r(u, v)$ are linear functions.

Proof. Define $U : \mathbb{R}^2 \to \mathbb{R}^3$ by $U := (p(u, v), q(u, v), r(u, v))$ and construct an affine maximal surface $x : \mathbb{R}^2 \to \mathbb{R}^3$ via (5.21). Then, from the affine theorema egregium, for x we have the relation

$$R = 2\chi = 2(J + L_1) = 2J \geq 0.$$

According to the discussion in Section 5.2.3, the Blaschke metric G of x has the expression $G = F((du)^2 + (dv)^2)$ and, therefore, we have $-\frac{1}{2}\Delta \ln F = \chi \geq 0$, i.e. $\ln F$ is necessarily a superharmonic function on the whole plane. Since $F \geq \delta > 0$ we have $F = \text{const}$. It follows that $\chi = J = 0$. Hence $x(\mathbb{R}^2)$ is an elliptic paraboloid. Let $\mathbb{R}^3$ be equipped with affine coordinates $\{x^1, x^2, x^3\}$, then after an appropriate affine transformation

$$\begin{pmatrix} \bar{x}^1 \\ \bar{x}^2 \\ \bar{x}^3 \end{pmatrix} = \begin{pmatrix} b_{11} & b_{12} & b_{13} \\ b_{21} & b_{22} & b_{23} \\ b_{31} & b_{32} & b_{33} \end{pmatrix} \begin{pmatrix} x^1 \\ x^2 \\ x^3 \end{pmatrix} + \begin{pmatrix} c_1 \\ c_2 \\ c_3 \end{pmatrix};$$

the surface $x(\mathbb{R}^2)$ can be expressed by the following equation

$$\bar{x}^3 = \frac{1}{2}[(\bar{x}^1)^2 + (\bar{x}^2)^2].$$

A direct calculation shows that the conormal vector field of the surface is

$$(-\bar{x}^1, -\bar{x}^2, 1)$$

and the Blaschke metric is

$$G = (d\bar{x}^1)^2 + (d\bar{x}^2)^2.$$

Now we consider the coordinates transformation

$$\bar{x}^1 = \bar{x}^1(u, v), \quad \bar{x}^2 = \bar{x}^2(u, v).$$

Our first observation is that the relation

$$F((du)^2 + (dv)^2) = (d\bar{x}^1)^2 + (d\bar{x}^2)^2$$

implies that we have either

$$\frac{\partial\bar{x}^1}{\partial u} = \frac{\partial\bar{x}^2}{\partial v}, \quad \frac{\partial\bar{x}^1}{\partial v} = -\frac{\partial\bar{x}^2}{\partial u};$$

or

$$\frac{\partial\bar{x}^1}{\partial u} = -\frac{\partial\bar{x}^2}{\partial v}, \quad \frac{\partial\bar{x}^1}{\partial v} = \frac{\partial\bar{x}^2}{\partial u}.$$

Hence the complex valued function

$$\varphi(z) = \bar{x}^1(u, v) + \sqrt{-1}\bar{x}^2(u, v),$$

with $z = u + \sqrt{-1}v$, is holomorphic or anti-holomorphic.
Next, from

$$\det\begin{vmatrix} p & q & r \\ p_u & q_u & r_u \\ p_v & q_v & r_v \end{vmatrix} = \text{const.}$$

and the observation

$$\begin{pmatrix} -\bar{x}^1 \\ -\bar{x}^2 \\ 1 \end{pmatrix} = \begin{pmatrix} b_{11} & b_{12} & b_{13} \\ b_{21} & b_{22} & b_{23} \\ b_{31} & b_{32} & b_{33} \end{pmatrix}\begin{pmatrix} p \\ q \\ r \end{pmatrix},$$

we derive

$$\det\begin{vmatrix} \dfrac{\partial\bar{x}^1}{\partial u} & \dfrac{\partial\bar{x}^2}{\partial u} \\[2mm] \dfrac{\partial\bar{x}^1}{\partial v} & \dfrac{\partial\bar{x}^2}{\partial v} \end{vmatrix} = \text{const.}$$

This implies that

$$\left|\varphi'(z)\right|^2 = \left(\frac{\partial \bar{x}^1}{\partial u}\right)^2 + \left(\frac{\partial \bar{x}^2}{\partial v}\right)^2 = \text{const.} \tag{5.24}$$

Now, using Liouville's theorem, which states that every bounded holomorphic (or anti-holomorphic) function on $\mathbb{C}$ must be constant, from (5.24) we get $\varphi'(z) = \text{const.}$ and thus $\varphi(z)$ is a linear function.

Therefore, $p(u, v)$, $q(u, v)$ and $r(u, v)$ are linear functions. $\qquad\square$

Remark 5.9. In Theorem 5.8, the condition

$$\det \begin{vmatrix} p & q & r \\ p_u & q_u & r_u \\ p_v & q_v & r_v \end{vmatrix} \geq \delta > 0$$

is essential. Take, for example, $p = 1$, $q = e^u \cos v$ and $r = e^u \sin v$, then it holds

$$\det \begin{vmatrix} p & q & r \\ p_u & q_u & r_u \\ p_v & q_v & r_v \end{vmatrix} = e^{2u} > 0,$$

but q and r are not linear functions.

5.2.5 Affine Bernstein problems

From the point of view of local differential geometry the formula (5.20) gives all affine maximal surfaces. On the other hand, the following global problem is interesting:

Problem. *Is there a complete, affine maximal surface which is not an elliptic paraboloid?*

About complete affine maximal surfaces there are two famous conjectures, one is a conjecture of S. S. Chern (see [60, 61]), the other is a conjecture of E. Calabi [42].

Chern's conjecture. *Let $f : \mathbb{R}^2 \to \mathbb{R}$ be a strongly convex function. If the graph*

$$M = \left\{(x^1, x^2, f(x^1, x^2)) \mid (x^1, x^2) \in \mathbb{R}^2\right\}$$

is an affine maximal surface then M must be an elliptic paraboloid.

Calabi's conjecture. *A locally strongly convex affine complete surface $x : M \to \mathbb{R}^3$ with affine mean curvature $L_1 \equiv 0$ is an elliptic paraboloid.*

Both of the two preceding conjectures were extended to higher dimensions in [187] and the generalized version usually is also called the *Chern* and *Calabi conjecture*, respectively. Notice that the two conjectures above differ in the assumption on the completeness of the affine maximal hypersurface considered. Conceptually, both conjectures are called *affine Bernstein conjectures*, and they were long standing open problems for several decades.

Remark 5.10. We recall different notions of completeness in affine hypersurface theory from Section 3.4. In Chern's conjecture one assumes Euclidean completeness, in Calabi's conjecture one assumes affine completeness. In Section 3.4 we showed that both completeness assumptions are not equivalent. In 2000, Trudinger and Wang [367] solved Chern's conjecture in dimension $n = 2$. Later, Li and Jia [174] solved Calabi's conjecture for $n = 2$. Li and Jia used a blow up analysis to show that, for an affine complete maximal surface, $\|\hat{B}\|_G$ (the tensor norm of the Weingarten form $\hat{B}$) is bounded above, then they used a result of Martinez and Milan [217] to complete the proof of Calabi's conjecture (for details see Section 5.4 below). So far the higher dimensional affine Bernstein problems are unsolved. In this monograph we first give a proof of Calabi's conjecture, and then give two different proofs of Chern's Conjecture in dimension $n = 2$.

5.3 Differential inequalities for $\Delta^{(B)}J$, $\Delta^{(B)}\Phi$, and $\Delta^{(C)}\Phi$

5.3.1 Notations from E. Calabi

We recall a notational convention from Section 2.7 and if necessary mark invariants of the Blaschke geometry by "B" as before. Let $x : M \to A^3$ be an affine maximal surface, and let M be simply connected. We choose a complex isothermal coordinate $z = u + \sqrt{-1}v$ with respect to the Blaschke metric such that $G = 2F|dz|^2$ (cf. Section 3.9). Then the components of its conormal vector field U are harmonic functions; thus there are holomorphic vector valued functions $Z(z) = (Z^1(z), Z^2(z), Z^3(z))$ such that

$$U = \sqrt{-1}\,(Z - \bar{Z}). \tag{5.25}$$

In this subsection we shall express quantities of the affine maximal surface in terms of the holomorphic curve $Z(z)$ and calculate $\Delta^{(B)}J$, mainly following Calabi (cf. [44]). As in Section 3.9 we put

$$\alpha := \frac{1}{2}(A_{111} + \sqrt{-1}A_{222}),$$

$$\beta := \frac{1}{2}\left(\frac{B_{11} - B_{22}}{2} - \sqrt{-1}B_{12}\right).$$

Then the Fubini–Pick form from (2.47) and the Weingarten form from (2.58) can be represented as in (3.127) and (3.128):

$$A = \alpha(dz)^3 + \bar{\alpha}(d\bar{z})^3, \tag{5.26}$$

$$\hat{B} = \beta(dz)^2 + 2L_1 F dz d\bar{z} + \bar{\beta}(d\bar{z})^2. \tag{5.27}$$

From (3.137), (3.141), and (3.142) it follows that

$$F = -\sqrt{-1}\,\mathrm{Det}(U, U_z, U_{\bar{z}}) = \mathrm{Det}(Z - \bar{Z}, Z', \bar{Z}') > 0, \tag{5.28}$$

$$\alpha = -\sqrt{-1}\,\mathrm{Det}(U, U_z, U_{zz}) = \mathrm{Det}(Z - \bar{Z}, Z', Z''), \tag{5.29}$$

$$\beta = -F^{-1}\frac{\partial\alpha}{\partial\bar{z}} = \mathrm{Det}(\bar{Z}', Z', Z'') \cdot [\mathrm{Det}(Z - \bar{Z}, Z', \bar{Z}')]^{-1}, \tag{5.30}$$

where we denote $Z' = \frac{\partial Z}{\partial z}$, $Z'' = \frac{\partial^2 Z}{\partial z^2}$, etc. For the following proposition we recall the notions of ∇' and ∇'' from Section 3.9.

Proposition 5.11. *Let $x : M \to A^3$ be a locally strongly convex affine maximal surface and $z = u + \sqrt{-1}v$ be a local complex isothermal parameter with respect to the Blaschke metric. Then the vector-valued cubic differential form*

$$\Psi = (\alpha Y + \beta x_z)(dz)^3$$

and the scalar valued differential form of degree six

$$\Xi = (\beta \nabla'\alpha - \alpha \nabla'\beta)(dz)^6$$

are holomorphic on M.

Remark 5.12. By extending Calabi's method, H. Li was able to construct a sextic holomorphic form for locally strongly convex affine surfaces with constant affine mean curvature which, for affine maximal surface, reduces to Calabi's preceding result. We refer to [189] for details and interesting applications.

To prove Proposition 5.11, we need the following,

Lemma 5.13. *Let a, b, c, d and e be vectors in $\mathbb{R}^3$ (or in $\mathbb{C}^3$). Then*

$$\mathrm{Det}(a, b, c) \cdot \mathrm{Det}(d, b, e) - \mathrm{Det}(a, b, e) \cdot \mathrm{Det}(d, b, c) = \mathrm{Det}(a, b, d) \cdot \mathrm{Det}(c, b, e).$$

Proof. One verifies this, using Sections 1.1.3 and 1.2.2. □

Proof of Proposition 5.11. It is sufficient to prove that Ψ and Ξ satisfy the following formulas in terms of Z:

$$\Psi = -\sqrt{-1}[Z', Z''](dz)^3,$$

$$\Xi = \mathrm{Det}(Z', Z'', Z''')(dz)^6.$$

First, from (3.137), (3.138), and (5.19), we get

$$Y = -\sqrt{-1}F^{-1}[U_z, U_{\bar{z}}],$$

$$x_z = \frac{1}{2}\left(\frac{\partial x}{\partial u} - \sqrt{-1}\frac{\partial x}{\partial v}\right)$$

$$= \frac{1}{2}\left(\left[U, \frac{\partial U}{\partial v}\right] + \sqrt{-1}\left[U, \frac{\partial U}{\partial u}\right]\right) = \sqrt{-1}[U, U_z].$$

Hence

$$\alpha Y + \beta x_z = -\sqrt{-1}\frac{\mathrm{Det}(Z-\bar{Z},Z',Z'')}{\mathrm{Det}(Z-\bar{Z},Z',\bar{Z}')}[Z',\bar{Z}']$$

$$-\sqrt{-1}\frac{\mathrm{Det}(\bar{Z}',Z',Z'')}{\mathrm{Det}(Z-\bar{Z},Z',\bar{Z}')}[(Z-\bar{Z}),Z'].$$

Using Lemma 5.13, we obtain

$$\mathrm{Det}(Z-\bar{Z},Z',Z'')[Z',\bar{Z}'] - \mathrm{Det}(Z',\bar{Z}',Z'')[(Z-\bar{Z}),Z'] = \mathrm{Det}(Z-\bar{Z},Z',\bar{Z}')[Z',Z''].$$

Consequently:

$$\Psi = -\sqrt{-1}[Z',Z''](dz)^3.$$

Second,

$$\beta\nabla'\alpha - \alpha\nabla'\beta =$$

$$-\frac{\mathrm{Det}(Z',\bar{Z}',Z'')\cdot\mathrm{Det}(Z-\bar{Z},Z',Z''')}{\mathrm{Det}(Z-\bar{Z},Z',\bar{Z}')} + \frac{\mathrm{Det}(Z-\bar{Z},Z',Z'')\cdot\mathrm{Det}(-\bar{Z}',Z',Z''')}{\mathrm{Det}(Z-\bar{Z},Z',\bar{Z}')}.$$

From Lemma 5.13 we obtain

$$\mathrm{Det}(Z-\bar{Z},Z',Z'')\cdot\mathrm{Det}(Z',\bar{Z}',Z''') - \mathrm{Det}(Z',\bar{Z}',Z'')\cdot\mathrm{Det}(Z-\bar{Z},Z',Z''')$$

$$= \mathrm{Det}(Z-\bar{Z},Z',\bar{Z}')\cdot\mathrm{Det}(Z',Z'',Z''').$$

Hence

$$\Xi = \mathrm{Det}(Z',Z'',Z''')(dz)^6. \qquad \square$$

Lemma 5.14. *The affine Weierstraß formula* (5.20) *can be rewritten in terms of the holomorphic curve Z as follows (cf. [44]):*

$$x = -\sqrt{-1}\left([Z,\bar{Z}] + \int[Z,dZ] - \int[\bar{Z},d\bar{Z}]\right).$$

5.3.2 Computation of $\Delta^{(B)}J$

For an affine maximal surfaces, E. Calabi computed the Laplacian of $J = \chi = F^{-3}\alpha\bar{\alpha}$ in terms of complex coordinates (cf. [44]). To present this computation, we again recall the notions of the derivations ∇', ∇'' and basic formulas from Section 3.9 up to (3.144):

$$\frac{1}{2}\Delta^{(B)}J = F^{-4}\nabla''\nabla'(\alpha\bar{\alpha})$$

$$= F^{-4}\nabla''((\nabla'\alpha)\bar{\alpha} + \alpha(\overline{\nabla''\alpha}))$$

$$= F^{-4}\{(\nabla'\alpha)(\overline{\nabla'\alpha}) + (\nabla'\nabla''\alpha + 3F\chi\alpha)\bar{\alpha} + \nabla''(-\alpha F\bar{\beta})\} \qquad (5.31)$$

$$= \|\nabla'\alpha\|^2 + 3J^2 + F^{-4}[-\nabla''(F\beta)\bar{\alpha} - F(-F\beta\bar{\beta} + \alpha(\overline{\nabla'\beta}))]$$

$$= \|\nabla'\alpha\|^2 + 3J^2 + \|\beta\|^2 - 2F^{-3}\mathrm{Re}(\bar{\alpha}\nabla'\beta),$$

where $\|\nabla'\alpha\|^2 = F^{-4}(\nabla'\alpha)(\overline{\nabla'\alpha})$ and $\|\beta\|^2 = F^{-2}\beta\bar{\beta}$. Denote by φ the local coefficient of Ξ (cf. Proposition 5.11), i.e.

$$\varphi = \beta\nabla'\alpha - \alpha\nabla'\beta = \mathrm{Det}(Z', Z'', Z''').$$

When $\alpha \neq 0$, we have

$$\bar{\alpha}\nabla'\beta = \frac{\bar{\alpha}}{\alpha}(\beta\nabla'\alpha - \varphi),$$

$$\|\nabla'\alpha\|^2 + \|\beta\|^2 - 2F^{-3}\mathrm{Re}\left(\frac{\bar{\alpha}}{\alpha}(\beta\nabla'\alpha - \varphi)\right) \geq \left\|\nabla'\alpha - \frac{F\alpha\bar{\beta}}{\bar{\alpha}}\right\|^2 - 2\|\varphi\|.$$

Inserting the last inequality into (5.31) we get

$$\frac{1}{2}\Delta^{(B)}J \geq 3J^2 - 2\|\varphi\|. \tag{5.32}$$

(5.32) holds at each point where $\alpha \neq 0$. Let p be a point such that $\alpha = 0$. If there is a neighborhood D of p such that $\alpha \equiv 0$ in D, then $\beta \equiv 0$ and so $\varphi \equiv 0$ in D. Hence (5.32) holds. If there is a point in every neighborhood of p such that $\alpha \neq 0$, then it follows from a continuity argument that (5.32) holds at p. Thus (5.32) holds everywhere.

5.3.3 Computation of $\Delta^{(B)}(\|\hat{B}\|^2)$

For an affine maximal surface, E. Calabi also computed the Laplacian of $\frac{1}{2}(J + \|\hat{B}\|^2) = \frac{1}{2}\|\alpha\|^2 + \|\beta\|$. Since

$$\begin{aligned}
\Delta^{(B)}\|\beta\| &= \|\beta\|^{-1}F^{-3}\nabla''\nabla'(\beta\bar{\beta}) - \frac{1}{2}\|\beta\|^{-3}F^{-5}(\nabla'(\beta\bar{\beta}))(\nabla''(\beta\bar{\beta})) \\
&= \|\beta\|^{-1}F^{-3}[5F^{-2}\alpha\bar{\alpha}\beta\bar{\beta} + 2F^{-1}\mathrm{Re}((\nabla'\alpha)\bar{\beta}^2) + (\nabla'\beta)\overline{\nabla'\beta}] \\
&\quad - \frac{1}{2}\|\beta\|^{-3}F^{-5}[\beta\bar{\beta}(\nabla'\beta)\overline{\nabla'\beta} + F^{-2}\alpha\bar{\alpha}\beta^2\bar{\beta}^2 + 2F^{-1}\mathrm{Re}(\alpha(\nabla'\beta)\bar{\beta}^3)] \\
&= \frac{9}{2}\|\alpha\|^2\|\beta\| + \frac{1}{2}\|\beta\|^{-1}\|\nabla'\beta\|^2 + 2\|\beta\|^{-1}F^{-4}\mathrm{Re}((\nabla'\alpha)\bar{\beta}^2) \\
&\quad - F^{-6}\|\beta\|^{-3}\mathrm{Re}(\alpha(\nabla'\beta)\bar{\beta}^3),
\end{aligned}$$

from (5.31) it follows that

$$\begin{aligned}
\Delta^{(B)}\left(\frac{1}{2}J + \|\beta\|\right) &= \Delta^{(B)}\left(\frac{1}{2}\|\alpha\|^2 + \|\beta\|\right) \\
&= 3\|\alpha\|^4 + \|\nabla'\alpha + F\beta^{3/2}\bar{\beta}^{-1/2}\|^2 \\
&\quad + \frac{1}{2}\|\beta\|^{-1}\{\|\nabla'\beta - 2\|\beta\|\alpha - F^{-1}\bar{\alpha}\beta^2\bar{\beta}^{-1}\|\}^2 \\
&\quad + \|\beta\|^{-2}\|\alpha\bar{\beta}^{3/2} - \bar{\alpha}\beta^{3/2}\|^2 \\
&\geq 3\|\alpha\|^4 + \|\beta\|^{-2}\|\alpha\bar{\beta}^{3/2} - \bar{\alpha}\beta^{3/2}\|^2 \geq 0.
\end{aligned}$$

Making use of this inequality we can prove the following theorem, which was first obtained by A. Martinez and F. Milan [217]; this result on affine complete affine maximal surface is an important step to the complete solution of Calabi's conjecture, stated below in Theorem 5.20; there we give a proof of A.-M. Li and F. Jia, see [174].

Theorem 5.15. *Let $x : M \to A^3$ be a locally strongly convex, affine complete, affine maximal surface. If there is a constant $N > 0$ such that the affine Gauß–Kronecker curvature $L_2 = \lambda_1\lambda_2$ satisfies $|\lambda_1\lambda_2| < N$ everywhere on M then $x(M)$ is an elliptic paraboloid.*

Proof. The conditions $|\lambda_1\lambda_2| < N$ and $\lambda_1 + \lambda_2 = 0$ imply that $\|\beta\|$ is bounded above by a positive constant C. On the other hand, we have

$$\Delta^{(B)}(\tfrac{1}{2}J + \|\beta\|) \geq 3J^2 \geq 6\left(\tfrac{1}{2}J + \|\beta\|\right)^2 - 12\|\beta\|^2$$
$$\geq 6\left(\tfrac{1}{2}J + \|\beta\|\right)^2 - 12C^2. \tag{5.33}$$

From Theorem 3.39 it follows that $\tfrac{1}{2}J + \|\beta\|$ is bounded from above. Being a bounded subharmonic function on a complete surface with $R \geq 0$, the function $\tfrac{1}{2}J + \|\beta\|$ must be a constant. It follows from (5.33) that $J = 0$ everywhere on M, therefore $x(M)$ is an elliptic paraboloid. □

5.3.4 Estimations of $\Delta^{(C)}\Phi$ and $\Delta^{(B)}\Phi$

As before we consider an n-dimensional affine maximal hypersurface. First we derive a differential inequality for $\Delta^{(C)}\Phi$; here Φ is defined by (2.165). Recall that, for any function F, in terms of the Calabi metric $\mathcal{H}$ (see Section 2.7.2) and the Blaschke metric G, we have

$$\frac{\|\operatorname{grad} F\|_{\mathcal{H}}^2}{\rho} = \|\operatorname{grad} F\|_G^2, \tag{5.34}$$

$$\Delta^{(C)}F = \rho\Delta^{(B)}F - \frac{n-2}{2}\mathcal{H}(\operatorname{grad} F, \operatorname{grad} \ln \rho). \tag{5.35}$$

Note that the equation $\Delta^{(B)}\rho = 0$ can be rewritten as (cf. also Lemma 5.27):

$$\Delta^{(C)}\rho = -\frac{n-2}{2}\frac{\|\operatorname{grad} \rho\|_{\mathcal{H}}^2}{\rho}, \quad \rho = \left[\det\left(\frac{\partial^2 f}{\partial x^i \partial x^j}\right)\right]^{-1/(n+2)}. \tag{5.36}$$

Now we prove the following result which is useful for our study of the affine Bernstein problems in later subsections.

Proposition 5.16. *Let f be a strongly convex C^∞-function defined in a domain $\Omega \subset \mathbb{R}^n$. Assume that f satisfies the p.d.e. (5.36). Then the following estimate holds for Φ and any*

$0 \le \delta < 1$:

$$\Delta^{(C)}\Phi \ge \frac{2\delta}{\rho^2}\sum(\rho_{,ij})^2 + \frac{n(1-\delta)}{2(n-1)}\frac{\|\text{grad } \Phi\|^2_{\mathcal{H}}}{\Phi}$$

$$- \frac{n(n-2)+(3n-2)\delta}{n-1}\mathcal{H}(\text{grad }\Phi, \text{grad }\ln\rho) \qquad (5.37)$$

$$+ \left[\frac{n^2-(n^2+4n-4)\delta}{2(n-1)} - \frac{(n+2)^2(n-1)}{8n}\right]\Phi^2.$$

Proof. We choose a local orthonormal frame field $\{e_i\}_{i=1}^n$ of the Calabi metric $\mathcal{H}$. Then, using usual notations for covariant derivatives with respect to $\mathcal{H}$, we have

$$\Phi = \sum\frac{(\rho_j)^2}{\rho^2}, \quad \Phi_{,i} = 2\sum\frac{\rho_{,j}\rho_{,ji}}{\rho^2} - 2\frac{\rho_{,i}}{\rho}\Phi, \qquad (5.38)$$

and, by using (5.36),

$$\Delta^{(C)}\Phi = \frac{2}{\rho^2}\sum((\rho_{,ji})^2 + \rho_{,j}\rho_{,jii}) - 8\sum\frac{\rho_{,j}\rho_{,i}\rho_{,ji}}{\rho^3} + (n+4)\Phi^2. \qquad (5.39)$$

Moreover, using the Ricci identity and (5.36), we have

$$\frac{2}{\rho^2}\sum\rho_{,j}\rho_{,jii} = (n-2)\left[\Phi^2 - 2\rho^{-3}\sum\rho_{,i}\rho_{,j}\rho_{,ij}\right] + 2\sum R_{ij}(\ln\rho)_{,i}(\ln\rho)_{,j}, \qquad (5.40)$$

where R_{ij} denote the components of the Ricci curvature tensor Ric of the Calabi metric. Inserting (5.40) into (5.39), we obtain

$$\Delta^{(C)}\Phi = 2\sum\frac{(\rho_{,ij})^2}{\rho^2} - 2(n+2)\rho^{-3}\sum\rho_{,i}\rho_{,j}\rho_{,ij}$$

$$+ 2\sum R_{ij}(\ln\rho)_{,i}(\ln\rho)_{,j} + 2(n+1)\Phi^2. \qquad (5.41)$$

It is easy to see that, if $\Phi(p) = 0$, then at p we have

$$\Delta^{(C)}\Phi = \frac{2}{\rho^2}\sum(\rho_{,ij})^2 \ge 0. \qquad (5.42)$$

Now we consider $p \in M$ with $\Phi(p) \ne 0$. To estimate the term $2\sum(\rho_{,ij})^2$, we further choose $\{e_i\}_{i=1}^n$ such that around p, it holds that

$$\rho_{,1} = \|\text{grad }\rho\|_{\mathcal{H}} > 0, \quad \rho_{,i} = 0 \text{ for all } i > 1.$$

We apply the inequality of Cauchy–Schwarz and derive

$$2\sum(\rho_{,ij})^2 \ge 2(\rho_{,11})^2 + 4\sum_{i>1}(\rho_{,1i})^2 + \frac{2}{n-1}(\Delta^{(C)}\rho - \rho_{,11})^2 \qquad (5.43)$$

$$= \frac{2n}{n-1}(\rho_{,11})^2 + 4\sum_{i>1}(\rho_{,1i})^2 + \frac{2(n-2)}{n-1}\frac{(\rho_{,1})^2\rho_{,11}}{\rho} + \frac{(n-2)^2}{2(n-1)}\frac{(\rho_{,1})^4}{\rho^2}$$

$$= \frac{2n}{n-1}(\rho_{,11})^2 + 4\sum_{i>1}(\rho_{,1i})^2 + \frac{2(n-2)}{n-1}\rho^{-1}\sum\rho_{,i}\rho_{,j}\rho_{,ij} + \frac{(n-2)^2}{2(n-1)}\rho^2\Phi^2.$$

Therefore, for any constant $0 \le \delta < 1$, if writing

$$\sum (\rho_{,ij})^2 = \delta \sum (\rho_{,ij})^2 + (1 - \delta) \sum (\rho_{,ij})^2,$$

and inserting (5.43) into (5.41), we obtain

$$
\begin{aligned}
\Delta^{(C)}\Phi \ge{}& \frac{2\delta}{\rho^2} \sum (\rho_{,ij})^2 + \frac{2n(1-\delta)}{n-1}\frac{(\rho_{,11})^2}{\rho^2} + \frac{4(1-\delta)}{\rho^2} \sum_{i>1} (\rho_{,1i})^2 \\
& - \left[\frac{2(n-2)(n-2+\delta)}{n-1} + 8 \right] \rho^{-3} \sum \rho_{,i}\rho_{,j}\rho_{,ij} \\
& + \left[\frac{(n-2)^2(1-\delta)}{2(n-1)} + 2(n+1) \right] \Phi^2 + 2 \sum R_{ij}(\ln \rho)_{,i}(\ln \rho)_{,j}.
\end{aligned}
\tag{5.44}
$$

From (5.38) we have

$$2\rho^{-3} \sum \rho_{,i}\rho_{,j}\rho_{,ij} = \sum (\ln \rho)_{,i}\Phi_{,i} + 2\Phi^2, \tag{5.45}$$

$$
\begin{aligned}
4\rho^{-4} \sum_i \Big(\sum_j \rho_{,j}\rho_{,ji} \Big)^2 &= \sum (\Phi_{,i})^2 + 8\Phi\rho^{-3} \sum \rho_{,i}\rho_{,j}\rho_{,ij} - 4\Phi^3 \\
&= \sum (\Phi_{,i})^2 + 4\Phi \sum (\ln \rho)_{,i}\Phi_{,i} + 4\Phi^3.
\end{aligned}
\tag{5.46}
$$

We can rewrite (5.46) as

$$\frac{(\rho_{,11})^2}{\rho^2} = \frac{1}{4\Phi} \sum (\Phi_{,i})^2 + \Phi^2 + \sum (\ln \rho)_{,i}\Phi_{,i} - \rho^{-2} \sum_{i>1} (\rho_{,1i})^2. \tag{5.47}$$

Then, inserting (5.45) and (5.47) into (5.44), we obtain

$$
\begin{aligned}
\Delta^{(C)}\Phi \ge{}& \frac{2\delta}{\rho^2} \sum (\rho_{,ij})^2 + \frac{2(n-2)(1-\delta)}{n-1}\rho^{-2} \sum_{i>1} (\rho_{,1i})^2 \\
& - \frac{n(n-2) + (3n-2)\delta}{n-1} \sum (\ln \rho)_{,i}\Phi_{,i} + \frac{n(1-\delta)}{2(n-1)} \sum \frac{(\Phi_{,i})^2}{\Phi} \\
& + \frac{n^2 - (n^2 + 4n - 4)\delta}{2(n-1)}\Phi^2 + 2 \sum R_{ij}(\ln \rho)_{,i}(\ln \rho)_{,j}.
\end{aligned}
\tag{5.48}
$$

Now, at the point p, we estimate the last term in (5.48):

$$\mathcal{U} := 2 \sum R_{ij}(\ln \rho)_{,i}(\ln \rho)_{,j} = 2\mathcal{H}(\mathrm{Ric}\,(\mathrm{grad}\,\ln \rho), \mathrm{grad}\,\ln \rho).$$

Since $\mathcal{U}$ is affinely invariant, up to an affine coordinate transformation, we can assume that $f_{ij}(p) = \delta_{ij}$ and

$$\frac{\partial \rho}{\partial x^1}(p) = \|\mathrm{grad}\,\rho\|_{\mathcal{H}}(p) > 0, \quad \frac{\partial \rho}{\partial x^i}(p) = 0 \text{ for all } i > 1. \tag{5.49}$$

It follows that

$$\mathcal{U}(p) = 2 \sum f^{ij}f^{kl}\frac{\partial \ln \rho}{\partial x^i}\frac{\partial \ln \rho}{\partial x^k}R_{jl} = 2R_{11}(p)\left[\frac{\partial \ln \rho}{\partial x^1}(p)\right]^2 = 2R_{11}(p)\Phi(p). \tag{5.50}$$

By using (2.163), (5.49), and the fact $\frac{\partial \ln \rho}{\partial x^j}(p) = -\frac{1}{n+2}\sum_i f_{iij}(p)$, we have

$$R_{11}(p) = \frac{1}{4}\sum (f_{ij1})^2 - \frac{1}{4}f_{111}\sum f_{ii1}.$$

On the other hand, similar to the derivation of (5.43), we get

$$\sum (f_{ij1})^2 \geq (f_{111})^2 + \sum_{i>1}(f_{ii1})^2 \geq (f_{111})^2 + \frac{1}{n-1}\left(\sum f_{ii1} - f_{111}\right)^2$$

$$\geq \frac{n}{n-1}(f_{111})^2 - \frac{2}{n-1}f_{111}\sum f_{ii1} + \frac{1}{n-1}\left(\sum f_{ii1}\right)^2,$$

and therefore, at p,

$$R_{11}(p) \geq \frac{n}{4(n-1)}(f_{111})^2 - \frac{(n+1)}{4(n-1)}f_{111}\sum f_{ii1} + \frac{1}{4(n-1)}\left(\sum f_{ii1}\right)^2$$

$$\geq -\frac{n-1}{16n}\left(\sum f_{ii1}\right)^2 = -\frac{(n-1)(n+2)^2}{16n}\Phi(p), \tag{5.51}$$

where we used the inequality

$$f_{111}\sum f_{ii1} \leq \frac{n}{n+1}(f_{111})^2 + \frac{n+1}{4n}\left(\sum f_{ii1}\right)^2.$$

From (5.50) and (5.51), we obtain the estimate

$$2\mathcal{H}(\mathrm{Ric}\,(\mathrm{grad}\,\ln\rho), \mathrm{grad}\,\ln\rho) \geq -\frac{(n+2)^2(n-1)}{8n}\Phi^2. \tag{5.52}$$

A combination of (5.48) and (5.52) immediately completes the proof of (5.37). □

Remark 5.17. If we take $\delta = 0$ in the inequality (5.37), it becomes

$$\Delta^{(C)}\Phi \geq \frac{n}{2(n-1)}\frac{\|\mathrm{grad}\,\Phi\|_{\mathcal{H}}^2}{\Phi} - \frac{n(n-2)}{n-1}\mathcal{H}\,(\mathrm{grad}\,\Phi, \mathrm{grad}\,\ln\rho)$$

$$+\left[\frac{n^2}{2(n-1)} - \frac{(n+2)^2(n-1)}{8n}\right]\Phi^2. \tag{5.53}$$

In particular, for $n = 2$, we have

$$\Delta^{(C)}\Phi \geq \frac{\|\mathrm{grad}\,\Phi\|_{\mathcal{H}}^2}{\Phi} + \Phi^2; \tag{5.54}$$

and for $n = 3$, we have

$$\Delta^{(C)}\Phi \geq \frac{3}{4}\frac{\|\mathrm{grad}\,\Phi\|_{\mathcal{H}}^2}{\Phi} - \frac{3}{2}\mathcal{H}(\mathrm{grad}\,\Phi, \mathrm{grad}\,\ln\rho) + \frac{1}{6}\Phi^2. \tag{5.55}$$

For a more general formula see [177] and [181]. An equivalent inequality of (5.53) in terms of the Blaschke metric is the following, which was first proved by Li and Jia in [173].

Proposition 5.18. *Let* $f(x^1, \ldots, x^n)$ *be a strongly convex* C^∞*-function satisfying the p.d.e. (5.16). In terms of the Blaschke metric, the Laplacian of the function* Φ, *which was defined in (2.165), satisfies the following inequality:*

$$\Delta^{(B)}\Phi \geq \frac{n}{2(n-1)} \frac{\|\text{grad } \Phi\|_G^2}{\Phi} - \frac{n^2 - n - 2}{2(n-1)} G(\text{grad } \Phi, \text{grad } \ln \rho)$$

$$+ \left[2 - \frac{(n-2)^2(n-1)}{8n} - \frac{n^2 - 2}{2(n-1)} \right] \frac{\Phi^2}{\rho}.$$

Remark 5.19. For $n = 2$ the foregoing Proposition implies the inequality $\Delta^{(B)}\Phi \geq 0$.

5.4 Proof of Calabi's conjecture in dimension 2

In this section we give a proof of Calabi's conjecture of dimension $n = 2$ due to Li and Jia (cf. [174]).

Theorem 5.20. *Let* $x : M \rightarrow \mathbb{R}^3$ *be a locally strongly convex affine maximal surface which is complete with respect to the Blaschke metric. Then it is an elliptic paraboloid.*

For the proof we use Theorem 5.15 and a useful Lemma of Hofer [123], which was applied several times in both symplectic geometry and complex geometry.

Lemma 5.21 ([123, p. 535]). *Let* (X, d) *be a complete metric space with metric* d, *and* $\bar{B}_a(p, d) := \{x \mid d(p, x) \leq a\}$ *be a ball with center* p *and radius* a. *Let* Ψ *be a nonnegative continuous function defined on* $\bar{B}_{2a}(p, d)$. *Then there is a point* $q \in \bar{B}_a(p, d)$ *and a positive number* $\varepsilon < \frac{a}{2}$ *such that*

$$\Psi(x) \leq 2\Psi(q), \text{ for all } x \in \bar{B}_\varepsilon(q, d) \text{ and } \varepsilon\Psi(q) \geq \frac{a}{2}\Psi(p).$$

Proof. Consider the function $F(x) = \Psi(x)d(x, \partial B_a(p, d))$ defined on $\bar{B}_a(p, d)$. Assume that F attains its maximum at q. Set

$$\varepsilon = \frac{1}{2}d(q, \partial B_a(p, d)).$$

Then we easily see that

$$\Psi(x) \leq 2\Psi(q),$$

for all $x \in \bar{B}_\varepsilon(q, d)$ and $\varepsilon\Psi(q) \geq \frac{a}{2}\Psi(p)$. $\qquad\square$

Let $x : M \rightarrow \mathbb{R}^3$ be a locally strongly convex affine maximal surface, which is complete with respect to the Blaschke metric. We choose local isothermal parameters (u, v) such that the Blaschke metric of M is given by

$$G = F((du)^2 + (dv)^2),$$

where F is a positive function of (u, v). The equation $\Delta^{(B)}U = -2L_1U$ with $L_1 = 0$ implies that U is harmonic with respect to G. Let $\theta := u + \sqrt{-1}\,v$. Define α and β as in (5.29) and (5.30) (by identification of θ with z there), respectively. As before, let $\|\cdot\|_G$ denote the norm with respect to the Blaschke metric, then we have

$$\|\alpha\|^2 := \left(\frac{F}{2}\right)^{-3}\cdot\alpha\bar{\alpha} = \frac{1}{2}\|A\|_G^2, \quad \|\beta\|^2 := \left(\frac{F}{2}\right)^{-2}\cdot\beta\bar{\beta} = \frac{1}{2}\|\hat{B}\|_G^2. \tag{5.56}$$

We want to show that there is a constant $N > 0$ such that $\|\hat{B}\|_G^2 \le N$ everywhere. The proof of this uses the following Steps 1–3 and additionally Lemma 5.22 below.

We assume that $\|\hat{B}\|_G^2$ is not bounded above. Then there is a sequence of points $\{p_\ell\}_{\ell\in\mathbb{N}} \subset M$ such that $\|\hat{B}\|_G^2(p_\ell) \to \infty$ as $\ell \to \infty$. We may assume that M is simply connected, otherwise we consider its universal covering space. As M is noncompact and (M, G) is complete with

$$\chi = J + L_1 \ge 0,$$

it is conformally equivalent to the complex plane $\mathbb{C}$. Then we may assume the isothermal parameters (u, v) are globally defined on M. Let $\bar{B}_1(p_\ell, G)$ be the geodesic ball with center p_ℓ and radius 1. Consider a family of functions $\Psi^{(\ell)}$, $\ell \in \mathbb{N}$, defined by

$$\Psi^{(\ell)} : \bar{B}_2(p_\ell, G) \to \mathbb{R},$$

$$\Psi^{(\ell)} := \|\text{grad}\ln F\|_G + \|A\|_G + \|\hat{B}\|_G^{1/2}.$$

In terms of $(u, v) := (u^1, u^2)$, and noting that

$$A = \sum A_{ij}^k du^i du^j \frac{\partial}{\partial u^k}, \quad \hat{B} = \sum B_{ij}du^i du^j,$$

we have

$$\|\text{grad}\ln F\|_G^2 = \frac{1}{F}\left[\left(\frac{\partial\ln F}{\partial u}\right)^2 + \left(\frac{\partial\ln F}{\partial v}\right)^2\right],$$

$$\|A\|_G^2 = \frac{1}{F}\sum(A_{ij}^k)^2, \quad \|\hat{B}\|_G^2 = \frac{1}{F^2}\sum(B_{ij})^2.$$

Using Lemma 5.21 we find a sequence of points q_ℓ and positive numbers ε_ℓ such that

$$\Psi^{(\ell)}(x) \le 2\Psi^{(\ell)}(q_\ell), \quad \text{for all } x \in \bar{B}_{\varepsilon_\ell}(q_\ell, G), \tag{5.57}$$

$$\varepsilon_\ell\Psi^{(\ell)}(q_\ell) \ge \frac{1}{2}\Psi^{(\ell)}(p_\ell) \to \infty. \tag{5.58}$$

The restriction of the surface x to the balls $\bar{B}_{\varepsilon_\ell}(q_\ell, G)$ defines a family $M^{(\ell)}$ of affine maximal surfaces. For every ℓ, we normalize $M^{(\ell)}$ as follows:

Step 1. For each ℓ, we denote by $u^{(\ell)}$, $v^{(\ell)}$ the restriction of the isothermal parameters (u, v) of M to $M^{(\ell)}$. First we take a parameter transformation:

$$\hat{u}^{(\ell)} = c^{(\ell)}u^{(\ell)}, \quad \hat{v}^{(\ell)} = c^{(\ell)}v^{(\ell)}, \quad c^{(\ell)} > 0, \tag{5.59}$$

where $c^{(\ell)}$ is a constant. Choosing $c^{(\ell)}$ appropriately and using an obvious notation $\hat{F}$, we may assume that, for every ℓ, we have $\hat{F}(q_\ell) = 1$. Note that, under the parameter transformation (5.59), $\Psi^{(\ell)}$ is invariant.

Step 2. We use the Weierstraß representation for affine maximal surfaces (see Section 5.2.2) to define, for every ℓ, a new surface $\tilde{M}^{(\ell)}$ from $M^{(\ell)}$ via its conormal by

$$\tilde{U}^{(\ell)} = \lambda^{(\ell)} U^{(\ell)}, \quad \lambda^{(\ell)} > 0;$$

we introduce new parameters $\tilde{u}^{(\ell)}, \tilde{v}^{(\ell)}$ by

$$\tilde{u}^{(\ell)} = b^{(\ell)} \hat{u}^{(\ell)}, \quad \tilde{v}^{(\ell)} = b^{(\ell)} \hat{v}^{(\ell)}, \quad b^{(\ell)} > 0,$$

where $\lambda^{(\ell)}$ and $b^{(\ell)}$ are constants. From the foregoing conormal equation one easily verifies that each $\tilde{M}^{(\ell)}$ again is a locally strongly convex affine maximal surface. We now choose $\lambda^{(\ell)} = (b^{(\ell)})^{\frac{2}{3}}$, $b^{(\ell)} = \Psi^{(\ell)}(q_\ell)$. Using again obvious notations $\tilde{F}$ and $\tilde{\Psi}^{(\ell)}$, one can see that

$$\tilde{F} = \hat{F}, \quad \tilde{\Psi}^{(\ell)} = \frac{1}{b^{(\ell)}} \Psi^{(\ell)}.$$

In fact, the first equation is trivial. Now we calculate the second one. We can easily get

$$\|\text{grad} \ln \tilde{F}\|_{\tilde{G}} = \frac{1}{b} \|\text{grad} \ln \hat{F}\|_{\hat{G}}.$$

From (5.56), (5.28)–(5.30) and our choice $\lambda^3 = b^2$ we have

$$\|\tilde{B}\|_{\tilde{G}}^2 = 2\|\tilde{\beta}\|^2 = 2\frac{1}{b^4}\|\beta\|^2 = \frac{1}{b^4}\|\hat{B}\|_{G}^2,$$

$$\|\tilde{A}\|_{\tilde{G}}^2 = 2\|\tilde{\alpha}\|^2 = 2\frac{1}{b^2}\|\alpha\|^2 = \frac{1}{b^2}\|A\|_{G}^2.$$

Then the second equation follows.

We denote $\tilde{B}_a(q_\ell, \tilde{G}) := \{x \in \tilde{M}^{(\ell)} \mid \tilde{d}^{(\ell)}(x, q_\ell) \le a\}$, where $\tilde{d}^{(\ell)}(x, q_\ell)$ denotes the geodesic distance from q_ℓ to x with respect to the Blaschke metric $\tilde{G}$ on $\tilde{M}^{(\ell)}$. Then $\tilde{\Psi}^{(\ell)}$ is defined on the geodesic ball $\tilde{B}_{\tilde{\varepsilon}_\ell}(q_\ell, \tilde{G})$ with

$$\tilde{\varepsilon}_\ell = \varepsilon_\ell \Psi^{(\ell)}(q_\ell) \ge \frac{1}{2} \Psi^{(\ell)}(p_\ell) \to \infty.$$

From (5.57) we have

$$\tilde{\Psi}^{(\ell)}(q_\ell) = 1 \quad \text{and} \quad \tilde{\Psi}^{(\ell)}(x) \le 2 \text{ for all } x \in \tilde{B}_{\tilde{\varepsilon}_\ell}(q_\ell, \tilde{G}).$$

Step 3. For any ℓ we introduce new parameters $\xi_1^{(\ell)}, \xi_2^{(\ell)}$ as follows:

$$\xi_1^{(\ell)} = \tilde{u}^{(\ell)} - \tilde{u}^{(\ell)}(q_\ell), \quad \xi_2^{(\ell)} = \tilde{v}^{(\ell)} - \tilde{v}^{(\ell)}(q_\ell).$$

Then, at q_ℓ, $(\xi_1^{(\ell)}, \xi_2^{(\ell)}) = (0,0)$ for any ℓ, and we can identify the parametrization $(\xi_1^{(\ell)}, \xi_2^{(\ell)}) =: (\xi_1, \xi_2)$ for any index ℓ. Let $\tilde{x}^{(\ell)}$ denote the position vector of $\tilde{M}^{(\ell)}$. An appropriate unimodular affine transformation gives

$$\tilde{x}^{(\ell)}(0) = 0, \tag{5.60}$$

$$\tilde{x}_{\xi_1}^{(\ell)}(0) = e_1 = (1,0,0), \tag{5.61}$$

$$\tilde{x}_{\xi_2}^{(\ell)}(0) = e_2 = (0,1,0), \tag{5.62}$$

$$\tilde{Y}^{(\ell)}(0) = (0,0,1). \tag{5.63}$$

Consider the open geodesic balls around the origin

$$B_{\tilde{\varepsilon}_\ell}(0, \tilde{G}) := \{(\xi_1, \xi_2) \in \mathbb{R}^2 \mid \tilde{d}^{(\ell)}(0,\xi) < \tilde{\varepsilon}_\ell\}$$

and the sequence $\tilde{M}^{(\ell)}$ of affine maximal surfaces $\tilde{x}^{(\ell)} : B_{\tilde{\varepsilon}_\ell}(0, \tilde{G}) \to \mathbb{R}^3$. They satisfy (5.60)–(5.63) and the conditions

$$\tilde{F}^{(\ell)}(0) = 1, \tag{5.64}$$

$$\tilde{\Psi}^{(\ell)}(0) = 1 \text{ and } \tilde{\Psi}^{(\ell)}(\xi) \le 2 \text{ for all } \xi \in B_{\tilde{\varepsilon}_\ell}(0, \tilde{G}), \tag{5.65}$$

$$\tilde{\varepsilon}_\ell \to \infty. \tag{5.66}$$

It follows from (5.17) and (5.60)–(5.64) that, for any ℓ, it holds the relation

$$(\tilde{U}_{\xi_1}^{(\ell)}, \tilde{U}_{\xi_2}^{(\ell)}, \tilde{U}^{(\ell)})(0) = I,$$

where I is the unit matrix.

Now, we need the following Lemma:

Lemma 5.22. *Let M be an affine maximal surface defined in a neighborhood of the origin $O \in \mathbb{R}^2$. Assume that, with the notation from above,*

(i) $$F(0) = 1, \quad (U_{\xi_1}, U_{\xi_2}, U)(0) = I,$$

(ii) $$\left(\frac{1}{F}\sum\left(\frac{\partial \ln F}{\partial \xi_i}\right)^2\right)^{\frac{1}{2}} + \left(\frac{1}{F}\sum (A_{ij}^k)^2\right)^{\frac{1}{2}} + \left(\frac{1}{F}\left(\sum (B_{ij})^2\right)^{\frac{1}{2}}\right)^{\frac{1}{2}} \le 2.$$

Denote $\bar{B}_{1/\sqrt{2}}(0) := \{(\xi_1, \xi_2) \in \mathbb{R}^2 \mid (\xi_1)^2 + (\xi_2)^2 \le \frac{1}{2}\}$. Then there is a constant $C_1 > 0$, such that, for $(\xi_1, \xi_2) \in \bar{B}_{1/\sqrt{2}}(0)$, the following estimates hold:
(1) $\frac{4}{9} \le F \le 4$;
(2) $\|U\| + \|U_{\xi_1}\| + \|U_{\xi_2}\| \le C_1$, *where $\| \cdot \|$ denotes the canonical norm in $\mathbb{R}^3$;*
(3) *denote $r_0 = \frac{1}{3}$, and let $B_{r_0}(0, G)$ be the geodesic ball with center O and radius r_0 with respect to the Blaschke metric G; then*

$$\bar{B}_{r_0}(0, G) \subset \left\{(\xi_1, \xi_2) \in \mathbb{R}^2 \mid (\xi_1)^2 + (\xi_2)^2 < \frac{1}{4}\right\} \subset \bar{B}_{1/\sqrt{2}}(0).$$

Proof. (1) Consider an arbitrary curve

$$\Gamma = \left\{ (\xi_1, \xi_2) \mid \xi_1 = a_1 s, \; \xi_2 = a_2 s; \; (a_1)^2 + (a_2)^2 = 1, \; s \ge 0 \right\}.$$

From the assumption we have

$$\frac{1}{F} \left(\frac{\partial \ln F}{\partial s} \right)^2 \le 2, \quad F(0) = 1.$$

Solving this differential inequality, we get

$$\left(\frac{1}{1 + \frac{\sqrt{2}}{2} s} \right)^2 \le F(s) \le \left(\frac{1}{1 - \frac{\sqrt{2}}{2} s} \right)^2.$$

Then the assumption gives $s \le \frac{1}{\sqrt{2}}$, and the assertion (1) follows.

(2) Note that the Christoffel symbols are given by $\frac{\partial \ln F}{\partial \xi_i}$ as follows:

$$\Gamma_{11}^1 = \Gamma_{12}^2 = \Gamma_{21}^2 = -\Gamma_{22}^1 = \frac{1}{2} \frac{\partial \ln F}{\partial \xi_1}; \quad \Gamma_{22}^2 = \Gamma_{12}^1 = \Gamma_{21}^1 = -\Gamma_{11}^2 = \frac{1}{2} \frac{\partial \ln F}{\partial \xi_2}.$$

Along the curve Γ the structure equation $U_{,ij} = -\sum A_{ij}^k U_{,k} - B_{ij} U$ gives an ODE, which can be written in matrix form:

$$\frac{dX}{ds} = XD, \tag{5.67}$$

where $X = (U_{\xi_1}, U_{\xi_2}, U)$, and D is a matrix whose elements depend on B_{ij}, A_{ij}^k and $\frac{\partial \ln F}{\partial \xi_i}$. From (5.67) it follows that

$$\frac{dX^t}{ds} = D^t X^t, \tag{5.68}$$

where we use an obvious notation for the transpose of a matrix. Then

$$\frac{d(X^t X)}{ds} = D^t X^t X + X^t XD. \tag{5.69}$$

Denote $f := \mathrm{tr}\,(X^t X)$. Taking the trace of (5.69) we get

$$\frac{df}{ds} = \mathrm{tr}\,(D^t X^t X) + \mathrm{tr}\,(X^t XD) \le Cf, \tag{5.70}$$

where C is a constant. In deriving the last inequality we have used (1) and the condition (ii). Solving (5.70) with the condition (i) we get (2).

Finally, from (1) we immediately get (3). $\qquad\qquad\square$

We now continue with the proof of Calabi's conjecture: Since $\tilde{\varepsilon}_\ell \to \infty$, we have the relation $\bar{B}_{\sqrt{2}/2}(O) \subset B_{\tilde{\varepsilon}_\ell}(O, \tilde{G})$ for ℓ big enough. In fact, by (1) of Lemma 5.22, the geodesic distance from O to the boundary of $\bar{B}_{\sqrt{2}/2}(O)$ with respect to the Blaschke metric on $\tilde{M}^{(\ell)}$ is less than $\sqrt{2}$.

Using (2) of Lemma 5.22 and a standard elliptic estimate (cf. Theorem 5.23 below), we get a C^k-estimate of $\tilde{U}^{(\ell)}$, independent of ℓ, for any k. It follows that there is a ball $\{(\xi_1, \xi_2) \in \mathbb{R}^2 \mid (\xi_1)^2 + (\xi_2)^2 \leq r^2\}$ and a subsequence of $\tilde{U}^{(\ell)}$ (still indexed by ℓ) such that it converges to a vector valued function $\tilde{U}$ on the ball, and correspondingly converges in all derivatives, where $r^2 < \frac{1}{2}$ is close to $\frac{1}{2}$. Thus, as limit, we get an affine maximal surface $\tilde{M}$, defined on a ball, which, by (3) of Lemma 5.22, contains the geodesic ball $\bar{B}_{r_0}(O, \tilde{G})$.

Now we extend the surface $\tilde{M}$ as follows: For every $p = (\xi_1^0, \xi_2^0) \in \partial \bar{B}_{r_0}(O, \tilde{G})$ we first make the parameter transformation: $\tilde{\xi}_i = b(\xi_i - \xi_i^0)$ such that, at p, $(\tilde{\xi}_1, \tilde{\xi}_2) = (0, 0)$, and for the limit surface $\tilde{M}$ we have $\tilde{F}(p) = 1$. We choose a frame $\{e_1, e_2, e_3\}$ at p such that $e_1 = \tilde{x}_{\tilde{\xi}_1}$, $e_2 = \tilde{x}_{\tilde{\xi}_2}$, $e_3 = \tilde{Y}$. We have

$$\tilde{F}^{(\ell)}(p) \to \tilde{F}(p) = 1, \quad \left(\tilde{U}^{(\ell)}_{\tilde{\xi}_1}, \tilde{U}^{(\ell)}_{\tilde{\xi}_2}, \tilde{U}^{(\ell)} \right)(p) \to I \text{ as } \ell \to \infty. \tag{5.71}$$

It is easy to see that, under the conditions (5.71) and (ii) in Lemma 5.22, the estimates (1), (2), and (3) in Lemma 5.22 hold again. By the same argument as above we conclude that there is a ball around p and a subsequence $\{\ell_k\}$, such that $\tilde{U}^{(\ell_k)}$ converges to a $\tilde{U}'$ on the ball, and correspondingly all derivatives. As limit, we get an affine maximal surface $\tilde{M}'$, which contains a geodesic ball of radius r_0 around p. Then we return to the original parameters (ξ_1, ξ_2) and the original frame $\{e_1, e_2, e_3\}$ at O. Here, we note that the geodesic distance is independent of the choice of the parameters and the frames.

It is obvious that $\tilde{M}$ and $\tilde{M}'$ agree on their common part. We repeat this procedure to extend $\tilde{M}$ to be defined on $\bar{B}_{2r_0}(O, \tilde{G})$, and then repeat it again. In this way we may extend $\tilde{M}$ to be an affine complete maximal surface defined in a domain $\Omega \subset \mathbb{R}^2$, which still is denoted by $\tilde{M}$; using (5.64) and (5.65) we get

$$\|\tilde{B}\|_{\tilde{G}} \leq 4, \quad \tilde{\Psi}(0) = 1.$$

By Theorem 5.15, $\tilde{M}$ must be an elliptic paraboloid, given by

$$x^3 = \frac{1}{2}((x^1)^2 + (x^2)^2), \tag{5.72}$$

where $\{x^1, x^2, x^3\}$ are the coordinates in $\mathbb{R}^3$ with respect to the frame $\{e_1, e_2, e_3\}$. For a paraboloid we have $\tilde{G} = (dx^1)^2 + (dx^2)^2$ and $\|\tilde{A}\|_{\tilde{G}} = 0$, $\|\tilde{B}\|_{\tilde{G}}^2 = 0$, $\tilde{R} = 0$ identically. Thus

$$\|\text{grad} \ln \tilde{F}\|_{\tilde{G}}(0) = 1. \tag{5.73}$$

We consider $\ln \tilde{F}$ as a function of (x^1, x^2). Since the scalar curvature vanishes identically, $\tilde{R} = 0$, from the formula

$$\Delta \ln \tilde{F} = -\tilde{R},$$

we conclude that $\ln \tilde{F}$ is a harmonic function. As $\|\text{grad} \ln \tilde{F}\| \leq 2$, $\ln \tilde{F}$ must be a linear function. In view of (5.73), without loss of generality, we may assume that $\ln \tilde{F} = x^1$.

We introduce complex coordinates and write $w := \xi_1 + \sqrt{-1}\,\xi_2$, $z := x^1 + \sqrt{-1}\,x^2$ on $\tilde{M}$, then the coordinate transformation function $w = w(z)$ is holomorphic or anti-holomorphic. We consider the case that $w(z)$ is holomorphic. For the case that $w(z)$ is anti-holomorphic, the discussion is similar.

Now we have $\tilde{G} = |dz|^2 = \tilde{F}|dw|^2$. It follows that

$$|w'(z)|^2 = \tilde{F}^{-1} = e^{-x^1}.$$

Let $Q(z) = e^{z/2}$. Then we have $|w'(z)Q(z)| = 1$. The maximum modulus theorem for holomorphic functions implies that $w'(z)Q(z) = c$, where c is a constant with $|c| = 1$. So $w'(z) = ce^{-z/2}$ and $w(z) = -2ce^{-z/2} + w_0$, where w_0 is a constant. We note that the function $e^{-z/2}$ is periodic in the variable x^2 with period 4π; the coordinate transformation $z \mapsto w(z)$ obviously defines a covering map $\pi : \mathbb{R}^2 \to \Omega$ of multiplicity infinity; this is impossible because both $\mathbb{R}^2$ and Ω are parametrization domains of $\tilde{M}$, respectively, and that $\pi : \mathbb{R}^2 \to \Omega$ is a local diffeomorphism.

The preceding contradiction shows that $\|B\|_G$ must be bounded from above on M. By Theorem 5.15, M is an elliptic paraboloid.

We have completed the proof of Theorem 5.20. $\qquad\qquad\square$

5.5 Chern's affine Bernstein conjecture

In this section we study Chern's conjecture on complete affine maximal hypersurfaces. In Section 5.5.2 we give a partial result on Chern's conjecture in arbitrary dimensions, and in Section 5.5.3 we present a proof of Chern's conjecture in dimension 2.

5.5.1 Some tools from p.d.e. theory

To solve Chern's conjecture (and also when we consider the problems in Chapter 6), we need the standard elliptic estimates from p.d.e. theory as well as the linearized Monge–Ampère equation theory of Caffarelli–Gutiérrez. For a reader who is unfamiliar with such tools we collect some details, for references see [101] and [34, 35].

First we recall the classical Schauder interior estimate.

Theorem 5.23 (cf. [101]). *Let Ω be a bounded $C^{2,\alpha}$-domain in $\mathbb{R}^n$ and L be a uniformly elliptic operator in Ω, defined by*

$$Lu := \sum a_{ij}u_{ij} + \sum b_i u_i + c,$$

where a_{ij} satisfy $\lambda|x|^2 \le \sum a_{ij}x^i x^j \le \Lambda|x|^2$ for constants $\Lambda > \lambda > 0$, and where the coefficients a_{ij}, b_i, $c \in C^\alpha(\Omega)$ satisfy

$$\|a_{ij}\|_{C^\alpha(\Omega)} \le C_1, \quad \|b_i\|_{C^\alpha(\Omega)} \le C_1, \quad \|c\|_{C^\alpha(\Omega)} \le C_1,$$

for some constant C_1. Assume that $u \in L^\infty(\overline{\Omega}) \cap C^{2,\alpha}(\Omega)$ is a solution of the p.d.e.

$$Lu = f \text{ in } \Omega$$

with $\|f\|_{C^\alpha(\Omega)} < \infty$. Then, for any Ω' with $\overline{\Omega'} \subset \Omega$, we have

$$\|u\|_{C^{2,\alpha}}(\Omega') \leq C(\|u\|_{L^\infty(\overline{\Omega})} + \|f\|_{C^\alpha(\Omega)}),$$

where C is a positive constant depending only on $n, \alpha, \lambda, \Lambda, C_1$ and $\mathrm{dist}\,(\Omega', \partial\Omega)$.

We also need the following $C^{2,\alpha}$-estimate of Monge–Ampère equations due to L. A. Caffarelli.

Theorem 5.24 ([34], see also [369]). *Let Ω be a bounded convex domain in $\mathbb{R}^n$, and let $\alpha \in (0,1)$ be a constant. Assume that $f \in C^\alpha(\Omega)$ has a positive lower bound and bounded $C^\alpha(\Omega)$-norm. If $u \in L^\infty(\overline{\Omega}) \cap C^{2,\alpha}(\Omega)$ is a convex generalized solution of the following boundary value problem*

$$\begin{cases} \det(u_{ij}) = f & \text{in } \Omega, \\ u = 0 & \text{on } \partial\Omega, \end{cases} \tag{5.74}$$

then, for any Ω' with $\overline{\Omega'} \subset \Omega$, we have

$$\|u\|_{C^{2,\alpha}(\Omega')} \leq C,$$

where C is a positive constant depending only on $n, \alpha, \Omega', \Omega, \inf_\Omega f$ and $\|f\|_{C^\alpha(\Omega)}$.

Now we describe the linearization of the Monge–Ampère equation. Let ϕ be a convex function defined on $\mathbb{R}^n$ and consider the Monge–Ampère equation

$$\det(\phi_{ij}) = f. \tag{5.75}$$

For a given function u defined on $\mathbb{R}^n$ we consider $\det(\phi_{ij} + tu_{ij})$. Then it is easy to see that

$$\det(\phi_{ij} + tu_{ij}) = \det(\phi_{ij}) + t\sum \Phi^{ij}u_{ij} + \cdots + t^n \det(u_{ij}),$$

where (Φ^{ij}) is the matrix of cofactors of the Hessian matrix (ϕ_{ij}). The coefficient of t in this expansion is called *the linearization* of the Monge–Ampère equation (5.75) and will be denoted by $L_\phi(u)$. This defines an elliptic partial differential operator (possibly degenerate) that can be written as

$$L_\phi(u) := \sum \Phi^{ij}u_{ij} = \sum(\Phi^{ij}u)_{ij} = \sum(\Phi^{ij}u_i)_j.$$

For a nonnegative solution u of the homogeneous equation $L_\phi u = 0$, Caffarelli and Gutiérrez [35] proved the Hölder estimates and the Harnack inequality of u, provided the function ϕ satisfies the condition

$$(C_1)^{-1} \leq \det(\phi_{ij}) \leq C_1 \tag{5.76}$$

for a positive constant C_1. Particularly, we have

Theorem 5.25 ([35]; see also [369, pp. 480–483 and p. 491]). *Let $\Omega \subset \mathbb{R}^n$ be a normalized convex domain. Let ϕ be a strongly convex function defined in Ω, satisfying (5.76) and $\phi = 0$ on $\partial\Omega$. Let $u \in C^2(\Omega)$ be a nonnegative solution of $L_\phi u = 0$. Then, for any section $S_\phi(x_0, h) \subset \Omega$, there exist positive constants $\alpha \in (0, 1)$ and C, depending on n, C_1, and h, such that*

$$\|u\|_{C^\alpha(S_\phi(x_0,h/2))} \leq C \|u\|_{L^\infty(\Omega)}. \tag{5.77}$$

Remark 5.26. Caffarelli and Gutiérrez's condition in [35] is weaker than (5.76), but Theorem 5.25 is enough for our applications to affine geometry.

5.5.2 A partial result on Chern's conjecture in arbitrary dimensions

Before giving the main result Theorem 5.28 below, first we prove:

Lemma 5.27. *The affine maximal hypersurface equation (5.16) is equivalent to one of the following two equations:*

$$\Delta^{(C)}\rho = -\frac{n-2}{2}\frac{\|\operatorname{grad}\rho\|_{\mathcal{H}}^2}{\rho} = -\frac{n-2}{2}\rho\Phi. \tag{5.78}$$

and

$$\sum U^{ij}w_{ij} = 0, \tag{5.79}$$

where $w = [\det(f_{kl})]^{-\frac{n+1}{n+2}}$, and $U^{ij} = [\det(f_{kl})]^{-1}f^{ij}$ is the cofactor of (f_{ij}).

Proof. The equivalence between (5.16) and (5.78) is easy: see also (5.36). Now we prove the equivalence between (5.78) and (5.79). By definition of $\Delta^{(C)}$ in Section 2.7.2, we have

$$\Delta^{(C)}\rho = \sum f^{ij}\rho_{ij} + \frac{n+2}{2\rho}\sum f^{ij}\rho_i\rho_j = \sum f^{ij}\rho_{ij} + \frac{n+2}{2}\frac{\|\operatorname{grad}\rho\|_{\mathcal{H}}^2}{\rho}. \tag{5.80}$$

On the other hand, since $w = [\det(f_{kl})]^{-\frac{n+1}{n+2}} = \rho^{n+1}$, we get

$$\sum U^{ij}w_{ij} = \sum U^{ij}(\rho^{n+1})_{ij} = (n+1)\sum U^{ij}(\rho^n\rho_i)_j$$
$$= (n+1)\rho^n \sum U^{ij}\left(\rho_{ij} + n\frac{\rho_i\rho_j}{\rho}\right)$$
$$= (n+1)\rho^{-2}\left(\sum f^{ij}\rho_{ij} + n\frac{\|\operatorname{grad}\rho\|_{\mathcal{H}}^2}{\rho}\right).$$

Using (5.80), the equivalence between (5.78) and (5.79) follows. □

Theorem 5.28. *Let M be a Euclidean complete affine maximal hypersurface given by*

$$M := \{(x^1, \ldots, x^{n+1}) \mid x^{n+1} = f(x), \ x = (x^1, \ldots, x^n) \in \Omega\},$$

where $\Omega \subset \mathbb{R}^n$ is a convex domain and f is a strongly convex smooth function defined on Ω. Then, if there is a constant $b > 1$ such that

$$b^{-1} \leq \det(\partial_i \partial_j f) \leq b$$

holds everywhere in Ω, f must be a quadratic polynomial.

Proof. Given any $p \in M$, adding a linear function if necessary, we may assume that

$$f(p) = 0, \quad \frac{\partial f}{\partial x^i}(p) = 0, \quad i = 1, \ldots, n.$$

Choose a sequence $\{C_k\}$ of positive numbers such that $C_k \to \infty$ as $k \to \infty$. Then, for any C_k, the section

$$S_f(p, C_k) := \{(x^1, x^2, \ldots, x^n) \in \mathbb{R}^n \mid f(x^1, x^2, \ldots, x^n) < C_k\}$$

is a bounded convex domain in $\mathbb{R}^n$. It is well known (cf. [111, Sect. 1.8]) that there exists a unique ellipsoid E_k which attains the minimum volume among all ellipsoids that contain $S_f(p, C_k)$ and that are centered at the barycenter of $S_f(p, C_k)$ such that

$$n^{-\frac{3}{2}} E_k \subset S_f(p, C_k) \subset E_k .$$

Let T_k be an affine transformation such that

$$T_k(E_k) = B_d(0) := \{(x^1, x^2, \ldots, x^n) \in \mathbb{R}^n \mid (x^1)^2 + \cdots + (x^n)^2 < d^2\},$$

and define functions $\{f^{(k)}\}$ by

$$f^{(k)}(x) := \frac{1}{C_k} f(T_k^{-1} x).$$

Then

$$B_{n^{-\frac{3}{2}}d}(0) \subset \Omega_k \subset B_d(0),$$

where

$$\Omega_k := \{(x^1, x^2, \ldots, x^n) \in \mathbb{R}^n \mid f^{(k)}(x^1, x^2, \ldots, x^n) < 1\}.$$

We choose T_k such that the inequalities

$$b^{-1} \leq \det\left(\frac{\partial^2 f^{(k)}}{\partial x^j \partial x^i}\right) \leq b \tag{5.81}$$

still hold. Using the comparison principle of determinants (cf. [111, p. 16]) we conclude that there is a constant $b_1 > 1$ such that

$$(b_1)^{-1} \leq d \leq b_1.$$

Taking subsequences, we may assume that $\{\Omega_k\}$ converges to a convex domain Ω and $\{f^{(k)}\}$ converges to a convex function f^∞, locally uniformly in Ω.

Consider the affine maximal hypersurface equation defined in the section $S_f(p,1) := \{x \in \Omega \mid f(x) < 1\}$

$$\sum U^{ij}w_{ij} = 0, \quad w = [\det(f_{ij})]^{-\frac{n+1}{n+2}},$$

where U^{ij} is the cofactor of (f_{ij}). As $\det(f_{ij})$ is bounded from above and below by constants (cf. (5.81)), by Theorem 5.25 we have a Hölder estimate of $\det(f_{ij})$ on the section $S_f(p, \frac{1}{2})$. Then using Theorem 5.24 we have

$$\|f\|_{C^{2,\alpha}(\Omega^*)} \le C_1',$$

for any Ω^* with $\overline{\Omega^*} \subset S_f(p, \frac{1}{2})$. It follows that $U^{ij} \in C^{\alpha}(\Omega^*)$, and we can apply Theorem 5.23 to the p.d.e.

$$\sum U^{ij}w_{ij} = 0, \quad -\frac{n+1}{n+2}\sum f^{ij}f_{ijk} = \frac{w_k}{w}$$

to obtain the estimate $\|f\|_{C^{4,\alpha}(\Omega_1^*)} \le C_2'$, for any Ω_1^* with $\overline{\Omega_1^*} \subset \Omega^*$. We note that the preceding arguments hold for each function $f^{(k)}$ and repeat the same procedure (the standard *bootstrap method*) to conclude that $\{f^{(k)}\}$ smoothly converges to f^{∞} in a neighborhood of $T_k(p) \in \Omega$, for k large enough. It follows that the functions $\Phi^{(k)}(T_k(\cdot))$ corresponding to $f^{(k)}$ are uniformly bounded.

Assume that $\Phi(p) \ne 0$; a direct calculation gives

$$\Phi^{(k)}(T_k(p)) = C_k \Phi(p) \to \infty.$$

Thus we get a contradiction, and hence $\Phi(p) = 0$. Since p is arbitrary, we have $\Phi = 0$ everywhere on M. It follows that $\det(f_{ij}) = $ const. Then the theorem of Jörgens–Calabi–Pogorelov (see Theorem 3.46) implies that f must be a quadratic polynomial. □

For a different proof of the preceding theorem, see also Z. Y. Zhang and F. Jia [412].

5.5.3 Proof of Chern's conjecture in dimension 2

First, we recall the notion of a *normalized convex set* in the sense of [111]: let $\Omega \subset \mathbb{R}^n$ be a bounded convex set with nonempty interior and E be the ellipsoid of minimal volume containing Ω and centered at the center of mass of Ω; then by Theorem 1.8.2 of [111], it holds $\alpha_n E \subset \Omega \subset E$, where $\alpha_n = n^{-3/2}$ and αE denotes the α-dilation of E with respect to its center. Notice that there is an affine transformation T such that $T(E) = B_1(O)$, here $B_t(x)$ denotes the Euclidean ball with center x and radius t. Hence we have

$$B_{\alpha_n}(O) \subset T(\Omega) \subset B_1(O).$$

Here, $T(\Omega)$ is called a normalization of Ω, and T is called an affine transformation that normalizes Ω. By ([111], p.48), the convex set Ω is called *normalized* if its center of mass is the origin O and $B_{\alpha_n}(O) \subset \Omega \subset B_1(O)$.

Next, given a positive constant C and a convex domain $\Omega \subset \mathbb{R}^n$, we denote by $\mathcal{S}(\Omega, C)$ the class of strongly convex C^∞-functions f, defined on Ω, such that

$$\inf_\Omega f(x) = 0, \quad f = C \text{ on } \partial\Omega.$$

Finally, before giving the proof of Chern's conjecture in dimension 2, we state the following lemma, for which we will give two proofs in subsequent Sections 5.5.5 and 5.6, respectively.

Lemma 5.29. *Let $\Omega_k \subset \mathbb{R}^2$ be a sequence of smooth normalized convex domains, converging to a convex domain Ω. Let $f^{(k)} \in \mathcal{S}(\Omega_k, C)$ and $q^k \in \Omega_k$, which satisfy $f^{(k)}(q^k) = 0$. Assume that the functions $f^{(k)}$ satisfy the p.d.e. (5.78). Then there exists a subsequence $f^{(i_\ell)}$ that locally uniformly converges to a convex function $f^\infty \in C^0(\Omega)$ such that, for the point p_0 satisfying $f^\infty(p_0) = 0$, we have the distance $\mathrm{dist}(p_0, \partial\Omega) > 0$. Moreover, there is an open neighborhood N of p_0 on which $f^{(i_\ell)}$ converges to f^∞, and also all their derivatives converge, so that f^∞ is smooth and strongly convex in N.*

Proof of Chern's Conjecture. Let $x : M \to \mathbb{R}^3$ be a locally strongly convex affine maximal surface, given as graph of a smooth, strongly convex function f, defined for all $(x^1, x^2) \in \mathbb{R}^2$. Thus, f satisfies the p.d.e. (5.78). Given any $p \in M$, by adding a linear function if necessary, we may assume that

$$f(p) = 0, \quad \frac{\partial f}{\partial x^i}(p) = 0, \ i = 1, 2.$$

Choose a sequence $\{C_k\}$ of positive numbers such that $C_k \to \infty$ as $k \to \infty$. Then, for any k, the section

$$S_f(p, C_k) := \left\{ (x^1, x^2) \in \mathbb{R}^2 \mid f(x^1, x^2) < C_k \right\}$$

is a bounded convex domain in $\mathbb{R}^2$. According to Theorem 1.8.2 in [111], there exists a unique ellipsoid E_k which attains the minimum volume among all ellipsoids that contain $S_f(p, C_k)$ and that are centered at the center of mass of $S_f(p, C_k)$ such that

$$2^{-\frac{3}{2}} E_k \subset S_f(p, C_k) \subset E_k.$$

Let T_k be an affine transformation such that

$$T_k(E_k) = B_1(0) := \left\{ (x^1, x^2) \in \mathbb{R}^2 \mid (x^1)^2 + (x^2)^2 < 1 \right\}.$$

Define the functions

$$f^{(k)}(x) := \frac{1}{C_k} f(T_k^{-1} x).$$

Then we have

$$B_{2^{-\frac{3}{2}}}(0) \subset \Omega_k \subset B_1(0),$$

where

$$\Omega_k := \{ (x^1, x^2) \in \mathbb{R}^2 \mid f^{(k)}(x^1, x^2) < 1 \}.$$

Taking subsequences if necessary, we may assume that $\{\Omega_k\}$ converges to a convex domain Ω and $\{f^{(k)}\}$ converges to a locally uniformly convex function f^∞ in Ω.

By Lemma 5.29, for k large enough, the function f^∞ is smooth and strongly convex in a neighborhood of $T_k(p) \in \Omega$. It follows that the functions $\Phi^{(k)}(T_k(\cdot))$ are uniformly bounded. By the same argument as in the proof of Theorem 5.28 we conclude that f must be a quadratic polynomial.

This completes the proof of Chern's Conjecture in dimension 2. □

5.5.4 Estimates for the determinant of the Hessian

In this subsection, by adopting notations of the previous two subsections, we shall estimate the determinant of the Hessian of certain convex functions. We use $\|\cdot\|_E$ to denote the norm of a vector with respect to the canonical Euclidean metric in $\mathbb{R}^n$. We give upper bounds for the determinant of the Hessian for a convex function in *sections*; such bounds were first obtained by Trudinger and Wang [367] for affine maximal hypersurfaces.

Lemma 5.30. *Let f be a strongly convex C^∞-function that is defined in a convex domain $\Omega \subset \mathbb{R}^n$ which satisfies $\inf_\Omega f(x) = 0$ and $f|_{\partial\Omega} = \text{const.} =: C > 0$.*

Further assume that f satisfies the p.d.e. (2.145) and that there is a constant $b > 0$ such that, for $\text{grad} f = \sum_i \frac{\partial f}{\partial x^i} \frac{\partial}{\partial x^i}$, we have

$$\max_\Omega \|\text{grad} f\|_E \le b, \quad \max_\Omega |L_1| \le b.$$

Then, for any $C' < C$ and with the associated section

$$S_f(C') := \{x \in \Omega \mid f(x) \le C'\},$$

we have the estimate

$$\det(f_{ij}) \le b_0 \quad \text{in } S_f(C'),$$

where $b_0 = b_0(n, b, C, \frac{C'}{C})$ is an appropriate constant.

Proof. Consider the function

$$F := \frac{\exp\left\{\frac{-m}{C-f} + \varepsilon \sum (\xi_k)^2\right\}}{\rho}$$

defined on $S_f(C)$, where m and ε are positive constants to be determined later. Clearly, F attains its supremum at some interior point p^* of $S_f(C)$. We choose a local orthonormal frame field $\{E_i\}$ of the Blaschke metric G on M near p^*. Then, at p^*, we have $\text{grad} F = \sum F_{,i} E_i = 0$ and $\Delta^{(B)} F = \sum F_{,ii} \le 0$, or equivalently,

$$- g f_{,i} + \varepsilon \left[\sum (\xi_k)^2\right]_{,i} - (\ln \rho)_{,i} = 0, \tag{5.82}$$

$$- g' \sum (f_{,i})^2 - g \Delta^{(B)} f + \varepsilon \Delta^{(B)} \left(\sum (\xi_k)^2\right) + \frac{\Phi}{\rho} - \frac{\Delta^{(B)}\rho}{\rho} \le 0, \tag{5.83}$$

where

$$g = \frac{m}{(C-f)^2}, \quad g' := \frac{2m}{(C-f)^3}, \quad \Phi = \|\text{grad } \ln \rho\|_{\mathcal{H}}^2$$

and "," denotes the covariant derivation with respect to the orthonormal frame field $\{E_i\}$ of the Blaschke metric. Inserting (2.146), (2.147), (2.157), and (5.82) into (5.83), we get

$$2\varepsilon \frac{\sum u^{ii}}{\rho} + (n+1)\frac{\Phi}{\rho} + nL_1 - g' \sum (f_{,i})^2 - \frac{n}{\rho}g \tag{5.84}$$
$$- (n+2)\varepsilon G\Big(\text{grad } \ln \rho, \text{grad } \sum (\xi_k)^2\Big) \le 0.$$

Using (5.82) and the inequality of Cauchy–Schwarz we have

$$\frac{\Phi}{\rho} = \|\text{grad } \ln \rho\|_G^2 = \sum [(\ln \rho)_{,i}]^2$$
$$= g^2 \sum (f_{,i})^2 + \varepsilon^2 \sum [(\sum (\xi_k)^2)_{,i}]^2 - 2g\varepsilon \sum f_{,i}(\sum (\xi_k)^2)_{,i}$$
$$\ge \frac{1}{2}g^2 \sum (f_{,i})^2 - \varepsilon^2 \sum [(\sum (\xi_k)^2)_{,i}]^2$$
$$= \frac{1}{2}g^2 \sum (f_{,i})^2 - 4\varepsilon^2 \rho^{-1} \sum f_{ij}\xi_i\xi_j$$
$$\ge \frac{1}{2}g^2 \sum (f_{,i})^2 - 4\varepsilon^2 \|\text{grad } f\|_E^2 \frac{\sum u^{ii}}{\rho}.$$

Similarly, using again the inequality of Cauchy–Schwarz, we get

$$(n+2)\varepsilon G\Big(\text{grad } \ln \rho, \text{grad } \sum (\xi_k)^2\Big) \le \frac{(n+2)^2 \varepsilon^2}{2n} \|\text{grad } \sum (\xi_k)^2\|_G^2 + n\|\text{grad } \ln \rho\|_G^2$$
$$\le \frac{2(n+2)^2 \varepsilon^2}{n} \|\text{grad } f\|_E^2 \frac{\sum u^{ii}}{\rho} + n\frac{\Phi}{\rho}.$$

Put the above estimates into (5.84); we obtain

$$\left[2\varepsilon - \Big(4 + \frac{2(n+2)^2}{n}\Big)\varepsilon^2 \|\text{grad } f\|_E^2\right] \frac{\sum u^{ii}}{\rho} + \Big(\frac{1}{2}g^2 - g'\Big)\sum (f_{,i})^2 - \frac{n}{\rho}g + nL_1 \le 0. \tag{5.85}$$

Now, we choose $\varepsilon = \varepsilon(n, \max_\Omega \|\text{grad } f\|_E)$ and m, such that

$$m = 4C, \quad \left[4 + \frac{2(n+2)^2}{n}\right]\varepsilon \cdot \|\text{grad } f\|_E^2 \le 1.$$

We note that $\frac{1}{2}g^2 - g' \ge 0$ and $\sum u^{ii} = \sum f_{ii} \ge n\rho^{-\frac{n+2}{n}}$, from (5.85) we get

$$\varepsilon\Big(\frac{1}{\rho}\Big)^{2(n+1)/n} - g\frac{1}{\rho} + L_1 \le 0. \tag{5.86}$$

For any $c > 0$, Young's inequality (see [13, p. 15]) implies that

$$g\frac{1}{\rho} \le \frac{n+2}{2(n+1)}(cg)^{2(n+1)/(n+2)} + \frac{n}{2(n+1)}\Big(\frac{1}{c\rho}\Big)^{2(n+1)/n}. \tag{5.87}$$

From (5.86) and (5.87), we have constants

$$a_i = a_i\left(n, \max_{\Omega} \|\mathrm{grad}\, f\|_E, \max_{\Omega} |L_1|\right), \quad i = 1, 2;$$

$$a_3 = a_3\left(n, \max_{\Omega} \|\mathrm{grad}\, f\|_E, \max_{\Omega} |L_1|, C, \frac{C'}{C}\right),$$

such that

$$\frac{1}{\rho(p^*)} \le a_1 g(p^*)^{\frac{n}{n+2}} + a_2 \le a_3.$$

It follows that, for any $x \in S_f(C')$, there holds

$$\frac{\exp\left\{\frac{-4C}{C-f} + \varepsilon \sum(\xi_k)^2\right\}}{\rho}(x) \le \frac{\exp\left\{\frac{-4C}{C-f} + \varepsilon \sum(\xi_k)^2\right\}}{\rho}(p^*) \le a_4$$

for some constant $a_4 = a_4\left(n, \max_{\Omega} \|\mathrm{grad}\, f\|_E, \max_{\Omega} |L_1|, C, \frac{C'}{C}\right)$.
Then Lemma 5.30 follows. □

If we consider the Legendre transformation function u of f, then a similar estimate to Lemma 5.30 also holds.

Lemma 5.31. *Let u be the Legendre transformation function of a strongly convex C^∞-function f which is defined in a convex domain $\Omega \subset \mathbb{R}^n$. Let Ω^* be the domain of the Legendre transformation, which satisfies $\inf_{\Omega^*} u(\xi) = 0$ and $u|_{\partial\Omega^*} = \mathrm{const.} =: C > 0$.*
Further assume that f satisfies the p.d.e. (2.146) and that there is a constant $b > 0$ such that

$$\max_{\Omega^*} \|\mathrm{grad}\, u\|_E \le b, \quad \max_{\Omega^*} |L_1| \le b.$$

Then, for any $C' < C$ and the section $S_u(C') := \{\xi \in \Omega^* | u(\xi) \le C'\}$, we have the estimate

$$\det(u_{ij}) \le b_0 \quad \text{in } S_u(C'),$$

where $b_0 = b_0(n, b, C, \frac{C'}{C})$ is an appropriate constant.

Proof. Consider the function

$$F := \rho \exp\left\{\frac{-m}{C - u} + \varepsilon \sum_k (x^k)^2\right\}$$

defined on the section $S_u(C)$. Then by a similar argument as in the proof of Lemma 5.30 we can prove the assertion. We omit the details. □

In the following we restrict to functions f of two variables. In this case we can estimate the determinant in a convex domain $\Omega \subset \mathbb{R}^2$, while usual estimates hold only in *sections*, just like in Lemmas 5.30 and 5.31. The following Lemmas 5.32 and 5.33 were first proved by Li and Jia for surfaces with constant affine mean curvature [175].

Lemma 5.32. *Let f be a strongly convex C^∞-function defined on a bounded convex domain $\Omega \subset \mathbb{R}^2$ which satisfies the p.d.e. (2.146). As before denote by Ω^* the Legendre transformation domain of f. Then, for any Ω'^* with $\overline{\Omega'^*} \subset \Omega^*$, the following estimate holds:*

$$\det(f_{ij}) \geq b_0 \quad \text{in } \Omega'^*,$$

where b_0 is a constant depending only on $\mathrm{dist}\,(\Omega'^*, \partial\Omega^*)$, $\mathrm{diam}\,(\Omega)$, $\mathrm{diam}\,(\Omega^*)$ and $\max_\Omega \{|L_1|\}$.

Proof. For any $\xi_0 = (\xi_1^0, \xi_2^0) \in \Omega^*$, we choose $\delta > 0$ such that $\delta < \mathrm{dist}\,(\xi_0, \partial\Omega^*)$. We also choose a fixed point $x_0 = (x_0^1, x_0^2) \in \Omega$. Consider the function

$$F := -\frac{m}{\delta^2 - \theta} + \rho\,(1 + \varepsilon P)$$

defined on $B_\delta^*(\xi_0)$, where

$$\theta := \sum (\xi_k - \xi_k^0)^2, \quad P := \sum (x^k - x_0^k)^2,$$

and m and ε are positive constants to be determined later. Without loss of generality, we assume that both ξ_0 and x_0 are the origin O, respectively. Clearly, F attains its supremum at some interior point ξ^* of $B_\delta^*(O)$. We choose a local orthonormal frame field $\{E_1, E_2\}$ of the Blaschke metric G on M near ξ^*. Then, at ξ^*, we have the relations $\mathrm{grad}\,F = \sum F_{,i}E_i = 0$ and $\Delta^{(B)}F = \sum F_{,ii} \leq 0$, or equivalently,

$$-\frac{m}{(\delta^2 - \theta)^2}\theta_{,i} + \varepsilon\rho P_{,i} + (1 + \varepsilon P)\rho_{,i} = 0, \tag{5.88}$$

$$-\frac{2m}{(\delta^2 - \theta)^3}\sum(\theta_{,i})^2 - \frac{m}{(\delta^2 - \theta)^2}\Delta^{(B)}\theta + \varepsilon\rho\,\Delta^{(B)}P$$
$$+ (1 + \varepsilon P)\Delta^{(B)}\rho + 2\varepsilon\sum\rho_{,i}P_{,i} \leq 0. \tag{5.89}$$

Using (2.146), (2.148), and (2.157), from (5.89) we obtain

$$\frac{2m}{(\delta^2 - \theta)^2}\,G\,(\mathrm{grad}\,\ln\rho, \mathrm{grad}\,\theta) + 4\varepsilon\rho\;G\,(\mathrm{grad}\,\ln\rho, \mathrm{grad}\,P) + 2\varepsilon\sum f^{ii}$$
$$-\frac{8m}{(\delta^2 - \theta)^3}\frac{\sum u^{ij}\xi_i\xi_j}{\rho} - \frac{2m}{(\delta^2 - \theta)^2}\frac{\sum u^{ii}}{\rho} - 2\rho(1 + \varepsilon P)L_1 \leq 0. \tag{5.90}$$

Using (5.88), we get

$$(1 + \varepsilon P)G\,(\mathrm{grad}\,\ln\rho, \mathrm{grad}\,\theta) \tag{5.91}$$
$$= -\varepsilon G\,(\mathrm{grad}\,P, \mathrm{grad}\,\theta) + \frac{m}{\rho(\delta^2 - \theta)^2}\|\mathrm{grad}\,\theta\|_G^2,$$
$$(1 + \varepsilon P)G\,(\mathrm{grad}\,\ln\rho, \mathrm{grad}\,P) \tag{5.92}$$
$$= \frac{m}{\rho(\delta^2 - \theta)^2}G\,(\mathrm{grad}\,P, \mathrm{grad}\,\theta) - \varepsilon\|\mathrm{grad}\,P\|_G^2.$$

On the other hand, a direct calculation in terms of the coordinates $\{x^i\}$ gives:

$$\rho G(\text{grad}P, \text{grad}\,\theta) = \sum f^{ij}\frac{\partial P}{\partial x^i}\frac{\partial \theta}{\partial x^j} = 4\sum x_i \xi_i \tag{5.93}$$

$$\geq -2\,\text{diam}(\Omega^*)\cdot\text{diam}(\Omega),$$

$$\|\text{grad}\,P\|_G^2 = \frac{4}{\rho}\sum f^{ij}x^i x^j. \tag{5.94}$$

Inserting all the above calculations into (5.90), we obtain

$$-\frac{8m}{(\delta^2-\theta)^3}\frac{\sum u^{ij}\xi_i\xi_j}{\rho} - \frac{2m}{(\delta^2-\theta)^2}\frac{\sum u^{ii}}{\rho} + 2\varepsilon\sum f^{ii} - \frac{4\varepsilon^2}{1+\varepsilon P}\sum f^{ij}x^i x^j \tag{5.95}$$

$$-\frac{16m\varepsilon}{\rho(\delta^2-\theta)^2}\,\text{diam}(\Omega^*)\cdot\text{diam}(\Omega) - 2\rho(1+\varepsilon P)L_1 \leq 0.$$

Denote by λ_1,λ_2 the two eigenvalues of $(\frac{\partial^2 u}{\partial\xi_i\partial\xi_j}) = (u_{ij}) = (f^{ij})$ and choose ε such that $4\varepsilon[\text{diam}(\Omega)]^2 \leq 1$. Then, from (5.95), we get

$$\varepsilon(\lambda_1+\lambda_2) \leq \frac{10m\delta^2}{\rho(\delta^2-\theta)^3}\left(\frac{1}{\lambda_1}+\frac{1}{\lambda_2}\right)$$

$$+\frac{16m\varepsilon}{\rho(\delta^2-\theta)^2}\text{diam}(\Omega^*)\cdot\text{diam}(\Omega) + 2\rho(1+\varepsilon P)L_1,$$

i.e.

$$\varepsilon\rho\lambda_1\lambda_2 \leq \frac{10m\delta^2}{(\delta^2-\theta)^3} + \frac{16m\varepsilon}{(\delta^2-\theta)^2}\frac{\lambda_1\lambda_2}{\lambda_1+\lambda_2}\text{diam}(\Omega^*)\cdot\text{diam}(\Omega) \tag{5.96}$$

$$+2\rho^2\frac{\lambda_1\lambda_2}{\lambda_1+\lambda_2}(1+\varepsilon P)L_1.$$

Recall that $\lambda_1\lambda_2 = \rho^4$ and that $\lambda_1+\lambda_2 \geq 2\sqrt{\lambda_1\lambda_2}$, we get

$$\varepsilon\rho^5 \leq \frac{10m\,\text{diam}^2(\Omega^*)}{(\delta^2-\theta)^3} + \frac{8m\varepsilon\,\text{diam}(\Omega^*)\,\text{diam}(\Omega)}{(\delta^2-\theta)^2}\rho^2 + (1+\varepsilon\,\text{diam}(\Omega))|L_1|\rho^4.$$

Using Young's inequality, we have proved

$$\rho^5 \leq c_0 + \frac{mc_1}{(\delta^2-\theta)^{10/3}} + \frac{mc_2}{(\delta^2-\theta)^3}.$$

It follows that, at ξ^*,

$$\rho \leq C_0 + \frac{m^{1/5}C_1}{(\delta^2-\theta)^{2/3}} + \frac{m^{1/5}C_2}{(\delta^2-\theta)^{3/5}}, \tag{5.97}$$

where, here and also later below, $\{c_i, C_i\}$ are constants depending only on $\text{diam}(\Omega)$, $\text{diam}(\Omega^*)$ and $\max_\Omega\{|L_1|\}$. Then using Young's inequality again, at ξ^*,

$$F = -\frac{m}{\delta^2-\theta} + \rho(1+\varepsilon P)$$

$$\leq -\frac{m}{\delta^2-\theta} + C_3 + \frac{m^{1/5}C_4}{(\delta^2-\theta)^{2/3}} + \frac{m^{1/5}C_5}{(\delta^2-\theta)^{3/5}}$$

$$\leq -\frac{m}{\delta^2-\theta} + C_6 + C_7 m^{3/5} + C_8 m^{1/2} + \frac{C_9}{\delta^2-\theta} \leq 0,$$

where we have chosen $m = m(C_6, C_7, C_8, C_9, \text{diam}(\Omega^*))$ large enough so that the last inequality holds. It follows that

$$F(\xi) = -\frac{m}{\delta^2 - \theta} + \rho(1 + \varepsilon P) \le F(\xi^*) \le 0 \quad \text{on } B_\delta^*(0),$$

and consequently,

$$\rho \le C_{10} \quad \text{on } B_{\delta/2}^*(0). \tag{5.98}$$

Finally, using a covering argument, we complete the proof of Lemma 5.32. □

Using the same method we can prove the following:

Lemma 5.33. *Let f be a strongly convex C^∞-function defined on a bounded convex domain $\Omega \subset \mathbb{R}^2$ which satisfies the p.d.e. (2.146). Denote by Ω^* the Legendre transformation domain of f. Then, for any Ω' with $\overline{\Omega'} \subset \Omega$, the following estimate holds:*

$$\det(f_{ij}) \le b_0 \quad \text{in } \Omega',$$

where b_0 is a constant depending only on $\text{dist}(\Omega', \partial\Omega)$, $\text{diam}(\Omega)$, $\text{diam}(\Omega^*)$ and $\max_\Omega |L_1|$.

5.5.5 First proof of Lemma 5.29

The following proof was given by Li and Jia in [177] (see also [181, p. 129 and pp. 153–159]). We need an important result from the classical convex body theory (see [33]). Let M be a convex hypersurface in $\mathbb{R}^{n+1}$ and e be a subset of M. We denote by $\psi_M(e)$ the Euclidean spherical image of e. Denote by $\sigma_M(e)$ the area (measure) of the spherical image $\psi_M(e)$, and by $A(e)$ the area of the set e on M. The ratio $\sigma_M(e)/A(e)$ is called the specific curvature of e. In the case $n = 2$, the following theorem due to Alexandrov-Pogorelov-Heinz holds (see [33, p. 35]):

Theorem 5.34. *A convex surface whose specific curvature is bounded away from zero is strongly convex.*

Proof of Lemma 5.29. From the assumption, taking subsequences if necessary, we may assume that $f^{(k)}$ locally uniformly converges to a convex function f^∞ in Ω. To extend the definition of f^∞ from Ω to the boundary $\partial\Omega$, we define

$$f^\infty(x) = \lim_{y \to x, y \in \Omega} f^\infty(y), \quad \forall x \in \partial\Omega.$$

Let $O \in \Omega_k$ be the center of mass of Ω_k and $u^{(k)}$ the Legendre transformation function of $f^{(k)}$ relative to O. Denote $D := \{x \in \Omega \mid f^\infty(x) = 0\}$. We discuss two cases.

(i) If $D \cap \partial\Omega = \emptyset$, then there is a constant $h > 0$ such that the level set satisfies $\overline{S_{f^\infty}(p_0, h)} \subset \Omega$, and thus we have a uniform estimate for $\|\text{grad} f^{(k)}\|_E$ in $S_{f^{(k)}}(q^k, h)$.

From Lemma 5.30 it follows that there is a uniform estimate for det $(f_{ij}^{(k)})$ in $\overline{S_{f^{(k)}}(q^k, \frac{h}{2})}$. Choose the radius R for a Euclidean ball such that

$$B_R(q^k) \supset \Omega^{(k)} \supset S_{f^{(k)}}(q^k, \frac{C}{2}).$$

Then, when $u^{(k)}$ denotes the Legendre transformation function of $f^{(k)}$, and when, corresponding to $\Omega^{(k)}$, the Legendre transformation domain is $(\Omega^{(k)})^*$, we have the relation $B_r^*(0) \subset (\Omega^{(k)})^*$, where $r = \frac{C}{4R}$. Therefore,

(1) $u^{(k)}$ locally converges to a convex function u^∞ on $B_r^*(0)$;
(2) from Lemma 5.30 and the relations of the Legendre transformation functions we see that $\det(u_{ij}^{(k)}) = [\det(f_{ij}^{(k)})]^{-1}$ is bounded from below on $B_r^*(0)$.

Let us denote by $(\xi_1, \xi_2, u^{(k)}(\xi))$ the graph of $u^{(k)}$, as a convex surface, its Gauß curvature $K_G^{(k)}$ can be written as

$$K_G^{(k)}(\xi) = \frac{\det(u_{ij}^{(k)})}{(1 + \|\operatorname{grad} u^{(k)}\|_E^2)^2}.$$

Since $\operatorname{grad} u^{(k)}(B_r^*(0)) \subset \Omega \subset B_1(0)$, we have $\|\operatorname{grad} u^{(k)}\|_E \leq 1$ in $B_r^*(0)$. It follows that

$$K_G^{(k)} \geq \frac{d_1}{4} > 0.$$

Then the specific curvature of the convex surface $(\xi_1, \xi_2, u^\infty(\xi))$ is bounded away from zero. By Theorem 5.34, we conclude that u^∞ is *strongly convex* on $B_{r/2}^*(0)$. Adding a linear function if necessary, we may assume that

$$u^{(k)}(0) = 0, \quad \frac{\partial u^{(k)}}{\partial \xi_i}(0) = 0, \quad i = 1, 2.$$

Hence there is a constant b_0 such that $\overline{S_{u^{(k)}}(0, b_0)}$ is compact in $B_{r/2}^*(0)$. Then, for sufficient large k, $S_{u^{(k)}}(0, b_0) \subset\subset B_{r/2}^*(0)$. By Lemma 5.31 we have

$$\det(u_{ij}^{(k)}) \leq d_2 \quad \text{on } S_{u^{(k)}}\left(0, \frac{b_0}{2}\right).$$

Now we use tools from Section 5.5.1, namely results of Caffarelli–Gutiérrez and estimates from Caffarelli–Schauder (cf. Theorems 5.24 and 5.25), to conclude that $\{u^{(k)}\}$ smoothly converges to u^∞ in $\{\xi \in B_r^*(0) \mid u^\infty(\xi) \leq \frac{b_0}{4}\}$. Therefore, u^∞ is a smooth and strongly convex function in an open neighborhood of $(0, 0)$ in $B_{r/2}^*(0)$, and hence f^∞ is also a smooth and strongly convex function in an open neighborhood of $(0, 0)$.

(ii) If $D \cap \partial\Omega \neq \emptyset$, we take a point $p \in D \cap \partial\Omega$. Since the p.d.e. (5.78) (i.e. (5.16)) is equiaffinely invariant, we may choose a new coordinate system such that the term $\|\operatorname{grad} f^{(k)}\|_E$ is uniformly bounded in $\overline{S_{f^{(k)}}(p, h)}$. Then the same argument as case (i) shows that f^∞ is smooth in a neighborhood of p, and we get a contradiction. This excludes the case $D \cap \partial\Omega \neq \emptyset$.

We have completed the proof of Lemma 5.29. □

5.6 An analytic proof of Chern's conjecture in dimension 2

It is interesting to give a purely analytic proof of Chern's conjecture. To achieve this, it suffices to give an analytic proof of Lemma 5.29, without using tools from the theory of convex bodies. Here, in order to introduce our analytic proof, we apply a method that we call the *real affine technique*. In recent years, the method of real (and complex) affine technique developed here, has been successfully applied to the study of Abreu's equation and extremal Kähler metrics (cf. [50–55]).

The key points of the new proof of Lemma 5.29 are the estimates of both, the second order term $\sum f_{ii}$ and the third order derivatives of f; these will be given in Sections 5.6.1–5.6.3 below. Then, finally in Section 5.6.4, we will complete our analytic proof of Lemma 5.29.

5.6.1 Technical estimates

First we consider the case of general dimension n. Let Ω be a convex domain in $\mathbb{R}^n$. For a positive constant C, we denote by $\mathcal{S}(\Omega, C)$ the class of those smooth and strongly convex functions, as defined in Section 5.5.3.

Given $f \in \mathcal{S}(\Omega, C)$, we consider the graph hypersurface M of f equipped with its Calabi metric $\mathcal{H} = \sum f_{ij} dx^i dx^j$. Now we consider the function

$$F := \exp\left\{\frac{-m}{C-f} + \tau\right\} Q \|\mathrm{grad}\, h\|_{\mathcal{H}}^2 \tag{5.99}$$

defined on Ω, where m is a positive constant to be specified later; Q, τ and h are prescribed smooth functions defined on $\overline{\Omega}$ such that $Q > 0$, $\tau > 0$ and $F \neq 0$.

From now on, in this section, we simplify the notation in calculations and write $\Delta := \Delta^{(C)}$, $A := A^{(C)}$, $R := R^{(C)}$ and so on.

Clearly, the function F attains its supremum at some interior point $p^* \in \Omega$ at which we have $\|\mathrm{grad}\, h\|(p^*) > 0$. First we prove the following general result.

Lemma 5.35. *With respect to a local orthonormal frame field $\{e_i\}_{i=1}^n$ the following inequality holds at the point p^*, where F attains its supremum:*

$$2\left(\frac{1}{n-1} - \delta - 1\right) \frac{\sum_i (\sum_j h_{ij} h_{,j})^2}{\|\mathrm{grad}\, h\|_{\mathcal{H}}^2}$$

$$+ \left[\Delta\tau - g' \sum (f_{,i})^2 - g\Delta f + \frac{\Delta Q}{Q} - \frac{\sum (Q_{,i})^2}{Q^2}\right] \|\mathrm{grad}\, h\|_{\mathcal{H}}^2 \tag{5.100}$$

$$- \frac{(n+2)^2}{8\delta} \Phi \|\mathrm{grad}\, h\|_{\mathcal{H}}^2 + \frac{2}{n-1}\left[1 - \frac{1}{(n-1)\delta}\right](\Delta h)^2$$

$$+ 2\sum_j h_{,j}(\Delta h)_{,j} + 2(1-\delta)\sum_{m,l}\left(\sum_i A_{mli} h_{,i}\right)^2 \leq 0,$$

where δ is any real number with $0 < \delta < 1$, *and*

$$g := \frac{m}{(C-f)^2}, \quad g' := \frac{2m}{(C-f)^3}. \tag{5.101}$$

Proof. Since the left hand side of (5.100) does not depend on the choice of the frame $\{e_i\}_{i=1}^n$, similar to the proof of Proposition 5.16, at p^* we can assume that $e_1 = \frac{\partial}{\partial x^1}$ and

$$f_{ij}(p^*) = \delta_{ij}, \quad h_{,1}(p^*) = \|\text{grad } h\|_{\mathcal{H}}(p^*), \quad h_{,i}(p^*) = 0 \text{ for } i > 1. \tag{5.102}$$

Then, at p^*, we can write

$$\frac{\sum_i (\sum_j h_{,ij} h_{,j})^2}{\|\text{grad } h\|_{\mathcal{H}}^2} = \sum (h_{,1i})^2, \quad \sum_{m,l} (\sum_i A_{mli} h_{,i})^2 = \sum (A_{ml1})^2 (h_{,1})^2.$$

From the assumptions, at p^*, we have grad $F = 0$ and $\Delta F \le 0$, i.e.

$$F_{,i} = 0, \quad \forall i, \tag{5.103}$$

$$\sum F_{,ii} \le 0. \tag{5.104}$$

Now we calculate both expressions (5.103) and (5.104) explicitly. We have

$$\left(-gf_{,i} + \tau_{,i} + \frac{Q_{,i}}{Q}\right) \sum (h_{,j})^2 + 2 \sum h_j h_{ji} = 0, \quad \forall i, \tag{5.105}$$

$$2 \sum (h_{,ij})^2 + 2 \sum h_j h_{jii} + 2 \sum \left(-gf_{,i} + \tau_{,i} + \frac{Q_{,i}}{Q}\right) h_j h_{ji}$$

$$+ \left[-g' \sum (f_{,i})^2 - g\Delta f + \Delta \tau + \frac{\Delta Q}{Q} - \frac{\sum (Q_{,i})^2}{Q^2}\right] (h_{,1})^2 \le 0. \tag{5.106}$$

From (5.105) we get

$$\left(-gf_{,i} + \tau_{,i} + \frac{Q_{,i}}{Q}\right) h_{,1} = -2h_{,1i}, \quad \forall i. \tag{5.107}$$

Applying the inequality of Cauchy–Schwarz, for any $\delta > 0$, we obtain

$$\sum (h_{,ij})^2 \ge (h_{,11})^2 + \frac{1}{n-1} \left(\sum_{i>1} h_{,ii}\right)^2 + 2 \sum_{i>1} (h_{,1i})^2$$

$$= (h_{,11})^2 + \frac{1}{n-1} (\Delta h - h_{,11})^2 + 2 \sum_{i>1} (h_{,1i})^2 \tag{5.108}$$

$$\ge \left(\frac{n}{n-1} - \delta\right) (h_{,11})^2 + \frac{1}{n-1} \left[1 - \frac{1}{(n-1)\delta}\right] (\Delta h)^2 + 2 \sum_{i>1} (h_{,1i})^2.$$

To estimate the term $\sum h_j h_{jii}$ in (5.106), we apply the Ricci identity and obtain

$$2 \sum h_j h_{jii} = 2 \sum h_{,j} (\Delta h)_{,j} + 2 \sum R_{ij} h_{,i} h_{,j}. \tag{5.109}$$

The Ricci tensor of the Calabi metric satisfies formula (2.163); moreover, at p^*, we have $2\sum A_{jii} = (n+2)(\ln\rho)_j$; we use the inequality of Cauchy–Schwarz to obtain

$$2\sum R_{ij}h_{,i}h_{,j} = 2\sum (A_{1ij})^2(h_{,1})^2 - (n+2)\sum A_{11i}(\ln\rho)_{,i}(h_{,1})^2$$

$$\geq 2(1-\delta)\sum (A_{1ij})^2(h_{,1})^2 - \frac{(n+2)^2}{8\delta}\Phi(h_{,1})^2. \tag{5.110}$$

We insert (5.107)–(5.110) into (5.106) and immediately get the assertion (5.100). □

Next, we consider the estimate (5.100) in two special situations, namely:

$$(1)\ h = f \quad \text{and} \quad (2)\ h = \frac{\partial f}{\partial x^1}.$$

For both cases we further estimate the invariant expression.

$$2\sum h_{,j}(\Delta h)_j + 2(1-\delta)\sum A_{mli}A_{mlj}h_{,i}h_{,j},$$

at p^*, that appears in (5.100).

Case 1: $h = f$. In this case, (5.102) is also assumed. Using the formula (2.166) we have

$$2\sum f_{,j}(\Delta f)_j = (n+2)\left[\frac{\rho_{,11}}{\rho}(f_{,1})^2 - \frac{(\rho_{,1})^2}{\rho^2}(f_{,1})^2 + \sum f_{,1}f_{,1}\frac{\rho_{,i}}{\rho}\right]. \tag{5.111}$$

Note that

$$f_{,ij} = f_{ij} + A_{ij1}f_{,1}$$

and

$$\sum (f_{,ij})^2 = \sum (A_{1ij})^2(f_{,1})^2 + n + (n+2)\frac{\rho_{,1}f_{,1}}{\rho}. \tag{5.112}$$

Similar to (5.108) we get

$$\sum (f_{,ij})^2 \geq (\frac{n}{n-1} - \varepsilon)(f_{,11})^2 + \frac{1}{n-1}[1 - \frac{1}{(n-1)\varepsilon}](\Delta f)^2 + 2\sum_{i>1}(f_{,1i})^2$$

for any $\varepsilon > 0$. Thus a combination of (5.111) and (5.112) gives

$$2\sum f_{,j}(\Delta f)_j + 2(1-\delta)\sum (A_{1ij})^2(f_{,1})^2 \tag{5.113}$$

$$\geq (n+2)\left[\frac{\rho_{,11}}{\rho}(f_{,1})^2 - \frac{(\rho_{,1})^2}{\rho^2}(f_{,1})^2 + \sum f_{,1}f_{,1}\frac{\rho_{,i}}{\rho}\right]$$

$$+ 2(1-\delta)\left\{\left(\frac{n}{n-1} - \varepsilon\right)(f_{,11})^2 + \frac{1}{n-1}\left[1 - \frac{1}{(n-1)\varepsilon}\right](\Delta f)^2 + 2\sum_{i>1}(f_{,1i})^2\right\}$$

$$- 2n(1-\delta) - 2(n+2)(1-\delta)\frac{\rho_{,1}f_{,1}}{\rho}.$$

Now taking $0 < \delta < 1$, and using the following two inequalities

$$\sum f_{,1}f_{,1}\frac{\rho_{,i}}{\rho} \geq -\sigma \sum (f_{,1i})^2 - \frac{1}{4\sigma}\Phi(f_{,1})^2, \quad \sigma > 0,$$

$$-\frac{\rho_{,1}f_{,1}}{\rho} \geq -\bar\sigma\left(\frac{\rho_{,1}f_{,1}}{\rho}\right)^2 - \frac{1}{4\bar\sigma} \geq -\bar\sigma\Phi(f_{,1})^2 - \frac{1}{4\bar\sigma}, \quad \bar\sigma > 0,$$

from (5.113) we get

$$
\begin{aligned}
2\sum f_{,j}(\Delta f)_{,j} &+ 2(1-\delta)\sum (A_{1ij})^2 (f_{,1})^2 \\
&\ge \left[2(1-\delta)\left(\frac{n}{n-1}-\varepsilon\right)-(n+2)\sigma\right]\sum (f_{,1i})^2 \\
&\quad + (n+2)\frac{\rho_{,11}}{\rho}(f_{,1})^2 + \frac{2(1-\delta)}{n-1}\left[1-\frac{1}{(n-1)\varepsilon}\right](\Delta f)^2 \\
&\quad - (n+2)\left[1+\frac{1}{4\sigma}+2(1-\delta)\bar\sigma\right]\Phi(f_{,1})^2 - \left[2n+\frac{(n+2)}{2\bar\sigma}\right](1-\delta)
\end{aligned}
\tag{5.114}
$$

for $0<\delta<1$ and any $\varepsilon>0, \sigma>0, \bar\sigma>0$.

In particular, for $n=2$, choosing ε, σ and $\bar\sigma$ to be appropriately small, we have

$$
\begin{aligned}
2\sum f_{,j}(\Delta f)_{,j} &+ 2(1-\delta)\sum (A_{1ij})^2 (f_{,1})^2 \\
&\ge 4\frac{\rho_{,11}}{\rho}(f_{,1})^2 + 4(1-3\delta)\sum (f_{,1i})^2 - \frac{2}{\delta}(\Delta f)^2 - 4\left(\frac{1}{\delta}+2\right)\Phi(f_{,1})^2 - 8.
\end{aligned}
\tag{5.115}
$$

Case 2: $h = \xi_1 = \frac{\partial f}{\partial x^1}$. First we recall that, in terms of the coordinates $(x^1,\dots,x^n)$, the Christoffel symbols satisfy $\Gamma_{ij}^k = \frac{1}{2}\sum f^{kl}f_{lij}$; then we have

$$
h_{,ij} = f_{1ij} - \frac{1}{2}\sum f_{1k}f^{kl}f_{lij} = \frac{1}{2}f_{1ij},
$$

$$
\sum f^{ki}f^{lj}h_{,kl}h_{,ij} = \frac{1}{4}\sum f^{ik}f^{jl}f_{1ij}f_{1kl}.
$$

It follows that

$$
\begin{aligned}
\sum f^{ik}f^{jl}f^{pr}f^{qs}A_{ijp}A_{klq}h_{,r}h_{,s} &= \frac{1}{4}\sum f^{ik}f^{jl}f_{ijp}f_{klq}f^{pr}f_{1r}f^{qs}f_{1s} \\
&= \sum f^{ik}f^{jl}h_{,ij}h_{,kl}.
\end{aligned}
\tag{5.116}
$$

Next, we choose a local orthonormal frame field $\{e_i\}_{i=1}^n$ such that at p^*,

$$
h_{,1}(p^*) = \|\mathrm{grad}\, h\|_{\mathcal H}(p^*), \quad h_{,i}(p^*) = 0 \text{ for } i > 1.
\tag{5.117}
$$

By (2.171) we have $\Delta h = -\frac{n+2}{2}\mathcal H(\mathrm{grad}\,\ln\rho,\,\mathrm{grad}\,h)$. It follows that

$$
(\Delta h)^2 \le \frac{(n+2)^2}{4}\Phi\|\mathrm{grad}\,h\|_{\mathcal H}^2 = \frac{(n+2)^2}{4}\Phi(h_{,1})^2,
$$

and, for any $\sigma>0$,

$$
\begin{aligned}
2\sum h_{,j}(\Delta h)_{,j} &= -(n+2)\left[\frac{\rho_{,11}}{\rho}(h_{,1})^2 - \frac{(\rho_{,1})^2}{\rho^2}(h_{,1})^2 + \sum h_{,1i}h_{,1}\frac{\rho_{,i}}{\rho}\right] \\
&\ge -(n+2)\frac{\rho_{,11}}{\rho}(h_{,1})^2 - \sigma\sum (h_{,1i})^2 - \frac{(n+2)^2}{4\sigma}\Phi(h_{,1})^2.
\end{aligned}
$$

Similar to (5.108), for $\varepsilon < \frac{1}{n-1}$, we derive the following estimate:

$$\sum (h_{,ij})^2 \geq \left(\frac{n}{n-1} - \varepsilon\right)(h_{,11})^2 - \frac{(n+2)^2[1-(n-1)\varepsilon]}{4(n-1)^2\varepsilon}\Phi(h_{,1})^2 + 2\sum_{i>1}(h_{,1i})^2.$$

Consequently, by (5.116) and choosing ε and σ appropriately, we obtain

$$\begin{aligned}
2\sum h_{,j}(\Delta h)_{,j} &+ 2(1-\delta)\sum A_{mli}A_{mlj}h_{,i}h_{,j}\\
&\geq \frac{2n-5n\delta}{n-1}(h_{,11})^2 + (4-6\delta)\sum_{i>1}(h_{,1i})^2 - (n+2)\frac{\rho_{,11}}{\rho}(h_{,1})^2\\
&\quad - \left[\frac{(n+2)^2}{4\delta} + \frac{(n+2)^2}{(n-1)^2\delta}\right]\Phi(h_{,1})^2,
\end{aligned}$$

(5.118)

for any $0 < \delta < \frac{1}{3}$.

5.6.2 Estimates for $\sum f_{ii}$

The following proposition is an analogue of Pogorelov's famous estimate for eigenvalues of the Hessian.

Proposition 5.36. *Let f be a smooth and strongly convex function defined on a normalized convex domain $\Omega \subset \mathbb{R}^n$, which satisfies equation (5.78). Assume that $f \in \mathcal{S}(\Omega, C)$, and that there exist constants $\alpha \geq 1$ and $d_1 \geq 0$ such that*

$$\frac{\Phi}{\rho^\alpha} \leq d_1, \quad \frac{1}{\rho} \leq d_1, \quad \|\mathrm{grad} f\|_{\mathbb{E}} \leq d_1$$

on $\bar{\Omega}$. Then there is a real constant $d_2 > 0$, depending only on α, d_1 and C, such that on Ω

$$\exp\left\{-\frac{8(n-1)(1+d_1)C}{C-f}\right\}\frac{\sum f_{ii}}{\rho^\alpha} \leq d_2.$$

Proof. We consider the function F as defined in (5.95) and put

$$h := \xi_1, \quad \tau := W + \theta, \quad Q := \frac{1}{\rho^\alpha}$$

for $W := \frac{\Phi}{\rho^\alpha}$ and $\theta := \varepsilon \sum(\xi_k)^2$, where ε is a positive constant to be specified later.

Assume that F attains its supremum at some interior point $p^* \in \Omega$. Choose a local orthonormal frame field $\{e_i\}_{i=1}^n$ with respect to the Calabi metric $\mathcal{H}$ near p^*, such that (5.117) holds. From $(\mathrm{grad}\, F)(p^*) = 0$ we have, at p^*,

$$\left(-gf_{,i} - \alpha\frac{\rho_{,i}}{\rho} + W_{,i} + \theta_{,i}\right)\sum(h_{,j})^2 + 2\sum h_{,j}h_{,ji} = 0, \quad 1 \leq i \leq n.$$

(5.119)

Inserting (5.118) into (5.100) with $\delta := \frac{1}{7n}$, we get, at p^*,

$$\frac{3}{n-1} \sum (h_{,1i})^2 - (n+2)\frac{\rho_{,11}}{\rho}(h_{,1})^2 - \frac{2}{(n-1)^2\delta}(\Delta h)^2$$
$$+ \left[\Delta \theta + \Delta W - g' \sum (f_{,i})^2 - g\Delta f + \left(\frac{n\alpha}{2} - a_1\right)\Phi\right](h_{,1})^2 \le 0, \qquad (5.120)$$

where a_1 is a constant depending only on n.

To estimate $\sum (h_{,1i})^2$ in (5.120), we use (5.119) to obtain:

$$2 \sum (h_{,1i})^2 = \frac{1}{2} \sum \left(gf_{,i} + \alpha\frac{\rho_{,i}}{\rho} - W_{,i} - \theta_{,i}\right)^2 (h_{,1})^2$$

$$\ge \left\{\frac{1}{4} \sum (gf_{,i} - W_{,i})^2 - \sum (\theta_{,i})^2 - \alpha^2\Phi\right\}(h_{,1})^2$$

$$\ge \left\{\frac{g^2}{4(1+W)} \sum (f_{,i})^2 - \frac{\sum (W_{,i})^2}{4W} - \sum (\theta_{,i})^2 - \alpha^2\Phi\right\}(h_{,1})^2 \qquad (5.121)$$

$$\ge \left\{\frac{g^2}{4(1+d_4)} \sum (f_{,i})^2 - \frac{\sum (W_{,i})^2}{4W} - \sum (\theta_{,i})^2 - \alpha^2\Phi\right\}(h_{,1})^2.$$

In the preceding series of inequalities we used

$$(a+b)^2 \ge (1-\eta)a^2 - \left(\frac{1}{\eta} - 1\right)b^2$$

with $\eta = \frac{W}{1+W}$.

Next, we estimate the terms $\Delta\theta$, ΔW, Δh, $\sum (\theta_{,i})^2$ and $\frac{\rho_{,11}}{\rho}$, respectively. Using the basic formulas in Section 2.7.2 and the inequality of Cauchy–Schwarz, we have

$$(n+2)\frac{|\rho_{,11}|}{\rho} \le \frac{\delta \sum (\rho_{,ij})^2}{\rho^{\alpha+2}} + \frac{(n+2)^2}{4\delta}\rho^\alpha. \qquad (5.122)$$

$$(\Delta h)^2 = \left[\frac{n+2}{2}\mathcal{H}(\text{grad}\ln\rho, \text{grad } h)\right]^2 \le \frac{(n+2)^2}{4}\Phi(h_{,1})^2. \qquad (5.123)$$

$$\sum (\theta_{,i})^2 = 4\varepsilon^2 \sum u^{ij}f_i f_j \le 4\varepsilon^2(d_1)^2 \sum u^{ii}, \qquad (5.124)$$

$$\Delta \theta = 2\varepsilon \sum u^{ii} - \frac{n+2}{2}\mathcal{H}(\text{grad }\ln\rho, \text{grad }\theta)$$

$$\ge \varepsilon \sum u^{ii} - \frac{1}{4}\varepsilon(n+2)^2(d_1)^2\Phi, \qquad (5.125)$$

$$W_{,i} = W\left(\frac{\Phi_{,i}}{\Phi} - \alpha\frac{\rho_{,i}}{\rho}\right),$$

$$\Delta W = \frac{\Phi}{\rho^\alpha}\left[\sum \left(\frac{\Phi_{,i}}{\Phi} - \alpha\frac{\rho_{,i}}{\rho}\right)^2 + \frac{\Delta\Phi}{\Phi} - \frac{\sum (\Phi_i)^2}{\Phi^2} + \frac{n}{2}\alpha\Phi\right]. \qquad (5.126)$$

Inserting (5.37) into (5.126) and using the inequality of Cauchy–Schwarz, we have

$$\Delta W \geq \frac{n(1-\delta)}{4(n-1)} \frac{\sum (W_{,i})^2}{W} + \frac{2\delta \sum (\rho_{,ij})^2}{\rho^{\alpha+2}} - a_2 \frac{\Phi^2}{\rho^\alpha}, \tag{5.127}$$

where $a_2 = a_2(n, \alpha)$ is a positive constant.

To estimate $\sum u^{ii}$, we use the relations

$$\left(\frac{\partial^2 u}{\partial \xi_i \partial \xi_j} \right) = \left(\frac{\partial^2 f}{\partial x^i \partial x^j} \right)^{-1} \quad \text{and} \quad (h_{,1})^2 = \sum f_{1i} f_{1j} f^{ij} = f_{11},$$

and so we have $\sum u^{ii} = \sum f_{ii} \geq (h_{,1})^2$.

By (2.166), we have the estimate

$$-g\Delta f = -g\left[n + \frac{n+2}{2} \mathcal{H}(\text{grad} \ln\rho, \text{grad} f) \right]$$

$$\geq -ng - \frac{g^2}{8(n-1)(1+d_1)} \sum (f_{,i})^2 - a_3 \Phi, \tag{5.128}$$

where $a_3 = a_3(n, d_1)$ is a positive constant.

Now, we choose $m = 8(n-1)(1+d_1)C$. Then

$$\frac{3g^2}{8(n-1)(1+d_1)} \geq g' + \frac{g^2}{8(n-1)(1+d_1)}. \tag{5.129}$$

If we choose $\varepsilon = \frac{n-1}{12(d_1)^2}$, then we have

$$\Delta\theta - \frac{3}{2(n-1)} \sum (\theta_{,i})^2 \geq \frac{n-1}{24(d_1)^2} \sum u^{ii} - \frac{1}{4}\varepsilon(n+2)^2(d_1)^2\Phi. \tag{5.130}$$

Finally, we insert (5.121)–(5.130) into (5.120), and obtain, at p^*,

$$(h_{,1})^2 - a_4 \Phi - a_5 g - a_6 \rho^\alpha - a_7 \leq 0 \tag{5.131}$$

for positive constants a_4, a_5, a_6, a_7 depending only on n, α, d_1. It follows that, at any $x \in \Omega$,

$$\left\{ \exp\left\{ -\frac{8(n-1)(1+d_1)C}{C-f} \right\} \frac{\|\text{grad} h\|_{\mathcal{H}}^2}{\rho^\alpha} \right\}(x)$$

$$= \{F \exp\{-W - \theta\}\}(x) \leq F(p^*)$$

$$\leq a_8 \left\{ \exp\left[-\frac{8(n-1)(1+d_1)C}{C-f} \right] \frac{\|\text{grad} h\|_{\mathcal{H}}^2}{\rho^\alpha} \right\}(p^*) \tag{5.132}$$

$$\leq a_8 \left\{ a_4 \frac{\Phi}{\rho^\alpha} + a_5 \frac{g}{\rho^\alpha} \exp\left\{ -\frac{8(n-1)(1+d_1)C}{C-f} \right\} + a_6 + \frac{a_7}{\rho^\alpha} \right\}(p^*)$$

$$\leq a_9$$

for $h = \xi_1$ and a constant $a_9 = a_9(n, \alpha, d_1, C) > 0$. It is easily seen that (5.132) also holds for each $h = \xi_i$ ($2 \leq i \leq n$) corresponding to the same upper bound a_9.

Since $\|\text{grad} \xi_i\|_{\mathcal{H}}^2 = f_{ii}$ for each i, by (5.132) and the above remarks we have completed the proof of Proposition 5.36. □

5.6.3 Estimates for the third order derivatives

In the remaining part of Section 5.6, we restrict to $n = 2$. The purpose of this subsection is the proof of the following results on the estimates for the third order derivatives.

Proposition 5.37. *Let $\Omega \subset \mathbb{R}^2$ be a normalized convex domain and $0 \in \Omega$ be the center of mass of Ω. Let $f \in \mathcal{S}(\Omega, C)$ (cf. Section 5.5.3) and assume that f satisfies the p.d.e. (5.78); moreover assume that there is a constant $b > 0$ such that*

$$\frac{1}{\rho} \leq b \text{ in } \Omega. \tag{5.133}$$

Then there exists a real constant $\alpha > 0$ such that the following estimates hold:

$$\frac{\Phi}{\rho^\alpha} \leq d, \quad \frac{\|\operatorname{grad} f\|_{\mathcal{H}}^2}{\rho^\alpha} \leq d \quad \text{on } \Omega_{C/2}, \tag{5.134}$$

where $\Omega_{C/2} = \{x \in \Omega \mid f(x) < \frac{C}{2}\}$ and $d > 0$ is a real constant that depends only on $\alpha, b,$ and C.

Proof. Related to the function f we define two numbers $\mathcal{A}$ and $\mathcal{D}$ by

$$\mathcal{A} := \max_{\Omega} \left\{ \exp\left(\frac{-m}{C-f}\right) \frac{\Phi}{\rho^\alpha} \right\},$$

$$\mathcal{D} := \max_{\Omega} \left\{ \exp\left(\frac{-m}{C-f} + K\right) \frac{g^2}{\rho^\alpha} \|\operatorname{grad} f\|_{\mathcal{H}}^2 \right\},$$

where $g := \frac{m}{(C-f)^2}$, $K := \frac{N}{\mathcal{A}} \exp\{\frac{-m}{C-f}\} \frac{\Phi}{\rho^\alpha}$ and m, α, N are positive constants to be determined later.

Claim I. For any positive constants α, N and $m \geq 2C$, we have

$$\mathcal{A} \leq \max \left\{ \frac{4b^\alpha}{\alpha C}, \frac{4}{\alpha} \mathcal{D} \exp\{-N\} \right\}. \tag{5.135}$$

To show that Claim I holds, we consider the function

$$L := \exp\left\{\frac{-m}{C-f}\right\} \frac{\Phi}{\rho^\alpha} \text{ in } \Omega.$$

Obviously, the function L attains its supremum at some interior point $\bar{p} \in \Omega$. Thus we have $(\operatorname{grad} L)(\bar{p}) = 0$ and $(\Delta L)(\bar{p}) \leq 0$. Differentiation with respect to an orthonormal frame field $\{e_1, e_2\}$ and an application of (5.78) give, at $\bar{p}$:

$$\frac{\Phi_{,i}}{\Phi} - gf_{,i} - \alpha \frac{\rho_{,i}}{\rho} = 0, \quad i = 1, 2, \tag{5.136}$$

$$\frac{\Delta\Phi}{\Phi} - \frac{\|\operatorname{grad}\Phi\|_{\mathcal{H}}^2}{\Phi^2} + \alpha\Phi - g'\|\operatorname{grad} f\|_{\mathcal{H}}^2 - g\Delta f \leq 0. \tag{5.137}$$

Deriving (5.137) we use $gf_{,i} = \frac{\Phi_i}{\Phi} - \alpha\frac{\rho_i}{\rho}$, and the function g' is defined as in (5.101): $g' = \frac{2m}{(C-f)^3}$.

On the other hand, the affine maximal surface equation gives the following lower bound for $\frac{\Delta\Phi}{\Phi}$ (see (5.54)):

$$\frac{\Delta\Phi}{\Phi} \geq \frac{\|\operatorname{grad}\Phi\|_{\mathcal{H}}^2}{\Phi^2} + \Phi. \tag{5.138}$$

Now we choose any m such that $m \geq 2C$. Then $g' \leq g^2$, and a combination of the inequalities (5.137) and (5.138) gives, at $\bar{p}$,

$$(1 + \alpha)\Phi \leq g^2\|\operatorname{grad}f\|_{\mathcal{H}}^2 + g\Delta f.$$

Using the relation (2.166) and the inequality of Cauchy–Schwarz we have

$$g\Delta f \leq g^2\|\operatorname{grad}f\|_{\mathcal{H}}^2 + \Phi + 2g.$$

It follows that, at $\bar{p}$,

$$\alpha\Phi \leq 2g^2\|\operatorname{grad}f\|_{\mathcal{H}}^2 + 2g.$$

Therefore, using the fact that $x^2 < 2\exp\{x\}$ for $x \geq 0$ we get, at $\bar{p}$,

$$\exp\left\{\frac{-m}{C-f}\right\}\frac{\Phi}{\rho^\alpha} \leq \frac{2}{\alpha}\exp\left\{\frac{-m}{C-f}\right\}g^2\frac{\|\operatorname{grad}f\|_{\mathcal{H}}^2}{\rho^\alpha} + \frac{2b^\alpha}{\alpha C}. \tag{5.139}$$

If, at $\bar{p}$, we have $\exp\left\{\frac{-m}{C-f}\right\}g^2\frac{\|\operatorname{grad}f\|_{\mathcal{H}}^2}{\rho^\alpha} \leq \frac{b^\alpha}{C}$, then (5.139) implies that $A \leq \frac{4b^\alpha}{\alpha C}$. If, on the other side, at $\bar{p}$, it holds $\exp\left\{\frac{-m}{C-f}\right\}g^2\frac{\|\operatorname{grad}f\|_{\mathcal{H}}^2}{\rho^\alpha} > \frac{b^\alpha}{C}$, then again, by (5.139) and the relation $K(\bar{p}) = N$, we have

$$\exp\{N\}A \leq \frac{4}{\alpha}\left\{\exp\left\{\frac{-m}{C-f}\right\}g^2\frac{\|\operatorname{grad}f\|_{\mathcal{H}}^2}{\rho^\alpha}\right\}(\bar{p})\exp\{N\} \leq \frac{4}{\alpha}D.$$

This verifies the assertion of Claim I.

According to (5.135), we discuss two cases as below:

(i) if $D \leq \frac{b^\alpha}{C}\exp\{N\}$, then both A and D are bounded, which immediately implies the conclusion of Proposition 5.37;

(ii) if $D \geq \frac{b^\alpha}{C}\exp\{N\}$, we will show that, A and D, are still bounded. This will be carried out in the next Claim.

Claim II. If $D \geq \frac{b^\alpha}{C}\exp\{N\}$, namely $\exp\{N\}A \leq \frac{4}{\alpha}D$, then for appropriate positive constants α, N and $m \geq 2C$, the number $D = D(f)$ is bounded from above by a constant depending only on α, b, m and N.

To show that Claim II holds, consider the function F defined as in (5.99), i.e.

$$F = \exp\left\{\frac{-m}{C-f} + \tau\right\}Q\|\operatorname{grad}h\|_{\mathcal{H}}^2 \quad \text{in } \Omega,$$

with the choice of

$$\tau := K, \quad Q := \frac{g^2}{\rho^\alpha}, \quad h := f.$$

Assume that F attains its supremum at some interior point $p^* \in \Omega$. Around p^*, we choose a local orthonormal (with respect to the Calabi metric) frame field $\{e_1, e_2\}$ such that $f_{,1}(p^*) = \|\operatorname{grad} f\|_{\mathcal{H}} > 0$. From $F_{,i}(p^*) = 0$ we have, at p^*,

$$\left(-gf_{,i} + \frac{4}{C-f}f_{,i} - \alpha\frac{\rho_{,i}}{\rho} + K_{,i} \right) f_{,1} + 2f_{,1i} = 0, \quad i = 1, 2. \tag{5.140}$$

Inserting (5.115) into (5.100), where we choose $\delta = \frac{1}{10}$, and using the formula $\Delta f = 2 + 2\frac{\sum \rho_{,i} f_{,i}}{\rho}$, then in the present situation we get, at p^*,

$$2(f_{,11})^2 + 2(f_{,12})^2 + 4\frac{\rho_{,11}}{\rho}(f_{,1})^2 + (\alpha - 372)\Phi(f_{,1})^2 - 312$$

$$+ \left[\Delta K - g'(f_{,1})^2 - 2\left(g - \frac{4}{C-f} \right)\left(1 + \frac{\rho_{,1}}{\rho}f_{,1} \right) \right] (f_{,1})^2 \le 0. \tag{5.141}$$

In the following we estimate three terms appearing in (5.141), namely the terms: $(f_{,11})^2 + (f_{,12})^2, \Delta K$ and $4\frac{\rho_{,11}}{\rho}(f_{,1})^2$.

(1) Estimate for the term $(f_{,11})^2 + (f_{,12})^2$. Now we choose $N > 6$; from (5.140) we have

$$2(f_{,11})^2 = \frac{1}{2}\left[gf_{,1} - \frac{4}{C-f}f_{,1} + \alpha\frac{\rho_{,1}}{\rho} - K_{,1} \right]^2 (f_{,1})^2 \tag{5.142}$$

$$\ge \frac{3}{4N}\left[\left(g - \frac{4}{C-f} \right)f_{,1} + \alpha\frac{\rho_{,1}}{\rho} \right]^2 (f_{,1})^2 - \frac{9}{10}\frac{(K_{,1})^2}{K}(f_{,1})^2,$$

where we used the fact $K \le N$ and the elementary inequality

$$(a + b)^2 \ge (1 - \varepsilon)a^2 - \left(\frac{1}{\varepsilon} - 1 \right)b^2$$

with $a = (g - \frac{4}{C-f})f_{,1} + \alpha\frac{\rho_{,1}}{\rho}$, $b = -K_{,1}$ and $\varepsilon = \frac{10K}{18+10K}$.
Similarly, we have

$$2(f_{,12})^2 \ge \frac{3}{4N}\alpha^2\left(\frac{\rho_{,2}}{\rho} \right)^2 (f_{,1})^2 - \frac{9}{10}\frac{(K_{,2})^2}{K}(f_{,1})^2. \tag{5.143}$$

(2) Estimate for the term ΔK.

$$K_{,i} = K\left(\frac{\Phi_{,i}}{\Phi} - gf_{,i} - \alpha\frac{\rho_{,i}}{\rho} \right), \tag{5.144}$$

$$\Delta K \ge \frac{\sum (K_{,i})^2}{K} + K\left[(1 + \alpha)\Phi - g'(f_{,1})^2 - 2g - 2g\frac{\rho_{,1}}{\rho}f_{,1} \right] \tag{5.145}$$

$$\ge \frac{\|\operatorname{grad} K\|_{\mathcal{H}}^2}{K} - 2Kg\frac{\rho_{,1}}{\rho}f_{,1} - N\left[g'(f_{,1})^2 + 2g \right].$$

(3) Estimate for the term $4\frac{\rho_{,11}}{\rho}(f_{,1})^2$. To estimate the invariant $\frac{\rho_{,11}}{\rho}(f_{,1})^2 = \sum \frac{\rho_{,ij}}{\rho} f_{,i} f_{,j}$, we will change the local orthonormal frame field by choosing $\{\tilde{e}_1, \tilde{e}_2\}$ such that, with respect to it, we have $\rho_{,1} = \|\text{grad}\,\rho\|$. Then we have

$$\rho_{,11} + \rho_{,22} = 0, \quad \Phi_{,i} = 2\frac{\rho_{,1}\rho_{,1i}}{\rho^2} - 2\frac{\rho_{,i}(\rho_{,1})^2}{\rho^3},$$

so we can easily check that

$$\sum \frac{(\rho_{,ij})^2}{\rho^2} = \frac{2(\rho_{,11})^2 + 2(\rho_{,12})^2}{\rho^2} \leq \frac{\sum(\Phi_{,i})^2}{\Phi} + 4\Phi^2. \tag{5.146}$$

Notice that both sides of (5.146) are independent of the choice of frames, we return to the frame $\{e_1, e_2\}$. From (5.146) and (5.144) it follows that, at p^*,

$$4\frac{|\rho_{,11}|}{\rho} \leq \sqrt{32}\,\Phi + \sqrt{8}\,\frac{\|\text{grad}\,\Phi\|_{\mathcal{H}}}{\sqrt{\Phi}}$$

$$\leq \sqrt{32}\Phi + \sqrt{8\Phi}\left\{\left[\sum\left(\frac{K_{,i}}{K}\right)^2\right]^{1/2} + \left[\sum\left(gf_{,i} + \alpha\frac{\rho_{,i}}{\rho}\right)^2\right]^{1/2}\right\}. \tag{5.147}$$

Applying the inequality of Cauchy–Schwarz we get

$$\sqrt{8\Phi}\left[\sum\left(\frac{K_{,i}}{K}\right)^2\right]^{1/2} \leq \frac{1}{20}\frac{\|\text{grad}\,K\|_{\mathcal{H}}^2}{K} + \frac{40}{N}\mathcal{A}\exp\left\{\frac{m}{C-f}\right\}\rho^\alpha,$$

$$\sqrt{8\Phi}\left[\sum\left(gf_{,i} + \alpha\frac{\rho_{,i}}{\rho}\right)^2\right]^{1/2} \leq 24N\Phi + \frac{4}{3N}\frac{1}{(C-f)^2}(f_{,1})^2$$

$$+ \frac{1}{12N}\sum\left[\left(g - \frac{4}{C-f}\right)f_{,i} + \alpha\frac{\rho_{,i}}{\rho}\right]^2.$$

Since $\mathcal{A}\exp\{N\} \leq \frac{4}{\alpha}\mathcal{D}$, we have

$$\mathcal{A}\exp\{N\} \leq \frac{4}{\alpha}\exp\left\{\frac{-m}{C-f} + K\right\}\frac{g^2\|\text{grad}\,f\|_{\mathcal{H}}^2}{\rho^\alpha}(p^*).$$

As $K(p^*) \leq N$, it follows that

$$\frac{40}{N}\mathcal{A}\exp\left\{\frac{m}{C-f}\right\}\rho^\alpha(f_{,1})^2 \leq \frac{160}{N\alpha}g^2(f_{,1})^4 \quad \text{at } p^*.$$

Inserting the preceding estimates into (5.147) we finally get

$$4\frac{\rho_{,11}}{\rho}(f_{,1})^2 \geq -\frac{1}{20}\frac{\|\text{grad}\,K\|_{\mathcal{H}}^2}{K}(f_{,1})^2 - \frac{160}{N\alpha}g^2(f_{,1})^4$$

$$- \frac{4}{3N}\frac{1}{(C-f)^2}(f_{,1})^4 - 30N\Phi(f_{,1})^2 \tag{5.148}$$

$$- \frac{1}{12N}\sum\left[\left(g - \frac{4}{C-f}\right)f_{,i} + \alpha\frac{\rho_{,i}}{\rho}\right]^2(f_{,1})^2.$$

After finishing the previous three estimates, we use the inequalities (5.142), (5.143), (5.145), and (5.148) and insert them into (5.141); we get, at p^*,

$$\frac{2}{3N}\sum\left[\left(g-\frac{4}{C-f}\right)f_{,i}+\alpha\frac{\rho_{,i}}{\rho}\right]^2(f_{,1})^2+(\alpha-92N)\Phi(f_{,1})^2$$

$$-2(N+1)g(f_{,1})^2-2\left(Kg+g-\frac{4}{C-f}\right)\frac{\rho_{,1}}{\rho}(f_{,1})^3-312$$ (5.149)

$$-\left[(N+1)g'+\frac{160}{N\alpha}g^2+\frac{4}{3N}\frac{1}{(C-f)^2}\right](f_{,1})^4\le 0.$$

Now for any $N\ge 200$, we choose α such that

$$N+1=\frac{2\alpha}{3N},\quad m=2N(N+1)\alpha C,$$

i.e.

$$\alpha=\frac{3N(N+1)}{2},\quad m=3[N(N+1)]^2 C.$$ (5.150)

Then we have

$$\alpha-92N>200N,\quad (N+1)g'\le\frac{1}{N\alpha}g^2,\quad \frac{4}{3N}\frac{1}{(C-f)^2}<\frac{1}{N\alpha}g^2,$$

and from (5.149) we obtain, at p^*,

$$\frac{2}{3N}\sum\left[\left(g-\frac{4}{C-f}\right)f_{,i}+\alpha\frac{\rho_{,i}}{\rho}\right]^2(f_{,1})^2+200N\Phi(f_{,1})^2$$

$$-2(N+1)g(f_{,1})^2-2\left(Kg+g-\frac{4}{C-f}\right)\frac{\rho_{,1}}{\rho}(f_{,1})^3$$ (5.151)

$$-\frac{108}{N^2(N+1)}g^2(f_{,1})^4-312\le 0.$$

Our following discussions depend on the sign of $\sum\rho_{,i}f_{,i}$ at p^*:

(a) If $(\sum\rho_{,i}f_{,i})(p^*)>0$, then, at p^*, we have the estimate

$$\frac{2}{3N}\left[\left(g-\frac{4}{C-f}\right)f_{,1}+\alpha\frac{\rho_{,1}}{\rho}\right]^2(f_{,1})^2-(2+2N)\left(g-\frac{4}{C-f}\right)\frac{\rho_{,1}}{\rho}(f_{,1})^3$$

$$\ge\frac{2}{3N}\left(g-\frac{4}{C-f}\right)^2(f_{,1})^4\ge\frac{1}{3N}g^2(f_{,1})^4,$$

and, since $K(p^*)\le N$, it holds at p^*,

$$\left[2N\left(g-\frac{4}{C-f}\right)-2Kg\right]\frac{\rho_{,1}f_{,1}}{\rho}\ge-\frac{8N}{C-f}\frac{\rho_{,1}f_{,1}}{\rho}\ge-200N\Phi-\frac{3}{50\alpha}g^2(f_{,1})^2.$$

Inserting the above two estimates into (5.151) we get, at p^*:

$$\frac{1}{6N}g^2(f_{,1})^4-2(N+1)g(f_{,1})^2-312\le 0.$$ (5.152)

(b) If $(\sum \rho_{,i} f_{,i})(p^*) \leq 0$, then it is easily seen that

$$2\left(Kg + g - \frac{4}{C-f}\right)\frac{\rho_{,1}}{\rho}(f_{,1})^3 \leq 0,$$

and the inequality of Cauchy–Schwarz implies that, at p^*,

$$\frac{2}{3N}\sum\left[\left(g - \frac{4}{C-f}\right)f_{,i} + \alpha\frac{\rho_{,i}}{\rho}\right]^2 (f_{,1})^2 + (\alpha - 92N)\Phi(f_{,1})^2$$

$$\geq \frac{2}{3N}\sum\left[\left(g - \frac{4}{C-f}\right)f_{,i} + \alpha\frac{\rho_{,i}}{\rho}\right]^2 (f_{,1})^2 + \frac{\alpha}{4}\Phi(f_{,1})^2$$

$$\geq \frac{2}{8\alpha + 3N}\left(g - \frac{4}{C-f}\right)^2 (f_{,1})^4 \geq \frac{1}{8\alpha + 3N}g^2(f_{,1})^4.$$

Hence from (5.149) we obtain, at p^*,

$$\frac{1}{2(8\alpha + 3N)}g^2(f_{,1})^4 - 2(N+1)g(f_{,1})^2 - 312 \leq 0. \tag{5.153}$$

Thus, (5.152) and (5.153) show that in both of the above two cases, at p^*, we have an inequality of the type

$$a_0 g^2(f_{,1})^4 - a_1 g(f_{,1})^2 - 312 \leq 0, \tag{5.154}$$

where a_0 and a_1 are real positive constants depending only on C, α and N. This further implies that, at p^*,

$$g^2\|\operatorname{grad} f\|_{\mathcal{H}}^4 \leq a_2.$$

Consequently, we have

$$\mathcal{D}^2 = \left\{\exp\left\{\frac{-2m}{C-f} + 2K\right\}\frac{g^4}{\rho^{2\alpha}}\|\operatorname{grad} f\|^4\right\}(p^*)$$

$$\leq a_2 \exp\{2N\}b^{2\alpha}\left\{g^2\exp\left\{\frac{-2m}{C-f}\right\}\right\}(p^*) \leq a_3,$$

where a_3 is a real positive constant depending only on C, α, b and N. Therefore, $\mathcal{D}$ is bounded from above as we announced in Claim II.

From Claim I and Claim II we complete the proof of Proposition 5.37. □

5.6.4 Second proof of Lemma 5.29

In this proof, we will use the same notations as in our first proof in Section 5.5.5. Again we consider two subcases.

(i) If $D \cap \partial\Omega = \emptyset$, then there is a constant $h > 0$ such that the level set satisfies $\overline{S_{f\infty}(p_0, h)} \subset \Omega$, thus we have a uniform estimate for $\|\operatorname{grad} f^{(k)}\|_E$ in $S_{f^{(k)}}(q^k, h)$.

For simplicity, let f denote any fixed function of the sequence $\{f^{(k)}\}$ with $f(q) = 0$. By Lemma 5.30 and Propositions 5.36, 5.37, we have the uniform estimates

$$\frac{\Phi}{\rho^\alpha} \le d, \quad \frac{1}{\rho^\alpha} \le d, \quad \frac{\sum f_{ii}}{\rho^\alpha} \le d$$

in the domain

$$S_f\left(q, \frac{C}{2}\right) := \left\{x \in \Omega \mid f(x) < \frac{C}{2}\right\},$$

where d is a positive constant independent of k.

We may assume that q_k converges to p_0. Let $B_R(q_k)$ be a Euclidean ball such that $\Omega \subset B_{\frac{R}{2}}(q_k)$. Then the Legendre transformation domain Ω^* of Ω satisfies the relation $B_\delta^*(O) \subset \Omega^*$, where $\delta = \frac{C}{2R}$ and $B_\delta^*(O) = \{\xi \mid (\xi_1)^2 + (\xi_2)^2 < \delta^2\}$. By Lemma 5.32, we have

$$\det(f_{ij}) \ge b \quad \text{in} \quad B_{\delta/2}^*(O),$$

where b is a constant independent of k. Restricting to $B_\delta^*(O)$, we have

$$-\frac{C}{2R} - C \le u^{(k)} = \sum \xi_i x^i - f^{(k)} \le \frac{C}{2R}.$$

Therefore, the sequence $u^{(k)}$ locally uniformly converges to a convex function u in $B_\delta^*(O)$, and there are constants $0 < \lambda \le \Lambda < \infty$ such that the following estimates hold in $B_{\delta/2}^*(O)$:

$$\lambda \le \lambda_i^{(k)} \le \Lambda, \quad \text{for} \quad i = 1, 2, \ldots, n; \ k = 1, 2, \ldots,$$

where $\lambda_1^{(k)}, \ldots, \lambda_n^{(k)}$ denote the eigenvalues of the matrix $(f_{ij}^{(k)})$. Then, standard elliptic estimates from Section 5.5.1 imply the assertion of Lemma 5.29 in the case (i).

(ii) In case $D \cap \partial\Omega \ne \emptyset$, let $p \in D \cap \partial\Omega$. Since the p.d.e. (5.78) is equiaffinely invariant, we may choose a new coordinate system such that $\|\operatorname{grad} f^{(k)}\|_E$ is uniformly bounded in $\overline{S_{f^{(k)}}(p, h)}$. Then the same argument as in (i) above shows that f is smooth in a neighborhood of p, and we get a contradiction. This excludes the case $D \cap \partial\Omega \ne \emptyset$. This completes the proof of Lemma 5.29. □

5.7 An affine Bernstein problem with respect to the Calabi metric

Besides the previous two different versions of an *affine Bernstein conjecture*, due to two different completeness assumptions, namely affine completeness and Euclidean completeness, one can consider a third possibility of assuming the completeness with respect to the Calabi metric. Concerning this case, we now describe the first result that was obtained by Li and Jia in [176] (see also Theorem 5.7.1 in [181]), namely:

Theorem 5.38. *Let $f(x)$ be a strongly convex function that is defined in a domain $\Omega \subset \mathbb{R}^n$. If $M = \{(x, f(x)) \mid x \in \Omega\}$ is an affine maximal hypersurface, and if M is complete with respect to the Calabi metric $\mathcal{H}$, then for $n = 2, 3$, M must be an elliptic paraboloid.*

Proof. Let $p_0 \in M$ be an arbitrary fixed point and let $r(x) = d(p_0, x)$ denote the distance function from p_0 with respect to $\mathcal{H}$. By adding a linear function, we may assume that p_0 has coordinates $(0, \ldots, 0)$ and

$$f(0) = 0, \quad f_i(0) = 0.$$

Now, for any given $a > 0$, we consider the function

$$F(x) = (a^2 - r^2(x))^2 \Phi(x)$$

defined on $\bar{B}_a(p_0, \mathcal{H}) = \{p \in M \mid d(p_0, p) \leq a\}$. It is obvious that F attains its supremum at some interior point $p^* \in B_a(p_0, \mathcal{H})$. Then, by an appropriate affine parameter transformation if necessary, we may assume $f_{ij}(p^*) = \delta_{ij}$; and thus, at p^*, we have

$$F_{,i} = 0, \quad \sum F_{,ii} \leq 0,$$

where the subscript ',' again denotes the covariant differentiation with respect to the Calabi metric. We calculate both expressions explicitly

$$\frac{\Phi_{,i}}{\Phi} - \frac{2(r^2)_{,i}}{a^2 - r^2} = 0, \tag{5.155}$$

$$\frac{\Delta^{(C)}\Phi}{\Phi} - \frac{\sum(\Phi_{,i})^2}{\Phi^2} - \frac{2\|\operatorname{grad} r^2\|_{\mathcal{H}}^2}{(a^2 - r^2)^2} - \frac{2\Delta^{(C)}(r^2)}{a^2 - r^2} \leq 0. \tag{5.156}$$

We insert (5.155) into (5.156) and get

$$\frac{\Delta^{(C)}\Phi}{\Phi} \leq \frac{6\|\operatorname{grad} r^2\|_{\mathcal{H}}^2}{(a^2 - r^2)^2} + \frac{2\Delta^{(C)}(r^2)}{a^2 - r^2} \tag{5.157}$$

$$= \frac{24r^2}{(a^2 - r^2)^2} + \frac{4}{a^2 - r^2} + \frac{4r\Delta^{(C)}r}{a^2 - r^2}.$$

To estimate $r\Delta^{(C)}r$, we only need to consider the case $p^* \neq p_0$. Denote by $a^* := d(p_0, p^*)$, then we have $a^* > 0$. Let

$$\bar{B}_{a^*}(p_0) = \{p \in M \mid d(p_0, p) \leq a^*\}.$$

By (5.54) and (5.55) we have

$$\max_{p \in \bar{B}_{a^*}(p_0)} \Phi(p) = \max_{p \in \partial B_{a^*}(p_0)} \Phi(p).$$

On the other hand, we have $a^2 - r^2 = a^2 - a^{*2}$ on $\partial B_{a^*}(p_0)$; it follows that

$$\max_{p \in \bar{B}_{a^*}(p_0)} \Phi(p) = \Phi(p^*).$$

We will now estimate the Ricci curvature of the Calabi metric on $\bar{B}_{a^*}(p_0)$. Let $p \in \bar{B}_{a^*}(p_0)$ be an arbitrary point, without loss of generality we may assume that $f_{ij}(p) = \delta_{ij}$. We use the expression of the Ricci tensor of the Calabi metric, namely (2.163), and choose arbitrary real numbers $\{\lambda_i\}_{i=1}^n$ with $\sum(\lambda_i)^2 \neq 0$; then we get

$$\sum R_{ij}(p)\lambda_i\lambda_j = \frac{1}{4}\sum f^{kl}f^{hm}(f_{hil}f_{mkj} - f_{hij}f_{mkl})\lambda_i\lambda_j$$

$$= \frac{1}{4}\sum_{m,j}\left(\sum_i f_{mij}\lambda_i\right)^2 - \frac{1}{4}\sum_m\left(\sum_{i,j}f_{mij}\lambda_i\lambda_j\sum_k f_{mkk}\right)$$

$$\geq \frac{1}{4}\sum_{m,j}\left(\sum_i f_{mij}\lambda_i\right)^2 - \frac{1}{4}\left(\sum_i\lambda_i^2\right)^{-1}\sum_m\left(\sum_{i,j}f_{mij}\lambda_i\lambda_j\right)^2$$

$$- \frac{1}{16}\sum_m\lambda_j^2\sum_m\left(\sum_k f_{mkk}\right)^2$$

$$\geq -\frac{1}{16}\sum_m\lambda_j^2\sum_m\left(\sum_k f_{mkk}\right)^2,$$

where we have used the inequality of Cauchy–Schwarz twice.

On the other hand, by definition

$$\Phi(p) = \frac{1}{(n+2)^2}\sum f^{ij}f^{kl}f_{kli}f^{mh}f_{mhj} = \frac{1}{(n+2)^2}\sum_m\left(\sum_k f_{mkk}\right)^2;$$

it follows that the Ricci curvature of the Calabi metric satisfies

$$\text{Ric}(p) \geq -\frac{(n+2)^2}{16}\Phi(p) \geq -\frac{(n+2)^2}{16}\Phi(p^*).$$

Thus, by the Laplacian comparison theorem (see Appendix B), we obtain

$$r\Delta^{(C)}r \leq (n-1)\left(1 + \frac{n+2}{4}\sqrt{\Phi(p^*)}\cdot r\right). \tag{5.158}$$

For the case $n = 3$, by (5.55), the inequality of Cauchy–Schwarz and (5.155), we have

$$\frac{\Delta^{(C)}\Phi}{\Phi} \geq \frac{3}{4}\sum\frac{(\Phi_{,i})^2}{\Phi^2} - \frac{3}{2}\sum\frac{\Phi_{,i}}{\Phi}\cdot\frac{\rho_{,i}}{\rho} + \frac{1}{6}\Phi \tag{5.159}$$

$$\geq -6\sum\frac{(\Phi_{,i})^2}{\Phi^2} + \frac{1}{12}\Phi = -\frac{96r^2}{(a^2-r^2)^2} + \frac{1}{12}\Phi.$$

Inserting (5.158) into (5.157), and using (5.159), we get

$$\Phi \leq \frac{1440r^2}{(a^2-r^2)^2} + \frac{144}{a^2-r^2} + \frac{120\sqrt{\Phi(p^*)}\cdot r}{a^2-r^2}. \tag{5.160}$$

Multiplying both sides of (5.160) by $(a^2 - r^2)^2$ and then applying the inequality of Cauchy–Schwarz

$$\sqrt{\Phi(p^*)}a(a^2 - r^2) \leq 60a^2 + \frac{\Phi(p^*)(a^2-r^2)^2}{240},$$

from (5.160) we obtain that, at p^*,

$$(a^2 - r^2)^2 \Phi \leq 10^5 a^2. \qquad (5.161)$$

For the case $n = 2$, by (5.54) we have

$$\Delta^{(C)}\Phi \geq \frac{\|\text{grad } \Phi\|_{\mathcal{H}}^2}{\Phi} + \Phi^2. \qquad (5.162)$$

Similar to the case $n = 3$, we can verify that, for $n = 2$, it holds, at p^*,

$$(a^2 - r^2)^2 \Phi \leq 32a^2 < 10^5 a^2. \qquad (5.163)$$

Hence, for $n = 2, 3$, at any interior point x of $\bar{B}_a(p_0, \mathcal{H})$, we have

$$\Phi(x) \leq \frac{10^5}{a^2 \left(1 - \frac{r^2(x)}{a^2}\right)^2}. \qquad (5.164)$$

Let $a \to \infty$ in (5.164): we obtain

$$\Phi \equiv 0.$$

It follows that

$$\det\left(\frac{\partial^2 f}{\partial x^i \partial x^j}\right) = \text{const.},$$

and thus the Calabi metric $\mathcal{H}$ and the Blaschke metric G coincide up to a constant multiple.

Therefore, M is an affine complete parabolic affine hypersphere and, by Theorem 3.45, we conclude that M is an elliptic paraboloid. $\qquad \square$

6 Hypersurfaces with constant affine Gauß–Kronecker curvature

6.1 The affine Gauß–Kronecker curvature

6.1.1 Motivation

Consider a locally strongly convex hypersurface immersion $x : M \to A^{n+1}$ into real affine space, equipped with its equiaffine normalization. Let us denote by G its Blaschke–Berwald metric and by $\hat{B}$ its Weingarten form. We call x to be of *hyperbolic type* (*elliptic type*, resp.) if $\hat{B}$ is *negative definite* (*positive definite*, resp.). If $\hat{B}$ is negative definite, we define $\tilde{B} := -\hat{B}$, then as a symmetric and positive definite $(0, 2)$-tensor, $\tilde{B}$ is called the *Weingarten metric* on M. The affine Gauß–Kronecker curvature (abbreviation: affine G-K curvature) is given by

$$S_n := \frac{\det(\tilde{B})}{\det(G)}. \tag{6.1}$$

Note that we will apply the preceding definition of S_n only in the hyperbolic case. If x is of hyperbolic type, the definition of L_n in Section 2.2.3 implies that $S_n = (-1)^n L_n$. When x is an affine hypersphere, according to Proposition 3.2, we have $\hat{B} = \lambda G$ for some constant $\lambda = L_1$, and therefore the Gauß–Kronecker curvature is constant, too. Thus a classification of complete hypersurfaces with constant Gauß–Kronecker curvature extends a classification of complete affine hyperspheres. We state the following problem for locally strongly convex hypersurfaces:

Problem 6.1. Classify all affine complete hypersurfaces with nonzero constant affine Gauß–Kronecker curvature.

If the hypersurface is elliptic it follows from Corollary 2.17 and the Newton inequality for the curvature functions L_1 and S_n that the Ricci tensor of the metric G is positively bounded below. Thus (M, G) must be compact. But then L_n = const. implies that x must be an ellipsoid (see Theorem 4.5).

On the other hand, the hyperbolic case turns out to be much more difficult. The paper [180] by Li, Simon and Chen at the first time gave a partial answer to Problem 6.1. Now we explain their work. The main results are summarized in the three Theorems 6.5, 6.11, and 6.21. The basic geometric facts which they used are the following:
(i) the Weingarten form $\hat{B}$ can be interpreted as metric of the Gauß conormal image;
(ii) the equation S_n = const. gives a fourth order p.d.e., which is shown to be equivalent to the fact that the two affine Gauß images (the affine indicatrix and the affine conormal image) of x are proper affine hyperspheres; moreover, it is shown that the immersion $x : M \to A^{n+1}$ can be represented as a graph hypersurface of some

convex function f, and the normal image of $x : M \to A^{n+1}$ can be represented also as a graph hypersurface of another convex function f^*; the Legendre transformation function u^* of f^* satisfies a second order p.d.e. of Monge–Ampère type for the Gauß conormal image; as a well-known fact its solution corresponds to an affine hypersphere;

(iii) the immersed hypersurface $x : M \to A^{n+1}$ is constructed from its Gauß conormal image (in terms of u^*) in a second step, solving another Monge–Ampère equation for the Legendre transformation function u of f (Theorem 6.11).

6.1.2 Main results

In Theorems 6.11 and 6.21 we will prove that the hypersurface, represented by the solution f, is Euclidean complete and also complete with respect to the Weingarten metric $\tilde{B}$ that we defined before; in our terminology the hypersurface is called *Weingarten complete*; in a particular case we can prove that it is affine complete.

Now we sketch some more details for the equation $S_n = $ const. For the Gauß conormal image we refer to Section 2.3.

Assume that $x : M \to A^{n+1}$ is locally given as a graph hypersurface by

$$x^{n+1} = f(x^1, \ldots, x^n).$$

From the expression of G and $\hat{B}$ (see (2.139) and (2.150)), one can see that in terms of the coordinates $(x^1, \ldots, x^n)$ and the function f, the condition $S_n = $ const. is a complicated p.d.e. of fourth order. Then, if we consider the equiaffine conormal immersion $U : M \to \mathbb{R}^{n+1}$ and use a Legendre transformation, we can split, as indicated, the p.d.e. of fourth order into two (second order) Monge–Ampère equations. As stated in Section 1.15.3, the conormal immersion $U : M \to \mathbb{R}^{n+1}$ describes a convex hypersurface with centroaffine normalization, i.e. its position vector is transversal to $U(M)$ and is taken as centroaffine normal.

Recalling from Section 1.12 that, for any centroaffine hypersurface

$$\tilde{x} : M \to \mathbb{R}^{n+1},$$

we can define a function ψ as follows: Choose a local centroaffine frame field $\{\tilde{x}; e_1, \ldots, e_{n+1}\}$, namely such that

$$e_1, \ldots, e_n \in T_{\tilde{x}}M, \quad e_{n+1} = \tilde{x}.$$

We can write

$$d\tilde{x} = \sum \omega^i e_i,$$
$$de_i = \sum \omega_i^j e_j + \sum h_{ij} \omega^j \tilde{x}.$$

Let us define

$$\psi := \frac{\det(h_{ij})}{[\mathrm{Det}(e_1, \ldots, e_n, \tilde{x})]^2}.$$

It is easy to see that ψ is both a centroaffine and an equiaffine invariant. We call ψ the Tchebychev function. It follows from Section 4.4.8 of [341] that ψ is related to the centroaffine Tchebychev form T by (cf. Lemma 1.16)

$$d \ln \psi = -2nT. \tag{6.2}$$

The affine G-K curvature S_n of $x : M \to A^{n+1}$ and the Tchebychev function ψ_U of its conormal image $U(M)$ satisfy the following:

Proposition 6.2. $S_n = \psi_U$.

Proof. Using the fundamental formulas (see Proposition 2.10)

$$U_{ij} = \sum \Gamma_{ij}^k U_k + \tilde{B}_{ij} U$$

and the fact (cf. Lemma 2.6 (iii) in Section 2.3)

$$\det(G_{ij}) = [\text{Det}^*(U_1, \dots, U_n, U)]^2,$$

we directly get $S_n = \psi_U$. □

Remark 6.3. Proposition 6.2 follows also from [312, p. 389], where the author states the relation between the relative Tchebychev fields of x and $U = \tilde{x}$.

According to Proposition 6.2 and equation (6.2), we see that $S_n = $ const. if and only if the associated centroaffine Tchebychev form T of $U : M \to \mathbb{R}^{n+1}$ is zero, and therefore, due to the fact stated in Section 1.15.2-3, this is equivalent to the fact that the centroaffine hypersurface $U : M \to \mathbb{R}^{n+1}$ is an equiaffine hypersphere with affine mean curvature $L_1 = -1$ and equiaffine normal U (cf. Section 1.15.3). Hence we have the following

Corollary 6.4. *Let $x : M \to A^{n+1}$ be a locally strongly convex hypersurface such that the Weingarten form is nondegenerate. Then*

$$S_n = \text{const.} \Longleftrightarrow U(M) \text{ is an affine hypersphere}$$
$$\Longleftrightarrow Y(M) \text{ is an affine hypersphere.}$$

From Proposition 6.2, the affine G-K curvature describes at the same time geometric properties of $U : M \to \mathbb{R}^{n+1}$. More precisely, in what follows we will see that to construct the immersion $x : M \to A^{n+1}$ from the given constant affine G-K curvature we can proceed in two steps:

Step 1. Construct $U(M)$ from $\psi_U = S_n$ by solving a Monge-Ampère equation.
Step 2. Construct $x(M)$ from $U(M)$ by solving another Monge-Ampère equation.

6.2 Splitting of the fourth order PDE S_n = const into two (second order) Monge–Ampère equations

Let $x : M \to A^{n+1}$ be a locally strongly convex hypersurface of hyperbolic type. Assume that $S_n = 1$ and that the Weingarten metric $\tilde{B}$ defines a complete Riemann metric on M. Then, both $U(M)$ and $Y(M)$, are complete hyperbolic affine hyperspheres. By Theorem 3.54, $Y(M)$ is also Euclidean complete. It follows that $Y(M)$ is the graph of a strongly convex function $f^*(x) = f^*(x^1, \ldots, x^n)$ defined on the whole $\mathbb{R}^n$. Without loss of generality, we assume that $f^* \geq 0$ and the center of $Y(M)$ has coordinates $(0, \ldots, 0, -1)$.

Let u^* be the Legendre transformation function of f^*, and Ω^* its Legendre transformation domain. Then, according to (2.144), the affine conormal vector field U^* of $Y(M)$ can be identified with

$$U^*(\xi^*) = \left[\det \left(\frac{\partial^2 u^*}{\partial \xi_i^* \partial \xi_j^*} \right) \right]^{\frac{1}{n+2}} (-\xi_1^*, \ldots, -\xi_n^*, 1), \quad \forall \xi^* \in \Omega^*, \tag{6.3}$$

where $\xi_i^* := \frac{\partial f^*}{\partial x^i}$ and $u^*(\xi_1^*, \ldots, \xi_n^*) := \sum x^i \frac{\partial f^*}{\partial x^i} - f^*(x)$.

Put $b := (0, \ldots, 0, 1)$. Then $\langle U^*, b \rangle > 0$ everywhere on $Y(M)$.

From the Weingarten structure equations of $x : M \to A^{n+1}$ (cf. Section 1.15.2-2 and equation (1.12)), we have

$$\langle U, Y \rangle = 1, \quad \langle U, dx \rangle = 0, \quad \langle U, dY \rangle = 0.$$

It follows that both $x : M \to A^{n+1}$ and $Y : M \to V$ have the same conormal field $U : M \to \mathbb{R}^{n+1}$, thus we have $\langle U, b \rangle > 0$ on $x(M)$. So we can consider $x : M \to A^{n+1}$ as a graph of

$$x^{n+1} = f(x^1, \ldots, x^n)$$

for some strongly convex function f defined on some domain $\tilde{\Omega}$ of $\mathbb{R}^n$.

Consider the Legendre transformation function u of f:

$$u(\xi_1, \ldots, \xi_n) = \sum x^i \frac{\partial f}{\partial x^i} - f(x), \quad \xi_i = \frac{\partial f}{\partial x^i}, \quad 1 \leq i \leq n, \tag{6.4}$$

and denote by Ω the Legendre transformation domain of f, i.e. $u : \Omega \to \mathbb{R}$ and

$$\Omega = \left\{ (\xi_1, \ldots, \xi_n) \mid \xi_i = \frac{\partial f(x)}{\partial x^i}, \, x = (x^1, \ldots, x^n) \in \tilde{\Omega} \right\}.$$

Then, according to (2.144) again, the affine conormal vector field U of $x(M)$ can be identified with

$$U(\xi) = \left[\det \left(\frac{\partial^2 u}{\partial \xi_i \partial \xi_j} \right) \right]^{\frac{1}{n+2}} (-\xi_1, \ldots, -\xi_n, 1), \quad \forall \xi \in \Omega. \tag{6.5}$$

We take the hyperplane $\Pi : \xi_{n+1} = 1$ and consider the central projection

$$P : U(M) \to \Pi$$

of $U(M)$ from the origin into Π; this defines a one to one mapping onto an open convex domain $P(U(M))$, which can be identified with the Legendre transformation domain Ω by formula (6.5).

Since $x : M \to A^{n+1}$ and $Y : M \to V$ have the same conormal vector at corresponding points, i.e. $U = U^*$, we have

$$\xi_i = \xi_i^*,$$
$$\Omega = \Omega^*,$$
$$\det\left(\frac{\partial^2 u}{\partial\xi_i\partial\xi_j}\right) = \det\left(\frac{\partial^2 u^*}{\partial\xi_i^*\partial\xi_j^*}\right).$$

(6.6)

Moreover, as $Y : M \to V$ is a complete affine hypersphere with affine mean curvature $L_1 = -1$, from Proposition 3.50 and the proof of the Calabi conjecture in Section 3.8 we see that $\Omega = \Omega^*$ is a bounded convex domain, and u^* satisfies

$$\begin{cases} \det\left(\dfrac{\partial^2 u^*}{\partial\xi_i\partial\xi_j}\right) = (-u^*)^{-n-2} & \text{in } \Omega, \\ u^* = 0 & \text{on } \partial\Omega. \end{cases}$$

(6.7)

In summary, we have proved the following

Theorem 6.5. *Let $x : M \to A^{n+1}$ be a locally strongly convex hypersurface of hyperbolic type. If the affine G-K curvature satisfies $S_n = 1$ and the Weingarten metric is complete then*

(i) *the immersion $x : M \to A^{n+1}$ can be considered as a graph of some strongly convex function $x^{n+1} = f(x^1, \ldots, x^n)$;*

(ii) *the Legendre transformation domain Ω of f is a bounded convex domain;*

(iii) *the Legendre transformation function u of f satisfies*

$$\det\left(\frac{\partial^2 u}{\partial\xi_i\partial\xi_j}\right) = (-u^*)^{-n-2} \quad \text{in } \Omega,$$

(6.8)

where u^* is the unique convex solution of the p.d.e.

$$\begin{cases} \det\left(\dfrac{\partial^2 u^*}{\partial\xi_i\partial\xi_j}\right) = (-u^*)^{-n-2} & \text{in } \Omega, \\ u^* = 0 & \text{on } \partial\Omega. \end{cases}$$

(6.9)

6.3 Construction of Euclidean complete hypersurfaces with constant affine G-K curvature

In this section we study the following problem:

How can we construct hypersurfaces with constant affine G-K curvature?

Given a bounded convex domain $\Omega \subset \mathbb{R}^n$ with C^∞-boundary (we always assume the origin $O \in \Omega$) and a function $\varphi \in C^\infty(\partial\Omega)$, we will prove the existence of the solutions $\{u, u^*\}$, resp., of the following p.d.e.'s:

$$\begin{cases} \det\left(\dfrac{\partial^2 u}{\partial\xi_i\partial\xi_j}\right) = (-u^*)^{-n-2} \ \text{ in } \Omega, \\ u = \varphi \ \text{ on } \partial\Omega, \end{cases} \tag{6.10}$$

$$\begin{cases} \det\left(\dfrac{\partial^2 u^*}{\partial\xi_i\partial\xi_j}\right) = (-u^*)^{-n-2} \ \text{ in } \Omega, \\ u^* = 0 \ \text{ on } \partial\Omega. \end{cases} \tag{6.11}$$

For that purpose, we first disturb (6.11) by considering, for any $t < 0$, the following boundary value problem: the p.d.e. (6.10) together with

$$\begin{cases} \det\left(\dfrac{\partial^2 u^*}{\partial\xi_i\partial\xi_j}\right) = (-u^*)^{-n-2} \ \text{ in } \Omega, \\ u^* = t \ \text{ on } \partial\Omega. \end{cases} \tag{6.12}$$

By a well-known result of Caffarelli et al. (see [36]), there is a unique smooth convex solution u_t^* satisfying (6.12). As $(-u_t^*)^{-n-2} \in C^\infty(\bar{\Omega})$, there is a unique smooth convex solution u_t satisfying (6.10).

Lemma 6.6. *For any* $0 > t_2 > t_1$, *we have*
(i) $u_{t_2}^*(\xi) \geq u_{t_1}^*(\xi), \ \forall \xi \in \Omega$,
(ii) $u_{t_2}(\xi) \leq u_{t_1}(\xi), \ \forall \xi \in \Omega$.

Proof. (i) If the assertion is not true, notice that $(u_{t_2}^* - u_{t_1}^*)|_{\partial\Omega} > 0$, then $u_{t_2}^* - u_{t_1}^*$ attains its negative minimum at an interior point $p \in \Omega$. At p, we have

$$(-u_{t_2}^*)^{-n-2} - (-u_{t_1}^*)^{-n-2} < 0.$$

Then, at p,

$$\det\left(\frac{\partial^2 u_{t_2}^*}{\partial\xi_i\partial\xi_j}\right) - \det\left(\frac{\partial^2 u_{t_1}^*}{\partial\xi_i\partial\xi_j}\right) < 0.$$

From the observation

$$\det\left(\frac{\partial^2 u_{t_2}^*}{\partial\xi_i\partial\xi_j}\right) - \det\left(\frac{\partial^2 u_{t_1}^*}{\partial\xi_i\partial\xi_j}\right) = \int_0^1 \frac{d}{ds}\left\{\det\left(\frac{\partial^2 u_{t_1}^*}{\partial\xi_i\partial\xi_j} + s\frac{\partial^2 (u_{t_2}^* - u_{t_1}^*)}{\partial\xi_i\partial\xi_j}\right)\right\} ds,$$

we see that the preceding inequality can be written as

$$\sum A^{ij}(\xi) \frac{\partial^2 (u^*_{t_2} - u^*_{t_1})}{\partial \xi_i \partial \xi_j} < 0, \tag{6.13}$$

where

$$A^{ij}(\xi) = \int_0^1 a^{ij}(\xi, s)ds,$$

and $a^{ij}(\xi, s)$ are the cofactor components of the matrix

$$s\left(\frac{\partial^2 u^*_{t_2}}{\partial \xi_i \partial \xi_j}\right) + (1 - s)\left(\frac{\partial^2 u^*_{t_1}}{\partial \xi_i \partial \xi_j}\right) = \left(\frac{\partial^2 u^*_{t_1}}{\partial \xi_i \partial \xi_j} + s \frac{\partial^2 (u^*_{t_2} - u^*_{t_1})}{\partial \xi_i \partial \xi_j}\right).$$

On the other hand, since the matrix $(A^{ij}(\xi))$ is positive definite and $u^*_{t_2} - u^*_{t_1}$ attains its minimum at p, we have

$$\sum A^{ij}(\xi) \frac{\partial^2 (u^*_{t_2} - u^*_{t_1})}{\partial \xi_i \partial \xi_j} \geq 0.$$

This is a contradiction to (6.13).

(ii) For $0 > t_2 > t_1$, the conclusion of (i) implies that

$$(-u^*_{t_2})^{-n-2} \geq (-u^*_{t_1})^{-n-2}.$$

Then (6.10) gives

$$\det\left(\frac{\partial^2 u_{t_2}}{\partial \xi_i \partial \xi_j}\right) - \det\left(\frac{\partial^2 u_{t_1}}{\partial \xi_i \partial \xi_j}\right) \geq 0.$$

The preceding inequality can be written as

$$\sum A^{ij}(\xi) \frac{\partial^2 (u_{t_2} - u_{t_1})}{\partial \xi_i \partial \xi_j} \geq 0, \tag{6.14}$$

where as above

$$A^{ij}(\xi) = \int_0^1 a^{ij}(\xi, s)ds,$$

and $a^{ij}(\xi, s)$ are the cofactor components of the matrix

$$(1 - s)\left(\frac{\partial^2 u_{t_1}}{\partial \xi_i \partial \xi_j}\right) + s\left(\frac{\partial^2 u_{t_2}}{\partial \xi_i \partial \xi_j}\right).$$

Since $(u_{t_2} - u_{t_1})|_{\partial \Omega} = 0$ and $(A^{ij}(\xi)) > 0$, from (6.14) and the maximum principle we obtain $u_{t_2} \leq u_{t_1}$ in Ω, as claimed. □

Now we show that the functions u_t are bounded from below. In fact, for the unique convex solution u^* of (6.11), we define

$$\tilde{u} = u^* + C,$$

where C is a constant such that $C < \inf_{\partial\Omega} \varphi$. Then from the inequality

$$u_t^* < u^* < 0 \text{ in } \Omega$$

we have

$$\det\left(\frac{\partial^2 u_t}{\partial \xi_i \partial \xi_j}\right) - \det\left(\frac{\partial^2 \tilde{u}}{\partial \xi_i \partial \xi_j}\right) < 0 \text{ in } \Omega.$$

Note that $(u_t - \tilde{u})|_{\partial\Omega} > 0$, thus the maximum principle implies that $u_t > \tilde{u}$.

Combining the above fact with Lemma 6.6, which implies that u_t is monotonically decreasing as $t \to 0$, we have shown that, as $t \to 0$, u_t converges locally uniformly to some function u. We will prove that u is a smooth solution of (6.10)–(6.11).

First we prove that, for any point $\bar{\xi} \in \partial\Omega$, $u(\xi) \to u(\bar{\xi})$ as $\xi \to \bar{\xi}$.

Without loss of generality we assume $\bar{\xi} = (\bar{\xi}_1, \ldots, \bar{\xi}_n) \in \partial\Omega$ with $\bar{\xi}_1 > 0$, and $(1, 0, \ldots, 0)$ is the exterior unit normal of $\partial\Omega$ at $\bar{\xi}$, then $\xi_1 \leq \bar{\xi}_1$ holds for any $\xi \in \bar{\Omega}$. Now, for the unique convex solution u^* of (6.11) we define

$$\underline{u} = u^* + a_1 \xi_1 + c.$$

We claim that, for any $\varepsilon > 0$, there are constants a_1 and c such that $\underline{u}$ satisfies the following two conditions: (i) $\underline{u}|_{\partial\Omega} < \varphi$; (ii) $\underline{u}(\bar{\xi}) = \varphi(\bar{\xi}) - \varepsilon$.

In fact, to ensure (ii), after fixing a_1, we can choose c such that

$$c = \varphi(\bar{\xi}) - \varepsilon - a_1 \bar{\xi}_1.$$

To choose a_1, we use the continuity of φ so that we can find a positive number δ with $\bar{\xi}_1 > \delta > 0$, such that if $\delta \leq \xi_1 \leq \bar{\xi}_1$ then $|\varphi(\xi) - \varphi(\bar{\xi})| < \varepsilon$. Now, it is sufficient to choose a_1 such that

$$a_1 > 2(\bar{\xi}_1 - \delta)^{-1} \left| \sup_{\partial\Omega} |\varphi| \right|.$$

To see that (i) holds, we consider an arbitrary $\xi = (\xi_1, \ldots, \xi_n) \in \partial\Omega$. If $\delta \leq \xi_1 \leq \bar{\xi}_1$, we have

$$\underline{u}(\xi) = a_1 \xi_1 + \varphi(\bar{\xi}) - \varepsilon - a_1 \bar{\xi}_1$$
$$= a_1(\xi_1 - \bar{\xi}_1) + \varphi(\bar{\xi}) - \varphi(\xi) + \varphi(\xi) - \varepsilon \leq \varphi(\xi).$$

If, on the other hand, $\xi_1 < \delta$, we have

$$\underline{u}(\xi) = a_1(\xi_1 - \bar{\xi}_1) + \varphi(\bar{\xi}) - \varepsilon \leq a_1(\delta - \bar{\xi}_1) + 2 \left| \sup_{\partial\Omega} |\varphi| \right| + \varphi(\xi) < \varphi(\xi).$$

Hence $\underline{u}$, defined as above, satisfies (i) and (ii) above.

By definition of u and the maximum principle, we see that for any $t < 0$ it holds $u_t > u$ and therefore

$$u_{-1/2} \geq u \geq \underline{u}.$$

Since $u_{-1/2}(\xi) \to \varphi(\bar{\xi})$ and $\underline{u}(\xi) \to \varphi(\bar{\xi}) - \varepsilon$ as $\xi \to \bar{\xi}$, we have

$$|u(\xi) - \varphi(\bar{\xi})| \leq 2\varepsilon \text{ as } \xi \to \bar{\xi}.$$

By the arbitrariness of $\varepsilon > 0$, we conclude that $u|_{\partial\Omega} = \varphi$.

Next we prove that $u \in C^{\infty}(\Omega)$ and u satisfies (6.10)–(6.11). We need the estimation of u_t up to third order, independent of t. First of all, we will estimate $|u_t|$. Since u_t is a convex function we have $u_t \leq \max \varphi$. On the other hand, we have $u_t \geq \underline{u}$. Thus we get

$$|u_t| \leq C_0 \tag{6.15}$$

for some constant $C_0 > 0$.

Since u_t is convex, we have

$$|\text{grad } u_t|(\xi) \leq \frac{1}{d}(\max u_t - \min u_t) \leq \frac{2C_0}{d},$$

where d is the distance from ξ to $\partial\Omega$. For any fixed point $\xi \in \Omega$ there is a neighborhood N of ξ such that, on N, we have the estimate

$$|\text{grad } u_t| \leq C_1 \tag{6.16}$$

for some appropriate constant $C_1 > 0$.

To estimate the second and the third order derivatives we use the following results due to Pogorelov [285] and Calabi–Nirenberg (cf. [36]).

Lemma 6.7 (cf. [285]). *Given a bounded convex domain Ω and assume that $F \in C^2(\Omega)$ is a positive function, let u be a C^4-solution of the equation*

$$\det(u_{ij}) = F(\xi) \text{ in } \Omega; \quad u = \text{const. on } \partial\Omega.$$

Then there is an estimate of the second order derivatives of u on Ω depending on $\max\{|u|, |\text{grad } u|^2, |\text{grad } \ln F|^2, \sum[(\ln F)_{ij}]^2\}$ and the distance to $\partial\Omega$.

Lemma 6.8 (cf. [36]). *Given a bounded convex domain Ω and assume that $F \in C^{2,1}(\bar{\Omega})$ is a positive function, let u be a C^5-solution of the equation*

$$\det(u_{ij}) = F(\xi) \text{ in } \Omega; \quad u = \text{const. on } \partial\Omega$$

such that $u \in C^{1,1}(\bar{\Omega})$. Then there is an estimate of the third order derivatives of u depending on the $C^{1,1}$-norm of u on $\bar{\Omega}$, the $C^{2,1}$-norm of $\ln F$ on $\bar{\Omega}$ and the distance to $\partial\Omega$.

We fix a point $t_0 < 0$, for example let $t_0 = -\frac{1}{2}$. For any fixed point $\xi' \in \Omega$, let

$$b_i = \frac{\partial u_{-1/2}}{\partial \xi_i}(\xi'), \quad c = u_{-1/2}(\xi') - \sum b_i \xi_i'.$$

By the strong convexity of $u_{-\frac{1}{2}}$ we have

$$u_{-1/2}(\xi') - \left(\sum b_i \xi_i' + c\right) = 0,$$

$$u_{-1/2}(\xi) - \left(\sum b_i \xi_i + c\right) > 0, \quad \forall \xi \in \Omega \setminus \{\xi'\}.$$

We can choose a small positive number ε such that the set

$$\Omega_{-\frac{1}{2},\varepsilon} := \left\{\xi \in \Omega \mid u_{-\frac{1}{2}}(\xi) - \left(\sum b_i \xi_i + c\right) \le \varepsilon\right\}$$

is compact. We introduce the notations

$$\Omega_{t,\varepsilon} := \left\{\xi \in \Omega \mid u_t - \left(\sum b_i \xi_i + c\right) \le \varepsilon\right\}.$$

Since, for all $-\frac{1}{2} \le t < 0$,

$$u_t \le u_{-\frac{1}{2}} \quad \text{and} \quad u_t|_{\partial\Omega} = \varphi,$$

we have

$$\Omega_{t,\varepsilon} \supset \Omega_{-\frac{1}{2},\varepsilon} \ni \xi', \quad \text{for all } -\frac{1}{2} \le t < 0,$$

and the domains $\Omega_{t,\varepsilon}$ are compact. By the estimates (6.15) and (6.16) and the Lemmas 6.7 and 6.8 we can find a $C^{2,\alpha}$ estimate for u_t on $\Omega_{-\frac{1}{2},\frac{\varepsilon}{2}}$, independent of t. It follows that u satisfies (6.10)–(6.11) on a neighborhood of ξ'. Since ξ' is arbitrary, u satisfies (6.10)–(6.11) in Ω.

Then it follows from the standard theory of p.d.e.'s (cf. Section 5.5.1) that we have $u \in C^\infty(\Omega)$. Moreover, from the maximum principle we can easily show the uniqueness of the solution of (6.10)–(6.11). We have proved the following

Theorem 6.9. *Given a bounded convex domain Ω with C^∞-boundary and a function $\varphi \in C^\infty(\partial\Omega)$, there is a unique strongly convex solution $u \in C^\infty(\Omega)$ of the equations (6.10)–(6.11).*

From the solution u of (6.10)–(6.11) we consider its Legendre transformation function f:

$$f(x^1, \ldots, x^n) = \sum \xi_i \frac{\partial u}{\partial \xi_i} - u(\xi), \quad x^i = \frac{\partial u}{\partial \xi_i}, \quad \xi \in \Omega,$$

$$\tilde{\Omega} = \left\{\left(\frac{\partial u}{\partial \xi_1}, \ldots, \frac{\partial u}{\partial \xi_n}\right) \mid \xi \in \Omega\right\}.$$

Then we can construct a hypersurface M in A^{n+1} as graph of f:

$$M := \left\{(x, f(x)) \mid x = (x^1, \ldots, x^n) \in \tilde{\Omega}\right\}. \tag{6.17}$$

Obviously, M is a strongly convex hypersurface with affine G-K curvature satisfying $S_n = 1$. Moreover, by equation (6.11) and the proof of the Calabi conjecture (cf. Theorem 3.57), the affine normal image $Y(M)$ is a complete affine hypersphere (with affine mean curvature $L_1 = -1$) whose Blaschke metric is exactly the Weingarten form of M. It follows that the Weingarten metric on M is complete.

Next we prove that the above defined hypersurface M is Euclidean complete (i.e. it is closed with respect to the Euclidean topology of $\mathbb{R}^{n+1}$). We need the following lemma:

Lemma 6.10. *Let u be the strongly convex solution of (6.10)–(6.11). Then, for any point $\bar{\xi} \in \partial\Omega$, there exist constants $b_1, \ldots, b_n, d$, depending only on Ω and φ, such that the function $\bar{u}$, defined by*

$$\bar{u} = u^* + b_1 \xi_1 + \cdots + b_n \xi_n + d,$$

satisfies

$$\bar{u}(\bar{\xi}) = u(\bar{\xi}),$$
$$\bar{u}(\xi) > u(\xi), \quad \forall \xi \in \partial\Omega \setminus \{\bar{\xi}\}.$$

Proof. Given $\bar{\xi} \in \partial\Omega$, by a unimodular centroaffine transformation of $(\xi_1, \ldots, \xi_n)$, we can assume that

$$\bar{\xi} = (0, \ldots, 0, \bar{\xi}_n), \quad \bar{\xi}_n < 0,$$

and the vector $(0, \ldots, 0, -1)$ is the exterior unit normal of $\partial\Omega$ at $\bar{\xi}$. Then the boundary $\partial\Omega$ can be locally expressed as the graph of a strongly convex function by

$$\xi_n = \kappa(\xi_1, \ldots, \xi_{n-1}),$$

where $\kappa(\xi_1, \ldots, \xi_{n-1})$ is assumed to satisfy the condition:

$$\frac{\partial \kappa}{\partial \xi_i}(\bar{\xi}) = 0, \quad i = 1, 2, \ldots, n - 1.$$

We first choose

$$b_k := \frac{\partial u}{\partial \xi_k}(\bar{\xi}) = \frac{\partial \varphi}{\partial \xi_k}(\bar{\xi}), \quad k = 1, 2, \ldots, n - 1,$$

and then, to ensure $\bar{u}(\bar{\xi}) = u(\bar{\xi})$, we fix

$$b_n := (-\bar{\xi}_n)^{-1} (d - u(\bar{\xi})),$$

where the constant d is still to be specified.

Next, we consider the following function defined on $\partial\Omega$:

$$F(\xi) = \bar{u}(\xi) - u(\xi) = b_1 \xi_1 + \cdots + b_n \xi_n + d - \varphi(\xi), \quad \forall \xi \in \partial\Omega.$$

For a small $\varepsilon > 0$, let $N_\varepsilon(\bar{\xi})$ be the neighborhood of $\bar{\xi}$, defined by

$$N_\varepsilon(\bar{\xi}) := \{\xi \in \partial\Omega \mid \bar{\xi}_n \le \xi_n < \bar{\xi}_n + \varepsilon\}.$$

Then, on $N_\varepsilon(\bar{\xi})$, we can write

$$F(\xi) = \sum_{i=1}^{n-1} b_i\xi_i + b_n\kappa(\xi_1, \ldots, \xi_{n-1}) + d - \varphi(\xi),$$

and it is easily seen that, for $1 \le i, j \le n - 1$,

$$F(\bar{\xi}) = 0, \qquad \frac{\partial F}{\partial \xi_i}(\bar{\xi}) = 0,$$

$$\frac{\partial^2 F}{\partial \xi_i \partial \xi_j} = (d - u(\bar{\xi}))(-\bar{\xi}_n)^{-1}\frac{\partial^2 \kappa}{\partial \xi_i \partial \xi_j} - \frac{\partial^2 \varphi}{\partial \xi_i \partial \xi_j}.$$

Denote by r the smallest eigenvalue of $\left(\frac{\partial^2 \kappa}{\partial \xi_i \partial \xi_j}\right)$. If we choose

$$d > \frac{|\bar{\xi}_n| \cdot \max_{\overline{N_\varepsilon(\xi)}}\left\{\sum_{i,j}\left|\frac{\partial^2 \varphi}{\partial \xi_i \partial \xi_j}\right|\right\}}{\min_{\overline{N_\varepsilon(\xi)}}\{r\}} + \max_{\partial\Omega}|\varphi|$$

then on $N_\varepsilon(\bar{\xi})$ it holds $F \ge 0$; while, outside of $N_\varepsilon(\bar{\xi})$, we have

$$F(\xi) \ge \sum_{i=1}^{n-1} b_i\xi_i + [d\varepsilon - \varphi(\bar{\xi})\xi_n]|\bar{\xi}_n|^{-1} - \varphi(\xi), \quad \xi \in \partial\Omega \setminus N_\varepsilon(\bar{\xi}).$$

Since Ω is a bounded convex domain and $|\varphi|$ is bounded, we can choose d large enough such that $F \ge 0$ on $\partial\Omega$.

This completes the proof of the lemma. $\qquad\qquad\square$

To prove that the hypersurface M defined by (6.17) is Euclidean complete, it is sufficient to prove that the Legendre transformation domain of u is $\mathbb{R}^n$. For that purpose, we consider the function defined on $\mathbb{R}^n$:

$$L(x) = \sup_{\xi \in \bar{\Omega}}\left\{-u(\xi) + \sum x^i\xi_i\right\}.$$

Given any point $\bar{x} \in \mathbb{R}^n$, it is obvious that there is a point $\bar{\xi} \in \bar{\Omega}$ such that

$$L(\bar{x}) = -u(\bar{\xi}) + \sum \bar{x}^i\bar{\xi}_i.$$

We will consider two cases:

(i) $\bar{\xi} \in \Omega$. In this case, the function $\Phi(\xi) = -u(\xi) + \sum \bar{x}^i\xi_i$ attains its maximum at $\bar{\xi}$. Then we have

$$\frac{\partial\Phi}{\partial \xi_i}(\bar{\xi}) = -\frac{\partial u}{\partial \xi_i}(\bar{\xi}) + \bar{x}^i = 0,$$

and therefore $\bar{x}$ belongs to the Legendre transformation domain of u.

(ii) $\bar{\xi} \in \partial\Omega$. In this case, for any $\xi \in \Omega$, we have

$$L(\bar{x}) > -u(\xi) + \sum \bar{x}^i \xi_i.$$

By Lemma 6.10 we can choose constants $\{b_1, \ldots, b_n, d\}$ so that the function

$$\bar{u} := u^* + \sum b_i \xi_i + d$$

satisfies

$$\bar{u}(\bar{\xi}) = u(\bar{\xi}),$$
$$\bar{u}(\xi) > u(\xi), \quad \forall \xi \in \partial\Omega \setminus \{\bar{\xi}\}.$$

Notice that

$$\det\left(\frac{\partial^2 \bar{u}}{\partial \xi_i \partial \xi_j}\right) = \det\left(\frac{\partial^2 u}{\partial \xi_i \partial \xi_j}\right) \quad \text{in } \Omega,$$

by the maximum principle we have

$$\bar{u} \geq u \quad \text{in } \bar{\Omega}.$$

Then

$$-\bar{u}(\bar{\xi}) + \sum \bar{x}^i \bar{\xi}_i = -u(\bar{\xi}) + \sum \bar{x}^i \bar{\xi}_i$$
$$> -u(\xi) + \sum \bar{x}^i \xi_i \qquad\qquad (6.18)$$
$$\geq -\bar{u}(\xi) + \sum \bar{x}^i \xi_i, \quad \forall \xi \in \Omega.$$

On the other hand, as u^* determines a complete hyperbolic affine hypersphere, its Legendre transformation domain is $\mathbb{R}^n$ (see Proposition 3.48), and therefore the Legendre transformation domain of $\bar{u}$ is also $\mathbb{R}^n$. Then there is a point $\xi^* \in \Omega$ such that

$$\bar{x}^i = \frac{\partial \bar{u}}{\partial \xi_i}(\xi^*), \quad 1 \leq i \leq n.$$

We consider the function $h(\xi) := -\bar{u}(\xi) + \sum \bar{x}^i \xi_i$, defined on Ω. From the fact

$$\frac{\partial h}{\partial \xi_i}(\xi^*) = 0, \quad 1 \leq i \leq n; \quad \left(\frac{\partial^2 h}{\partial \xi_i \partial \xi_j}(\xi^*)\right) < 0,$$

we see that $h(\xi)$ attains its maximum at the interior point ξ^*. But this is a contradiction to (6.18). It follows that the Legendre transformation domain of u is $\mathbb{R}^n$.

In summary, we have proved the following

Theorem 6.11. *Given a bounded convex domain Ω in $\mathbb{R}^n$ with C^∞-boundary and a function $\varphi \in C^\infty(\partial\Omega)$, we can construct a locally strongly convex hypersurface M in A^{n+1} defined as the graph of a strongly convex function f such that the Legendre transformation domain relative to f is Ω, and the Legendre transformation function u of f satisfies $u|_{\partial\Omega} = \varphi$.*

Moreover, M is of constant affine G-K curvature, and it is both, Euclidean complete and Weingarten complete.

6.4 Completeness with respect to the Blaschke metric

Given a bounded convex domain $\Omega \subset \mathbb{R}^n$ with C^∞-boundary $\partial\Omega$, and a function $\varphi \in C^\infty(\partial\Omega)$, in the preceding section we proved that from the solution u of the p.d.e.'s (6.10)–(6.11), we can construct a Euclidean complete hypersurface M with constant affine G-K curvature. More precisely, M is given as the graph hypersurface of the Legendre transformation function f of u.

In this section we study the following problem:

Is the above M also complete with respect to the Blaschke metric G?

To deal with the completeness with respect to the Blaschke metric, we need a gradient estimate of the following type:

$$\frac{\|\operatorname{grad} f\|_G}{f} \leq C \quad \text{on } \Omega$$

for some constant $C > 0$, or equivalently (cf. (6.10)) an estimate as follows:

$$-\frac{u^*}{f^2} \sum u_{ij}\xi_i\xi_j \leq C \quad \text{on } \Omega.$$

Once having such estimates, we can finally obtain the main result (see Theorem 6.21 below).

First of all, we give several estimates for solutions u and u^* of the p.d.e.'s (6.10)–(6.11), together with their Legendre transformation functions f and f^*, respectively. These estimates are of independent interest because the functions involved are closely related to affine hyperspheres.

Lemma 6.12. *Let Ω be a bounded convex domain in $\mathbb{R}^n$ and u^* be a strongly convex function which satisfies the p.d.e.*

$$\det\left(\frac{\partial^2 u^*}{\partial\xi_i\partial\xi_j}\right) = (-u^*)^{-n-2} \quad \text{in } \Omega.$$

Then, regarding f^* as a function of $\{\xi_1, \ldots, \xi_n\}$, it holds

$$\sum u^{*ij}(u^*f^*)_{ij} = 0. \tag{6.19}$$

Proof. By definition, we have $\frac{\partial f^*}{\partial\xi_i} = \sum \frac{\partial f^*}{\partial x^k} \cdot \frac{\partial x^k}{\partial\xi_i} = \sum \xi_k(u^*)_{ki}$ and

$$(u^*f^*)_i = (u^*)_i f^* + u^* \sum \xi_k(u^*)_{ki}, \tag{6.20}$$

$$(u^*f^*)_{ij} = (u^*)_{ij}f^* + (u^*)_i \sum \xi_k(u^*)_{kj} + (u^*)_j \sum \xi_k(u^*)_{ki}$$
$$+ u^*\left[(u^*)_{ij} + \sum \xi_k(u^*)_{kij}\right]. \tag{6.21}$$

Using

$$\sum (u^*)^{ij} (u^*)_{ijk} = \frac{\partial \ln \det((u^*)_{ij})}{\partial \xi_k} = -(n+2) \frac{(u^*)_k}{u^*}$$

and (6.21), we obtain

$$\sum (u^*)^{ij} (u^* f^*)_{ij} = n f^* + 2 \sum (u^*)_i \xi_i + n u^* - (n+2) \sum (u^*)_i \xi_i$$
$$= 0.$$

(6.22)

This proves the lemma. $\qquad\square$

Lemma 6.13. *Let* $\Omega \subset \mathbb{R}^n$ *be a bounded convex domain with* C^∞-*boundary and* u^* *be a strongly convex function which satisfies the p.d.e.*

$$\begin{cases} \det \left(\dfrac{\partial^2 u^*}{\partial \xi_i \partial \xi_j} \right) = (-u^*)^{-n-2} & \text{in } \Omega, \\ u^* = 0 & \text{on } \partial\Omega. \end{cases}$$

Then there are constants $d_2^* > d_1^* > 0$ *such that*

$$d_1^* \le -u^* f^* \le d_2^* \quad \text{holds in } \Omega.$$

(6.23)

Proof. For any $t < 0$ we consider the convex solution u_t^* of the p.d.e.

$$\begin{cases} \det \left(\dfrac{\partial^2 u^*}{\partial \xi_i \partial \xi_j} \right) = (-u^*)^{-n-2} & \text{in } \Omega, \\ u^* = t & \text{on } \partial\Omega. \end{cases}$$

Since, by Lemma 6.12, $-u_t^* f_t^*$ attains its maximum and minimum on the boundary $\partial\Omega$, we now estimate $-u_t^* f_t^* |_{\partial\Omega}$.

For any point $\bar{\xi} \in \partial\Omega$, the assumptions on Ω imply the existence of two balls $B_r(\eta)$ and $B_R(\eta')$ such that

$$B_r(\eta) \subset \Omega \subset B_R(\eta'), \quad \partial B_r(\eta) \cap \partial\Omega = \{\bar{\xi}\} = \partial B_R(\eta') \cap \partial\Omega,$$

where $r = r(\bar{\xi}) > 0$, $R = R(\bar{\xi}) > 0$ and

$$B_r(\eta) := \left\{ \xi \in \mathbb{R}^n \mid |\xi - \eta|^2 < r^2 \right\},$$
$$B_R(\eta') := \left\{ \xi \in \mathbb{R}^n \mid |\xi - \eta'|^2 < R^2 \right\}$$

for some $\eta = \eta(\bar{\xi}) = (\eta_1, \ldots, \eta_n)$, $\eta' = \eta'(\bar{\xi}) = (\eta_1', \ldots, \eta_n') \in \mathbb{R}^n$.

Since $\partial\Omega$ is compact, we can further choose the constants r and R such that they depend only on Ω (i.e. r and R do not depend on $\bar{\xi}$).

Consider the function

$$\bar{u}_t(\xi) := -\alpha(t) \sqrt{r^2 - |\xi - \eta|^2 + \frac{t^2}{\alpha^2(t)}}, \quad \xi \in B_r(\eta),$$

where $\alpha(t) > 0$ is the smallest positive root of the equation

$$\alpha^{2n+2}r^2 + \alpha^{2n}t^2 = 1.$$

It is easy to check that

$$\begin{cases} \det\left(\dfrac{\partial^2 \bar{u}_t}{\partial \xi_i \partial \xi_j}\right) = (-\bar{u}_t)^{-n-2} & \text{in } B_r(\eta), \\ \bar{u}_t = t & \text{on } \partial B_r(\eta). \end{cases}$$

So we have

$$\det\left(\frac{\partial^2 \bar{u}_t}{\partial \xi_i \partial \xi_j}\right) - \det\left(\frac{\partial^2 u_t^*}{\partial \xi_i \partial \xi_j}\right) = (-\bar{u}_t)^{-n-2} - (-u_t^*)^{-n-2} \text{ in } B_r(\eta), \tag{6.24}$$

and it is obvious to see that

$$u_t^* \leq \bar{u}_t \text{ on } \partial B_r(\eta). \tag{6.25}$$

As in the proof of Lemma 6.6, we can rewrite (6.24) as

$$\sum A^{ij}(\xi)(\bar{u}_t - u_t^*)_{ij} = b(\xi)(\bar{u}_t - u_t^*) \text{ in } B_r(\eta), \tag{6.26}$$

where $b(\xi) > 0$, and the matrix $(A^{ij}(\xi))$ is positive definite. From (6.25), (6.26) and applying the maximum principle, we derive the following inequality:

$$u_t^* \leq \bar{u}_t \text{ in } B_r(\eta). \tag{6.27}$$

Since $\bar{u}_t(\bar{\xi}) = u_t^*(\bar{\xi}) = t$, letting $\frac{\partial}{\partial \nu}$ denote $\sum \xi_i \frac{\partial}{\partial \xi_i}$, (6.27) implies that

$$\frac{\partial u_t^*}{\partial \nu}(\bar{\xi}) \geq \frac{\partial \bar{u}_t}{\partial \nu}(\bar{\xi}) = \frac{\alpha^2(t)}{-t}(\bar{\xi} - \eta) \cdot \bar{\xi}.$$

It follows that

$$(-u_t^* f_t^*)(\bar{\xi}) = -t f_t^*(\bar{\xi}) = -t \left[\frac{\partial u_t^*}{\partial \nu}(\bar{\xi}) - u_t^*(\bar{\xi}) \right] \tag{6.28}$$

$$\geq \alpha^2(t)(\bar{\xi} - \eta(\bar{\xi})) \cdot \bar{\xi} + t^2.$$

By the convexity of Ω and the compactness of $\partial\Omega$, there will exist two constants $d_2 > d_1 > 0$ such that

$$d_2 \geq (\xi - \eta(\xi)) \cdot \xi \geq d_1, \text{ for all } \xi \in \partial\Omega. \tag{6.29}$$

Since $-u_t^* f_t^*$ attains its minimum on $\partial\Omega$, from (6.28) and (6.29) we have

$$-u_t^* f_t^* \geq \alpha^2(t)d_1 + t^2 \text{ in } \Omega.$$

Let $t \to 0^-$ we get

$$-u^* f^* \geq \frac{d_1}{r^{2/(n+1)}} =: d_1^* > 0 \text{ in } \Omega.$$

On the other hand, if we consider the function

$$\underline{u}_t(\xi) := -\beta(t)\sqrt{R^2 - |\xi - \eta'|^2 + \frac{t^2}{\beta^2(t)}}, \quad \xi \in B_R(\eta'),$$

where $\beta(t) > 0$ is the smallest positive root of the equation

$$\beta^{2n+2}R^2 + \beta^{2n}t^2 = 1,$$

then we get

$$\begin{cases} \det\left(\dfrac{\partial^2 \underline{u}_t}{\partial \xi_i \partial \xi_j}\right) = (-\underline{u}_t)^{-n-2} & \text{in } B_R(\eta'), \\ \underline{u}_t = t & \text{on } \partial B_R(\eta'). \end{cases}$$

Now we have

$$\det\left(\frac{\partial^2 \underline{u}_t}{\partial \xi_i \partial \xi_j}\right) - \det\left(\frac{\partial^2 u_t^*}{\partial \xi_i \partial \xi_j}\right) = (-\underline{u}_t)^{-n-2} - (-u_t^*)^{-n-2} \quad \text{in } \Omega, \tag{6.30}$$

which can be rewritten as

$$\sum A^{ij}(\xi)(\underline{u}_t - u_t^*)_{ij} = b(\xi)(\underline{u}_t - u_t^*) \quad \text{in } \Omega, \tag{6.31}$$

where $b(\xi) > 0$, and the matrix $(A^{ij}(\xi))$ is positive definite.

From the fact $\underline{u}_t|_{\partial\Omega} \leq \max(\underline{u}_t|_{\partial B_R(\eta')}) = t$, we get

$$\underline{u}_t \leq u_t^* \quad \text{on } \partial\Omega. \tag{6.32}$$

From (6.31), (6.32), and the maximum principle we derive the inequality

$$\underline{u}_t \leq u_t^* \quad \text{in } \Omega. \tag{6.33}$$

Since $\underline{u}_t(\bar{\xi}) = u_t^*(\bar{\xi}) = t$, for $\frac{\partial}{\partial v} = \sum \xi_i \frac{\partial}{\partial \xi_i}$, (6.33) implies that

$$\frac{\partial u_t^*}{\partial v}(\bar{\xi}) \leq \frac{\partial \underline{u}_t}{\partial v}(\bar{\xi}) = \frac{\beta^2(t)}{-t}(\bar{\xi} - \eta) \cdot \bar{\xi}.$$

Together with (6.29) it follows that

$$\begin{aligned} (-u_t^* f_t^*)(\bar{\xi}) = -t f_t^*(\bar{\xi}) &= -t\left[\frac{\partial u_t^*}{\partial v}(\bar{\xi}) - u_t^*(\bar{\xi})\right] \\ &\leq \beta^2(t)(\bar{\xi} - \eta(\bar{\xi})) \cdot \bar{\xi} + t^2 \\ &\leq \beta^2(t)d_2 + t^2. \end{aligned} \tag{6.34}$$

By the arbitrariness of $\bar{\xi} \in \partial\Omega$ and from the fact that $-u_t^* f_t^*$ attains its maximum on $\partial\Omega$, we have

$$-u_t^* f_t^* \leq \beta^2(t)d_2 + t^2 \quad \text{in } \Omega.$$

Let $t \to 0^-$, we finally get

$$-u^* f^* \leq \frac{d_2}{R^{2/(n+1)}} =: d_2^* \quad \text{in } \Omega. \qquad \square$$

Remark 6.14. Since u^* is a bounded function, it is easy to see that the estimate (6.23) is equivalent to the estimate

$$d_1^{*\prime} \le -u^* \frac{\partial u^*}{\partial v} \le d_2^{*\prime} \quad \text{in } \Omega \tag{6.35}$$

for some constants $d_2^{*\prime} > d_1^{*\prime} > 0$.

Remark 6.15. Let u^* and f^* be the functions given in Lemma 6.12. If we consider the hypersurface $M^* = \{(\xi, u^*(\xi)) \mid \xi \in \Omega\}$, then $-u^* f^*$ is the affine support function of M^* relative to the origin.

In fact, the affine conormal of M^* is

$$U^* = \left[\det \left(\frac{\partial^2 u^*}{\partial \xi_i \partial \xi_j} \right) \right]^{-\frac{1}{n+2}} \left(-\frac{\partial u^*}{\partial \xi_1}, \dots, -\frac{\partial u^*}{\partial \xi_n}, 1 \right),$$

and therefore, the affine support function of M^* relative to the origin, at the point $x^* = (\xi, u^*(\xi))$, is (cf. Definition 2.12)

$$\Lambda^*(x^*) = \langle U^*, -x^* \rangle = \left[\det \left(\frac{\partial^2 u^*}{\partial \xi_i \partial \xi_j} \right) \right]^{-\frac{1}{n+2}} \left(\sum \frac{\partial u^*}{\partial \xi_i} \xi_i - u^* \right) = -u^* f^*.$$

Lemma 6.16. *Let $\Omega \subset \mathbb{R}^n$ be a bounded convex domain with C^∞-boundary and u^* be a strongly convex function which satisfies the p.d.e.*

$$\begin{cases} \det \left(\dfrac{\partial^2 u^*}{\partial \xi_i \partial \xi_j} \right) = (-u^*)^{-n-2} \quad \text{in } \Omega, \\ u^* = 0 \quad \text{on } \partial \Omega. \end{cases}$$

Then there are constants $a_2^* > a_1^* > 0$ such that

$$a_1^* \sqrt{d(\xi)} \le -u^*(\xi) \le a_2^* \sqrt{d(\xi)} \quad \text{for all } \xi \in \Omega, \tag{6.36}$$

where $d(\xi) := \text{dist}(\xi, \partial \Omega)$ denotes the (Euclidean) distance from ξ to $\partial \Omega$.

Proof. First we prove the existence of a constant $a_1^* > 0$ such that

$$-u^*(\xi) \ge a_1^* \sqrt{d(\xi)} \quad \text{for all } \xi \in \Omega.$$

From the proof of Lemma 6.13, for the smooth bounded convex domain Ω, we recall the fact that there are constants $R > r > 0$ such that, for any $\bar{\xi} \in \partial \Omega$, we have two balls $B_R(\eta')$ and $B_r(\eta)$, with center $\eta' = \eta'(\bar{\xi}) \in \Omega$ and $\eta = \eta(\bar{\xi}) \in \Omega$, respectively, which possess the following properties:

$$B_r(\eta) \subset \Omega \subset B_R(\eta'), \quad \partial B_r(\eta) \cap \partial \Omega = \{\bar{\xi}\} = \partial B_R(\eta') \cap \partial \Omega,$$

For any $\tilde{\xi} \in \Omega$, let $\bar{\xi} \in \partial \Omega$ be the point such that $d(\tilde{\xi}) = |\bar{\xi} - \tilde{\xi}|$. We will consider two cases.

(i) If $d(\tilde{\xi}) < r$, we consider the ball $B_r(\eta(\tilde{\xi})) \ni \tilde{\xi}$ and define the function

$$\bar{u}(\xi) := -r^{-\frac{1}{n+1}} \sqrt{r^2 - |\xi - \eta|^2}, \quad \xi \in B_r(\eta).$$

Adopting the same arguments as in the proof of Lemma 6.13, we can show that $u^* \le \bar{u}$ in $B_r(\eta)$. So, by the relation $r = |\tilde{\xi} - \eta| + d(\tilde{\xi})$, we obtain

$$-u^*(\tilde{\xi}) \ge r^{-\frac{1}{n+1}} \sqrt{r^2 - |\tilde{\xi} - \eta|^2} \ge r^{\frac{n-1}{2n+2}} \sqrt{d(\tilde{\xi})}.$$

(ii) If $d(\tilde{\xi}) \ge r$, we note that the set

$$\bar{\Omega}_r := \{\xi \in \Omega \mid d(\xi) \ge r\}$$

is compact and define

$$a' = \frac{\min_{\xi \in \bar{\Omega}_r}(-u^*(\xi))}{\sqrt{\max_{\xi \in \bar{\Omega}_r} d(\xi)}},$$

then we have

$$-u^*(\tilde{\xi}) \ge a' \sqrt{d(\tilde{\xi})}.$$

Regarding (i) and (ii), we choose $a_1^* = \min\{r^{\frac{n-1}{2n+2}}, a'\}$ and have

$$-u^*(\xi) \ge a_1^* \sqrt{d(\xi)} \quad \text{for any point } \xi \in \Omega.$$

On the other hand, for any $\tilde{\xi} \in \Omega$ and $\bar{\xi} \in \partial\Omega$ such that $d(\tilde{\xi}) = |\tilde{\xi} - \bar{\xi}|$, we consider the ball $B_R(\eta'(\bar{\xi})) \ni \tilde{\xi}$ and define the function

$$\underline{u}(\xi) := -R^{-\frac{1}{n+1}} \sqrt{R^2 - |\xi - \eta'|^2}, \quad \xi \in B_R(\eta').$$

Using the same arguments as in the proof of Lemma 6.13, we can show that $u^* \ge \underline{u}$ in Ω. From the relation $R = |\bar{\xi} - \eta'| + d(\tilde{\xi})$, we obtain

$$-u^*(\tilde{\xi}) \le R^{-\frac{1}{n+1}} \sqrt{R^2 - |\bar{\xi} - \eta'|^2}$$

$$= R^{-\frac{1}{n+1}} \sqrt{(d(\tilde{\xi}))^2 + 2|\bar{\xi} - \eta'| \cdot d(\tilde{\xi})}$$

$$\le \sqrt{3} R^{\frac{n-1}{2n+2}} \sqrt{d(\tilde{\xi})}.$$

We choose $a_2^* = \sqrt{3} R^{\frac{n-1}{2n+2}}$ and complete the proof of Lemma 6.16. $\qquad\square$

For a later purpose, we first prove an algebraic inequality.

Lemma 6.17 (cf. Lemma 5.7 of [172]). *Let (a_{ij}) be a symmetric positive definite $n \times n$ matrix, (b_{ij}) be a symmetric $n \times n$ matrix. Then*

$$\sum a^{ik} a^{jl} b_{ij} b_{kl} \ge \left(\frac{n}{n-1} - \delta\right) \left(\frac{b_{11}}{a_{11}}\right)^2 - \frac{1 - (n-1)\delta}{(n-1)^2 \delta} \left(\sum a^{ij} b_{ij}\right)^2,$$

for any $0 < \delta < \frac{1}{n-1}$.

Proof. Denote by $\mu_1, \ldots, \mu_n$ the solutions of the equation

$$\det(b_{ij} - \mu a_{ij}) = 0,$$

i.e. $\{\mu_i\}_{i=1}^n$ denote the eigenvalues of the the matrix (b_{ij}) relative to the positive definite matrix (a_{ij}). Let (a^{ij}) be the inverse matrix of (a_{ij}), then we have

$$\sum \mu_i = \sum a^{ij} b_{ij}, \quad \sum (\mu_i)^2 = \sum a^{ik} a^{jl} b_{ij} b_{kl}. \tag{6.37}$$

For any $\xi = (\xi_1, \ldots, \xi_n) \neq 0$, we have

$$\min_{1 \leq i \leq n}\{\mu_i\} \leq \frac{\sum b_{ij} \xi_i \xi_j}{\sum a_{ij} \xi_i \xi_j} \leq \max_{1 \leq i \leq n}\{\mu_i\}. \tag{6.38}$$

Denoting $\mu^2 = \max_{1 \leq i \leq n}\{(\mu_i)^2\}$ and choosing $\xi = (1, 0, \ldots, 0)$ in (6.38), we get

$$\left(\frac{b_{11}}{a_{11}}\right)^2 \leq \mu^2.$$

Using the inequality of Cauchy–Schwarz we obtain, for any $\delta > 0$,

$$\sum (\mu_i)^2 \geq \mu^2 + \frac{1}{n-1}\left(\sum \mu_i - \mu\right)^2$$

$$= \frac{n}{n-1}\mu^2 + \frac{1}{n-1}\left(\sum \mu_i\right)^2 - \frac{2}{n-1}\mu \sum \mu_i$$

$$\geq \left(\frac{n}{n-1} - \delta\right)\mu^2 + \left[\frac{1}{n-1} - \frac{1}{(n-1)^2\delta}\right]\left(\sum \mu_i\right)^2.$$

Then by (6.37) the assertion immediately follows. □

For studying the affine completeness of the hypersurface M as defined in (6.17), we prove the following

Lemma 6.18. *Let u and u^* be solutions of the p.d.e.'s (6.10)–(6.11). Let f be the Legendre transformation function of u, which is still regarded as a function on Ω. Assume that there are constants $d_2 > d_1 > 0$ such that*

$$d_1 \leq -u^* f \leq d_2 \quad in \ \Omega,$$

then there is a constant $Q > 0$ such that

$$-(u^*)^3 \sum u_{ij} \xi_i \xi_j \leq Q \quad in \ \Omega. \tag{6.39}$$

Proof. Since the hypersurface M, defined in (6.17), is Euclidean complete (cf. Theorem 6.11), for any constant $C > 1$, the section

$$S_C := \{(x, f(x)) \in M \mid f(x) \leq C\}$$

is compact. We define

$$\tilde{\Omega}_C := \{x \in \tilde{\Omega} \mid f(x) \le C\},$$

$$\Omega_C := \left\{\xi \in \Omega \mid \left(\frac{\partial u}{\partial \xi_1}, \dots, \frac{\partial u}{\partial \xi_n}\right) \in \tilde{\Omega}_C\right\},$$

and, in a standard way, we will identify the three compact sets S_C, $\tilde{\Omega}_C$ and Ω_C.

Without loss of generality we assume $f(0) = 1$, $\frac{\partial f}{\partial x_i}(0) = 0$, $1 \le i \le n$. This implies that both, $\tilde{\Omega}_C$ and Ω_C, include the origin.

Recall that the Blaschke metric G of M can be written as (cf. Section 2.7.1-3)

$$G = \sum G_{ij} d\xi_i d\xi_j,$$

where $G_{ij} = \rho u_{ij} := \rho \frac{\partial^2 u}{\partial \xi_i \partial \xi_j}$, $\rho = [\det(u_{ij})]^{1/(n+2)}$.

Now we consider the function

$$F := \exp\left\{-\frac{m}{C-f}\right\} \frac{\sum u_{ij}\xi_i\xi_j}{f^3} \tag{6.40}$$

defined on $S_C \cong \Omega_C$, where m is a positive constant to be determined later. Obviously, F attains its maximum at some interior point $\bar{\xi} \in \Omega_C$, $\bar{\xi} \ne 0$. Then, at $\bar{\xi}$, we have

$$\operatorname{grad} F = 0 \quad \text{and} \quad \Delta^{(B)}F \le 0,$$

or, equivalently,

$$F_i = 0, \quad 1 \le i \le n; \quad \sum u^{ij}F_{ij} \le 0, \tag{6.41}$$

where $F_i = \frac{\partial F}{\partial \xi_i}$, $F_{ij} = \frac{\partial^2 F}{\partial \xi_i \partial \xi_j}$, and so on.

For simplicity and without loss of generality, we may assume that $\bar{\xi}$ is $(\bar{\xi}_1, 0, \dots, 0)$, $\bar{\xi}_1 > \delta > 0$. To calculate the derivatives of F, we use the relation $\frac{\partial f}{\partial \xi_i} = \sum \xi_k u_{ki}$, so that the equations (6.41) become, at $\bar{\xi}$,

$$\bar{\xi}_1 u_{11i} = g(\bar{\xi}_1)^2 u_{11} u_{1i} - 2u_{1i}, \quad 1 \le i \le n, \tag{6.42}$$

$$(\bar{\xi}_1)^2 \sum u^{ij} u_{ij11} + 4\bar{\xi}_1 \sum u^{ij} u_{ij1} - g(\bar{\xi}_1)^3 u_{11} \sum u^{ij} u_{ij1} + 2n \tag{6.43}$$

$$- ng(\bar{\xi}_1)^2 u_{11} - g^2(\bar{\xi}_1)^4 (u_{11})^2 - \left[\frac{2m}{(C-f)^3} - \frac{3}{f^2}\right](\bar{\xi}_1)^4 (u_{11})^2 \le 0,$$

where

$$g := \frac{m}{(C-f)^2} + \frac{3}{f}; \tag{6.44}$$

and in deriving (6.43) we have used (6.42).

Next, we need to estimate the terms $\sum u^{ij} u_{ij1}$ and $\sum u^{ij} u_{ij11}$ in (6.43). If we take logarithms of both sides of the first equation in (6.10) and differentiate with respect to ξ_1 we find

$$\sum u^{ij} u_{ij1} = -(n+2)\frac{u_1^*}{u^*}. \tag{6.45}$$

We may assume that u^* attains its minimum at the origin, then $u_1^* \geq 0$ on Ω. Therefore, from (6.35) and the condition $-u^* f > d_1 > 0$, we have

$$0 \leq \sum u^{ij} u_{ij1} = (n + 2) \frac{-u^* u_1^*}{(u^*)^2} \leq c_1 f^2, \tag{6.46}$$

for some constant $c_1 > 0$.

Differentiating (6.45) with respect to ξ_1 again, we get

$$\sum u^{ij} u_{ij11} + \sum (u^{ij})_1 u_{ij1} = -(n + 2) \frac{u_{11}^*}{u^*} + (n + 2) \left(\frac{u_1^*}{u^*} \right)^2 \geq 0.$$

Differentiation of $\sum u^{ij} u_{jk} = \delta_k^i$ yields $(u^{ij})_1 = -\sum u^{ik} u_{kl1} u^{lj}$, so we have

$$\sum (u^{ij})_1 u_{ij1} = -\sum u^{ik} u^{lj} u_{kl1} u_{ij1}.$$

Hence

$$\sum u^{ij} u_{ij11} \geq -\sum (u^{ij})_1 u_{ij1} = \sum u^{ik} u^{jl} u_{kl1} u_{ij1}. \tag{6.47}$$

An application of Lemma 6.17 gives, for any $\delta < \frac{1}{n-1}$,

$$\sum u^{ij} u_{ij11} \geq \left(\frac{n}{n-1} - \delta \right) \left(\frac{u_{111}}{u_{11}} \right)^2 - \frac{1 - (n-1)\delta}{(n-1)^2 \delta} \left(\sum u^{ij} u_{ij1} \right)^2. \tag{6.48}$$

Inserting (6.42), (6.46), and (6.48) into (6.43), we get

$$\left[\left(\frac{1}{n-1} - \delta \right) g^2 - \frac{2m}{(C-f)^3} \right] (\xi_1)^4 (u_{11})^2$$

$$- g \left(\xi_1 c_1 f^2 + n + \frac{4n}{n-1} \right) (\xi_1)^2 u_{11} - \frac{1 - \delta(n-1)}{\delta(n-1)^2} c_1^2 (\xi_1)^2 f^4 \leq 0. \tag{6.49}$$

We further choose $\delta < \frac{1}{2(n-1)}$ and $m = (n-1)^2 C$. Then using the inequality

$$\left(\frac{1}{n-1} - \delta \right) g^2 - \frac{2m}{(C-f)^3} \geq \left[\frac{1}{2(n-1)} - \delta \right] g^2, \tag{6.50}$$

we finally obtain the following inequality:

$$ag^2 (\xi_1)^4 (u_{11})^2 - g(b_1 f^2 + b_2)(\xi_1)^2 u_{11} - b_3 f^4 \leq 0, \tag{6.51}$$

where b_1, b_2, b_3 are positive constants depending only on n and Ω, and where

$$a = \frac{1}{2(n-1)} - \delta > 0.$$

From (6.51) we find

$$(\xi_1)^2 u_{11} \leq \frac{g(b_1 f^2 + b_2) + g\sqrt{ab_3 f^4}}{ag^2} \leq b_4 f^3,$$

for some constant $b_4 > 0$, which implies that $F(\bar{\xi}) \le b_4$. Since F attains its maximum at $\bar{\xi}$, we have

$$\frac{\sum u_{ij}\xi_i\xi_j}{f^3} \le \exp\left\{\frac{(n-1)^2 C}{C-f}\right\} \cdot b_4 \quad \text{for all } \xi \in \Omega_C.$$

Let $C \to \infty$, we obtain

$$\frac{\sum u_{ij}\xi_i\xi_j}{f^3} \le \exp\{(n-1)^2\} \cdot b_4 \quad \text{in } \Omega.$$

The above estimate together with the condition $-u^* f \le d_2$ gives

$$-(u^*)^3 \sum u_{ij}\xi_i\xi_j \le Q \quad \text{in } \Omega \tag{6.52}$$

for some constant $Q > 0$. This finishes the proof of Lemma 6.18. □

Lemma 6.19. *Let u and u^* be the solutions of the p.d.e.'s (6.10)–(6.11). Let f be the Legendre transformation function of u, which is still regarded as a function on Ω. Then there is a constant $d_2 > 0$ such that*

$$-u^* f \le d_2 \quad \text{in } \Omega. \tag{6.53}$$

Proof. Similar to the proof of Lemma 6.10, we can show that, for any point $\bar{\xi} \in \partial\Omega$, there are appropriate constants $a = (a_1, \ldots, a_n)$ and d depending only on Ω and φ, such that the function

$$\underline{u}(\xi) = u^*(\xi) + a \cdot \xi + d$$

satisfies the following conditions:

$$\underline{u}(\bar{\xi}) = u(\bar{\xi}),$$
$$\underline{u}(\xi) \le u(\xi), \quad \forall \xi \in \partial\Omega,$$

and obviously

$$\det\left(\frac{\partial^2 \underline{u}}{\partial\xi_i\partial\xi_j}\right) = (-u^*)^{-n-2} \quad \text{in } \Omega.$$

It follows from the maximum principle that

$$\underline{u} \le u \quad \text{in } \Omega.$$

Since u is strongly convex, for any t with $0 \le t \le 1$, we have

$$\frac{\partial u}{\partial \nu}(t\bar{\xi}) < \frac{u(\bar{\xi}) - u(t\bar{\xi})}{|\bar{\xi} - t\bar{\xi}|} \le \frac{\underline{u}(\bar{\xi}) - \underline{u}(t\bar{\xi})}{|\bar{\xi} - t\bar{\xi}|}$$

$$= \frac{a \cdot (\bar{\xi} - t\bar{\xi}) - u^*(t\bar{\xi})}{|\bar{\xi} - t\bar{\xi}|} \le |a| + \frac{-u^*(t\bar{\xi})}{d(t\bar{\xi})},$$

where $\frac{\partial}{\partial v} = \sum \xi_i \frac{\partial}{\partial \xi_i}$. Using Lemma 6.16, we get

$$\frac{\partial u}{\partial v}(t\bar{\xi}) \leq \frac{C}{-u^*(t\bar{\xi})},$$

for some constant $C > 0$. Since $\bar{\xi}$ is arbitrary, we have

$$\frac{\partial u}{\partial v} \leq \frac{C}{-u^*} \quad \text{in } \Omega. \tag{6.54}$$

It follows that

$$-u^* f \leq C + uu^* \leq d_2 \quad \text{in } \Omega$$

for some constant $d_2 > 0$. $\qquad\square$

When comparing Lemma 6.18 and Lemma 6.19, we realize that, in order to verify the assertion (6.39) of Lemma 6.18, it remains to derive an estimate of the type

$$-u^* f \geq d_1 > 0 \quad \text{in } \Omega,$$

for an appropriate real constant d_1. The following proof works for the special case that $\Omega = B_1(O)$; so far we do not have a proof for arbitrary convex domains.

When Ω is the unit ball $B_1(O)$, it is easy to see that the solution of (6.11) is

$$u^*(\xi) = -\sqrt{1 - \sum(\xi_i)^2},$$

and the p.d.e. (6.10) becomes

$$\begin{cases} \det\left(\dfrac{\partial^2 u}{\partial \xi_i \partial \xi_j}\right) = \left[1 - \sum(\xi_i)^2\right]^{-\frac{n+2}{2}} & \text{in } B_1(O), \\ u = \varphi & \text{on } \partial B_1(O). \end{cases} \tag{6.55}$$

The p.d.e. (6.55) was studied in [172] in connection with spacelike hypersurfaces in the Minkowski space with constant Gauß–Kronecker curvature. In particular, we have the following lemma:

Lemma 6.20 (cf. Lemma 5.6 of [172]). *Let u be the convex solution of (6.55), f be the Legendre transformation function of u. Then there is a constant $d_1 > 0$ such that*

$$\sqrt{1 - \sum(\xi_i)^2} f \geq d_1 \quad \text{in } B_1(O).$$

Proof. For any r, $\frac{1}{2} \leq r < 1$, and any point $\bar{\xi} \in \partial B_r(O)$, we can show that, similar to the proof of Lemma 6.10, there are appropriate constants $b_1, \ldots, b_n$, depending only on $\bar{\xi}$; and d, depending only on φ, but not on $\bar{\xi}$ and r, such that the function

$$\bar{u}(\xi) = -\sqrt{1 - \sum(\xi_i)^2} + \sum b_i \xi_i + d$$

satisfies the following conditions:

$$\bar{u}(\bar{\xi}) = u(\bar{\xi}),$$
$$\bar{u}(\xi) > u(\xi), \ \ \forall\, \xi \in \partial B_r(O)\backslash\{\bar{\xi}\}.$$

Since we have

$$\det\left(\frac{\partial^2 u}{\partial\xi_i\partial\xi_j}\right) = \left[1 - \sum(\xi_i)^2\right]^{-\frac{n+2}{2}} = \det\left(\frac{\partial^2\bar{u}}{\partial\xi_i\partial\xi_j}\right) \ \ \text{in } B_r(O),$$

the maximum principle implies that

$$u(\xi) \le \bar{u}(\xi) \ \ \text{in } B_r(O).$$

Without loss of generality, we may assume that $\bar{\xi} = (r, 0, \dots, 0)$. Then, at $\bar{\xi}$,

$$\frac{\partial u}{\partial\xi_1} \ge \frac{\partial\bar{u}}{\partial\xi_1} = \frac{\bar{\xi}_1}{\sqrt{1 - (\bar{\xi}_1)^2}} + b_1.$$

So, at $\bar{\xi}$,

$$\bar{\xi}_1\frac{\partial u}{\partial\xi_1} - u(\bar{\xi}) \ge \frac{1}{\sqrt{1 - (\bar{\xi}_1)^2}} - d,$$

i.e.

$$f(\bar{\xi})\sqrt{1 - \sum(\bar{\xi}_i)^2} \ge 1 - d\sqrt{1 - \sum(\bar{\xi}_i)^2}. \tag{6.56}$$

By the arbitrariness of r and $\bar{\xi} \in \partial B_r(O)$, we see that the inequality (6.56) holds at any point $\xi \in B_1(O)\backslash B_{1/2}(O)$.

From (6.56) we see that, if $\sum(\xi_i)^2 \ge \max\{\frac{1}{4}, 1 - \frac{1}{4d^2}\}$ then

$$f(\xi)\sqrt{1 - \sum(\xi_i)^2} \ge \frac{1}{2}.$$

If, on the other hand, $\sum(\xi_i)^2 < \max\{\frac{1}{4}, 1 - \frac{1}{4d^2}\}$ then recalling $f \ge 1$ we have

$$f(\xi)\sqrt{1 - \sum(\xi_i)^2} \ge \min\left\{\frac{\sqrt{3}}{2}, \frac{1}{2d}\right\}.$$

Hence, if we choose $d_1 = \min\{\frac{1}{2}, \frac{1}{2d}\}$ then the assertion of the lemma follows. $\quad\square$

Now we can prove the main result of this section.

Theorem 6.21. *Given $\varphi \in C^\infty(B_1(O))$, from the solution u of (6.55) we can construct a hypersurface M with the following properties:*
(i) *$S_n = 1$;*
(ii) *M is Euclidean complete;*
(iii) *M is complete with respect to the Blaschke metric.*

Proof. The first two properties are part of Theorem 6.11. We now prove (iii). By definition, we have the computation

$$\frac{\|\text{grad}\,f\|_G^2}{f^2} = \frac{-u^*}{f^2} \sum u^{ij} \frac{\partial f}{\partial \xi_i} \frac{\partial f}{\partial \xi_j}$$

$$= -\frac{u^*}{f^2} \sum u^{ij} \xi_k u_{ik} \xi_l u_{jl}$$

$$= -\frac{u^*}{f^2} \sum u_{kl} \xi_k \xi_l.$$

Thus, in the special case when $\Omega = B_1(O)$, the previous Lemmas 6.18, 6.19, and 6.20 give the estimate

$$\frac{\|\text{grad}\,f\|_G}{f} \le C \ \text{ in } B_1(O), \tag{6.57}$$

for some constant $C > 0$.

For the hypersurface defined by (6.17), i.e.

$$M := \{(x, f(x)) \mid x = (x^1, \dots, x^n) \in \tilde{\Omega}\},$$

we consider an arbitrary unit speed geodesic $\gamma(s)$, with respect to the Blaschke metric G, starting from the point $p_0 = (0, \dots, 0, 1)$:

$$\sigma : [0, L] \to \tilde{\Omega}; \ \ \gamma(s) = (\sigma(s), f(\sigma(s))),$$

here we have assumed $f(x) \ge f(0) = 1$. From (6.57), we have

$$\frac{df(\sigma(s))}{ds} = G(\text{grad}\,f, \sigma'(s)) \le \|\text{grad}\,f\|_G \le Cf(\sigma(s)).$$

It follows that

$$s \ge C^{-1} \ln f(\sigma(s)),$$

so we have $L = +\infty$ and M is complete with respect to the Blaschke metric. $\qquad\square$

7 Geometric inequalities

Geometric inequalities have a wide range of applications within geometry itself as well as in many fields in mathematics and physics; as exemplary fields we mention differential equations, functional analysis, the calculus of variations in the large, the theory of functions of a complex variable, the geometry of numbers, discrete geometry, etc. Lutwak gives references in [204]; see also [31].

Affine isoperimetric inequalities have always occupied a central position in the field of geometric convexity. Experts are referred to [196–206]. It is the purpose of this chapter to give an introduction to this interesting field and to prove some of the most important geometric inequalities in affine geometry. The commentary of Leichtweiß in [23, pp. 21–36] shows the influence of Blaschke to this field.

7.1 The affine isoperimetric inequality

In this section we derive an affine analogue of the classical isoperimetric inequality. Again, we introduce a Euclidean inner product on V and denote the Euclidean space by $E^{n+1} =: E$. We consider an ovaloid $x : M \to E$. We denote by $S(M)$ the total affine area of M with respect to the Blaschke metric, by $S_E(M)$ the total area of M with respect to the Euclidean metric induced from E^{n+1}, and by $\mathrm{Vol}(M)$ the $(n + 1)$-dimensional volume of the convex body in E^{n+1} bounded by $x(M)$. The volume $\mathrm{Vol}(M)$ is equiaffinely invariant.

The so-called affine isoperimetric inequality states that for an arbitrary ovaloid M in A^{n+1},

$$(S(M))^{n+2} \le (n + 1)^{n+2} \cdot \sigma_{n+1}^2 \cdot (\mathrm{Vol}(M))^n$$

where

$$\sigma_{n+1} = \pi^{(n+1)/2} \left(\Gamma \left(\frac{n + 3}{2} \right) \right)^{-1}$$

is the $(n + 1)$-dimensional volume of the unit ball in E^{n+1}; Γ denotes the Gamma-function. The equality holds if and only if M is an ellipsoid.

This inequality is due to Blaschke [20] for $n = 2$. Santalo [294] and Deicke [65] generalized it to higher dimensions. In the form stated above the inequality is due to Petty [278].

7.1.1 Steiner symmetrization

Let P be a hyperplane in E^{n+1} and let l be a straight line orthogonal to the hyperplane P. For any chord d of the ovaloid parallel to l, we consider a (straight) line segment d^* with the same length as d, which lies in the same (straight) line as d and has its midpoint lying on P. The union of all such straight line segments defines a new convex body. We denote by M^* the boundary surface of the new convex body. Obviously, M^* is also a C^∞-ovaloid in E^{n+1}. The construction of M^* from the given ovaloid is called the *Steiner symmetrization* of $x(M)$. In the following we will give an analytic expression for the Steiner symmetrization.

Choose a Cartesian coordinate system $(x^1, \ldots, x^{n+1})$ in E^{n+1} such that the equation of the hyperplane P is $x^{n+1} = 0$. Draw straight lines which are parallel to the x^{n+1}-axis and tangent to $x(M)$. All the intersection points of these lines and P enclose a convex subdomain Ω of P and the set of all the tangent points divides $x(M)$ into two hypersurfaces Σ and $\tilde{\Sigma}$. Σ and $\tilde{\Sigma}$ can be represented, respectively, by convex functions f and $\tilde{f}$ defined on Ω, i.e.

$$\Sigma: \quad x^{n+1} = f(x^1, \ldots, x^n),$$
$$\tilde{\Sigma}: \quad x^{n+1} = -\tilde{f}(x^1, \ldots, x^n), \quad (x^1, \ldots, x^n) \in \Omega,$$

where the Hessian matrices of f and $\tilde{f}$ are positive definite. M^* is also divided into two hypersurfaces Σ^* and $\tilde{\Sigma}^*$. They can be represented by

$$\Sigma^*: \quad x^{n+1} = \frac{1}{2}\left[f(x^1, \ldots, x^n) + \tilde{f}(x^1, \ldots, x^n)\right],$$
$$\tilde{\Sigma}^*: \quad x^{n+1} = -\frac{1}{2}\left[f(x^1, \ldots, x^n) + \tilde{f}(x^1, \ldots, x^n)\right].$$

We have the following theorem.

Theorem 7.1. *Let $x(M)$ be a C^∞-ovaloid and let P be a hyperplane in E^{n+1}. Let M^* be the ovaloid obtained from $x(M)$ by the Steiner symmetrization. Then the total affine areas and the Euclidean $(n + 1)$-dimensional volumes, respectively, satisfy the following relations:*

$$S(M^*) \geq S(M), \tag{7.1}$$

$$\mathrm{Vol}(M^*) = \mathrm{Vol}(M) \tag{7.2}$$

with equality in (7.1) holding if and only if all midpoints of the chords of $x(M)$, orthogonal to P, lie on the same hyperplane.

To prove this theorem, we need the following lemma.

Lemma 7.2. *Let T and Q be two positive definite $n \times n$ matrices. Then*

$$\left[\det\left(\frac{T + Q}{2}\right)\right]^{1/(n+2)} \geq \frac{1}{2}\left[(\det(T))^{1/(n+2)} + (\det(Q))^{1/(n+2)}\right], \tag{7.3}$$

with equality holding if and only if $T = Q$.

Proof. Since T and Q are both positive definite, there is a nonsingular matrix E such that simultaneously

$$T = E \begin{pmatrix} \lambda_1 & & & 0 \\ & \lambda_2 & & \\ & & \ddots & \\ 0 & & & \lambda_n \end{pmatrix} E^t, \qquad Q = E \begin{pmatrix} \mu_1 & & & 0 \\ & \mu_2 & & \\ & & \ddots & \\ 0 & & & \mu_n \end{pmatrix} E^t,$$

where E^t denotes the transposed matrix and where the numbers $\lambda_1, \lambda_2, \ldots, \lambda_n$ and $\mu_1, \mu_2, \ldots, \mu_n$ are all positive. Thus it is sufficient to prove the following inequality:

$$\left[\left(\frac{\lambda_1 + \mu_1}{2} \right) \left(\frac{\lambda_2 + \mu_2}{2} \right) \cdots \left(\frac{\lambda_n + \mu_n}{2} \right) \right]^{1/(n+2)}$$
$$\geq \frac{1}{2} [(\lambda_1 \lambda_2 \cdots \lambda_n)^{1/(n+2)} + (\mu_1 \mu_2 \cdots \mu_n)^{1/(n+2)}].$$

In fact, from Minkowski's inequality [13, p. 26] we have

$$\left[\left(\frac{\lambda_1 + \mu_1}{2} \right) \left(\frac{\lambda_2 + \mu_2}{2} \right) \cdots \left(\frac{\lambda_n + \mu_n}{2} \right) \right]^{1/n} \geq \frac{1}{2} \left[(\lambda_1 \lambda_2 \cdots \lambda_n)^{1/n} + (\mu_1 \mu_2 \cdots \mu_n)^{1/n} \right],$$

with equality holding if and only if

$$\frac{\lambda_1}{\mu_1} = \frac{\lambda_2}{\mu_2} = \cdots = \frac{\lambda_n}{\mu_n}.$$

Hence

$$\left[\left(\frac{\lambda_1 + \mu_1}{2} \right) \left(\frac{\lambda_2 + \mu_2}{2} \right) \cdots \left(\frac{\lambda_n + \mu_n}{2} \right) \right]^{1/(n+2)} \tag{7.4}$$
$$\geq \left[\frac{1}{2} \left((\lambda_1 \lambda_2 \cdots \lambda_n)^{1/n} + (\mu_1 \mu_2 \cdots \mu_n)^{1/n} \right) \right]^{n/(n+2)}$$
$$\geq \frac{1}{2} \left[(\lambda_1 \lambda_2 \cdots \lambda_n)^{1/(n+2)} + (\mu_1 \mu_2 \cdots \mu_n)^{1/(n+2)} \right],$$

where the last inequality is due to the convexity of the function x^α ($\alpha < 1$). The equality in (7.4) holds if and only if

$$\lambda_1 \lambda_2 \cdots \lambda_n = \mu_1 \mu_2 \cdots \mu_n$$

and equality holds in the foregoing expression. Thus we have proved that the inequality (7.3) holds, and it becomes an equality if and only if

$$\lambda_1 = \mu_1, \lambda_2 = \mu_2, \ldots, \lambda_n = \mu_n. \qquad \square$$

Proof of Theorem 7.1. It is easy to see that the volume elements of the Blaschke metrics of Σ and $\tilde{\Sigma}$ are, respectively,

$$[\det(\partial_i \partial_j f)]^{1/(n+2)} dx^1 \wedge dx^2 \wedge \cdots \wedge dx^n$$

and

$$[\det(\partial_j\partial_i\tilde{f})]^{1/(n+2)}dx^1 \wedge dx^2 \wedge \cdots \wedge dx^n.$$

Thus

$$S(M) = \int_\Omega \left\{[\det(\partial_j\partial_i f)]^{1/(n+2)} + [\det(\partial_j\partial_i\tilde{f})]^{1/(n+2)}\right\}dx^1 \wedge dx^2 \wedge \cdots \wedge dx^n.$$

Similarly

$$S(M^*) = 2\int_\Omega \left\{\det\left[\frac{1}{2}(\partial_j\partial_i f) + \frac{1}{2}(\partial_j\partial_i\tilde{f})\right]\right\}^{1/(n+2)}dx^1 \wedge dx^2 \wedge \cdots \wedge dx^n.$$

By Lemma 7.2 we get

$$S(M^*) \geq S(M),$$

and $S(M^*) = S(M)$ if and only if at any point in Ω

$$\partial_j\partial_i f = \partial_j\partial_i\tilde{f}.$$

Thus $S(M^*) = S(M)$ holds if and only if

$$\tilde{f} = f - \sum a_i x^i - b$$

for some constants $b, a_1, \ldots, a_n$. This is equivalent to the fact that the midpoints of the chords of M which are parallel to the x_{n+1}-axis, satisfy

$$x^{n+1} = \frac{1}{2}(f - \tilde{f}) = \frac{1}{2}\left(\sum a_i x^i + b\right),$$

i.e. the midpoints lie on a hyperplane in E^{n+1}. The equality (7.2) is obvious from the Steiner symmetrization. $\qquad\square$

7.1.2 A characterization of ellipsoids

Theorem 7.3. *Let M be an ovaloid in A^{n+1}. If for any direction v the midpoints of the chords of M parallel to v lie on some hyperplane, then M is an ellipsoid.*

Proof. We shall use induction to prove the theorem. The case $n = 1$ is nontrivial: let Γ be a simple closed convex curve and suppose that the midpoints of the parallel chords of Γ in any direction lie on some straight line segment. Let AB be any chord of Γ and denote by CD the chord consisting of the midpoints of the chords of Γ parallel to AB. Obviously, CD passes through the midpoint of AB. Let O be the midpoint of CD. There must be a chord of Γ parallel to AB containing O. Without loss of generality we assume that AB passes through O. We assert that the chord EF consisting of the midpoints of chords parallel to CD coincides with AB. Obviously, EF passes through O and does

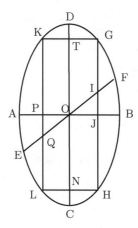

Fig. 7.1: An illustration to the proof of Theorem 7.3.

not coincide with CD. If our assertion does not hold, we draw an arbitrary chord GH parallel to CD which meets EF at I and meets AB at $J, I \neq J$. We draw chords GK and HL such that they are parallel to AB. They meet CD at T and N, respectively, and Γ at K and L, respectively (cf. Fig. 7.1). Then $GTNH$ is a parallelogram and $GT = HN$. Connect KL, it meets AB at P and meets EF at Q. Obviously, $GT = TK, HN = NL$. Thus $GK = HL$ and $GKLH$ is a parallelogram. Consequently, $KP = GJ, PL = JH$ and KL is parallel to CD. On the other hand, since EF is comprised of the midpoints of the chords parallel to CD, we have $GI = IH, KQ = QL$. Hence

$$KP < PL = JH < GJ = KP.$$

This contradiction gives our assertion: EF and AB coincide. Introducing again a Euclidean inner product, by a unimodular affine transformation, we may assume that AB is orthogonal to CD. This implies that Γ is symmetric with respect to AB and CD. Thus it is centrally symmetric with respect to O. Now let l_1 be any chord of Γ through O and l_2 be the chord consisting of the midpoints of the chords parallel to l_1. From the argument above, l_1 consist of the midpoints of the chords parallel to l_2. This shows that the line through O in the direction of l_1 is the affine normal of Γ (cf. Sect. 2.2.4) in both end points of l_1. Since l_1 is arbitrary, we conclude that all affine normals of Γ intersect at O. It follows that Γ is an ellipse. Hence the theorem is true for $n = 1$. Assume that the assertion is true for $n \leq k$. Let $x(M)$ now be an ovaloid in A^{k+2} satisfying the assumptions in Theorem 7.3. Obviously, for any $(m + 1)$-dimensional subspace in A^{k+2}, where $1 \leq m \leq k$, intersecting $x(M)$, the intersection is an m-dimensional ovaloid in A^{n+1} which satisfies the assumptions in the theorem, and thus is an m-dimensional ellipsoid.

We are going to prove that M is an ellipsoid. Let AB be any chord of M. The midpoints of chords of M parallel to AB lie on a hyperplane Π. Then the intersection $M \cap \Pi$ of Π and M is an ellipsoid in Π. Denote by O the center of $M \cap \Pi$. There must be a chord of M parallel to AB and through O. Without loss of generality we assume that AB passes through O. Then $AO = BO$. We assert that M is centrally symmetric with respect to O. In fact let CD be any chord passing through O. Denote by η the 2-dimensional plane spanned by AB and CD. Then $M \cap \eta$ is an ellipse. Obviously, $M \cap \Pi \cap \eta$ has only two points, say E and F, and EF is a chord through O. Since O is the center of $M \cap \Pi$, we have $EO = FO$. Consider the ellipse $M \cap \eta$. The facts $AO = BO$ and $EO = FO$ imply that O is the center of $M \cap \eta$. Hence $CO = DO$. Since CD is an arbitrary chord through O, our assertion is proved.

Now we make use of the geometric meaning of the affine normal (see Section 2.2.4) to prove that M is an ellipsoid. Let P be any point on M, and PQ be a chord passing through O. Then $PO = QO$. Let $T_P M$ be the tangent hyperplane of M at P, and Π^* be any hyperplane parallel to $T_P M$, which intersects M. By inductive assumption $M \cap \Pi^*$ is an ellipsoid in Π^*. Denote by O^* the intersection point of Π^* and OP. We assert that O^* is the center of $M \cap \Pi^*$. In fact let KL be any chord of $M \cap \Pi^*$ through O^*. Denote by ξ the 2-dimensional plane spanned by KL and OP. Then $M \cap \xi$ is an ellipse and O is the center of $M \cap \xi$ by the central symmetry of M with respect to O. Hence OP is the affine normal of $M \cap \xi$ and therefore $KO^* = LO^*$ (see Section 2.2.4). Since KL is an arbitrary chord through O^*, O^* must be the center of $M \cap \Pi^*$. By the geometric meaning of the affine normal we know that OP is the affine normal of M. Since P is arbitrary, we conclude that all the affine normals of M intersect at O, hence M is an ellipsoid. □

7.1.3 The affine isoperimetric inequality

We recall the notation for $S(M)$ and $S_E(M)$ from the beginning of Section 7.1. We denote by dV (see equation (5.2)) and dS_E the corresponding volume elements. For the proof of the affine isoperimetric inequality we use the Steiner symmetrization.

Theorem 7.4. *Let M be an ovaloid in A^{n+1}. Then*

$$(S(M))^{n+2} \leq (n + 1)^{n+2} \sigma_{n+1}^2 (\text{Vol}(M))^n, \tag{7.5}$$

with equality holding if and only if M is an ellipsoid, where

$$\sigma_{n+1} = \pi^{(n+1)/2} \left(\Gamma \left(\frac{n + 3}{2} \right) \right)^{-1}$$

is the $(n + 1)$-dimensional volume of the unit ball in E^{n+1}.

Proof. First, we prove

$$(S(M))^{n+2} \leq (n + 1) \sigma_{n+1} (S_E(M))^{n+1}. \tag{7.6}$$

Denote by K the Euclidean Gauß–Kronecker curvature of M, which is positive everywhere. Then the Riemannian volume element $\omega(I)$ of the first fundamental form (see

Section 1.5) and the volume element $dV = \omega(e)$ of the Blaschke metric satisfy the following relation on M:

$$dV = \omega(e) = K^{1/(n+2)}\omega(I).$$

We use Hölder's inequality [13, p. 21] and have

$$\left(\frac{\int_M K^{1/(n+2)} dS_E}{\int_M dS_E}\right)^{n+2} \leq \frac{\int_M K dS_E}{\int_M dS_E} = \frac{O_n}{\int_M dS_E},$$

where O_n is the n-dimensional volume of the Euclidean unit hypersphere in E^{n+1} and $O_n = (n+1)\sigma_{n+1}$. This proves (7.6).

Secondly, we consider a standard n-hypersphere Σ satisfying

$$\text{Vol}(\Sigma) = \text{Vol}(M).$$

Then

$$S_E(\Sigma) = (n+1)(\sigma_{n+1})^{1/(n+1)}(\text{Vol}(M))^{n/(n+1)}.$$

By the convergence theorem of W. Gross (cf. [20, §115]), we can construct a new ovaloid M^* from $x : M \to E^{n+1}$ by a number of suitable Steiner symmetrizations, such that

$$S_E(M^*) < S_E(\Sigma) + \varepsilon$$

for any small positive number ε. From Theorem 7.1 and (7.6) we have

$$S(M) \leq S(M^*) \leq [(n+1)\sigma_{n+1}(S_E(M^*))^{n+1}]^{1/(n+2)}$$
$$\leq [(n+1)\sigma_{n+1}(S_E(\Sigma) + \varepsilon)^{n+1}]^{1/(n+2)}.$$

Since ε may be any small positive number, we can conclude

$$(S(M))^{n+2} \leq (n+1)\sigma_{n+1}(S_E(\Sigma))^{n+1}$$
$$= (n+1)^{n+2}\sigma_{n+1}^2(\text{Vol}(M))^n.$$

Hence (7.5) holds for any ovaloid in A^{n+1}. In case of equality

$$(S(M))^{n+2} = (n+1)^{n+2}\sigma_{n+1}^2(\text{Vol}(M))^n$$

the midpoints of the parallel chords of M in any direction lie on a hyperplane. Otherwise there is a hyperplane P such that M^* obtained from $x : M \to E^{n+1}$ by Steiner symmetrization satisfies

$$S(M^*) \geq S(M),$$
$$\text{Vol}(M^*) = \text{Vol}(M).$$

Thus, $(S(M^*))^{n+2} > (n+1)^{n+2}\sigma_{n+1}^2(\text{Vol}(M))^n$, which contradicts (7.5). Hence, according to Theorem 7.3, $x : M \to E^{n+1}$ must be an ellipsoid. $\square$

7.2 Inequalities for higher affine mean curvatures

Let $x : M \to A^3$ be an ovaloid. The affine theorema egregium $\chi = J + L_1$ (see (2.125)) gives

$$\int_M L_1 dV \leq \int_M \chi dV = 4\pi, \tag{7.7}$$

where, as before, $dV = w(e)$ denotes the volume element with respect to the Blaschke metric. When $L_2 > 0$ on M, it follows from Lemma 4.4 that both affine principal curvatures are positive: $\lambda_1 > 0$ and $\lambda_2 > 0$. Hence

$$L_1 \geq \sqrt{L_2}$$

and

$$\int_M \sqrt{L_2} dV \leq \int_M L_1 dV \leq 4\pi. \tag{7.8}$$

Here, $\int_M \sqrt{L_2} dV$ is the volume of the ovaloid $Y(M)$ with respect to the centroaffine metric (for the geometry of $Y(M)$ see e.g. [341], section 4.6.1). We are going to generalize (7.7) and (7.8) to higher dimensions, mainly following Calabi [43].

7.2.1 Mixed volumes

For later applications, we introduce briefly the notion of the mixed volumes of convex bodies and the Alexandrov–Fenchel inequality (cf. [31, p. 143]).

Let N_1 and N_2 be two convex bodies and let λ be a real number. We define new convex bodies $N_1 + N_2$ and λN_1 as follows:

$$N_1 + N_2 = \{a + b \mid a \in N_1, \ b \in N_2\},$$
$$\lambda N_1 = \{\lambda a \mid a \in N_1\}.$$

Theorem 7.5 (Minkowski). *Let* $N_1, N_2, \ldots, N_s$ *be convex bodies in* A^n *and let* $\lambda_1, \lambda_2, \ldots, \lambda_s$ *be* s *nonnegative real numbers. Then the n-dimensional volume* $\mathrm{Vol}(\sum_{i=1}^{s} \lambda_i N_i)$ *of the convex body* $\sum_{i=1}^{s} \lambda_i N_i$ *is a homogeneous polynomial of degree n in the variables* $\lambda_1, \lambda_2, \ldots, \lambda_s$:

$$\mathrm{Vol}\left(\sum_{i=1}^{s} \lambda_i N_i\right) = \sum_{i_1, \ldots, i_s} \mathrm{Vol}(N_{i_1}, N_{i_2}, \ldots, N_{i_s}) \lambda_{i_1} \lambda_{i_2} \cdots \lambda_{i_s}, \tag{7.9}$$

where the coefficients $\mathrm{Vol}(N_{i_1}, \ldots, N_{i_s})$ *are symmetric with respect to* $N_{i_1}, \cdots, N_{i_s}$.

In Theorem 7.5, the coefficients $\mathrm{Vol}(N_{i_1}, N_{i_2}, \ldots, N_{i_s})$ are called the *mixed volumes* of $N_{i_1}, \ldots, N_{i_s}$. Obviously, $\mathrm{Vol}(N_1, N_1, \ldots, N_1)$ is the volume of N_1, i.e.

$$\mathrm{Vol}(N_1, N_1, \ldots, N_1) = \mathrm{Vol}(N_1).$$

Equation (7.9) can be written in the form

$$\text{Vol}\left(\sum_{i=1}^{s}\lambda_i N_i\right) = \sum_{\substack{1\leq p_1<\cdots<p_r\leq s \\ p_1+\cdots+p_r=n}} \frac{n!}{p_1! p_2!\cdots p_r!}\text{Vol}(N_{i_1},p_1;\cdots;N_{i_r},p_r)\lambda_{i_1}^{p_1}\lambda_{i_2}^{p_2}\cdots\lambda_{i_r}^{p_r},$$

where the sum is taken over all possible ordered decompositions of the number n into natural summands $p_1 + \cdots + p_r = n$, while

$$\text{Vol}(N_{i_1},p_1;\cdots;N_{i_r},p_r) := \text{Vol}(\underbrace{N_{i_1},N_{i_1},\ldots,N_{i_1}}_{p_1},\ldots,\underbrace{N_{i_r},\ldots,N_{i_r}}_{p_r}).$$

The notion of mixed volumes has the following properties:
(i)   independence of the order:

$$\text{Vol}(N_1,N_2,\ldots,N_n) = \text{Vol}(N_{i_1},N_{i_2},\ldots,N_{i_n}),$$

where $(i_1,i_2,\ldots,i_n)$ is any permutation of the numbers $(1,\ldots,n)$;
(ii)   positive multilinearity: for $\alpha_0 \geq 0, \alpha_1 \geq 0$

$$\text{Vol}(\alpha_0 N_0 + \alpha_1 N_1, N_2,\ldots,N_n) = \alpha_0\text{Vol}(N_0,N_2,\ldots,N_n) + \alpha_1\text{Vol}(N_1,N_2,\ldots,N_n);$$

(iii)   invariance with respect to parallel translation:

$$\text{Vol}(a_1 + N_1, a_2 + N_2,\ldots,a_n + N_n) = \text{Vol}(N_1,N_2,\ldots,N_n),$$

where $(a_1,\ldots,a_n) \in A^n$;
(iv)   invariance with respect to a unimodular affine transformation $T : A^n \rightarrow A^n$:

$$\text{Vol}(TN_1, TN_2,\ldots,TN_n) = \text{Vol}(N_1,N_2,\ldots,N_n);$$

(v)   monotonicity: if N_i, N_i^* are convex bodies in A^n satisfying $N_i \subset N_i^*, 1 \leq i \leq n$, then

$$\text{Vol}(N_1,N_2,\ldots,N_n) \leq \text{Vol}(N_1^*,N_2^*,\ldots,N_n^*).$$

It follows immediately that mixed volumes are nonnegative.

In many geometric problems, one considers mixed volumes which depend only on two convex bodies. The following is the famous Alexandrov–Fenchel inequality [31, §20.1].

Theorem 7.6. *Let $N_1,\ldots,N_n$ be convex bodies in A^n, then it holds the relation*

$$[\text{Vol}(N_1,N_2,\ldots,N_n)]^2 \geq \text{Vol}(N_1,N_1,N_3,\ldots,N_n)\text{Vol}(N_2,N_2,N_3,\ldots,N_n). \tag{7.10}$$

As direct corollaries of (7.10) we further have

$$[\text{Vol}(N_1,N_2,\ldots,N_n)]^n \geq \text{Vol}(N_1)\text{Vol}(N_2)\cdots\text{Vol}(N_n), \tag{7.11}$$

$$[\text{Vol}(N,k;L,n-k)]^2 \geq \text{Vol}(N,k-1;L,n-k+1)\cdot\text{Vol}(N,k+1;L,n-k-1). \tag{7.12}$$

Note. Section 20.5 of [31] contains a discussion of known results when equalities hold in (7.10)–(7.12). See also [318] and the paper [319], where further references are given.

7.2.2 Integral inequalities for curvature functions

As several times before, we consider A^{n+1} to carry additionally a Euclidean structure. Again, we denote the Euclidean space by $E = E^{n+1}$, and identify the vector spaces: $V = V^*$. Let $x : M \to E$ be an ovaloid. We shall relate the curvature integrals $\int_M L_j dV$ to certain mixed volumes in E. Introduce the following collection of n-forms on M:

$$\phi_k = \frac{1}{(n+1)!} \mathrm{Det}(\underbrace{-dY, \ldots, -dY}_{k}, \underbrace{dx, \ldots, dx}_{n-k}, -x),$$

$$\Psi_k = \frac{1}{(n+1)!} \mathrm{Det}(\underbrace{-dY, \ldots, -dY}_{k}, \underbrace{dx, \ldots, dx}_{n-k}, Y), \quad \text{for } k = 0, 1, \ldots, n,$$

where x is the position vector of M and Y is the affine normal vector of M. Suppose that the affine Weingarten form (B_{ij}) of M is positive definite everywhere, i.e. $L_n > 0$ on M (Lemma 4.4). As this form is the centroaffine metric of the spherical normal indicatrix $Y : M \to V$, the hypersurface $Y(M)$ is an ovaloid. Fixing an origin $O \in E$, we can also identify V and E and consider Y to be a hypersurface $Y : M \to E$. Put $x_0 := x$ and $x_1 := -Y$. We consider the one-parameter family of ovaloids in E defined by

$$x_t = x_0 + t x_1 : M \to E,$$

for $t \geq 0$ sufficiently small, so that x_t is an immersion. Then the volume enclosed by $M_t := x_t(M)$ is

$$\mathrm{Vol}(M_t) = \frac{1}{(n+1)!} \int_M \mathrm{Det}(dx_t, dx_t, \ldots, dx_t, -x_t) \tag{7.13}$$

$$= \int_M \sum_{k=0}^{n} \binom{n}{k} t^k (t\Psi_k + \phi_k).$$

Let $C(M_t)$ denote the convex body enclosed by M_t. Since $Y_i = -\sum B_i^j x_j$, $x(M)$ and $-Y(M)$ have the same Euclidean exterior unit normal vector at corresponding points. From the definition at the beginning of Section 7.2.1 we have

$$C(M_t) = C(M) + t C(-Y(M)).$$

Hence $\mathrm{Vol}(M_t)$ can be expressed in terms of the mixed volumes, namely

$$\mathrm{Vol}(M_t) = \sum_{k=0}^{n+1} \binom{n+1}{k} t^k V_k, \tag{7.14}$$

where

$$V_k := \mathrm{Vol}(\underbrace{M, \ldots, M}_{n-k}, \underbrace{-Y(M), \ldots, -Y(M)}_{k}).$$

Comparing (7.13) with (7.14) we have

$$\binom{n+1}{k}V_k = \int_M \left[\binom{n}{k-1}\Psi_{k-1} + \binom{n}{k}\phi_k\right] \text{ for } k = 1, 2, \ldots, n.$$

$$V_0 = \int_M \phi_0 .$$

Since

$$\frac{1}{(n+1)!} d\left[\mathrm{Det}(\underbrace{-dY,\ldots,-dY}_{k}, \underbrace{dx,\ldots,dx}_{n-k}, x, Y)\right] = \Psi_k - \phi_{k+1},$$

we get

$$\int_M \phi_k = \int_M \Psi_{k-1}.$$

Hence

$$V_k = \int_M \Psi_{k-1}.$$

On the other hand, from

$$\sum_{k=0}^{n} \binom{n}{k} t^k \Psi_k = \frac{1}{(n+1)!} \mathrm{Det}(d(x-tY),\ldots,d(x-tY),Y)$$

$$= \frac{1}{n+1} \sum_k \binom{n}{k} L_k t^k dV,$$

it follows that

$$V_{k+1} = \int_M \Psi_k = \frac{1}{n+1} \int_M L_k dV. \tag{7.15}$$

In particular,

$$V_1 = \frac{1}{n+1} S(M) \quad \text{and} \quad V_0 = \mathrm{Vol}(M).$$

Using these formulas, we can write the isoperimetric inequality (7.5) in the form

$$V_1^{n+1} \le \sigma_{n+1}^2 V_0^n, \tag{7.16}$$

where equality holds if and only if the ovaloid is an ellipsoid. The inequality (7.12) now reads

$$V_k^2 \ge V_{k-1} \cdot V_{k+1}. \tag{7.17}$$

We consider the function $\ln V_k$, $0 \le k \le n+1$, as a function of k. Then the inequality (7.17) states that the graph, formed by discrete points $(k, \ln V_k)$ with $0 \le k \le n+1$, is concave, while the inequality (7.16) shows that the point $(\frac{n+2}{2}, \ln \sigma_{n+1})$ does not lie below the chord joining two points $(0, \ln V_0)$ and $(1, \ln V_1)$. Because of the concavity condition, the chord joining any two points of the graph, at least one of which has an abscissa $k \leq \frac{n+2}{2}$, cannot pass above the point $(\frac{n+2}{2}, \ln \sigma_{n+1})$. In other words, for any two integers k, k' with $0 \leq k < k' \leq n + 1$ and $k < \frac{n+2}{2}$,

$$V_{k'}^{n+2-2k} V_k^{2k'-n-2} \leq \sigma_{n+1}^{2(k'-k)}. \tag{7.18}$$

In particular, when n is even and either k or k' equals $\frac{n+2}{2}$, it follows immediately that

$$V_{(n+2)/2} \leq \sigma_{n+1}. \tag{7.19}$$

If equalities in (7.18) and (7.19) hold then

$$V_1^{n+2} = \sigma_{n+1}^2 V_0^n, \tag{7.20}$$

and M is an ellipsoid (see (7.16)). Thus, from (7.15) we have the following theorem (see [43]):

Theorem 7.7. *Let M be an ovaloid in E. If $L_n > 0$ everywhere then for any two integers k, j with $0 \leq k < j \leq n + 1$ and $k < \frac{n+2}{2}$, we have*

$$\left(\int_M L_{j-1} dV \right)^{n+2-2k} \cdot \left(\int_M L_{k-1} dV \right)^{2j-n-2} \leq O_n^{2(j-k)},$$

with equality holding if and only if M is an ellipsoid, where

$$O_n = 2\pi^{(n+1)/2} \left(\Gamma \left(\frac{n+1}{2} \right) \right)^{-1}$$

is the n-dimensional volume of the Euclidean unit ball in E^{n+1}. In particular, when n is even,

$$\int_M L_{n/2} dV \leq O_n,$$

with equality holding if and only if M is an ellipsoid.

7.2.3 Total centroaffine area

Now we will generalize the inequality (7.8) to higher dimensions. As in Section 7.2.2, if we assume $\hat{B} = (B_{ij})$ to be positive definite everywhere, then $Y(M)$ is also an ovaloid; again we take $-Y$ to be the centroaffine normal vector of $Y(M)$ such that $\hat{B}$ is the centroaffine metric of $Y(M)$. Choose a local equiaffine frame field $\{x; e_1, \ldots, e_n, e_{n+1}\}$ such that $G_{ij} = \delta_{ij}$. Then the volume of $Y(M)$ with respect to the centroaffine metric is (see [341, Sects. 4.6 and 6.3])

$$\int_M \sqrt{\det(B_{ij})} \, \omega^1 \wedge \cdots \wedge \omega^n = \int_M \sqrt{L_n} \, dV.$$

From Hölder's inequality [13, p. 21],

$$\left(\frac{\int_M \sqrt{L_n}dV}{\int_M dV}\right)^2 \le \frac{\int_M L_n dV}{\int_M dV},$$

we have

$$\left(\int_M \sqrt{L_n}dV\right)^2 \le \int_M dV \cdot \int_M L_n dV$$

$$= (n+1)^2 V_1 V_{n+1}$$

$$= (n+1)^2 V_0^{-n} V_1^{n+2} \prod_{j=1}^{n} \left(\frac{V_{j-1}V_{j+1}}{V_j^2}\right)^{n+1-j}$$

$$\le (n+1)^2 V_0^{-n} V_1^{n+2}$$

$$\le (n+1)^2 \sigma_{n+1}^2 = O_n^2,$$

where we have used (7.17) and (7.20). As a consequence,

$$\int_M \sqrt{L_n}dV \le O_n.$$

Equality in this last relation implies that

$$V_1^{n+2} = \sigma_{n+1}^2 V_0^n.$$

As in (7.20) we conclude that $x(M)$ is an ellipsoid. We have proved the following theorem.

Theorem 7.8. *Let $x : M \to E$ be an ovaloid. If $L_n > 0$ on M, then the volume $\int_M \sqrt{L_n}dV$ of $Y(M)$ with respect to the centroaffine metric satisfies*

$$\int_M \sqrt{L_n}dV \le O_n$$

and equality holds if and only if $x(M)$ is an ellipsoid.

A. Basic concepts from differential geometry

As stated in the introduction we expect the reader to be familiar with basic facts from Euclidean hypersurface theory, Riemannian manifolds, and affine connections. We summarize these basic facts (except the Euclidean hypersurface theory; see Section 1.5 in this volume). In particular, we recall several ideas about tensor algebras, differential manifolds and some fundamental formulas and theorems in Riemann geometry. For details and proofs we refer the readers to [25, 49, 63, 106, 124, 140, 266, 345, 392, 409].

A.1 Tensors and exterior algebra

A.1.1 Tensors

Let V be an n-dimensional vector space over the real number field $\mathbb{R}$, V^* denotes its dual space, and $\langle \, , \rangle : V^* \times V \to \mathbb{R}$ the canonical scalar product. Then V^* is the space whose elements are linear functions from V to $\mathbb{R}$; classically, they are called covectors. Let $\{e_1, \ldots, e_n\}$ be a basis of V. Then there exists a unique dual basis $\{f^1, \ldots, f^n\}$ of V^* such that

$$f^i(e_j) = \delta^i_j,$$

where δ^i_j is the Kronecker symbol; this means that

$$\delta^i_j = \begin{cases} 1, & i = j, \\ 0, & i \neq j. \end{cases}$$

Definition A.1. A *tensor* of *contravariant order* p and *covariant order* q over V, or simply a (p, q)-tensor, is a multilinear map

$$\phi : \underbrace{V \times \cdots \times V}_{q} \times \underbrace{V^* \times \cdots \times V^*}_{p} \to \mathbb{R},$$

that is, a function ϕ that is linear in each variable separately.

Thus covectors are $(0,1)$-tensors, vectors in V are $(1,0)$-tensors, bilinear forms over V are $(0,2)$-tensors.
For $x \in V$ and $f^* \in V^*$ we considered above the standard scalar product $\langle \, , \rangle : V^* \times V \to \mathbb{R}$ such that $\langle f^*, x \rangle = f^*(x)$. Then $\langle \, , \rangle : V^* \times V \to \mathbb{R}$ is a $(1,1)$-tensor over V.
Denote by $\mathfrak{T}^q_p(V)$ the set of all (p, q)-tensors on V. With the obvious addition and scalar multiplication $\mathfrak{T}^q_p(V)$ is a vector space over $\mathbb{R}$.

Definition A.2. The *tensor product* $\alpha \otimes \beta$ of α in $\mathfrak{T}^q_p(V)$ and β in $\mathfrak{T}^s_r(V)$ is an element in $\mathfrak{T}^{q+s}_{p+r}(V)$, defined by

$$(\alpha \otimes \beta)(x_1, \ldots, x_q, x_{q+1}, \ldots, x_{q+s}, y^1, \ldots, y^p, y^{p+1}, \ldots, y^{p+r})$$
$$= \alpha(x_1, \ldots, x_q, y^1, \ldots, y^p) \, \beta(x_{q+1}, \ldots, x_{q+s}, y^{p+1}, \ldots, y^{p+r}),$$

where $x_i \in V$ and $y^j \in V^*$.

It follows from the definition that the tensor multiplication satisfies the following rules:
(i) $(\alpha_1 + \alpha_2) \otimes \beta = \alpha_1 \otimes \beta + \alpha_2 \otimes \beta$,
     $\alpha \otimes (\beta_1 + \beta_2) = \alpha \otimes \beta_1 + \alpha \otimes \beta_2$;

(ii) $(m\alpha) \otimes \beta = \alpha \otimes (m\beta) = m(\alpha \otimes \beta)$, for $m \in \mathbb{R}$;

(iii) $(\alpha \otimes \beta) \otimes \gamma = \alpha \otimes (\beta \otimes \gamma) =: \alpha \otimes \beta \otimes \gamma$.

Let $\{e_1, \dots, e_n\}$ be a basis of V, and $\{f^1, \dots, f^n\}$ its dual basis. By multilinearity, any (p, q)-tensor ϕ is completely determined by its n^{p+q} values $\phi(e_{i_1}, \dots, e_{i_q}, f^{j_1}, \dots, f^{j_p})$. Obviously,

$$f^{j_1} \otimes \cdots \otimes f^{j_q} \otimes e_{j_1} \otimes \cdots \otimes e_{j_p}(e_{k_1}, \dots, e_{k_q}, f^{l_1}, \dots, f^{l_p}) = \delta^{j_1}_{k_1} \cdots \delta^{j_q}_{k_q} \cdot \delta^{l_1}_{j_1} \cdots \delta^{l_p}_{j_p}. \tag{A.1}$$

Using (A.1) we can easily prove the following theorem.

Theorem A.3. *Let $\{e_1, \dots, e_n\}$ be a basis of V, and $\{f^1, \dots, f^n\}$ its dual basis. Then the set*

$$\{f^{j_1} \otimes \cdots \otimes f^{j_q} \otimes e_{j_1} \otimes \cdots \otimes e_{j_p} \mid 1 \le i_1, \dots, i_q \le n,\ 1 \le j_1, \dots, j_p \le n\}$$

is a basis of $\mathcal{T}^q_p(V)$ and $\dim(\mathcal{T}^q_p(V)) = n^{p+q}$.

From Theorem A.3, any (p, q)-tensor ϕ can be written in a unique way in the form

$$\phi = \sum_{i_1, \dots, i_q; j_1, \dots, j_p} \phi^{j_1 \cdots j_p}_{i_1 \cdots i_q} f^{i_1} \otimes \cdots \otimes f^{i_q} \otimes e_{j_1} \otimes \cdots \otimes e_{j_p}, \tag{A.2}$$

where

$$\phi^{j_1 \cdots j_p}_{i_1 \cdots i_q} = \phi(e_{i_1}, \dots, e_{i_q}, f^{j_1}, \dots, f^{j_p}). \tag{A.3}$$

The numbers $\phi^{j_1 \cdots j_p}_{i_1 \cdots i_q}$ are called the *components* of the tensor ϕ with respect to the basis $\{e_1, \dots, e_n\}$. The following gives the transformation law for the components of a tensor under a change of the basis in V. Let $\{\tilde{e}_i\}$ and $\{\tilde{f}^i\}$ be another pair of dual bases in V and V^*, respectively, so that

$$\tilde{e}_i = \sum a^j_i e_j, \quad \tilde{f}^i = \sum b^i_k f^k,$$

where

$$\sum a^k_i b^j_k = \delta^j_i.$$

Then multilinearity implies

$$\tilde{\phi}^{j_1 \cdots j_p}_{i_1 \cdots i_q} = \sum a^{k_1}_{i_1} \cdots a^{k_q}_{i_q} b^{j_1}_{l_1} \cdots b^{j_p}_{l_p} \phi^{l_1 \cdots l_p}_{k_1 \cdots k_q}.$$

It is a standard convention in differential geometry (the so-called *Einstein convention*) that indices repeated below and above indicate summation over the range of these indices, unless stated otherwise. Very often, the summation sign is dropped, and only the place of the same indices, once below and once above, indicates summation.

Let ϕ be a (p, q)-tensor and $k, m \in \mathbb{N}$ fixed. Then we can obtain a new $(p-1, q-1)$-tensor $C^m_k(\phi)$ from ϕ by setting, for $\{e_i\}, \{f^i\}$ being dual bases as above:

$$C^m_k(\phi)(x_1, \dots, x_{q-1}, y^1, \dots, y^{p-1}) = \sum_{i=1}^n \phi(x_1, \dots, x_{k-1}, e_i, x_{k+1}, \dots, x_{q-1}, y^1, \dots, y^{m-1}, f^i, y^{m+1}, \dots, y^{p-1}).$$

Obviously, $C^m_k(\phi)$ is well defined. $C^m_k(\phi)$ is called the *contraction* of ϕ over the k-th subscript and the m-th superscript.

Now we list the tensor operations in terms of the components relative to a basis $\{e_1, \dots, e_n\}$:

$$(\phi + \psi)^{j_1 \cdots j_q}_{i_1 \cdots i_p} = \phi^{j_1 \cdots j_q}_{i_1 \cdots i_p} + \psi^{j_1 \cdots j_q}_{i_1 \cdots i_p}; \quad (m\phi)^{j_1 \cdots j_q}_{i_1 \cdots i_p} = m\phi^{j_1 \cdots j_q}_{i_1 \cdots i_p}, \tag{A.4}$$

for $\phi, \psi \in \mathcal{T}_q^p(V)$, $m \in \mathbb{R}$;

$$(\phi \otimes \psi)_{i_1 \cdots i_{s+p}}^{j_1 \cdots j_{r+q}} = \phi_{i_1 \cdots i_s}^{j_1 \cdots j_r} \cdot \psi_{i_{s+1} \cdots i_{s+p}}^{j_{r+1} \cdots j_{r+q}}, \tag{A.5}$$

for $\phi \in \mathcal{T}_r^s(V)$, $\psi \in \mathcal{T}_q^p(V)$;

$$(C_k^m(\phi))_{i_1 \cdots i_{p-1}}^{j_1 \cdots j_{q-1}} = \sum_{\ell=1}^{n} \phi_{i_1 \cdots i_{k-1} \ell i_{k+1} \cdots i_p}^{j_1 \cdots j_{m-1} \ell j_{m+1} \cdots j_q}, \tag{A.6}$$

for $\phi \in \mathcal{T}_q^p(V)$.

A.1.2 Exterior algebra [106, Chap. 5]

Denote by G_p the permutation group formed by the permutations of p letters $(1, \ldots, p)$. Given any permutation $\sigma \in G_p$ and any $(0, p)$-tensor ϕ we define a new $(0, p)$-tensor $\sigma(\phi)$ by

$$\sigma(\phi)(x_1, \ldots, x_p) = \phi(x_{\sigma(1)}, \ldots, x_{\sigma(p)}). \tag{A.7}$$

Definition A.4. A $(0, p)$-tensor ϕ is called *symmetric* if $\sigma(\phi) = \phi$ for every permutation $\sigma \in G_p$; ϕ is called *skew-symmetric*, or *alternating*, if we have $\sigma(\phi) = (\operatorname{sgn} \sigma)\phi$ for every σ in G_p, where $\operatorname{sgn} \sigma$ is either 1 or -1 according to whether σ is even or odd, respectively.

We introduce the following two linear transformations on the vector space $\mathcal{T}_0^p(V)$:

$$S(\phi) = \frac{1}{p!} \sum_{\sigma \in G_p} \sigma(\phi), \tag{A.8}$$

$$A(\phi) = \frac{1}{p!} \sum_{\sigma \in G_p} \operatorname{sgn} \sigma \cdot \sigma(\phi), \quad \phi \in \mathcal{T}_0^p(V), \tag{A.9}$$

the summation being over all elements of G_p. For any ϕ, $S(\phi)$ is symmetric, and $A(\phi)$ is skew-symmetric. S and A are called the *symmetrization operator* and the *alternation operator*, respectively.

Denote by $\Lambda^p(V)$ the set of all skew-symmetric tensors of type $(0, p)$. We set $\Lambda^0(V) := \mathbb{R}$. For $p = 1$, we have $\Lambda^1(V) = V^*$. The set $\Lambda^p(V)$ forms a subspace of $\mathcal{T}_0^p(V)$. Note that $\Lambda^p(V) = 0$ for $p > n$, because of the skew-symmetry. The tensor product of alternating tensors is not, in general, an alternating tensor.

Definition A.5. The mapping from $\Lambda^p(V) \times \Lambda^q(V)$ to $\Lambda^{p+q}(V)$, defined by

$$\wedge: (\alpha, \beta) \mapsto \frac{(p+q)!}{p! \, q!} A(\alpha \otimes \beta)$$

is called the *exterior product* (or *wedge product*) of α and β, which is denoted by $\alpha \wedge \beta := \wedge(\alpha, \beta)$.

Theorem A.6. *The exterior multiplication $\wedge$ satisfies the following rules:*
(i) $(\alpha_1 + \alpha_2) \wedge \beta = \alpha_1 \wedge \beta + \alpha_2 \wedge \beta$,
    $\alpha \wedge (\beta_1 + \beta_2) = \alpha \wedge \beta_1 + \alpha \wedge \beta_2$;
(ii) $\alpha \wedge (\beta \wedge \gamma) = (\alpha \wedge \beta) \wedge \gamma =: \alpha \wedge \beta \wedge \gamma$;
(iii) $\alpha \wedge \beta = (-1)^{pq} \beta \wedge \alpha$,

where $\alpha, \alpha_1, \alpha_2 \in \Lambda^p(V)$, $\beta, \beta_1, \beta_2 \in \Lambda^q(V)$, $\gamma \in \Lambda^r(V)$.

Proof. We refer to [106, Chap. 5]. $\qquad \square$

Corollary A.7. *(i)* *If $\alpha, \beta \in \Lambda^1(V)$ then $\alpha \wedge \beta = -\beta \wedge \alpha$, $\alpha \wedge \alpha = 0$.*
(ii) *If $f^1, \ldots, f^p \in V^*$ and $x_1, \ldots, x_p \in V$ then we have*

$$f^1 \wedge f^2 \wedge \cdots \wedge f^p(x_1, \ldots, x_p) = p! A(f^1 \otimes \cdots \otimes f^p)(x_1, \ldots, x_p)$$

$$= \sum_{\sigma \in G_p} \operatorname{sgn} \sigma \cdot f^1(x_{\sigma(1)}) \cdots f^p(x_{\sigma(p)})$$

$$= \det \begin{vmatrix} f^1(x_1) & \cdots & f^1(x_p) \\ \vdots & & \vdots \\ f^p(x_1) & \cdots & f^p(x_p) \end{vmatrix}.$$

(iii) *If $\{e_1, \ldots, e_n\}$ is a basis of V and $\{f^1, \ldots, f^n\}$ its dual basis, then*

$$f^{i_1} \wedge \cdots \wedge f^{i_p}(e_{j_1}, \ldots, e_{j_p}) = \delta^{i_1 \cdots i_p}_{j_1 \cdots j_p},$$

where $\delta^{i_1 \cdots i_p}_{j_1 \cdots j_p}$ denotes the generalized Kronecker symbol, namely,

$$\delta^{i_1 \cdots i_p}_{j_1 \cdots j_p} = \begin{cases} 1, & \text{if } (j_1, \ldots, j_p) \text{ is an even permutation of } (i_1, \ldots, i_p); \\ -1, & \text{if } (j_1, \ldots, j_p) \text{ is an odd permutation of } (i_1, \ldots, i_p); \\ 0, & \text{otherwise.} \end{cases}$$

Theorem A.8. *Let $\{f^1, \ldots, f^n\}$ be a basis of V^*. Then the set*

$$\{f^{i_1} \wedge \cdots \wedge f^{i_p} \mid 1 \leq i_1 < i_2 < \cdots < i_p \leq n\}$$

is a basis of $\Lambda^p(V)$, $1 \leq p \leq n$, and $\dim(\Lambda^p(V)) = \binom{n}{p}$.

A student not so familiar with multilinear algebra should give a proof as exercise; apply Corollary A.7 (iii).

From Theorem A.8, any $\alpha \in \Lambda^p(V)$ can be written in a unique way in the form

$$\alpha = \sum_{i_1 < \cdots < i_p} a_{i_1 \cdots i_p} f^{i_1} \wedge \cdots \wedge f^{i_p} \tag{A.10}$$

where the summation is over all combinations $1 \leq i_1 < i_2 < \cdots < i_p \leq n$. An alternative representation of α can be written in a unique way in the form

$$\alpha = \sum b_{i_1 \cdots i_p} f^{i_1} \wedge \cdots \wedge f^{i_p} \tag{A.11}$$

where $1 \leq i_1, \ldots, i_p \leq n$, and the coefficients $b_{i_1 \cdots i_p}$ are skew-symmetric with respect to the subscripts.

Consider the direct sum of the vector spaces $\Lambda^p(V)$:

$$\Lambda(V) := \Lambda^0(V) \oplus \Lambda^1(V) \oplus \cdots \oplus \Lambda^n(V).$$

We define a product on $\Lambda(V)$ by extending the exterior product to be bilinear. Then $\Lambda(V)$ is an associative algebra of dimension 2^n, called the exterior algebra of V.

The following two theorems are fundamental for the method of moving frames; therefore we state them including their proofs.

Theorem A.9. *Let $f^1, f^2, \ldots, f^p \in V^*$. Then $f^1, \ldots, f^p$ are linearly dependent if and only if*

$$f^1 \wedge f^2 \wedge \cdots \wedge f^p = 0.$$

Proof. If $f^1, \ldots, f^p$ are linearly dependent, then we can express one of them, say f^p, as a linear combination of the others: $f^p = \sum_{i=1}^{p-1} a_i f^i$. Thus,

$$f^1 \wedge f^2 \wedge \cdots \wedge f^p = f^1 \wedge f^2 \wedge \cdots \wedge \sum_{i=1}^{p-1} a_i f^i = 0.$$

Conversely, if $f^1, \ldots, f^p$ are linearly independent, we can find $f^{p+1}, \ldots, f^n$ such that $\{f^1, \ldots, f^n\}$ form a basis of V^*. Let $\{e_1, \ldots, e_n\}$ be its dual basis in V. Then by Corollary A.7 (iii),

$$f^1 \wedge f^2 \wedge \cdots \wedge f^p(e_1, \ldots, e_p) = 1.$$

Therefore $f^1 \wedge f^2 \wedge \cdots \wedge f^p \neq 0.$ □

Theorem A.10 (Cartan's Lemma). *Let $f^1, \ldots, f^p \in V^*$ be p linearly independent linear functions on V, and let $g_1, \ldots, g_p$ be p linear functions on V. If*

$$\sum_{i=1}^{p} f^i \wedge g_i = 0, \tag{A.12}$$

then

$$g_i = \sum_{j=1}^{p} b_{ij} f^j, \quad for\ i = 1, \ldots, p,$$

with

$$b_{ij} = b_{ji}.$$

Proof. Choose $f^{p+1}, \ldots, f^n$ such that $\{f^1, \ldots, f^n\}$ is a basis of V^*. Then we can write

$$g_i = \sum_{j=1}^{p} b_{ij} f^j + \sum_{m=p+1}^{n} b_{im} f^m.$$

We insert this expression into (A.12) and obtain

$$\sum_{1 \leq i < j \leq p} (b_{ij} - b_{ji}) f^i \wedge f^j + \sum_{i \leq p < m} b_{im} f^i \wedge f^m = 0.$$

As the exterior products $f^i \wedge f^j$ are linearly independent, we arrive at the assertion. □

Since $V = (V^*)^*$, one can similarly define $\Lambda^p(V^*)$ and the exterior algebra $\Lambda(V^*)$, and the above theorems hold. The elements of $\Lambda^p(V^*)$ are usually called the multivectors of degree p or briefly p-vectors.

Remark A.11. Sometimes in the literature, $\Lambda^p(V^*)$ denotes the set of all skew-symmetric $(0, p)$-tensors, and $\Lambda^p(V)$ denotes the set of all p-vectors.

A.2 Differentiable manifolds

A.2.1 Differentiable manifolds and submanifolds

Definition A.12. Let M be a topological Hausdorff space with a countable basis of open sets. M is said to be a *topological manifold of dimension n* if for each point p of M there is an open neighborhood U of p which is homeomorphic to an open subset U' of $\mathbb{R}^n$. Letting $\varphi : U \to U'$ be this homeomorphism, we call (U, φ) a *coordinate chart* and U a *coordinate neighborhood* of p. For $q \in U$, we call the numbers $x^1(q), \ldots, x^n(q)$, given by $\varphi(q) =: (x^1(q), \ldots, x^n(q))$, the *local coordinates* of $q \in M$; the mappings $x^j \circ \varphi : U \to \mathbb{R}$ are called *coordinate functions*.

Definition A.13. An n-dimensional topological manifold M is called an *n-dimensional differentiable manifold* of class C^k if there is a collection of coordinate charts $(U_\alpha, \varphi_\alpha)$, where α belongs to some index set $\mathcal{U}$, such that the following conditions hold:

(I) $M = \bigcup U_\alpha$, i.e. $\{U_\alpha\}_{\alpha \in \mathcal{U}}$ is an open covering of M;

(II) for each α, β with $U_\alpha \cap U_\beta \neq \emptyset$, the map $\varphi_\beta \circ \varphi_\alpha^{-1}$ is an injective C^k-map from $\varphi_\alpha(U_\alpha \cap U_\beta)$ into $\mathbb{R}^n$;

(III) the collection $\{(U_\alpha, \varphi_\alpha)\}_{\alpha \in \mathcal{U}}$ is maximal relative to (II); this means, if φ is any homeomorphism mapping an open set U of M onto an open set in $\mathbb{R}^n$ such that, for each $(U_\alpha, \varphi_\alpha)$ with $U_\alpha \cap U \neq \emptyset$, and $\varphi \circ \varphi_\alpha^{-1}$ and $\varphi_\alpha \circ \varphi^{-1}$ are C^k maps from $\varphi_\alpha(U \cap U_\alpha)$ and $\varphi(U \cap U_\alpha)$ into $\mathbb{R}^n$, respectively, then it must be $(U, \varphi) \in \{(U_\alpha, \varphi_\alpha)\}_{\alpha \in \mathcal{U}}$.

Remark and Examples A.14. (i) $\varphi \in C^k(U)$ means that, for $\varphi : U \to \mathbb{R}$, the partial derivatives up to order k exist and are continuous, where $k = 1, 2, \ldots, \infty$; $\varphi \in C^0$ (resp. $\varphi \in C^\omega$) means that φ is continuous (resp. φ is real analytic).

(ii) A family $\{(U_\alpha, \varphi_\alpha)\}_{\alpha \in \mathcal{U}}$ of C^k-charts, satisfying (I) and (II) in Definition A.13, is called a C^k-*atlas of* M. The atlas is called *maximal*, if (III) is additionally satisfied. A maximal C^k-atlas on M is called a C^k-*structure* on M. Obviously, any C^k-atlas uniquely determines a maximal C^k-atlas on M which contains the given atlas; see [25, 63].

(iii) We give some examples of differentiable manifolds:

   (1) $\mathbb{R}^n$ is a C^∞-manifold. In fact, for an open covering with one chart we take $\mathbb{R}^n$ itself and let φ be the identity mapping:

   (2) the n-dimensional sphere S^n is a C^∞-manifold. Using stereographic projection from the north pole $(0, \ldots, 0, 1)$ we determine a coordinate neighborhood (U_N, φ_N). In the same way we determine by stereographic projection from the south pole $(0, \ldots, 0, -1)$ a neighborhood (U_S, φ_S). Then these two coordinate charts determine a C^∞-structure on S^n;

   (3) open submanifolds: let M be a C^k-manifold and D an open set of M. Suppose that $\{(U_\alpha, \varphi_\alpha)\}_{\alpha \in \mathcal{U}}$ defines the C^k-structure of M. Let $U'_\alpha = U_\alpha \cap D$, and let φ'_α be the restriction of φ_α to U'. Then $\{(U'_\alpha, \varphi'_\alpha)\}$ determine a C^k-structure on D. D, with this C^k-structure, is called an open submanifold of M.

Definition A.15. Let M and N be C^∞-manifolds of dimensions m and n, respectively. A map $F : M \to N$ is called a C^∞-*mapping* if for every $p \in M$ there exist coordinate neighborhoods (U, φ) of p and (V, ψ) of $F(p)$ with $F(U) \subset V$ such that $\psi \circ F \circ \varphi^{-1} : \varphi(U) \to \psi(V)$ is C^∞; this means that $F|_U : U \to V$ may be written in local coordinates $x^1, \ldots, x^m$ and $y^1, \ldots, y^n$ as a mapping from $\varphi(U)$ into $\psi(V)$ by C^∞-functions

$$y^1 = y^1(x^1, \ldots, x^m), \ldots, y^n = y^n(x^1, \ldots, x^m).$$

A C^∞-map $f : M \to \mathbb{R}$ is called a C^∞-*function* on M. Denote by $C^\infty(M)$ the set of all C^∞-functions on M.

Notation. $f_i = \partial_i f$, etc.

Definition A.16. A C^∞-mapping $F : M \to N$ between C^∞-manifolds is called a *diffeomorphism* if it is a homeomorphism and F^{-1} is C^∞. M and N are called *diffeomorphic* if there exists a diffeomorphism $f : M \to N$.

Now we proceed to define the concept of a tangent vector on M. Recall that a vector at a point in Euclidean space can be considered as an operator on C^∞-functions (directional derivative). This motivates the definition of tangent vectors on a manifold.

Definition A.17. Let M be a differentiable manifold and $p \in M$.
(i) Let $f_i : U_i \to \mathbb{R}$ be differentiable at $p \in U_i$, where $U_i \subset M$ is open for $i = 1, 2$. We define the equivalence relation as follows: $(f_1, U_1) \sim (f_2, U_2)$, if there exists an open set U, such that $p \in U \subset U_1 \cap U_2$ and $f_1|U \equiv f_2|U$. The equivalence class $[f] := [(f, V)]$ for $f : V \to \mathbb{R}$, $p \in V$, is called a (differentiable) germ of functions at p. Denote by $C^\infty(p)$ the set of equivalence classes; this set is an $\mathbb{R}$-algebra with unit element $1 \in C^\infty(p)$ where the operations are defined in an obvious way. It is a standard convention in the literature to denote an equivalence class again by f instead of $[f]$.
(ii) A *tangent vector* to M at p is a map $X_p : C^\infty(p) \to \mathbb{R}$ satisfying the two conditions

$$X_p(af + bg) = a(X_p f) + b(X_p g), \tag{A.13}$$
$$X_p(fg) = (X_p f)g(p) + f(p)(X_p g) \tag{A.14}$$

for all $f, g \in C^\infty(p)$ and all $a, b \in \mathbb{R}$.

Denote by $T_p M$ the set of all tangent vectors to M at p. $T_p M$ is a vector space with the following rules of addition and scalar multiplication; $T_p M$ is called the *tangent space* of M at p:

$$(X_p + Y_p)(f) = X_p(f) + Y_p(f), \quad X_p, Y_p \in T_p M, \ f \in C^\infty(p),$$
$$(aX_p)(f) = a(X_p(f)), \quad\quad\quad X_p \in T_p M, \ f \in C^\infty(p), \ a \in \mathbb{R}.$$

We remark that the definition of tangent vectors uses only the set $C^\infty(p)$; thus, if U is any open set of M containing p, then $T_p U$ and $T_p M$ can be naturally identified.

Definition A.18. Let M and N be smooth manifolds, $F : M \to N$ a smooth map. The *differential* of F at $p \in M$ is the map $F_* : T_p M \to T_{F(p)} N$ defined as follows. For any $X \in T_p M$ and $g \in C^\infty(F(p))$, we define

$$F_*(X)(g) := X(g \circ F).$$

We must prove that the map $F_*(X) : C^\infty(F(p)) \to \mathbb{R}$ is a tangent vector at $F(p)$, i.e. it satisfies (A.13) and (A.14). (A.13) is obvious. For (A.14) we have

$$F_*(X)(fg) = X(fg \circ F) = X[(f \circ F)(g \circ F)]$$
$$= X(f \circ F)g(F(p)) + f(F(p))X(g \circ F)$$
$$= F_*(X)(f)g(F(p)) + f(F(p))F_*(X)(g)$$

for $X \in T_p M$ and $f, g \in C^\infty(F(p))$, and so (A.14) is satisfied.

Theorem A.19. *(i) Let M and N be smooth manifolds, $F : M \to N$ be a smooth map. Then we have a homomorphism $F_* : T_p M \to T_{F(p)} N$.*
(ii) Let M, N and W be smooth manifolds. Let $F : M \to N$ and $G : N \to W$ be smooth maps. Then $(G \circ F)_* = G_* \circ F_*$.

Proof. (i) For $X, Y \in T_p M$, $a, b \in \mathbb{R}$, and $f \in C^\infty(F(p))$ we have

$$F_*(aX + bY)f = (aX + bY)(f \circ F) = aX(f \circ F) + bY(f \circ F)$$
$$= aF_*(X)f + bF_*(Y)f = [aF_*(X) + bF_*(Y)]f.$$

We leave the proof of (ii) to the reader. $\qquad\qquad\qquad\qquad\qquad\qquad\qquad\qquad\qquad\qquad\square$

Corollary A.20. *Let M be a smooth manifold of dimension n, and let (U, φ) be a coordinate chart containing p ∈ M. Then $\varphi_* : T_pM \to T_{\varphi(p)}\mathbb{R}^n$ is an isomorphism, therefore $\dim(T_pM) = n$.*

We identify $T_{\varphi(p)}\mathbb{R}^n$ with $\mathbb{R}^n$. Let $\frac{\partial}{\partial x^1}, \ldots, \frac{\partial}{\partial x^n}$ be the canonical basis vectors of $\mathbb{R}^n$. Then the vectors $\varphi_*^{-1}\left(\frac{\partial}{\partial x^1}\right), \ldots, \varphi_*^{-1}\left(\frac{\partial}{\partial x^n}\right)$ form a basis of T_pM, we call it a *coordinate frame*. Later we will drop the symbol φ_*^{-1} for convenience, that is, we will denote by $\frac{\partial}{\partial x^1}, \ldots, \frac{\partial}{\partial x^n}$ the coordinate frame relative to the local coordinate system $x^1, \ldots, x^n$. Sometimes one calls such a local basis a *Gauß basis field*; we simply write $\partial_i = \frac{\partial}{\partial x^i}$ if the coordinates $x^1, \ldots, x^n$ are given.

Let $F : M \to N$ be a smooth map, and let (U, φ) and (V, ψ) be coordinate charts on M and N with $F(U) \subset V$. Assume that F is given in local coordinates by

$$y^1 = y^1(x^1, \ldots, x^n), \ldots, y^m = y^m(x^1, \ldots, x^n).$$

A simple computation shows that, at $p \in U$,

$$F_*\left(\frac{\partial}{\partial x^j}\right) = \sum \frac{\partial y^i}{\partial x^j} \frac{\partial}{\partial y^i}, \quad \text{for } j = 1, 2, \ldots, n. \tag{A.15}$$

In particular, for two local coordinate systems $x^1, \ldots, x^n$ and $y^1, \ldots, y^n$ on a manifold M, we have the transformation rule

$$\frac{\partial}{\partial x^i} = \sum \frac{\partial y^j}{\partial x^i} \frac{\partial}{\partial y^j}. \tag{A.16}$$

The *rank* of the map F at p is defined to be the rank of the Jacobian matrix $(\frac{\partial y^j}{\partial x^i})$ in (A.15). It is easy to show that the rank is independent of the choice of coordinates.

Definition A.21. Let M and N be smooth manifolds of dimensions m and n, respectively. Let $F : M \to N$ be a smooth map. If, at each point p of M, F_* is injective, i.e. if the rank of F at p is m, then F is called an *immersion* of M into N and $F(M)$ is called an *immersed submanifold* of N. If F is an immersion, and if moreover F is an injective map, i.e. if $F(p) \neq F(q)$ for $p \neq q$, then F is called an *embedding* (or *imbedding*) of M into N and $F(M)$ is called an *embedded submanifold* of N. A C^∞-immersion from an open interval of $\mathbb{R}$ into M is called a C^∞-*curve* in M. If we consider a curve from a closed interval into M, we mean the restriction of a differentiable map of an open interval containing this closed interval.

A.2.2 Tensor fields on manifolds

Let M be a C^∞-manifold and assume $p \in M$. Denote by T_p^*M the dual space of T_pM. For any smooth function $f \in C^\infty(p)$ one can define an element $df_p \in T_p^*M$, called the *differential* of f at p, by the formula

$$df_p(X_p) = X_pf, \quad \text{for } X_p \in T_pM. \tag{A.17}$$

Let (U, φ) be a coordinate chart with local coordinates $x^1, \ldots, x^n$. Then the differentials $d(x^i \circ \varphi)$, $1 \leq i \leq n$, satisfy

$$d(x^i \circ \varphi)(\varphi_*^{-1}(\partial_j)) = \varphi_*^{-1}(\partial_j)(x^i \circ \varphi) = \partial_j(x^i) = \delta_j^i \tag{A.18}$$

for any $p \in U$. It is usual to drop the symbol φ. (A.18) means that $dx^1, \ldots, dx^n$ is the dual basis of $\partial_1, \ldots, \partial_n$ for any $p \in U$.

Let $\Phi_p \in \mathfrak{T}_s^r(T_pM)$; that means, Φ_p is an (s, r)-tensor over the $\mathbb{R}$-vector space T_pM. With respect to a local coordinate system at p and its Gauß basis $\partial_1, \ldots, \partial_n$, Φ_p has a representation:

$$\Phi_p = \sum \Phi_{j_1 \cdots j_r}^{i_1 \cdots i_s}(p) dx^{j_1} \otimes \cdots \otimes dx^{j_r} \otimes \partial_{i_1} \otimes \cdots \otimes \partial_{i_s}|_p.$$

We call $\Phi : p \mapsto \Phi_p$ a differentiable (s, r)-tensor field on M, if, for given local coordinates, the coefficients $\Phi^{i_1 \cdots i_s}_{j_1 \cdots j_r}$ are differentiable functions. It is easy to see that the coefficients are differentiable with respect to an arbitrary coordinate system, if they are differentiable with respect to one system. Denote by $\mathcal{T}^r_s(M)$ the set of all smooth (s, r)-tensor fields on M, and in particular

$$\mathcal{T}^r(M) := \mathcal{T}^r_0(M), \quad \mathcal{T}_s(M) := \mathcal{T}^0_s(M), \quad \mathcal{T}^0_0(M) =: C^\infty(M).$$

Then $\mathcal{T}(M) := \bigotimes^\infty_{r,s=0} \mathcal{T}^r_s(M)$ together with the tensor product is an associative algebra over $\mathbb{R}$. Usually one calls $(1,0)$-tensor fields also vector fields, and $(0,1)$ tensor fields also covector fields or 1-forms. We denote by $\Gamma(M) := \mathcal{T}^0_1(M)$ the set of all smooth vector fields on M. The set $\Gamma(M)$ is a real vector space under the natural addition and multiplication. For $X, Y \in \Gamma(M)$, the bracket $[X, Y]$ is a vector field on M defined by

$$[X, Y]_p f = X_p(Yf) - Y_p(Xf), \quad \text{for all } f \in C^\infty(M). \tag{A.19}$$

Theorem A.22. *The set $\Gamma(M)$ with the bracket-operation $[\,,\,]$ is a Lie algebra, that is, the operation $[\,,\,]$*
(i)   *is bilinear over $\mathbb{R}$;*
(ii) *is skew-symmetric, which means*

$$[X, Y] = -[Y, X], \quad \text{for all } X, Y \in \Gamma(M);$$

(iii) *satisfies the Jacobi-identity*

$$[X, [Y, Z]] + [Y, [Z, X]] + [Z, [X, Y]] = 0.$$

Let $F : M \to N$ be a smooth map between smooth manifolds. F induces a map $F^* : \mathcal{T}^r(N) \to \mathcal{T}^r(M)$, defined for $\phi \in \mathcal{T}^r(N)$ by

$$(F^*\phi)_p(X_{1p}, \ldots, X_{rp}) = \phi_p(F_*(X_{1p}), \ldots, F_*(X_{rp})). \tag{A.20}$$

We call $F^*\phi$ the *pull back* to M of the covariant tensor field ϕ on N.

Definition A.23. An alternating $(0, r)$-tensor field on M is called an *exterior differential form* of degree r, or simply an *r-form*.

Denote by $\Lambda^r(M)$ the set of all smooth exterior differential form of degree r. Set (see (A.11))

$$\Lambda(M) = \Lambda^0(M) \oplus \Lambda^1(M) \oplus \cdots \oplus \Lambda^n(M),$$

where $\Lambda^0(M) := C^\infty(M)$. With respect to the exterior multiplication $\wedge$, where $(\alpha \wedge \beta)_p := \alpha_p \wedge \beta_p$ for $\alpha, \beta \in \Lambda(M), p \in M$, $\Lambda(M)$ is an associative algebra, called the exterior algebra on M.
    The proof of the following theorem is immediate.

Theorem A.24. *Let $F : M \to N$ be a smooth mapping. Then $F^* : \Lambda(N) \to \Lambda(M)$ is an algebra homomorphism.*

We introduce the operator $d : \Lambda(M) \to \Lambda(M)$ of exterior differentiation [25, p. 218].

Theorem A.25. *Let M be a smooth manifold of dimension n. Then there exists a unique linear mapping $d : \Lambda(M) \to \Lambda(M)$, called the exterior differential, such that:*
(i)   *$d : \Lambda^k(M) \to \Lambda^{k+1}(M)$;*
(ii) *$d(f) := df$ (ordinary differential) for $f \in C^\infty(M)$;*

(iii) *if $\alpha \in \Lambda^r(M)$ and $\beta \in \Lambda^s(M)$, then*

$$d(\alpha \wedge \beta) = (d\alpha) \wedge \beta + (-1)^r \alpha \wedge d\beta;$$

(iv) $d^2 = 0$.

Remark A.26. We frequently use the following local expression for forms: let $\omega \in \Lambda^k(M)$ and (U, φ) be a local coordinate chart with local coordinates $\{x^1, \ldots, x^n\}$. Then the restriction of ω to U can be expressed as (we use the notation ω also for the restriction)

$$\omega = \sum_{1 \leq i_1 < \cdots < i_k \leq n} a_{i_1 \cdots i_k} dx^{i_1} \wedge \cdots \wedge dx^{i_k},$$

where the coefficients $a_{i_1 \cdots i_k}$ are C^∞-functions on U. Note that

$$d\omega = \sum_{i_1 < \cdots < i_k} \sum_j \partial_j(a_{i_1 \cdots i_k}) dx^j \wedge dx^{i_1} \wedge \cdots \wedge dx^{i_k}.$$

Proposition A.27. *Let $\theta \in \Lambda^1(M)$ and $X, Y \in \Gamma(M)$, then*

$$d\theta(X, Y) = X(\theta(Y)) - Y(\theta(X)) - \theta([X, Y]).$$

Remark A.28. For the analogous result, when $\theta \in \Lambda^p(M)$, see [124, Satz 12.5].

Theorem A.29. *Let M and N be smooth manifolds and let $F : M \rightarrow N$ be a smooth mapping. Then*

$$d \circ F^* = F^* \circ d.$$

Proof. Since both F^* and d are linear it is sufficient to establish $F^* \circ d = d \circ F^*$ on forms of the type $\omega = f dx^{i_1} \wedge \cdots \wedge dx^{i_k}$ in a local coordinate neighborhood U. We proceed by induction on the degree of the forms. For $f \in C^\infty(N)$ we have

$$F^*(df) = d(F^* f),$$

namely:

$$F^*(df)(X_p) = df(F_* X_p) = (F_* X_p)f = X_p(f \circ F) = X_p(F^* f) = d(F^* f)(X_p),$$

for $p \in U, X_p \in T_p M$.

Assume the theorem to be true for all forms of degree less than k and let ω be a form of degree k of the type above. Let $\omega_1 = f dx^{i_1}$ and $\omega_2 = dx^{i_2} \wedge \cdots \wedge dx^{i_k}$. Then $\omega = \omega_1 \wedge \omega_2$, and $d\omega_2 = 0$. We have

$$dF^*(\omega_1 \wedge \omega_2) = d[(F^* \omega_1) \wedge (F^* \omega_2)]$$
$$= (dF^* \omega_1) \wedge (F^* \omega_2) - (F^* \omega_1) \wedge (dF^* \omega_2)$$
$$= F^*(d\omega_1) \wedge F^* \omega_2 = F^*(d\omega_1 \wedge \omega_2) = F^* d(\omega_1 \wedge \omega_2). \qquad \square$$

A.2.3 Integration on manifolds ([103, Sect. 1.6]; [124, § 19])

Definition A.30. A differentiable manifold M of dimension n is called *orientable* if there is a continuous exterior form of degree n which is nowhere zero on M. M is called *oriented* if such an n-form is given.

Two forms define the same orientation if they differ from each other by a factor which is positive on M.

Obviously, a connected orientable manifold has exactly two possible orientations.

Definition A.31. The *support* of a real function f (resp. exterior differential form ω) on M, denoted by supp(f) (resp. supp(ω)), is the topological closure of the set of points of M at which f (resp. ω) is not equal to zero (resp. nontrivial).

Definition A.32. An open covering of M is called *locally finite* if any compact subset of M meets only a finite number of its elements.

Theorem A.33. *Let $\mathcal{Z}$ be a base for the topology of M. Then there is a locally finite open covering of M whose elements are in $\mathcal{Z}$.*

Theorem A.34 (Partition of unity). *Let $\{N_\alpha\}$ be an open covering of a smooth manifold M. Then there are smooth functions $\{g_\alpha\}$ satisfying the following conditions:*
(i)   *for each α we have $0 \le g_\alpha \le 1$, and supp(g_α) is compact and contained in N_α;*
(ii) *every point of M has a neighborhood which meets only finitely many sets of the family $\{$supp(g_α)$\}$;*
(iii) $\sum g_\alpha = 1$.

The family $\{g_\alpha\}$ is called a partition of unity subordinate to the covering $\{N_\alpha\}$ (see [140, vol. I, App. 3], or [87, Chap. 8.3]; see also [63] or [30, Sect. 3.4]).

We now define the *integral* of an n-form σ with compact support supp(σ). Choose an open covering $\{N_\alpha\}$ of M such that each N_α is a coordinate neighborhood. Let $\{g_\alpha\}$ be a corresponding partition of unity. Then

$$\sigma = \Big(\sum_\alpha g_\alpha \Big)\sigma = \sum_\alpha (g_\alpha \sigma).$$

Obviously, supp($g_\alpha\sigma$) is contained in supp(g_α), and supp(g_α) lies in a coordinate neighborhood N_α. In terms of a local coordinate system $(x^1,\ldots,x^n)$ on N_α, we can write

$$g_\alpha \sigma = f(x^1,\ldots,x^n)dx^1 \wedge \cdots \wedge dx^n,$$

where $f(x^1,\ldots,x^n)$ is a differentiable function on N_α. We define the integral of $(g_\alpha\sigma)$ by:

$$\int_M g_\alpha\sigma := \int_{N_\alpha} g_\alpha\sigma = \int_{N_\alpha} f(x^1,\ldots,x^n)dx^1 \wedge \cdots \wedge dx^n. \tag{A.21}$$

where the right-hand side is a usual Riemannian integral.

It is routine to show that the integral is well defined, which means one has to show that the definition is independent of the choice of the coordinate neighborhood N_α which contains the support supp(g_α); this follows from the rules for the transformation of multiple integrals.

In a next step, we define the integral of σ over M. According to Theorem A.33 we know that any point p in supp(σ) has an open coordinate neighborhood V_p which meets only finitely many sets of $\{$supp(g_α)$\}$. These sets V_p, for all p in supp(σ), form an open covering of supp(σ). Because supp(σ) is compact, it has a finite open subcovering. Therefore, there are only a finite number of $g_\alpha\sigma$ which are not identically zero. We define

$$\int_M \sigma = \sum_\alpha \int_M g_\alpha\sigma,$$

where the right-hand side is a finite sum. Because each $g_\alpha\sigma$ has a support lying in a coordinate neighborhood, we can evaluate $\int_M \sigma$ according to (A.21).

Again, it is routine to show that the definition of the integral is independent of the choice of the covering $\{N_\alpha\}$ and the subordinate partition of unity $\{g_\alpha\}$.

Now we introduce the notion of a manifold with boundary. Consider

$$H^n := \{x = (x^1,\ldots,x^n) \in \mathbb{R}^n \mid x^n \ge 0\}$$

with the relative topology of $\mathbb{R}^n$.

Definition A.35. A C^∞-*manifold with boundary* is a Hausdorff space M with a countable basis of open sets and a differentiable structure $\mathcal{U}$ in the following sense: $\mathcal{U} = \{U_\alpha, \varphi_\alpha\}$ consists of a family of open subsets U_α of M each with a homeomorphism φ_α onto an open subset of H^n such that
(i) $M = \cup U_\alpha$;
(ii) if $U_\alpha \cap U_\beta \neq \emptyset$, then $\varphi_\beta \circ \varphi_\alpha^{-1}$ and $\varphi_\alpha \circ \varphi_\beta^{-1}$ are diffeomorphisms of $\varphi_\alpha(U_\alpha \cap U_\beta)$ and $\varphi_\beta(U_\alpha \cap U_\beta)$;
(iii) $\mathcal{U}$ is maximal with respect to the properties (i) and (ii).

The points of M which have a neighborhood homeomorphic to an open set of $\mathbb{R}^n$ are called *interior points*. The other points are called *boundary points*. We denote by ∂M the set of boundary points.

   One verifies that if $\partial M \neq \emptyset$, then ∂M is a C^∞-manifold of dimension $(n-1)$ without boundary and the inclusion $i : \partial M \to M$ is an imbedding; if M is oriented, then ∂M is orientable, and an orientation of M induces a natural orientation of ∂M. Let $p \in \partial M$, and let U be a coordinate neighborhood of p such that $x^i(p) = 0$ for all i and for all points of $U \cap M, x^n \geq 0$. If $dx^1 \wedge \cdots \wedge dx^n$ defines the orientation for M, then $(-1)^n dx^1 \wedge \cdots \wedge dx^{n-1}$ determines the induced orientation for ∂M.

Theorem A.36 (Stokes' Formula). *Let M be an oriented, compact C^∞-manifold of dimension n and let ω be a smooth $(n-1)$-form on M. Then*

$$\int_M d\omega = \int_{\partial M} \omega.$$

When $\partial M = \emptyset$, the integral over M vanishes. It is standard to write $\int_{\partial M} \omega$ instead of $\int_{\partial M} i^* \omega$.

For the proof see [25].

A.3 Affine connections and Riemannian geometry: Basic facts

In this section we state some fundamental concepts in terms of moving frames; see [63, 103, 120, 121, 124, 140].

A.3.1 Affine connections

We start with the standard definition of affine connections from the so-called invariant view point.

Definition A.37 (Affine connection - invariant view point). A mapping

$$\nabla : \Gamma(M) \times \Gamma(M) \to \Gamma(M),$$

is called an *affine connection*, if ∇ satisfies the following conditions:
(i) ∇ is $C^\infty(M)$-linear in the first argument;
(ii) ∇ is $\mathbb{R}$-linear in the second argument;
(iii) for $X, Y \in \Gamma(M)$ and $f \in C^\infty(M)$, ∇ satisfies

   (Leibniz' rule) $\nabla_X fY = (Xf)Y + f\nabla_X Y.$

A.3.1.1 Covariant differentiation

One can extend ∇ to a mapping:

$$\nabla : \Gamma(M) \times \mathfrak{T}(M) \to \mathfrak{T}(M),$$

where $\mathfrak{T}(M)$ is the algebra of tensor fields on M, satisfying the following properties:
(i) ∇ is $C^\infty(M)$-linear in the first argument;
(ii) $\nabla_X f = (Xf)$ for $f \in C^\infty(M) \subset \mathfrak{T}(M)$;
(iii) ∇_X is type preserving, which means: for $T \in \mathfrak{T}_s^r(M)$ we have
      $\nabla_X T \in \mathfrak{T}_s^r(M)$;
(iv) $\nabla_X(T \otimes \hat{T}) = (\nabla_X T) \otimes \hat{T} + T \otimes \nabla_X \hat{T}$;
(v) ∇_X commutes with every contraction.

From the foregoing we define the *covariant derivative* ∇T of $T \in \mathfrak{T}(M)$: if $T \in \mathfrak{T}_s^r(M)$ then $\nabla T \in \mathfrak{T}_{s+1}^r(M)$, namely

$$(\nabla_X T)(X_1,\ldots,X_s) := \nabla_X(T(X_1,\ldots,X_s)) - \sum_{i=1}^{s} T(X_1,\ldots,\nabla_X X_i,\ldots,X_s),$$

where $(\nabla_X T)(X_1,\ldots,X_s) \in \mathfrak{T}_0^r(M)$, and

$$(\nabla T)(X_1,\ldots,X_s;X) := (\nabla_X T)(X_1,\ldots,X_s).$$

We call T *parallel* with respect to the affine connection ∇ if the covariant derivative ∇T vanishes identically on M.

It is well known that the operation of ∇_X on $\mathfrak{T}(M)$ is completely determined by its operation on the algebra of functions $C^\infty(M)$ and the module of vector fields $\Gamma(M)$; see [140, vol. I, Chap. 3.2].

Obviously, the concepts of affine connection and covariant differentiation generalize the concept of directional derivative in $\mathbb{R}^n$.

A.3.1.2 Frame fields

Let $U \subset M$ be an open subset. An n-tuple

$$\{e_1,\ldots,e_n\} \subset \Gamma(U)$$

is called a *local frame field* on U if $e_1,\ldots,e_n$ are linearly independent at any $p \in U$. A *local coframe field* $\{\omega^1,\ldots,\omega^n\}$ on U is given by an n-tuple $\{\omega^1,\ldots,\omega^n\} \subset \mathfrak{T}_0^1(U)$ such that $\omega^1,\ldots,\omega^n$ are linearly independent at any $p \in U$. Two frame fields are called *dual* on U if they are dual at any $p \in U$.

A.3.1.3 Affine connections – Cartan's viewpoint

Let ∇ be an affine connection on M, and $U \subset M$ be open with a local frame field $\{e_1,\ldots,e_n\}$ and its dual coframe field $\{\omega^1,\ldots,\omega^n\}$; let $X \in \Gamma(U)$. Then the relation

$$\nabla_X e_i = \sum \omega_i^j(X) e_j$$

defines n^2 covector fields ω_i^j on U:

$$\omega_i^j(X) = \omega^j(\nabla_X e_i).$$

Thus, an affine connection assigns to each local frame field n^2 one-forms; they are called *connection forms*.

It is obvious that a change of the frame determines the transformation of the connection forms; moreover, n^2 one-forms which satisfy this transformation rule determine an affine connection.

The local representation of ∇ with respect to the frame defines the so-called *Christoffel symbols* Γ_{ij}^k

$$\nabla_{e_j} e_i = \sum \Gamma_{ij}^k e_k.$$

The Christoffel symbols and the connection forms are related by

$$\omega_i^j = \sum \Gamma_{ik}^j \omega^k, \text{ or } \omega_i^j(e_k) = \Gamma_{ik}^j.$$

From A.3.1.1 one can express the covariant derivative ∇T of a tensor via the connection forms, or the Christoffel symbols, respectively. Denote by

$$T_{j_1 \cdots j_s, l}^{i_1, \ldots, i_r}$$

the components of ∇T with respect to the local frame field $\{e_1, \ldots, e_n\}$ on U; then

$$T_{j_1 \cdots j_s, l}^{i_1 \cdots i_r} = e_l T_{j_1 \cdots j_s}^{i_1 \cdots i_r} - \sum T_{j_1 \cdots j_{k-1} m j_{k+1} \cdots j_s}^{i_1 \cdots i_r} \Gamma_{j_k l}^m + \sum T_{j_1 \cdots j_k}^{i_1 \cdots i_{k-1} m i_{k+1} \cdots i_r} \Gamma_{ml}^{i_k},$$

or equivalently

$$\sum T_{j_1 \cdots j_s, l}^{i_1 \cdots i_r} \omega^l = dT_{j_1 \cdots j_s}^{i_1 \cdots i_r} - \sum T_{j_1 \cdots j_{k-1} m j_{k+1} \cdots j_s}^{i_1 \cdots i_r} \omega_{j_k}^m + \sum T_{j_1 \cdots j_s}^{i_1 \cdots i_{k-1} m i_{k+1} \cdots i_r} \omega_m^{i_k}.$$

A.3.1.4 Torsion tensor and curvature tensor

For a given affine connection ∇ on M, one defines
(i)   the *torsion tensor* T: $T(X, Y) := \nabla_X Y - \nabla_Y X - [X, Y]$;
(ii) the *curvature tensor* R: $R(X, Y)Z := \nabla_X \nabla_Y Z - \nabla_Y \nabla_X Z - \nabla_{[X,Y]} Z$.

In terms of a local frame field such that $e_j = \partial_j$, these tensors have the following representations:

$$T(e_j, e_k) = \sum T_{jk}^i e_i, \quad R(e_k, e_l)e_j = \sum R_{jkl}^i e_i,$$

where

$$T_{jk}^i = \Gamma_{jk}^i - \Gamma_{kj}^i,$$

$$R_{jkl}^i = \partial_k \Gamma_{lj}^i - \partial_l \Gamma_{kj}^i + \sum \left\{ \Gamma_{lj}^m \Gamma_{km}^i - \Gamma_{kj}^m \Gamma_{lm}^i \right\}.$$

Moreover, T and R locally induce 2-forms T^i and Ω_j^i in the following way:

$$T(X, Y) =: \sum T^i(X, Y)e_i,$$

$$R(X, Y)e_j =: \sum \Omega_j^i(X, Y)e_i.$$

The induced forms T^i and Ω_j^i and the local coefficients of T and R satisfy the following relations:

$$\Omega_j^i = \frac{1}{2} \sum R_{jkl}^i \omega^k \wedge \omega^l,$$

$$T^i = \frac{1}{2} \sum T_{jk}^i \omega^j \wedge \omega^k,$$

and the local coefficients satisfy the skew-symmetry relations:

$$R_{jkl}^i = -R_{jlk}^i; \quad T_{jk}^i = -T_{kj}^i.$$

The 2-forms T^i and Ω_j^i are called *torsion forms* and *curvature forms*, respectively. From the definitions, the forms and the local coefficients satisfy

$$T_{jk}^i = T(e_j, e_k)(\omega^i), \quad T^i = \sum T_{jk}^i dx^j \otimes dx^k,$$
$$R_{jkl}^i = \Omega_j^i(e_k, e_l), \quad \Omega_j^i = \sum R_{jkl}^i dx^k \otimes dx^l.$$

A.3.1.5 Cartan structure equations
The forms from Sections A.3.1.3–A.3.1.4 are related by the Cartan structure equations:

$$d\omega^j = \sum \omega^k \wedge \omega_k^j + T^j,$$
$$d\omega_i^j = \sum \omega_i^k \wedge \omega_k^j + \Omega_i^j.$$

A.3.1.6 Torsion-free connections
An affine connection ∇ is called *torsion-free* (or *symmetric*) if $T \equiv 0$ on M. Equivalently, for any local representation, we have $\Gamma_{ij}^k = \Gamma_{ji}^k$ (symmetry).

A.3.1.7 Bianchi identities
The curvature tensor satisfies the following two (cyclic) identities, which we will need only for torsion-free connections:

(i) $R(X, Y)Z + R(Y, Z)X + R(Z, X)Y = 0$; locally:

$$R_{jkl}^i + R_{klj}^i + R_{ljk}^i = 0;$$

(ii) $(\nabla_X R)(Y, Z) + (\nabla_Y R)(Z, X) + (\nabla_Z R)(X, Y) = 0$; locally:

$$R_{jkl,m}^i + R_{jlm,k}^i + R_{jmk,l}^i = 0.$$

Both relations follow from exterior differentiation of the Cartan structure equations.

A.3.1.8 Ricci identities
For a given affine connection ∇, we define the second covariant derivative of an (r, s)-tensor field T to be an $(r, s + 2)$-tensor field, repeating the procedure form A.3.1.1:

$$(\nabla^2 T)(X_1, \ldots, X_s; X; Y) := (\nabla_Y(\nabla T))(X_1, \ldots, X_s; X).$$

Then $\nabla^2 T = \nabla(\nabla T)$ satisfies

$$(\nabla^2 T)(X_1, \ldots, X_s; X; Y) = \nabla_Y(\nabla_X T)(X_1, \ldots, X_s) - (\nabla_{\nabla_Y X} T)(X_1, \ldots, X_s).$$

Similarly, one can define the covariant derivatives of higher order, satisfying similar relations. From these and the definition of the curvature tensor, one gets the so-called Ricci identities which generalize the classical theorem of H. A. Schwarz for the partial derivatives. We write them in local notation; (see [89, p. 30]):

$$T_{j_1 \cdots j_s, kl}^{i_1 \cdots i_r} - T_{j_1 \cdots j_s, lk}^{i_1 \cdots i_r} = \sum_{\alpha=1}^{s} T_{j_1 \cdots j_{\alpha-1} p j_{\alpha+1} \cdots j_s}^{i_1 \cdots i_r} R_{j_\alpha kl}^p - \sum_{\beta=1}^{r} T_{j_1 \cdots j_s}^{i_1 \cdots i_{\beta-1} q i_{\beta+1} \cdots i_r} R_{qkl}^{i_\beta};$$

here the components on the left hand side belong to $\nabla^2 T$.

A.3.1.9 Connections and volume forms
Let ∇ be an affine connection. A volume form ω is an exterior n-form on M (see Section A.2.3). ω is called *parallel* with respect to ∇ if the covariant derivative $\nabla\omega$ vanishes identically on M:

$$\nabla\omega = 0.$$

A.3.1.10 Ricci curvature

Let ∇ be an affine connection. The *Ricci tensor field* Ric is a (0,2)-tensor field defined as follows:

$$\text{Ric}(X, Y) := \text{tr}\{Z \to R(Z, X)Y\}.$$

In terms of local coordinates it is usual to write

$$R_{ij} := \text{Ric}(\partial_i, \partial_j), \quad \text{thus } R_{ij} = \sum R^k_{jki}.$$

It is easy to prove the following: The Ricci tensor of ∇ is symmetric, $\text{Ric}(X, Y) = \text{Ric}(Y, X)$, if and only if ∇ admits a parallel volume form ω (see [311, p. 99]).

A.3.1.11 Covariant Hessian

The *covariant Hessian* of a function $f \in C^\infty(M)$ with respect to a given connection ∇ is defined to be

$$(\text{Hess } f)(X, Y) := Y(Xf) - (\nabla_Y X)f;$$

obviously the Hessian is a (0,2)-tensor field, the second covariant derivative of the function. It is symmetric if and only if ∇ is torsion free.

A.3.2 Riemann manifolds

Definition A.38. Let M be a connected, oriented C^∞-manifold of dimension $\dim M = n \geq 2$ and G a symmetric (0,2)-tensor field. If G is nondegenerate (positive definite), then (M, G) is called a *semi-Riemann (Riemann) manifold*. Semi-Riemann manifolds are also called *pseudo-Riemann manifolds*. G is called the (Riemann) metric.

With respect to a local frame field $\{e_1, \dots, e_n\}$ and its dual frame field $\{\omega^1, \dots, \omega^n\}$ we can express the metric G by

$$G = \sum G_{ij}\omega^i \otimes \omega^j.$$

We denote by (G^{ij}) the inverse matrix of (G_{ij}) such that $\sum G^{ik}G_{kj} = \delta^i_j$. It is a standard convention in the local notation to raise and lower indices; these reversible processes relate $(r + 1, 1)$-fields and $(r, s + 1)$-fields in the following way. We choose $r = 1, s = 2$ for the sake of simplicity:

$$\phi^k_{ij} = \sum \phi_{ijp}G^{pk}, \quad \phi_{ijk} = \sum \phi^p_{ij}G_{pk}.$$

With a few exceptions we consider Riemanian metrics (where G is positive definite).

A local frame field $\{e_1, \dots, e_n\}$ is called *orthonormal* (with respect to the metric G) if it satisfies $G(e_i, e_j) = \delta_{ij}$. Obviously one can apply the Gram-Schmidt-orthogonalization to an arbitrary local frame field to construct an orthonormal local frame field.

A Riemannian metric G on M induces at each $p \in M$ an inner product on the tangent vector space; thus T_pM carries a Euclidean structure (analogously, a semi-Riemannian metric induces a pseudo-Euclidean structure). Denote by

$$\|X\| := |G(X, X)|^{1/2}$$

the induced norm, and by

$$L := \int_a^b \left\|\frac{dC}{dt}\right\| dt$$

the arc length of the C^∞-curve $C : [a, b] \to M$, where $\frac{d}{dt}$ denotes the tangent vector at $t \in \mathbb{R}$ and

$$\frac{dC}{dt} := C_*\left(\frac{d}{dt}\right).$$

Within a coordinate neighborhood U with coordinate system $(x^1, \ldots, x^n)$ the curve is given by $(x^1(t), \ldots, x^n(t))$, and so the arc length from $C(a)$ to $C(t)$ is given by

$$s = L(t) = \int_a^t \sqrt{\sum G_{ij}\frac{dx^i}{dt}\frac{dx^j}{dt}}\, dt.$$

One can easily verify that the value of the integral is independent of the parametrization. The arc length along the curve from $C(a)$ to $C(t)$, denoted by $s = L(t)$, gives a canonical parameter. We define the *distance* of two points $p, q \in M$ to be the infimum

$$d(p, q) := \inf L(c)$$

of the arc lengths of all piecewise differentiable curves connecting p and q. Then (M, d) is a metric space and the topology of (M, d) is the topology of the given manifold M (cf. [25], p.187–189).

A.3.2.1 Riemann volume

On an oriented Riemann manifold M there is a nowhere vanishing n-form defined by

$$dV = \sqrt{\det(G_{ij})}\, \omega^1 \wedge \cdots \wedge \omega^n.$$

It is easy to show that dV is independent of the choice of frames that is, dV is a global n-form, called the *oriented Riemann volume element*. We can use dV to define the integral of functions. Let f be a C^∞-function on M with compact support. The integral of f over M is denoted by $\int_M f dV$.

A.3.2.2 Fundamental theorem of Riemann geometry

For a given Riemann manifold (M, G) there exists a unique connection ∇ satisfying two properties:
(a) $\nabla_Z G(X, Y) = G(\nabla_Z X, Y) + G(X, \nabla_Z Y)$, for $X, Y, Z \in \Gamma(M)$;
(b) ∇ is torsion-free.

This unique connection ∇ is called the *Levi–Civita connection* of the Riemann metric G; relation (a) is called the *Ricci lemma*. We rewrite (a) and (b) in terms of the connection forms:

(a')
$$dG_{ij} = \sum G_{ik}\omega_j^k + \sum G_{kj}\omega_i^k,$$

(b')
$$d\omega^i = \sum \omega^j \wedge \omega_j^i.$$

The Levi–Civita connection is characterized by the following *Koszul formula*:

(c) $2G(\nabla_Y Z, X) = YG(Z, X) + ZG(X, Y) - XG(Y, Z) - G(Y, [Z, X]) + G(Z, [X, Y]) + G(X, [Y, Z])$.

In local coordinates one can express the Christoffel symbols Γ_{ij}^k of ∇ by

(c') $\Gamma_{ij}^k = \sum \Gamma_{ijm}G^{mk}$, where $2\Gamma_{ijm} := \{-\partial_m G_{ij} + \partial_i G_{jm} + \partial_j G_{im}\}.$

See [266, Chap. 3, Thm. 11].

A.3.2.3 Riemannian curvature tensor

For a given Riemannian manifold (M, G), we define the *Riemannian curvature tensor*, a $(0,4)$-tensor, by

(a) $$R(X_1, X_2, X_3, X_4) := G(R(X_3, X_4)X_2, X_1),$$

where the $(1,3)$ curvature tensor of the Levi–Civita connection is given as in Section A.3.1.4.

Local description:

(a') $$R_{ijkl} = \sum G_{im} R^m_{jkl}.$$

One can reformulate the Bianchi identities from Section A.3.1.7 in terms of the Riemannian curvature tensor. Moreover, one has skew-symmetry relations (see Section A.3.1.4). We summarize all relations in local notation, including the connection forms:

(b) $$\Omega_{ij} := \sum G_{im}\Omega_j^m, \quad \Omega_{ij} = -\Omega_{ji};$$

(c) $$R_{ijkl} = -R_{jikl} = -R_{ijlk}, \quad R_{ijkl} = R_{klij}.$$

Bianchi identities:

(d)
$$R_{ijkl} + R_{iklj} + R_{iljk} = 0,$$
$$R_{ijkl,m} + R_{ijmk,l} + R_{ijlm,k} = 0.$$

A.3.2.4 Ricci lemma

We reformulate formula (a') from Section A.3.2.2 in terms of covariant differentiation with respect to the Levi–Civita connection (see Section A.3.1.1 above):

$$\nabla G = 0;$$

locally

$$G_{ij,k} = 0.$$

As a consequence, G^{ij} satisfies: $G^{ij}_{,k} = 0$.

A.3.2.5 Curvature

For a Riemannian manifold (M, G) and its Levi–Civita connection ∇, the curvature tensor R and the Ricci tensor Ric are well defined. As (M, G) has a volume element which is parallel with respect to ∇, the Ricci tensor is symmetric. One defines the *scalar curvature*

$$R := tr_G \text{ Ric} := \sum G^{ij} R_{ij}$$

and the *normalized scalar curvature* χ

$$n(n - 1)\chi := R.$$

For $n = 2$, the normalized scalar curvature is also called the *Gauß curvature*.

Denote by g the genus of a closed, oriented 2-dimensional manifold (where closed means: compact without boundary). The following is the famous Gauß-Bonnet theorem [140, vol. II, note 20, p. 358], connecting the Riemannian structure and the topology of M.

Theorem A.39. *Let (M, G) be a closed, oriented and 2-dimensional Riemann manifold. Then*

$$\int_M \chi\, dV = 4\pi(1 - g).$$

Next we define the sectional curvature of a Riemannian manifold of dimension $n \geq 2$.

Definition A.40. At $p \in M$ we consider a two-dimensional subspace Π of T_pM, spanned by an orthonormal basis $X_1, X_2 \in \Pi$. Then the *sectional curvature* $K(p, \Pi)$ is defined via the Riemannian curvature tensor by

$$K(p, \Pi) := R(X_1, X_2, X_1, X_2).$$

One easily verifies that $K(p, \Pi)$ depends only on (p, Π), and not on the choice of the orthonormal basis X_1, X_2 of Π.

If $\{e_1, \ldots, e_n\}$ is a local orthonormal frame, and X, Y are orthonormal, say $X = \sum x^j e_j, Y = \sum y^i e_i$, then

$$K(p, \Pi) = \sum R_{ijkl} x^i x^k y^j y^l.$$

A.3.3 Manifolds of constant curvature, Einstein manifolds

Let (M, G) be a Riemannian manifold of dimension n.

Definition A.41. A Riemannian manifold is said to be of *constant curvature* if all sectional curvatures at all points have the same constant value K. (M, G) is called *flat* if $K \equiv 0$.

One verifies that a Riemannian manifold of dimension $n \geq 3$ has constant curvature K if and only if the Riemannian curvature tensor satisfies the relation

$$R_{ijkl} = K(G_{ik}G_{jl} - G_{il}G_{jk}); \tag{A.22}$$

in this case the curvature forms Ω_i^j, relative to an orthonormal coframe field ω^i, are given by

$$\Omega_i^j = K\omega^i \wedge \omega^j. \tag{A.23}$$

For dimension $n = 2$, (A.22) is always satisfied, K being the (not necessarily constant) Gauß curvature.

Definition A.42. A Riemannian manifold is said to be an *Einstein manifold* if the Ricci tensor and the metric tensor are proportional; for dimension $n = 2$, this relation is true for any Riemannian metric.

Suppose that $R_{ij} = f\, G_{ij}$. Contracting this equality, we get $f = \frac{R}{n}$, that is, for an Einstein manifold

$$R_{ij} = \frac{R}{n}G_{ij} = (n - 1)\chi G_{ij}, \tag{A.24}$$

where χ is the normalized scalar curvature.

For an Einstein manifold of dimension $n \geq 3$, the scalar curvature R must be constant. In fact, from the Bianchi identity A.3.2.3 (d), by contraction we get $R_{,k} = 2\sum G^{ml}R_{mk,l}$. But contracting the covariant derivative of (A.24) gives $R_{,k} = n\sum G^{ml}R_{mk,l}$. Hence when $n \neq 2$, the scalar curvature R must be constant.

We are going to give some examples of Riemannian manifolds of constant curvature. Before doing so we define the concept of the induced metric for an immersed submanifold.

Definition A.43. Let (N, G) be a C^∞-Riemannian (or pseudo-Riemannian) manifold of dimension n, let M be a C^∞-manifold of dimension $m < n$, and let $F : M \to N$ be an immersion. F^*G is called the *induced metric* on M, where F^* was defined in (A.20).

A.3.4 Examples

(1) The Euclidean space $\mathbb{E}^n$ is a space of zero curvature.
(2) The standard n-sphere $\mathbb{S}^n$ of radius 1 in $\mathbb{E}^{n+1}$ with the induced metric has constant sectional curvature $K = 1$.
(3) We introduce the Minkowski inner product $\langle \, , \, \rangle$ in $\mathbb{R}^{n+1}$ by

$$\langle x, y \rangle := \sum_{i=1}^{n} x^i y^i - x^{n+1} y^{n+1}, \quad x, y \in \mathbb{R}^{n+1},$$

then, by the inertial theorem in linear algebra, the induced metric on the hyperboloid

$$\mathbb{H}^n : (x^1)^2 + \cdots + (x^n)^2 - (x^{n+1})^2 = -1$$

is a positive definite metric. A straightforward calculation shows that the hyperboloid $\mathbb{H}^n$ with the induced metric has constant sectional curvature -1.

Definition A.44. Two Riemannian manifolds (M, G) and $(\bar{M}, \bar{G})$ are said to be *isometric* (resp. *locally isometric*) if there exists a diffeomorphism (resp. local diffeomorphism) $F : M \to \bar{M}$ such that $F^*\bar{G} = G$.

Theorem A.45. *Every Riemannian manifold M of constant curvature $K = 1, 0, -1$ is locally isometric to one of the three examples above.*

For the proof see [140, vol. I, Chap. 5.3].

A.3.5 Exponential mapping and completeness

Let (M, G) be a C^∞-Riemannian manifold and let $C : [a, b] \to M$ be a C^∞-curve. A C^∞-vector field Y is said to be *parallel* along C if

$$\nabla_{dc/dt} Y = 0.$$

In terms of a local coordinate system $(x^1, \ldots, x^n)$, letting $C(t) = \{x^i(t)\}$, $Y(t) = Y(C(t)) = \sum y^i(t) \frac{\partial}{\partial x^i}$, $Y(t)$ is parallel along C if and only if

$$\frac{dy^k}{dt} + \sum \Gamma_{ij}^k y^i \frac{dx^j}{dt} = 0, \quad k = 1, \ldots, n. \tag{A.25}$$

By the existence and uniqueness theorem for ordinary differential equations we get

Theorem A.46. *Given a C^∞-curve $C : [a, b] \to M$ and a vector $Y(a) \in T_{C(a)}M$, there is a unique vector field Y along C which is parallel along C, with given initial value $Y(a)$.*
The vector field $Y(t)$ is said to be obtained from $Y(a)$ by parallel translation along C.

It is easy to see from (A.25) and A.3.2.2 that if $X(t)$ and $Y(t)$ are two parallel vector fields along $C(t)$ then $G(X(t), Y(t))$ is constant along C.

Definition A.47. A C^∞-curve $C(t)$ in M is called a *geodesic* if its field of tangent vectors is parallel along C.

In a local coordinate chart, the curve $C(t) = (x^1(t), \ldots, x^n(t))$ is a geodesic if and only if

$$\frac{d^2 x^i(t)}{dt^2} + \sum \Gamma^i_{jk}(C(t)) \frac{dx^j(t)}{dt} \frac{dx^k(t)}{dt} = 0, \quad i = 1, \ldots, n. \tag{A.26}$$

Applying the existence and uniqueness theorem for ordinary differential equations to (A.26) yields:

Theorem A.48. *For any point $p \in M$ there exist a neighborhood V of p, and real numbers $\varepsilon > 0, \delta > 0$ such that, if $q \in V$ and $X_q \in T_q M$ with $\|X_q\| < \varepsilon$, then there is a unique geodesic $x(t) = x(t, q, X_q)$ defined for $-\delta < t < \delta$ with $x(0) = q$ and $\frac{dx}{dt}\big|_{t=0} = X_q$. The mapping defined by $(t, q, X_q) \mapsto x(t, q, X_q)$ is C^∞ on the open set $|t| < \delta$, $q \in V$, $\|X_q\| < \varepsilon$.*

Remark A.49. Let $x(t)$ be a geodesic with $x(0) = q$, $\frac{dx}{dt}(0) = X_q$. If we change to a parameter $t = ct'$, where $c \neq 0$ is a constant, then $\bar{x}(t') = x(ct')$ is again a geodesic with $\bar{x}(0) = x(0) = q$ and $\frac{d\bar{x}}{dt'}\big|_{t'=0} = \frac{dx}{dt}\big|_{t=0} \cdot c = cX_q$. Hence, if ε is small enough, we may assume $\delta > 2$.

Definition A.50. The *exponential mapping* $\exp_q : T_q M \to M$ is defined as follows. For $X_q \in T_q M$ let $x(t)$ be the unique geodesic with $x(0) = q$ and $\frac{dx}{dt}\big|_{t=0} = X_q$. We define $\exp_q X_q := x(1)$.

We note that since $\left\| \frac{dx}{dt} \right\|$ is constant along the geodesic $x(t)$, its arc length from $x(0)$ to $x(1)$ is just $\|X_q\|$, that is, $\exp X_q$ is the point on the unique geodesic $x(t)$ determined by X_q, whose distance from q along the geodesic is $\|X_q\|$. Suppose $X_q \in T_q M$ for which $\exp_q X_q$ is defined, then $\exp_q tX_q$ is defined for $|t| < 1$ and $\exp_q tX_q = x(t), x(t)$ being the geodesic determined by X_q. In terms of a local coordinate system $(x^1, \ldots, x^n)$, letting $x^i(q) = a^i, X_q = \sum y^i \frac{\partial}{\partial x^i}(q)$, we have

$$x^i(1, a^1, \ldots, a^n; ty^1, \ldots, ty^n) = x^i(t, a^1, \ldots, a^n; y^1, \ldots, y^n).$$

Hence

$$\sum \frac{\partial x^i}{\partial y^j}(0) y^j = \frac{dx^i}{dt}(0, a^1, \ldots, a^n; y^1, \ldots, y^n) = y^i.$$

It follows that $\frac{\partial x^i}{\partial y^j}(0) = \delta^i_j$, that is, the Jacobian matrix of the exponential mapping $\exp_q$ at $X_q = 0$ is the unit matrix. Then, by the inverse function theorem, we get

Theorem A.51. *For every point $q \in M$ there is a neighborhood N of the zero vector 0 of the vector space $T_q M$ such that $\exp : N \to \exp\{N\}$ is a diffeomorphism.*

A.3.5.1 Normal coordinate system

For $q \in M$, choose an orthonormal basis $\{e_1, \ldots, e_n\}$ of $T_q M$, and let $X_q = \sum y^i e_i$. Then $(y^1, \ldots, y^n)$ can be considered as the coordinates of the point $\exp_q X_q$. The coordinate system $(y^1, \ldots, y^n)$ is called a *normal coordinate system*. Relative to the normal coordinate system $(y^1, \ldots, y^n)$ we have

(i) $G_{ij}(0) = \delta_{ij}$;

(ii) the equations of the geodesics through q take the form $y^i = b^i t$, b^i being constants;

(iii) $\Gamma^k_{ij}(0) = 0$, for $i, j, k = 1, \ldots, n$.

(i) and (ii) are trivial. Inserting $y^i = b^i t$ into the equations (A.26) of the geodesics we get $\sum \Gamma_{ij}^k(0)b^i b^j = 0$, which gives (iii).

On a Riemannian manifold (M, G), we introduce the distance $d(p, q)$ between two points p and q of M.

Definition A.52. A Riemanian manifold (M, G) is called *complete* if (M, d) is a complete metric space.

One verifies that if a piecewise differentiable curve from p to q has length equal to $d(p, q)$, then it is a geodesic when parametrized by arc length. A geodesic segment whose length is the distance between its endpoints is called a minimal geodesic.

Given any $v \in T_pM$, there is a unique geodesic $C : I \to M$, $0 \in I \subset \mathbb{R}$, such that

(i) $C(0) = p$, $\frac{dC}{dt}\big|_{t=0} = v$;

(ii) if $\tilde{C} : \tilde{I} \to M$ is another geodesic satisfying (i), then $\tilde{I} \subset I$; that means, the domain I of C is the largest possible.

A geodesic satisfying (i) and (ii) is called *maximal* or *geodesically inextendible*.

If the maximal geodesic is defined for $-\infty < t < +\infty$, we say that the *geodesic can be extended to infinity*.

Theorem A.53 (Hopf and Rinow). *The following four properties are equivalent:*
(i)   *the Riemannian manifold (M, G) is complete;*
(ii) *for some point $p \in M$, all geodesics from p are infinitely extendable:*
(iii) *All geodesics are infinitely extendable.*
(iv) *All bounded closed subsets of M are compact.*

Theorem A.54. *If M is complete, then any pair (p, q) of points can be joined by a minimal geodesic.*

For proofs of Theorems A.53 and A.54, see [6, 25, 266].

A.3.5.2 Jacobi fields
Let $C : [a, b] \to M$ be a geodesic. A vector field X along the geodesic is called a *Jacobi field* if it satisfies the following second order ordinary linear differential equation, called the *Jacobi equation*:

$$\nabla_{dc/dt}\nabla_{dc/dt}X + R\left(X, \frac{dc}{dt}\right)\frac{dc}{dt} = 0.$$

A Jacobi field along C is uniquely determined by the values of X and $\nabla_{dc/dt}X$ at one point $C(t_0)$ of C.

For Jacobi fields see [49, Sects. 4–5]; or [140, vol. II, Chap. VII].

A.4 Green's formula

Let (M, G) be an oriented C^∞-Riemannian manifold with Levi–Civita connection ∇, X a C^∞-vector field. Choose a local orthonormal frame field $\{e_i\}_{i=1}^n$ on M. Denote by $\{\omega^i\}_{i=1}^n$ its dual frame field. Consider $X = \sum X^i e_i$. The *divergence* of X, denoted by div$X$, is a function on M defined by

$$\mathrm{div}X = \sum X^i_{;i}. \tag{A.27}$$

Define an $(n-1)$-form ω by

$$\omega = \sum(-1)^{i-1}X^i\omega^1 \wedge \cdots \wedge \omega^{i-1} \wedge \omega^{i+1} \wedge \cdots \wedge \omega^n. \tag{A.28}$$

It is easy to check that ω is well-defined on M. A straightforward calculation gives

$$d\omega = \operatorname{div}X\,dV \tag{A.29}$$

where $dV = \omega^1 \wedge \cdots \wedge \omega^n$ is the oriented Riemannian volume element of M. If M is a compact manifold with boundary, then, by Stokes' formula,.

$$\int_M \operatorname{div}X dV = \int_{\partial M} G(X,\nu)dO, \tag{A.30}$$

where ν is the outer unit normal vector field to ∂M relative to the induced orientation of ∂M; dO is the oriented volume element of ∂M. One usually calls (A.30) the *divergence formula*.

Let f be a C^∞-function on M. The *gradient vector field* $\operatorname{grad} f$ and the *Laplacian* Δf of f are defined, respectively, by

$$\operatorname{grad} f := \sum G^{ij}f_{,i}e_j, \tag{A.31}$$

$$\Delta f := \sum G^{ij}f_{,ij} = \operatorname{tr}_G(\operatorname{Hess} f), \tag{A.32}$$

where Hess denotes the covariant Hessian from A.3.1.11. In terms of a local coordinate system $(x^1,\ldots,x^n)$, a direct calculation gives

$$\Delta f = \frac{1}{\sqrt{\det(G_{k\ell})}} \sum \frac{\partial}{\partial x^i}\left(G^{ij}\sqrt{\det(G_{k\ell})}\frac{df}{dx^j}\right). \tag{A.33}$$

A function f on M is said to be *harmonic* (resp. *subharmonic or superharmonic*) if $\Delta f = 0$ (resp. ≥ 0 or ≤ 0) everywhere.

Theorem A.55 (E. Hopf). *Let M be a closed C^∞-manifold, f be a subharmonic (superharmonic) function on M. Then f is a constant.*

For a proof see [140, vol. II, p. 338].

Finally, as a direct consequence of (A.30), we have the following

Theorem A.56 (Green's Formula). *Let f_1 and f_2 be smooth functions on a compact Riemannian manifold (M,g) with boundary. Then*

$$\int_M (f_1\Delta f_2 - f_2\Delta f_1)dV = \int_{\partial M} (f_1\,\nu(f_2) - f_2\,\nu(f_1))dO, \tag{A.34}$$

where ν is the outer unit normal vector field of $\partial M \hookrightarrow M$.

Proof. Just notice that $f_1\Delta f_2 - f_2\Delta f_1 = \operatorname{div}(f_1\operatorname{grad}f_2 - f_2\operatorname{grad}f_1)$, and use (A.30). $\qquad\square$

B. Laplacian comparison theorem

The purpose of this appendix is to prove the Laplacian comparison theorem.

Let (M, G) be a connected, complete Riemannian manifold of dimension n and let $p \in M$ be a fixed point. For any unit vector $X \in T_pM$, $\sigma(t) := \exp_p(tX)$, $0 \leq t < \infty$, is a smooth geodesic starting from p and parametrized by arc length. For a sufficiently small positive number t, $\sigma|_{[0,t]}$ realizes the distance between $\sigma(0)$ and $\sigma(t)$. However, if t is too large, this may no longer be true. Set

$$t^* = \sup \{t \in \mathbb{R}^+ \,|\, \sigma|_{[0,t]} \text{ is a minimizing geodesic segment joining } p \text{ and } \sigma(t)\},$$

then $t^* \in \mathbb{R}^+ \cup \{+ \infty\}$, where $\mathbb{R}^+$ denotes the set of positive real numbers. If $t^* <+ \infty$ we call $\sigma(t^*)$ the *cut point* of p along σ. If $t^* =+ \infty$ we say that there is no cut point of p along σ. It is well known that if $\sigma(t^*)$ is the cut point of p, then either $\sigma(t^*)$ is the first conjugate point (see [49]), or there exist at least two minimizing geodesic from p to $\sigma(t^*)$. The union of all cut points of p is called the *cut locus* and denoted by $C(p)$.

Let S_p be the set of unit tangent vectors in T_pM. We define a function $\mu : S_p \to \mathbb{R}^+ \cup \{+ \infty\}$ as follows: for each $X \in S_p$ consider the geodesic $\sigma(t) = \exp_p(tX)$, $0 \leq t < \infty$. If $\exp_p(t^*X)$ is the cut point of p along σ, we set $\mu(X) = t^*$. If there is no cut point along σ, we set $\mu(X) = \infty$. We introduce a topology in $\mathbb{R}^+ \cup \{+ \infty\}$ by taking intervals (a, b) and $(a, \infty] = (a, \infty) \cup \{\infty\}$ as a base for the open sets. Then the function μ is continuous. Set

$$E = \{tX \,|\, X \in S_p, \; 0 \leq t < \mu(X)\}. \tag{B.1}$$

Then (cf. [140])

(i) $\exp_p$ maps E diffeomorphically onto an open subset of M;

(ii) M is a disjoint union of $\exp_p(E)$ and the cut locus $C(p)$ of p.

Obviously, the open subset $\exp_p(E)$ of M is the largest open subset of M in which a normal coordinate system around p can be defined. It is also easy to see that the set $\exp_p(E)$ is a star domain with center p and that $C(p)$ is the boundary of $\exp_p(E)$.

We denote by $r = r(x)$ the *geodesic distance function* from p. In terms of a normal coordinate system $(x^1, \ldots, x^n)$ in $\exp_p(E)$,

$$r(x) = \sqrt{(x^1)^2 + \cdots + (x^n)^2}\,.$$

Hence $r(x)$ is differentiable in the subset $\exp_p(E) - \{p\}$. Obviously, at the differentiable points,

$$\|\operatorname{grad} r\| = 1. \tag{B.2}$$

We recall the definition of the covariant Hessian from Section A.3.1.11 and evaluate the Hessian and the Laplacian of r. Let $x_0 \in \exp_p(E)$ be an arbitrary point and $\sigma(t) : [0, a] \to M$ a minimizing geodesic parametrized by arc length and satisfying $\sigma(0) = p, \sigma(a) = x_0$. Let $X \in T_{x_0}M$ and $G(X, T(x_0)) = 0$, where $T = \sigma_*(\frac{\partial}{\partial t})$ is the tangent field of $\sigma(t)$. Since p is not conjugate to x_0, there is a unique Jacobi field $\tilde{X}(t)$ along $\sigma(t)$ which satisfies $\tilde{X}(0) = 0, \tilde{X}(a) = X$. Let $\beta(u), -\varepsilon < u < \varepsilon$, be a C^∞-curve on the unit hypersphere in T_pM with center at p such that $\beta(0) = T(0)$, $\frac{\partial \beta}{\partial u}|_{u=0} = \nabla_T \tilde{X}|_{t=0}$. Consider the geodesic variation $\alpha(t, u) = \exp_p(t\beta(u))$, $0 \leq t \leq a$, $-\varepsilon < u < \varepsilon$. When ε is small enough, we can assume that $\alpha(t, u)$ is differentiable. Then $\alpha_*(\frac{\partial}{\partial t}) = \frac{\partial}{\partial r}$ and the variational field $\tilde{X} = \alpha_*(\frac{\partial}{\partial u})$ satisfies that $\tilde{X}|_{\alpha(t,0)} = \tilde{X}(t)$ and $[\tilde{X}, \frac{\partial}{\partial r}] = 0$. Moreover, from the Gauß lemma it is easy to see that

$Vr = G\left(V, \frac{\partial}{\partial r}\right)$ for any tangent vector V on $\exp_p(E) - \{p\}$. The Hessian satisfies

$$\begin{aligned}
\text{Hess}(r)(X,X) &= [\tilde{X}(\tilde{X}r) - (\nabla_{\tilde{X}}\tilde{X})r]_{x_0} \\
&= \left[\tilde{X}G\left(\tilde{X}, \frac{\partial}{\partial r}\right) - G\left(\nabla_{\tilde{X}}\tilde{X}, \frac{\partial}{\partial r}\right)\right]_{x_0} \\
&= G\left(\tilde{X}, \nabla_{\tilde{X}}\frac{\partial}{\partial r}\right)\Big|_{x_0} \\
&= G(\tilde{X}, \nabla_{\partial/\partial r}\tilde{X})|_{x_0} \\
&= \int_0^a TG(\tilde{X}, \nabla_T\tilde{X})\,dt \\
&= \int_0^a \left[\|\nabla_T\tilde{X}\|^2 + G(\tilde{X}, \nabla_T\nabla_T\tilde{X})\right]dt.
\end{aligned}$$

Here we have used the fact that $\left[\tilde{X}, \frac{\partial}{\partial r}\right] = 0$. Since $\tilde{X}(t)$ is a Jacobi field, then

$$\nabla_T\nabla_T\tilde{X} = R(T,\tilde{X})T,$$

where R is the curvature tensor. Thus we obtain that

$$\text{Hess}(r)(X,X) = \int_0^a \left[\|\nabla_T\tilde{X}\|^2 + G(\tilde{X}, R(T,\tilde{X})T)\right]dt = \text{Ind}\,(\tilde{X},\tilde{X}) \tag{B.3}$$

where Ind denotes the *index form* defined on the space of all smooth vector fields orthogonal to T.

Theorem B.1 (Laplacian comparison theorem). *Let (M,G) be an n-dimensional complete Riemannian manifold with Ricci curvature bounded from below by $(n-1)K$ and $(\tilde{M},\tilde{G})$ an n-dimensional complete simply connected Riemannian manifold with constant curvature K. Denote the geodesic distance functions from $p \in M$ and $\tilde{p} \in \tilde{M}$ by r and $\tilde{r}$, respectively. Suppose that, at $x \in M, \tilde{x} \in \tilde{M}$, the distance functions r and $\tilde{r}$ are differentiable and satisfy $r(x) = \tilde{r}(\tilde{x})$. Then*

$$\Delta r(x) \le \tilde{\Delta}\tilde{r}(\tilde{x}), \tag{B.4}$$

where Δ and $\tilde{\Delta}$ denote the Laplace operators on M and $\tilde{M}$, respectively.

Proof. Let $\sigma : [0,a] \to M$ be the minimal geodesic from p to x, and $\tilde{\sigma} : [0,a] \to \tilde{M}$ the minimal geodesic from $\tilde{p}$ to $\tilde{x}$. Assume that σ and $\tilde{\sigma}$ are parametrized by arc lengths. Take an orthonormal basis $\left\{e_1 = \frac{\partial\sigma}{\partial t}(a), e_2, \ldots, e_n\right\}$ of $T_x M$ and extend $e_2, e_3, \ldots, e_n$ to Jacobi fields $E_2(t), E_3(t), \ldots, E_n(t)$ along $\sigma(t)$ such that $E_2(0) = E_3(0) = \cdots = E_n(0) = 0$. Noting that $\text{Hess}(r)(e_1, e_1) = 0$ and using (B.3), we have

$$\Delta r(x) = \sum_{i=2}^n \text{Hess}(r)(e_i, e_i) = \sum_{i=2}^n \text{Ind}\,(E_i, E_i).$$

Similarly,

$$\tilde{\Delta}\tilde{r}(\tilde{x}) = \sum_{i=2}^n \text{Hess}(\tilde{r})(\tilde{e}_i, \tilde{e}_i) = \sum_{i=2}^n \text{Ind}\,(\tilde{E}_i, \tilde{E}_i),$$

where $\left\{\tilde{e}_1 = \frac{\partial\tilde{\sigma}}{\partial t}(a), \tilde{e}_2, \ldots, \tilde{e}_n\right\}$ is an orthonormal basis of $T_{\tilde{x}}\tilde{M}$ and $\tilde{E}_i$ is the Jacobi field along $\tilde{\sigma}(t)$ satisfying $\tilde{E}_i(0) = 0$ and $\tilde{E}_i(a) = \tilde{e}_i$ for $i = 2, \ldots, n$.

Next, we extend $\{e_1, e_2, \ldots, e_n\}$ and $\{\tilde{e}_1, \tilde{e}_2, \ldots, \tilde{e}_n\}$ to be orthonormal, parallel frame fields $\{e_1(t) = \frac{\partial \sigma}{\partial t}, e_2(t), \ldots, e_n(t)\}$ along $\sigma(t)$ and $\{\tilde{e}_1(t) = \frac{\partial \tilde{\sigma}}{\partial t}, \tilde{e}_2(t), \ldots, \tilde{e}_n(t)\}$ along $\tilde{\sigma}(t)$, respectively. Since $\tilde{M}$ is a manifold of constant curvature K, we have

$$\tilde{E}_i(t) = \tilde{f}(t)\tilde{e}_i(t), \quad i = 2, \ldots, n,$$

where $\tilde{f}(t)$ is a solution of the Jacobi equation

$$\begin{cases} \dfrac{d^2 \tilde{f}}{dt^2} + K\tilde{f} = 0, \\ \tilde{f}(0) = 0, \ \tilde{f}(a) = 1. \end{cases} \tag{B.5}$$

For $i = 2, \ldots, n$, define the vector fields $V_i(t) = \tilde{f}(t)e_i(t)$. Then

$$V_i(0) = 0, \quad V_i(a) = e_i(a) = E_i(a).$$

Since Jacobi fields minimize the index form Ind among vector fields along a geodesic with the same values at endpoints (cf. [49]), we get

$$\Delta r(x) = \sum_{i=2}^{n} \text{Ind}\,(E_i, E_i) \leq \sum_{i=2}^{n} \text{Ind}\,(V_i, V_i)$$

$$= \sum_{i=2}^{n} \int_0^a (\|\nabla_{e_1} V_i\|^2 - G(R(e_1, V - i)e_1, V - i))dt$$

$$= \sum_{i=2}^{n} \int_0^a (\|\nabla_{\tilde{e}_1} \tilde{E}_i\|^2 - \tilde{f}^2 G(R(e_1, e_i)e_1, e_i))dt$$

$$\leq \sum_{i=2}^{n} \int_0^a (\|\nabla_{\tilde{e}_1} \tilde{E}_1\|^2 - \tilde{f}^2 (n-1)K)dt$$

$$= \sum_{i=2}^{n} \text{Ind}\,(\tilde{E}_i, \tilde{E}_i) = \tilde{\Delta} \tilde{r}(\tilde{x}). \qquad \square$$

Now we assume N to be a manifold with constant negative curvature K and r to be the geodesic distance function from p. Let $\{e_1 = \frac{\partial}{\partial r}, e_2, \ldots, e_n\}$ be a local orthonormal frame field around p which is parallel along the geodesics starting from p. Let $E_i(t), 2 \leq i \leq n$, be Jacobi fields along a geodesic $\sigma(t), 0 \leq t \leq r$, starting from p, which satisfy $E_i(0) = 0$ and $E_i(r) = e_i(\sigma(r))$. Then we have (cf. [49])

$$E_i(t) = \frac{1}{\sinh(\sqrt{-K}r)} \sinh(\sqrt{-K}t)e_i(t) \tag{B.6}$$

and

$$\Delta r = \sum_{i=1}^{n} \text{Hess}(r)(e_i, e_i) = \sum_{i=2}^{n} \text{Ind}\,(E_i, E_i)$$

$$= \sum_{i=2}^{n} \int_0^r [\|\nabla_T E_i\|^2 + G(E_i, \nabla_T \nabla_T E_i)]dt$$

$$= -\frac{(n-1)}{\sinh^2(\sqrt{-K}r)} \int_0^r [\sinh^2(\sqrt{-K}t) + \cosh^2(\sqrt{-K}t)]dt \tag{B.7}$$

$$= \frac{(n-1)\sqrt{-K}}{\sinh^2(\sqrt{-K}r)} \int_0^r d[\sinh(\sqrt{-K}t)\cosh(\sqrt{-K}t)]$$

$$= (n-1)\sqrt{-K} \coth(\sqrt{-K}r).$$

From Theorem B.1 and (B.7) we have the following theorem.

Theorem B.2. *Let M be an n-dimensional complete Riemannian manifold with Ricci curvature bounded from below by a negative constant $K < 0$. Then, for $x \in \exp_p(E)$, the geodesic distance function r satisfies*

$$\Delta r(x) \le (n-1)\sqrt{-K}\coth\left(\sqrt{-K}r(x)\right).$$

Corollary B.3. *Let M be as in Theorem B.2, then for $x \in \exp_p(E)$,*

$$\Delta\,(r(x))^2 \le 2\left[1 + (n-1)\left(1 + \sqrt{-K}\,r(x)\right)\right].$$

Proof. From Theorem B.2 we have

$$\Delta r^2 = 2\|\operatorname{grad} r\|^2 + 2r\Delta r$$
$$\le 2\left[1 + (n-1)r\sqrt{-K}\coth\left(\sqrt{-K}\,r\right)\right].$$

The conclusion follows from the fact that

$$r\sqrt{-K}\coth\left(\sqrt{-K}r\right) \le \left(1 + \sqrt{-K}\,r\right). \qquad \square$$

Bibliography

[1] J. A. Aledo, A. Martínez, and F. Milán: An extension of the affine Bernstein problem, *Results Math.* **60** (2011), 157–174.

[2] A. D. Aleksandrov and J. A. Volkov: Uniqueness theorems for surfaces in the large IV, *Amer. Math. Soc. Transl. II. Ser.* **21** (1962), 403–411.

[3] S. Amari: Differential geometric methods in statistics. Lecture Notes Statistics 28, Springer Berlin etc., 1985.

[4] S. Amari: Differential geometry of statistics – towards new developments, in: S. Amari et al. (eds.), *Differential Geometry in Statistical Inferences*, IMS Monograph Series 10, Hayward, 1987.

[5] M. Antonowicz: On the Bianchi–Bäcklund construction for affine minimal surfaces, *J. Phys. A: Math. Gen.* **20** (1987), 1989–1996.

[6] T. Aubin: *Nonlinear Analysis on Manifolds. Monge–Ampère equations*, Springer, New York, 1981.

[7] J. C. Baez: The Octonions, *Bull. Amer. Math. Soc.* **39**(2) (2002), 145–205.

[8] W. Barthel: Zur Affingeometrie auf Mannigfaltigkeiten. *J.-ber. Deutsch. Math.-Verein.* **68** (1966), 13–44 [Zbl. 134.388].

[9] W. Barthel: Strukturelle Betrachtungen zur affinen Differentialgeometrie, *Results Math.* **13** (1988), 409–419.

[10] W. Barthel: Zur Affinen Differentialgeometrie – Kurventheorie in der allgemeinen Affingeometrie. Geometry, *Proc. Congr. Thessaloniki/Greece 1987*, 5–19 (1988) [Zbl. 642.53011].

[11] W. Barthel: Die Bedeutung Christoffelscher Zusammenhänge in der affinen Differentialgeometrie, *E. B. Christoffel-Symp. Aachen 1979*, 568–589 (1981) [Zbl. 482.53007].

[12] W. Barthel, R. Volkmer, and I. Haubitz: Thomsensche Minimalflächen – analytisch und anschaulich, *Results Math.* **3** (1980), 129–154.

[13] E. F. Beckenbach and R. Bellmann: Inequalities, Springer-Verlag, Berlin Göttingen Heidelberg, 1961.

[14] M. Berger, P. Gauduchon, and E. Mazet: Let spectre d'une variété Riemannienne, Lecture Notes Math. 194, Springer, 1971.

[15] L. Berwald: Die Grundlagen der Hyperflächen im euklidischen Raum gegenüber inhaltstreuen Affinitäten, *Monatshefte Math.* **32** (1922), 89–106.

[16] G. Bianchi and P. M. Gruber: Characterizations of ellipsoids, *Arch. Math.* **49** (1987), 344–350.

[17] O, Birembaux and M. Djorić: Isotropic affine spheres. *Acta Math. Sinica, English Ser.* **28** (2012), 1955–1972.

[18] C. Blanc and F. Fiala: Le type d'une surface et sa courbure totale. *Comm. Math. Helv.* **14** (1941–42), 230–233.

[19] W. Blaschke: Vorlesungen über Differentialgeometrie, I, Berlin, Springer, 1921.

[20] W. Blaschke: Vorlesungen über Differentialgeometrie, II, Berlin, Springer, 1923.

[21] W. Blaschke: Kreis und Kugel, 2nd ed., Walter de Gruyter & Co., Berlin, 1956.

[22] W. Blaschke: Gesammelte Werke, vol. 4, Affine Differentialgeometrie. Differentialgeometrie der Kreis- und Kugelgruppen, Thales Verlag, Essen, 1985.

[23] W. Blaschke: Gesammelte Werke, vol. 5, Konvexgeometrie. Thales Verlag, Essen, 1985.

[24] N. Bokan, K. Nomizu, and U. Simon: Affine hypersurfaces with parallel cubic forms. *Tôhoku Math. J., II. Ser.* **42** (1990), 101–108.

[25] W. M. Boothby: An introduction to differentiable manifolds and Riemannian geometry, Academic Press, New York London, 1975.

[26] O. Boruvka: Lineare Differentialtransformationen 2. Ordnung, Verlag VEB Deutscher Wiss., Berlin, 1967.

[27] H. Brauner: Eine Kennzeichnung der Minimalflächen von G. Thomsen. *Rad. Jugosl. Akad. Znan. Umjet., Mat. Znan.* **7** (1988), 1–15 [Zbl.676.53007].

[28] F. Brickell: A new proof of Deicke's theorem on homogeneous functions. *Proc. Amer. Math. Soc.* **16** (1965), 190–191.

[29] F. Brickell: A theorem on homogeneous functions. *J. London Math. Soc.* **42** (1967), 325–329.

[30] F. Brickell and R. S. Clark: Differentiable Manifolds, Van Nostrand, London, 1970.

[31] Y. D. Burago and V. Zalgaller: *Geometric Inequalities*, Springer-Verlag, Berlin, 1988.

[32] W. Burau and U. Simon: Blaschkes Beiträge zur affinen Differentialgeometrie. In: [22], (1985), 11–34.

[33] H. Busemann: Convex surfaces, Interscience Publ., New York London, 1958.

[34] L. A. Caffarelli: Interior $W^{2,p}$ estimates for solutions of Monge–Ampère equations. *Ann. Math.* **131** (1990), 135–150.

[35] L. A. Caffarelli and C. E. Gutiérrez: Properties of the solutions of the linearized Monge–Ampère equation. *Amer. J. Math.* **119** (1997), 423–465.

[36] L. Caffarelli, L. Nirenberg, and J. Spruck: The Dirichlet problem for nonlinear second-order elliptic equations, I, Monge–Ampère equation. *Comm. Pure Appl. Math.* **37** (1984), 369–402. Erratum: *Comm. Pure Appl. Math.* **40** (1987), 659–662.

[37] L. Caffarelli, L. Nirenberg, and J. Spruck: The Dirichlet problem for nonlinear second-order elliptic equations, III, Functions of the eigenvalues of the Hessian. *Acta Math.* **155** (1985), 261–301.

[38] L. Caffarelli, J. J. Kohn, L. Nirenberg, and J. Spruck: The Dirichlet problem for nonlinear second-order elliptic equations, II, Complex Monge–Ampère, and uniformly elliptic, equations. *Comm. Pure Appl. Math.* **38** (1985), 209–252.

[39] E. Calabi: An extension of E. Hopf's maximum principle with an application to Riemannian Geometry, *Duke Math. J.* **25** (1958), 45–56.

[40] E. Calabi: Improper affine hyperspheres of convex type and a generalization of a theorem by K. Jörgens, *Mich. Math. J.* **5** (1958), 105–126.

[41] E. Calabi: Complete affine hypersurfaces I, *Symposia Math.* **10** (1972), 19–38.

[42] E. Calabi: Hypersurfaces with maximal affinely invariant area, *Amer. J. Math.* **104** (1982), 91–126.

[43] E. Calabi: Hypersurface geometry in a vector space, unpublished manuscript, 67 pp., without year.

[44] E. Calabi: Convex affine maximal surfaces, *Results Math.* **13** (1988), 209–223.

[45] E. Calabi: Examples of Bernstein problems for some nonlinear equations, *Proc. Symposia Pure Math.* **15** (1970), 223–230.

[46] E. Calabi: *Géométrie différentielle affine des hypersurfaces*. Sém. Bourbaki, 33e année, vol. 1980/81, Exp. 573, Lect. Notes Math. 901, pp. 189–204, Springer, Berlin New York, 1981.

[47] E. Calabi: Affine differential geometry and holomorphic curves, Lect. Notes Math. 1422, pp. 15–21, Springer, Berlin, 1990.

[48] I. Chavel: *Eigenvalues in Riemannian Geometry*, Academic Press, Inc., Orlando, FL, 1984.

[49] J. Cheeger and D. G. Ebin: *Comparison Theorems in Riemannian Geometry*, North-Holland Publishing Co., Amsterdam Oxford; American Elsevier Publishing Co., Inc., New York, 1975.

[50] B. H. Chen, Q. Han, A.-M. Li, and L. Sheng: Interior estimates for the n-dimensional Abreu's equation, *Adv. Math.* **251** (2014), 35–46.

[51] B. H. Chen, A.-M. Li, and L. Sheng: Affine techniques on extremal metrics on toric surfaces, arXiv: math.DG/1008.2606v4.

[52] B. H. Chen, A.-M. Li, and L. Sheng: Extremal metrics on toric surfaces. arXiv: math.DG/1008.2607v4.

[53] B. H. Chen, A.-M. Li, and L. Sheng: The Abreu equation with degenerated boundary conditions, *J. Diff. Equations* **252** (2012), 5235–5259.

[54] B. H. Chen, A.-M. Li, and L. Sheng: Interior regularization for solutions of Abreu's equation, *Acta Math. Sin. (Engl. Ser.)* **29** (2013), 33–38.

[55] B. H. Chen, A.-M. Li, and L. Sheng: Uniform K-stability for extremal metrics on toric varieties, *J. Diff. Equations* **257** (2014), 1487–1500.

[56] S. Y. Cheng and S. T. Yau: On the regularity of the Monge–Ampère equation $\det(\frac{\partial^2 u}{\partial x^i \partial x^j}) = F(x, u)$, *Commun. Pure and Appl. Math.* **30** (1977), 41–68.

[57] S. Y. Cheng and S. T. Yau: Complete affine hypersurfaces, Part I, The completeness of affine metrics, *Commun. Pure Appl. Math.* **39** (1986), 839–866.

[58] S. Y. Cheng and S. T. Yau: On the regularity of the solution of the n-dimensional Minkowski problem, *Communications Pure Appl. Math.* **29** (1976), 495–516.

[59] S. Y. Cheng and S. T. Yau: The real Monge–Ampère equation and affine flat structure, in: Proceedings of the 1980 Beijing Symposium on Differential Geometry and Differential Equations, pp. 339–370, Science Press, Beijing, 1982.

[60] S. S. Chern: Affine minimal hypersurfaces, *Proc. Japan–United States Sem. Tokyo* **1977** (1979), 17–30.

[61] S. S. Chern: The mathematical works of W. Blaschke, *Abh. Math. Sem. Hamburg* **39** (1973), 1–9.

[62] S. S. Chern: Integral formulas for hypersurfaces in Euclidean space and their applications to uniqueness theorem, *J. Math. and Mech.* **8** (1959), 947–955.

[63] S. S. Chern, W. H. Chen, and K. S. Lam: Lectures on Differential Geometry, Series on Univ. Math., Vol. 1, World Scientific Publ. Co., Inc., River Edge, NJ, 1999.

[64] S. S. Chern and C.-L. Terng: An analogue of Bäcklund's theorem in affine geometry, *Rocky Mt. J. Math.* **10** (1980), 105–124.

[65] A. Deicke: Über die Finsler-Räume mit $A_i = 0$, *Arch. Math.* **4** (1953), 45–51.

[66] R. Deszcz: Certain curvature characterizations of affine hypersurfaces, *Coll. Math.* **43** (1992), 21–39.

[67] F. Dillen: Locally symmetric complex affine hypersurfaces, *J. Geom.* **33** (1988), 27–38.

[68] F. Dillen: The complex version of a theorem by Berwald. Soochow, *J. Math.* **14** (1988), 41–50.

[69] F. Dillen: Equivalence theorems in affine differential geometry, *Geom. Dedicata* **32** (1989), 81–91.

[70] F. Dillen: Polynomials with constant Hessian determinant, *J. Pure Appl. Algebra* **71** (1991), 13–18.

[71] F. Dillen, A. Martinez, F. Milan, F. G. Santos, and L. Vrancken: On the Pick invariant, the affine mean curvature and the Gauss curvature of affine surfaces, *Results Math.* **20** (1991), 622–642.

[72] F. Dillen, K. Nomizu, and L. Vrancken: Conjugate connections and Radon's theorem in affine differential geometry, *Monatsh. Math.* **109** (1990), 221–235.

[73] F. Dillen and L. Verstraelen: Real and complex locally symmetric affine hypersurfaces. In: Affine Differentialgeometrie, in: Proceedings Conf. Oberwolfach 1986, TU Berlin (ISBN 3798311927) 420–432, 1988.

[74] F. Dillen and L. Verstraelen (eds.): Geometry and Topology of Submanifolds, IV. Proc. Conf. Diff. Geom. Vision, Leuven June 1991, World Scientific Singapore, 1992.

[75] F. Dillen, I. van de Woestijne, L. Verstraelen, and L. Vrancken (eds): Geometry and Topology of Submanifolds, V. Proc. Conf. Diff. Geom. Vision, Leuven July 1992, World Scientific Publ., River Edge, NJ, 1993.

[76]   F. Dillen and L. Vrancken: Affine Differential Geometry of Hypersurfaces. In: M. Boyom, J. -
        M. Morvan, L. Verstraelen (eds.), *Geometry and Topology of Submanifolds* II , pp. 144–164.
        World Scientific, Singapore etc., 1990.

[77]   F. Dillen and L. Vrancken: Affine minimal higher order parallel affine surfaces, in: T. M. Ras-
        sias (ed.), *The problem of Plateau. A tribute to J. Douglas and T. Rado*, pp. 76–86, World Sci-
        entific, Singapore, 1992.

[78]   F. Dillen and L. Vrancken: Generalized Cayley surfaces. in: Proceedings of the Conference on
        Global Analysis and Global Differential Geometry, Berlin 1990, Lecture Notes in Mathematics
        1481, pp. 36–47, Springer, Berlin, 1991.

[79]   F. Dillen and L. Vrancken: 3-dimensional affine hypersurfaces in $\mathbb{R}^4$ with parallel cubic form,
        Nagoya Math. J. **124** (1991), 41–53.

[80]   F. Dillen and L. Vrancken: Homogeneous affine hypersurfaces with rank one shape operators,
        Math. Z. **212** (1993), 61–72.

[81]   F. Dillen and L. Vrancken: Calabi-type composition of affine spheres, *Diff. Geom. Appl.* **4**
        (1994), 303–328.

[82]   F. Dillen and L. Vrancken: The classification of 3-dimensional locally strongly convex homoge-
        neous affine hypersurfaces, *Manuscripta Math.* **80** (1993), 165–180.

[83]   F. Dillen and L. Vrancken: Quasi-umbilical, locally strongly convex homogeneous affine hyper-
        surfaces, *J. Math. Soc. Japan.* **46** (1994), 477–502.

[84]   F. Dillen, L. Vrancken, and S. Yaprak: Affine hypersurfaces with parallel cubic form, *Nagoya
        Math. J.* **135** (1994), 153–164.

[85]   M. P. do Carmo: *Differentialgeometrie von Kurven und Flächen*, Vieweg, Braunschweig, 1983.

[86]   M. P. do Carmo: *Riemannian Geometry*, Birkhäuser, Basel etc., 1992.

[87]   J. Dugundji: *Topology*, Allyn and Bacon. Boston, 1966, 8th printing 1973.

[88]   N. W. Efimov: *Flächenverbiegung im Großen. Mit einem Nachtrag von E. Rembs und K. P.
        Grotemeyer*, Akademie-Verlag, Berlin, 1957.

[89]   L. P. Eisenhart: *Riemannian Geometry*, Princeton Univ. Press, Princeton, 1949.

[90]   L. P. Eisenhart: *Non-Riemannian Geometry*, AMS Colloquium Publications, vol. VIII, New York,
        1927.

[91]   F. Fabricius-Bjerre: On a conjecture of G. Bol, *Math. Scand.* **40** (1977), 194–196.

[92]   W. F. Firey: The determination of convex bodies from their mean radius of curvature functions,
        Mathematika **14** (1967), 1–13.

[93]   H. Flanders: Local theory of affine hypersurfaces, *J. Analyse Math.* **15** (1965), 353–387.

[94]   O. Forster: *Analysis*, 3 vols., Vieweg, Braunschweig-Wiesbaden, 1976, 1979.

[95]   D. Fried: Distality, completeness and affine structures, *J. Diff. Geom.* **24** (1986), 265–273.

[96]   L. Gårding: An inequality for hyperbolic polynomials, *J. Math. Mech.* **8** (1959), 957–965.

[97]   S. Gigena: The classification of affine complete hyperspheres, in: Proc. of the Eleventh Brazil-
        ian Math. Col. (Pocos de Caldos, (1977), vol. II, pp. 629–641. Inst. Mat. Pura Appl., Rio de
        Janeiro, 1978.

[98]   S. Gigena: On a conjecture of E. Calabi, *Geom. Dedicata* **11** (1981), 387–396.

[99]   S. Gigena: General affine invariance of hypersurfaces, *Math. Notae* **29** (1981/82), 135–145, in
        Spanish.

[100]   S. Gigena: Integral invariants of convex cones, *J. Diff. Geom.* **13** (1978), 191–222.

[101]   D. Gilbarg and N. S. Trudinger: *Elliptic Partial Differential Equations of Second Order*, reprint
        of the 1998 edition,Springer-Verlag, Berlin, 2001.

[102]   E. Glässner: Ein Affinanalogon zu den Scherkschen Minimalflächen, *Arch. Math.* **27** (1977),
        436–439.

[103]   S. I. Goldberg: *Curvature and Homology*, Academic Press, New York London, 1962.

[104] R. Grambow: Ableitung der Affinvarianten einer krummen Fläche aus den Bewegungsinvarianten, Dissertation Univ. Hamburg, 1922.

[105] W. H. Greub: *Linear Algebra*, 3rd ed., Springer, Berlin etc., 1967.

[106] W. H. Greub: *Multilinear Algebra*, Springer, Berlin etc., 1967.

[107] K.-P. Grotemeyer: Die Integralsätze der affinen Flächentheorie, *Arch. Math.* **3** (1952), 38–43.

[108] K.-P. Grotemeyer: Eine kennzeichnende Eigenschaft der Affinsphären, *Arch. Math.* **3** (1952), 307–310.

[109] H. W. Guggenheimer: *Differential geometry*, Mc Graw-Hill, New York etc., 1963.

[110] H. W. Guggenheimer: Applications of polarity, *Ann. Mat. Pura Appl., IV Ser.* **102** (1975), 369–383.

[111] C. E. Gutiérrez: *The Monge–Ampère Equation*, Birkhäuser, Boston, MA, 2001.

[112] G. H. Hardy, J. E. Littlewood, and G. Polya: *Inequalities*, Cambridge Univ. Press, Cambridge, 1934.

[113] E. Heil: Zur affinen Differentialgeometrie der Eilinien, Dissertation TH Darmstadt. 1965.

[114] E. Heil: Affine Scheitel von Ovalen, *Elemente Math.* **25** (1970), 84–85.

[115] E. Heil: Verschärfungen des Vierscheitelsatzes und ihre relativgeometrischen Verallgemeinerungen, *Math. Nachr.* **45** (1970), 228–241.

[116] E. Heil: Eigenvalue estimates for Hill's equation, *J. Diff. Equat.* **18** (1975), 179–187.

[117] E. Heil: Relative and affine normals, *Results Math.* **13** (1988), 240–254.

[118] E. Heil: Wieviele Affinnormalen gehen durch einen Punkt? in: Proc. of the Congress of Geometry, Thessaloniki/Greece 1987, pp. 54–66, 1988 [Zbl.642.53010].

[119] E. Heil and U. Kern: Ovaloids without Vertices – Interplay between Euclidean and Affine Differential Geometry, in: B. Fuchsteiner, W. A. J. Luxemburg (eds.), *Analysis and Geometry: Trends in Research and Teaching*, pp. 141–148, BI, Mannheim Zürich, 1992.

[120] S. Helgason: *Differential Geomtry and Symmetrics Spaces*, Academic Press, New York London, 1962.

[121] N. J. Hicks: *Notes on Differential Geometry*, Van Nostrand, New York etc., 1965.

[122] F. Hirzebruch: Elliptische Differentialoperatoren auf Mannigfaltigkeiten. In: *Arbeitsgemeinschaft für Forschung NRW*, Heft 157, pp. 33–56, Westdeutscher Verlag, Köln Opladen, 1966.

[123] H. Hofer: Pseudoholomorphic curves in symplectizations with applications to the Weinstein conjecture in dimension three, *Invent. Math.* **114** (1993), 515–563.

[124] H. Holmann and H. Rummler: *Alternierende Differentialformen*, BI Wissenschaftsverlag, Zürich, 1972.

[125] H. Hopf: *Differential Geometry in the Large*, Lecture Notes in Mathematics 1000, Springer-Verlag, Berlin, 1983.

[126] C. C. Hsiung and J. K. Shahin: Affine differential geometry of closed hypersurfaces, *Proc. London Math. Soc.* **17** (1967), 715–735.

[127] H. Huck, R. Roitzsch, U. Simon, W. Vortisch, R. Walden, B. Wegner, and W. Wendland: *Beweismethoden der Differentialgeometrie im Großen*, Lecture Notes in Math. 335, Springer, Berlin, 1973.

[128] Z. Hu and C. Li: The classification of 3-dimensional Lorentzian affine hypersurfaces with parallel cubic form, *Diff. Geom. Appl.* **29** (2011), 361–373.

[129] Z. Hu, C. Li, H. Li, and L. Vrancken: Lorentzian affine hypersurfaces with parallel cubic form, *Results Math.* **59** (2011), 577–620.

[130] Z. Hu, C. Li, H. Li, and L. Vrancken: The classification of 4-dimensional nondegenerate affine hypersurfaces with parallel cubic form, *J. Geom. Phys.* **61** (2011), 2035–2057.

[131] Z. Hu, H. Li, U. Simon, and L. Vrancken: On locally strongly convex affine hypersurfaces with parallel cubic form, Part I, *Diff. Geom. Appl.* **27** (2009), 188–205.

[132] Z. Hu, H. Li, and L. Vrancken: Characterizations of the Calabi product of hyperbolic affine hyperspheres, *Results Math.* **52** (2008), 299–314.

[133] Z. Hu, H. Li, and L. Vrancken: Locally strongly convex affine hypersurfaces with parallel cubic form, *J. Diff. Geom.* **87** (2011), 239–307.

[134] Z. Hu, C. Li, and C. Zhang: On quasi-umbilical locally strongly convex homogeneous affine hypersurfaces, *Diff. Geom. Appl.* **33** (2014), 46–74.

[135] Z. Hu, C. Li, and D. Zhang: A differential geometry characterization of the Cayley hypersurface, *Proc. Amer. Math. Soc.* **139** (2011), 3697–3706.

[136] W. Jelonek: Affine locally symmetric surfaces, *Geom. Dedicata* **44** (1992), 189–221.

[137] W. Jelonek: Affine surfaces with parallel shape operator, *Anales Polon. Math.* **56** (1992), 179–186.

[138] K. Jörgens: Über die Lösungen der Differentialgleichung $rt - s^2 = 1$, *Math. Ann.* **127** (1954), 130–134.

[139] F. Klein: Vergleichende Betrachtung über neuere geometrische Forschungen, *Math. Ann.* **43** (1893), 63–100.

[140] S. Kobayashi and K. Nomizu: *Foundations of differential geometry*, vols. I and II, Interscience Publishers John Wiley & Sons, Inc., New York London Sydney, 1963 and 1969.

[141] M. Kozlowski: Affine maximal surfaces, *Anzeiger der Österreichischen Akademie der Wissenschaften* **8** (1987), 137–139.

[142] M. Kozlowski: A class of affine maximal hypersurfaces, *Rend. Circ. Mat. Palermo, II. Ser.* **37**(3) (1988), 444–448.

[143] M. Kozlowski: Improper affine spheres, *Anzeiger Österr. Akad. Wiss., math.-naturwiss. Kl.* **6** (1988), 95–96.

[144] M. Kozlowski: One parameter families of improper affine spheres, *Anzeiger der Österr. Akademie der Wissenschaften, math.-naturwiss. Kl.* **126** (1989), 81–82.

[145] M. Kozlowski and U. Simon: Hyperflächen mit äquiaffiner Einsteinmetrik, *Mathematica*, Festschrift E. Mohr, TU Berlin, pp. 179–190, 1985.

[146] E. Kreyszig and A. Pendel: Spherical curves and their analogues in affine differential geometry, *Proc. Amer. Math. Soc.* **48** (1975), 423–428.

[147] M. Kriele and L. Vrancken: Lorentzian affine hyperspheres with constant affine sectional curvature, *Trans. Amer. Math. Soc.* **352** (2000), 1581–1599.

[148] T. Kurose: Dual connections and affine geometry, *Math. Z.* **203** (1990), 115–121.

[149] T. Kurose: On the Minkowski problem in affine geometry, *Results Math.* **20** (1991), 643–649.

[150] T. Kurose: Two results in the affine hypersurface theory, *J. Math. Soc. Japan* **41** (1989), 539–548.

[151] D. Laugwitz: Einige differentialgeometrische Charakterisierungen der Quadriken, *Ann. Mat. Pura Appl., IV. Ser.* **55** (1961), 307–314.

[152] D. Laugwitz: Über Eilinien im Großen in der zentralaffinen Differentialgeometrie, *Math. Z.* **79** (1962), 425–438.

[153] D. Laugwitz: Zur Differentialgeometrie der Hyperflächen in Vektorräumen und zur affingeometrischen Deutung der Theorie der Finsler-Räume, *Math. Z.* **67** (1957), 63–74.

[154] D. Laugwitz: *Differentialgeometrie in Vektorräumen, unter besonderer Berücksichtigung der unendlichdimensionalen Räume*, Vieweg & Sohn, Braunschweig, 1965.

[155] K. Leichtweiß: Über einige Eigenschaften der Affinoberfläche beliebiger konvexer Körper, *Results in Math.* **13** (1988), 255–282.

[156] K. Leichtweiß: Konvexgeometrie, *Math. Phys. Sem. Berichte* **32** (1985), 55–75.

[157] K. Leichtweiß: Über eine Formel Blaschkes zur Affinoberfläche, *Stud. Sci. Math. Hung.* **21** (1986), 453–474, [Zbl.561.53012].

[158] K. Leichtweiß: Zur Affinoberfläche konvexer Körper, *Manuscripta Math.* **56** (1986), 429–464.

[159] K. Leichtweiß: Über eine geometrische Deutung des Affinnormalenvektors einseitig gekümmter Hyperflächen, *Arch. Math.* **53** (1989), 613–621.

[160] K. Leichtweiß: Bemerkungen zur Definition einer erweiterten Affinoberfläche von E. Lutwak, *Manuscripta Math.* **65** (1989), 181–197.

[161] K. Leichtweiß: Bemerkungen zur Monotonie der Affinoberfläche von Eihyperflächen, *Math. Nachr.* **147** (1990), 47–60.

[162] K. Leichtweiß: On the history of the affine surface area for convex bodies, *Results Math.* **20** (1991), 650–656.

[163] K. Leichtweiß: On inner parallel bodies in the equiaffine geometry, in: B. Fuchsteiner, W. A. J. (eds.), *Analysis and Geometry*, pp. 113–124. BI, Mannheim etc., 1992, IBSN 3-411-15621-X.

[164] A.-M. Li: Uniqueness theorems in affine differential geometry, Part I, *Results Math.* **13** (1988), 283–307.

[165] A.-M. Li: Variational formulas for higher affine mean curvature, *Results Math.* **13** (1988), 318–326.

[166] A.-M. Li: *Affine maximal surface and harmonic functions*, Lect. Notes Math. 1369, pp. 142–151, Springer, Berlin, 1989.

[167] A.-M. Li: Some theorems in affine differential geometry, *Acta Math. Sinica, New Series* **5** (1989), 345–354.

[168] A.-M. Li: Affine completeness and Euclidean completeness, in: *Global Diff. Geom. Global Analysis*, Proceedings, Berlin 1990. Lecture Notes Math. 1481 , pp. 116–126, Springer, Berlin, 1991.

[169] A.-M. Li: Calabi conjecture on hyperbolic affine hyperspheres, *Math. Z.* **203** (1990), 483–491.

[170] A.-M. Li: Calabi conjecture on hyperbolic affine hyperspheres (2), *Math. Ann.* **293** (1992), 485–493.

[171] A.-M. Li: A characterization of ellipsoids, *Results Math.* **20** (1991), 657–659.

[172] A.-M. Li: Spacelike hypersurfaces with constant Gauss-Kronecker curvature in the Minkowski space, *Arch. Math.* **64** (1995), 534–551.

[173] A.-M. Li and F. Jia: Affine differential geometry and partial differential equations of fourth order, in: W. H. Chen, A. M. Li, U. Simon, et al. (eds.), *Geometry and Topology of Submanifolds*, World Scientific, Singapore, 2000.

[174] A.-M. Li and F. Jia: The Calabi conjecture on affine maximal surfaces, *Results Math.* **40** (2001), 265–272.

[175] A.-M. Li and F. Jia: Euclidean complete affine surfaces with constant affine mean curvature, *Ann. Glob. Anal. Geom.* **23** (2003), 283–304.

[176] A.-M. Li and F. Jia: A Bernstein property of affine maximal hypersurfaces, *Ann. Glob. Anal. Geom.* **23** (2003), 359–372.

[177] A.-M. Li and F. Jia: A Bernstein property of some fourth order partial differential equations, *Results Math.* **56** (2009), 109–139.

[178] A.-M. Li, H. Li, and U. Simon: Centroaffine Bernstein problems, *Diff. Geom. Appl.* **20** (2004), 331–356.

[179] A.-M. Li, L. Sheng, and U. Simon: Blaschke hypersurfaces with constant negative affine mean curvature, *Ann. Glob. Anal. Geom.* **47** (2015), 225–238.

[180] A.-M. Li, U. Simon, and B. Chen: A two-step Monge–Ampère procedure for solving a fourth order PDE for affine hypersurfaces with constant curvature, *J. reine angew. Math.* **487** (1997), 179–200.

[181] A.-M. Li, R. Xu, U. Simon, and F. Jia: *Affine Bernstein Problems and Monge–Ampère Equations*, World Scientific Publ. Co., Hackensack, NJ, 2010.

[182] A.-M. Li, R. Xu, U. Simon, and F. Jia: Notes on Chern's affine Bernstein conjecture, *Results Math.* **60** (2011), 133–155.

[183] A.-M. Li, K. Nomizu, and C. P. Wang: A generalization of Lelieuvre's formula, *Results Math.* **20** (1991), 682–690.

[184] A.-M. Li and G. Penn: Uniqueness theorems in affine differential geometry, Part II, *Results Math.* **13** (1988), 308–317.

[185] A.-M. Li and C. P. Wang: Canonical centroaffine hypersurfaces in $\mathbb{R}^{n+1}$, *Results Math.* **20** (1991), 660–681.

[186] A.-M. Li and G. Zhao: *Affine differential geometry*, Sichuan Educational Press, Chengdu, China, 1990, in Chinese.

[187] A.-M. Li, U. Simon, and G. Zhao: *Global Affine Differential Geometry of Hypersurfaces*, de Gruyter Expositions in Mathematics 11. Walter de Gruyter & Co., Berlin, 1993.

[188] A.-M. Li, U. Simon, and G. Zhao: Hypersurfaces with prescribed affine Gauss-Kronecker curvature, *Geom. Dedicata* **81** (2000), 141–166.

[189] H. Li: A sextic holomorphic form of affine surfaces with constant affine mean curvature, *Arch. Math. (Basel)* **82** (2004), 263–272.

[190] H. Li: Variational problems and PDEs in affine differential geometry, *Banach Center Publ.* **69** (2005), 9–41.

[191] M. Linden and H. Reckziegel: On affine maps between affinely connected manifolds, *Geom. Dedicata* **33** (1990), 91–98.

[192] J. Loftin: Affine spheres and convex $\mathbb{RP}^n$-manifolds, *Amer. J. Math.* **123** (2001), 255–274.

[193] J. Loftin: Riemannian metrics on locally projectively flat manifolds, *Amer. J. Math.* **124** (2002), 595–609.

[194] J. Loftin: Affine spheres and Kähler-Einstein metrics, *Math. Res. Lett.* **9** (2002), 425–432.

[195] J. Loftin: Survey on affine spheres, in: *Handbook of Geometric Analysis*, vol. II, pp. 161–191, Adv. Lect. Math. 13, Int. Press, Somerville, MA, 2010.

[196] E. Lutwak: On some affine isoperimetric inequalities, *J. Diff. Geom.* **23** (1986), 1–13.

[197] E. Lutwak: Centroid bodies and dual mixed volumes, *Proc. London Math. Soc.* **60** (1990), 365–391.

[198] E. Lutwak: Extended affine surface area, *Adv. Math.* **85** (1991), 39–68.

[199] E. Lutwak: Inequalities for Hadwiger's harmonic Quermassintegrals, *Math. Ann.* **280** (1988), 165–175.

[200] E. Lutwak: Intersection bodies and dual mixed volumes, *Adv. Math.* **71** (1988), 232–261.

[201] E. Lutwak: Mixed affine surface area, *J. Math. Anal. Appl.* **125** (1987), 351–360.

[202] E. Lutwak: Mixed projection inequalities, *Trans. Amer. Math. Soc.* **287** (1985), 91–106.

[203] E. Lutwak: On the Blaschke-Santalo inequality, *Annals of the New York Academy of Sciences* **440** (1985), 106–112.

[204] E. Lutwak: Selected affine isoperimetric inequalities. A survey article, in: P. M. Gruber and J. M. Wills (eds.), *Handbook Convex Geometry*, North Holland, Amsterdam, 1993.

[205] E. Lutwak: The Brunn-Minkowski-Firey theory I: Mixed volumes and the Minkowski problem, *J. Diff. Geom.* **38** (1993), 131–150.

[206] E. Lutwak: The Brunn-Minkowski-Firey theory II: Affine and geominimal surface area, *Adv. Math.* **118** (1996), 244–294.

[207] E. Lutwak: A minimax inequality for inscribed cones, *J. Math. Anal. Appl.* **176** (1993), 148–155.

[208] M. A. Magid: Timelike Thomsen surfaces, *Results Math.* **20** (1991), 691–697.

[209] M. A. Magid and K. Nomizu: On affine surfaces whose cubic forms are parallel relative to the affine metric, *Proc. Japan Acad.* **65** (1989), Ser. A; 215–218.

[210] M. A. Magid and P. J. Ryan: Flat affine spheres in $\mathbb{R}^3$, *Geom. Dedicata* **33** (1990), 277–288.

[211] M. A. Magid and P. J. Ryan: Affine 3-spheres with constant affine curvature, *Trans. Amer. Math. Soc.* **330** (1992), 887–901.

[212] F. Manhart: Zur relativen Differentialgeometrie der Hyperflächen, Dissertation TU Wien, 1982.

[213]   F. Manhart: Die Affinminimalrückungsflächen, *Arch. Math.* **44** (1985), 547–556.

[214]   F. Manhart: Uneigentliche Relativsphären im dreidimensionalen euklidischen Raum, welche Drehflächen sind. Sitzungsberichte Österr. Akad. Wiss., Abt. II, Math. Phys. Techn. Wiss. 195, pp. 231–289, 1986.

[215]   F. Manhart: Kennzeichnungen euklidischer Hypersphären durch isoperimetrische Ungleichungen, *Glas. Mat., III. Ser.* **24**(44) (1989), No. 4, 541–555 [Zbl.715.53007].

[216]   F. Manhart: 2-Flächen mit einer Relativmetrik verschwindender Krümmung, in: Proc. 3rd Congress of Geometry, May 1991. Thessaloniki, 284, 1991.

[217]   A. Martínez and F. Milán: On the affine Bernstein problem, *Geom. Dedicata* **37** (1991), 295–302.

[218]   A. Martínez and F. Milán: Affine isoperimetric problems and surfaces with constant affine mean curvature, *Manuscripta Math.* **75** (1992), 35–41.

[219]   A. Martínez and F. Milán: Convex affine surfaces with constant affine mean curvature, in: Proc. Conf. Global Diff. Geometry Global Analysis. Berlin 1990, Lecture Notes Math, 1481, pp. 139–144, Springer, Berlin, 1991.

[220]   A. Martínez and F. Milán: On affine-maximal ruled surfaces, *Math. Z.* **208** (1991), 635–644.

[221]   F. Milan: Pick invariant and affine Gauss-Kronecker curvature, *Geom. Dedicata* **45** (1993), 41–47.

[222]   F. Milan: Superficies afines con curvatura media afin constante, thesis Univ. Granada 1991.

[223]   S. Montiel and A. Ros: Compact hypersurfaces: the Alexandrov theorem for higher order mean curvatures, in: B. Lawson and K. Tenenblat (eds.), *Differential Geometry*, Pitman Monographs and Surveys Pure Appl. Math. 52, pp. 279–296, Longman Sci. Tech., Harlow, 1991.

[224]   E. Müller: Relative Minimalflächen, *Monatshefte Math. Physik* **31** (1921), 3–19.

[225]   H.-F. Münzner: Die Poincarésche Indexmethode und ihre Anwendungen in der affinen Flächentheorie im Großen, Dissertation FU Berlin, 1963.

[226]   H.-F. Münzner: Maximumprinzipähnliche Sätze der euklidischen und der affinen Flächentheorie, *Arch. Math.* **17** (1966), 569–576.

[227]   H.-F. Münzner: Analoge Kennzeichnungen von Relativsphären und Ellipsoiden, *Math. Z.* **92** (1966), 1–11.

[228]   H.-F. Münzner: Über eine spezielle Klasse von Nabelpunkten und analoge Singularitäten in der zentroaffinen Flächentheorie, *Commentarii Math. Helvet.* **41** (1966–67), 88–104.

[229]   S. B. Myers: Riemannian manifolds with positive mean curvature, *Duke Math. J.* **8** (1941), 401–404.

[230]   S. Nakajima: Über die Isoperimetrie der Ellipsoide und Eiflächen mit konstanter mittlerer Affinkrümmung im $(n + 1)$-dim. Raume, *Japanese J. Mathematics* **2** (1927), 193–196.

[231]   S. Nakajima: Über Relativ-Minimalflächen, *Tôhoku Math. J.* **29** (1928), 421–424.

[232]   F. Neumann: Centroaffine invariants of plane curves in connection with the theory of the second order linear differential equations, *Archivum Math. (Brno)* **4** (1968), 201–216 [MR 42, 2414].

[233]   L. Nirenberg: Monge–Ampère equations and some associated problems in geometry, in: Proc. ICM Vancouver 1974, vol. II, pp. 275–279, Canad. Math. Congress, Montreal, Quebec, 1975.

[234]   J. C. C. Nitsche: *Vorlesungen über Minimalflächen*, Springer, Berlin etc., 1975.

[235]   K. Nomizu: What is affine differential geometry? Differential Geometry Meeting, Univ. Münster 1982. Tagungsbericht, pp. 42–43, 1982.

[236]   K. Nomizu: On completeness in affine differential geometry, *Geom. Dedicata* **20** (1986), 43–49.

[237]   K. Nomizu: *Fundamentals of Linear Algebra*, McGraw-Hill, New York etc., 1966.

[238] K. Nomizu: A survey of recent results in affine differential geometry, in: L. Verstraelen and A. West (eds.) *Geometry and Topology of Submanifolds*, III Leeds-Conference 1990, pp. 227–256, World Scientific, Singapore etc., 1991.

[239] K. Nomizu: Introduction to Affine Differential Geometry, Part I, Lecture Notes, MPI preprint 88–37, 1988; Revised: Department of Mathematics, Brown University (1989).

[240] K. Nomizu and B. Opozda: On normal and conormal maps for affine hypersurfaces, *Tôhoku Math. J.* **44** (1992), 425–431.

[241] K. Nomizu and B. Opozda: On affine hypersurfaces with parallel nullity, *J. Math. Soc. Japan* **44** (1992), 693–699.

[242] K. Nomizu and B. Opozda: Locally symmetric connections on possibly degenerate hypersurfaces, *Bull Pol. Acad. Sci.* **40** (1992), 143–150.

[243] K. Nomizu and B. Opozda: Integral formulas for affine surfaces and rigidity theorems of Cohn–Vossen type, in: F. Dillen and L. Verstraelen (eds.), Geometry and Topology of Submanifolds, IV. Prov. Conf. Diff. Geom. and Vision. Leuven, Belgium, June 1991, pp. 133–142, World Scientific Publ. Co., River Edge, NJ, 1992.

[244] K. Nomizu and U. Pinkall: On a certain class of homogeneous projectively flat manifolds, *Tôhoku Math. J. II* (1987), 39, 407–427.

[245] K. Nomizu and U. Pinkall: On the geometry of affine immersions, *Math. Z.* **195** (1987), 165–178.

[246] K. Nomizu and U. Pinkall: Cubic form theorem for affine immersions, *Results Math.* **13** (1988), 338–362.

[247] K. Nomizu and U. Pinkall: Cayley surfaces in affine differential geometry, *Tôhoku Math. J.* **41** (1989), 589–596.

[248] K. Nomizu and U. Pinkall: On the geometry of projective immersions, *Tôhoku Math. J.* **39** (1987), 407–427.

[249] K. Nomizu and U. Pinkall: Immersions of low rank in differential geometry, in: B. Lawson, K. Tenenblat (eds.), *Differential Geometry*, pp. 297–302, Longman Sci. Tech., Harlow, 1991.

[250] K. Nomizu, U. Pinkall, and F. Podestá: On the geometry of affine Kähler immersions, *Nagoya Math. J.* **120** (1990), 205–222.

[251] K. Nomizu, U. Pinkall, U. Simon, and C. Scharlach: Affine Differentialgeometrie, Math. Forschungsinstitut Oberwolfach, Tagungsbericht 7, 1991.

[252] K. Nomizu, U. Pinkall, and U. Simon (eds.): *Affine Differential Geometry*, Proc. Math. Forschungsinstitut Oberwolfach, Febr. 1991; Print: Birkhäuser, Basel, 1991; Distribution: TU Berlin; ISBN 3798314608.

[253] K. Nomizu and F. Podestá: On Affine Kähler Structures, *Bull. Soc. Math. Belg. Tijdschr. Belg. Wisk. Gen. Ser. B* **41**(3) (1989), 275–281.

[254] K. Nomizu and F. Podestá: On the Cartan-Norden theorem for affine Kähler immersions, *Nagoya Math. J.* **121** (1991), 127–135.

[255] K. Nomizu and T. Sasaki: A new model of unimodular-affinely homogeneous surfaces, *Manuscripta Math.* **73** (1991), 39–44.

[256] K. Nomizu and T. Sasaki: On a theorem of Chern and Terng on affine surfaces, *Manuscripta Math.* **66** (1990), 303–307.

[257] K. Nomizu and T. Sasaki: On certain quartic form for affine surfaces, *Tôhoku Math. J.* **44** (1992), 25–33.

[258] K. Nomizu and T. Sasaki: On the classification of projectively homogeneous surfaces, *Results Math.* **20** (1991), 698–724.

[259] K. Nomizu and T. Sasaki: Centroaffine immersions of codimension two and projective hypersurface theory, *Nagoya Math. J.* **132** (1993), 63–90.

[260] K. Nomizu and T. Sasaki: *Affine Differential Geometry*, Cambridge University Press, Cambridge, 1994.

[261] K. Nomizu and U. Simon: Conjugate connections, in: F. Dillen and L. Verstraelen (eds.), *Geometry and Topology of Submanifolds* IV. Proc. Conf. Diff. Geom. and Vision, Leuven, Belgium, June 1991, pp. 152–172, World Scientific, Singapore etc., 1992.

[262] A. P. Norden: *Affinely-Connected Spaces*, Gosudarstv. Izdat. Tehn.-Teor. Lit. Moskow Leningrad, 1950.

[263] I. V. Oliker: Remarks on projectively invariant nonlinear partial differential equations, *Results Math.* **13** (1988), 363–366.

[264] V. Oliker and U. Simon: The Christoffel problem in relative differential geometry, in: Colloquia Mathematica Societatis Janos Bolyai 46. Topics in Differential Geometry, Debrecen (Hungary), pp. 973–1000, 1984 [Zbl.646.53006].

[265] V. Oliker and U. Simon: Affine geometry and polar hypersurfaces, in: B. Fuchsteiner and W. A. J. Luxemburg (eds.) *Analysis and Geometry: Trends in Research and Teaching* , pp. 87–112, BI, Mannheim Zürich, 1992 [ISBN 3-411-15621-X].

[266] B. O'Neill: *Semi-Riemannian Geometry with Applications to Relativity*, Academic Press, New York, 1983.

[267] B. Opozda: Some extensions of Radon's theorem. In: Proc. Conf. Global Diff. Geometry Global Analysis. Berlin 1990, Lecture Notes Math. 1481, pp. 185–191, Springer, Berlin, 1991.

[268] B. Opozda: Some equivalence theorems in affine hypersurface theory, *Monatsh. Math.* **113** (1992), 245–254.

[269] B. Opozda: Locally symmetric connections on surfaces, *Results Math.* **20** (1991), 725–743.

[270] B. Opozda: Some relations betwen Riemannian and affine geometry, *Geom. Dedicata* **47** (1993), 225–236.

[271] B. Opozda and L. Verstraelen: On a New Curvature Tensor in Affine Differential Geometry, in: M. Boyom, J.-M. Morvan, and L. Verstraelen (eds.), *Geometry and Topology of Submanifolds III*, pp. 271–293, World Scientific, Singapore etc., 1990.

[272] H. Pabel: Translationsflächen in der äquiaffinen Differentialgeometrie, *J. Geom.* **40** (1991), 148–164.

[273] B. Palmer: Spacelike constant mean curvature surfaces in pseudo-Riemannian space form, *Ann. Glob. Anal. Geom.* **8** (1990), 217–226.

[274] H.-P. Paukowitsch: Zur Kurventheorie *n*-dimensionaler affiner Räume, *Österreich. Akad. Wiss. math.-naturw. Kl. S.-ber. Abt. II* **185** (1976) (1977), 443–458, [Zbl.392.53003].

[275] H.-P. Paukowitsch: Begleitfiguren und Invariantensystem minimaler Differentiationsordnung von Kurven im reellen *n*-dimensionalen affinen Raum, *Monatsh. Math.* **85** (1978), 137–148.

[276] G. Penn and U. Simon: Deformations of hypersurfaces in equiaffine differential geometry, *Ann. Glob. Anal. Geom.* **5** (1987), 123–131.

[277] C. M. Petty: Ellipsoids. *Convexity and its applications. Collect. Surv.*, pp. 264–276, Birkhäuser, Basel, 1983.

[278] C. M. Petty: Affine isoperimetric problems, in: J. E. Goodman, E. Lutwak, J. Malkevitch, and R. Pollack (eds.), *Discrete geometry and convexity*, Proc. Conf. New York 1982, *Annals New York Acad. Sci.* **440** (1985), 113–127.

[279] C. M. Petty: Geominimal surface area, *Geom. Dedicata* **3** (1974), 77–97.

[280] G. Pick: Natürliche Geometrie ebener Transformationsgruppen, *Sitz. Ber. der kais. Akad. d. Wiss. in Wien* (1906), 1–21.

[281] G. Pick: Über affine Geometrie IV. Differentialinvarianten der Flächen gegenüber affinen Transformationen, *Ber. Verh. Sächs. Ges. Wiss. Leipzig, Math.-Phys. Kl.* **69** (1917), 107–136.

[282] U. Pinkall, A. Schwenk-Schellschmidt, and U. Simon: Geometric methods for solving Codazzi and Monge–Ampère equations, *Math. Ann.* **298** (1994), 89–100.

[283] F. Podestá: Affine Transformations in Affine Differential Geometry, *Results Math.* **16** (1989), 150–161.

[284] A. V. Pogorelov: On the improper convex affine hyperspheres, *Geom. Dedicata* **1** (1972), 33–46.

[285] A. V. Pogorelov: *The Minkowski multidimensional problem*, John Wiley & Sons, New York Toronto London, 1978.

[286] A. V. Pogorelov: Improper convex affine hyperspheres [in Russian], *Doklady Akad. Nauk SSSR* **202** (1972), 1008–1011 [English translation in: *Soviet Math. Doklady* **13** (1972), 240–244].

[287] A. V. Pogorelov: Complete affine minimal hypersurfaces, *Soviet Math. Dokl.* **38** (1989), 217–219.

[288] A. V. Pogorelov: Unique determination of affine-minimal hypersurfaces, *Soviet Math. Dokl.* **36** (1988), 617–618.

[289] J. Radon: Die Grundgleichungen der affinen Flachentheorie, *Leipziger Berichte* **70** (1918), 91–107.

[290] R. C. Reilly: Affine geometry and the form of the equation of a hypersurface, *Rocky Mt. J. Math.* **16** (1986), 553–565.

[291] R. C. Reilly: The relative differential geometry of nonparametric hypersurfaces, *Duke Math. J.* **43** (1976), 705–721.

[292] E. Salkowski: *Affine Differentialgeometrie*, Walter de Gruyter, Berlin Leipzig, 1934.

[293] L. A. Santalo: A geometrical characterization for the affine differential invariants of a space curve, *Bull. Amer. Math. Soc.* **52** (1946), 625–632.

[294] L. A. Santalo: Un invariante afin para los cuerpos convexos del espacia de *n* dimensiones, *Portugal math.* **8** (1949), 155–161.

[295] T. Sasaki: Hyperbolic affine hyperspheres, *Nagoya Math. J.* **77** (1980), 107–123.

[296] T. Sasaki: An affine isoperimetric inequality for a strongly convex closed hypersurface in the unimodular affine space A^{n+1}, *Kumamoto J. Sci. (Math.)* **16** (1984), 23–38.

[297] T. Sasaki: On the Green function of a complete Riemannian or Kähler manifold with asymptotically negative constant curvature and applications, *Advanced Studies in Pure Math. Geometry of Geodesics and Related Topics* **3** (1984), 387–421.

[298] T. Sasaki: A note on characteristic functions and projectively invariant metrics on a bounded convex domain, *Tokyo J. Math.* **8** (1985), 49–79.

[299] T. Sasaki: On the characteristic function of a strictly convex domain and the Fubini-Pick invariant, *Results Math.* **13** (1988), 367–378.

[300] T. Sasaki: *Projective differential geometry and linear homogeneous differential equations*, Lecture Notes Brown Univ., 94pp., 1989.

[301] T. Sasaki: On a projectively minimal hypersurface in the unimodular affine space, *Geom. Dedicata* **23** (1987), 237–251.

[302] T. Sasaki: On the projective geometry of hypersurfaces. In: Équations différentielles dans le champ complexe, Colloq. Fr.-Jap., Strasbourg/Fr. 1985, vol. III, pp. 115–161, 1988. Publ. Inst. Rech. Math. Av., Univ. Louis Pasteur, Strasbourg, 1988.

[303] T. Sasaki: *On the Veronese embedding and related system of differential equations*, Proc. Conf. Global Diff. Geometry Global Analysis. Berlin 1990, Lecture Notes Math. 1481, pp. 210–247, Springer, Berlin, 1991.

[304] H. Schaal: Zur lokalen affinen Differentialgeometrie ebener und gewundener Flächenkurven. I: Ebene Flächenkurven, *Math. Z.* **90** (1965), 71–94.

[305] H. Schaal: Zur lokalen affinen Differentialgeometrie ebener und gewundener Flächenkurven. II: Gewundene Flächenkurven, *Math. Z.* **90** (1965), 95–116.

[306] H. Schaal: Neue Erzeugungen der Minimalflächen von G. Thomsen, *Monatsh. Math.* **77** (1973), 433–461.

[307] H. Schaal: Die Affinminimalflächen von G, *Thomsen. Arch. Math.* **24** (1973), 208–217.

[308] H. Schaal: Die Ennepersche Minimalfläche als Grenzfall der Minimalflächen von G. Thomsen, *Arch. Math.* **24** (1973), 320–322.

[309] C. Scharlach: Affin-konforme Geometrie regulärer Hyperflächen, Diploma Thesis, TU Berlin 1989.

[310] C. Scharlach: Some results in centro affine differential geometry, in: F. Dillen and L. Verstraelen (eds.), *Geometry and Topology of Submanifolds*, IV. Proc. Conf. Diff. Geom. Vision, Leuven 1991, pp. 198–206, World Scientific, Singapore etc., 1992.

[311] P. A. Schirokow and A. P. Schirokow: *Affine Differentialgeometrie*, Leipzig, Teubner, 1962. [Zbl. 106.147; Russ. original Zbl. 85.367].

[312] R. Schneider: Zur affinen Differentialgeometrie im Großen I, *Math. Z.* **101** (1967), 375–406.

[313] R. Schneider: Zur affinen Differentialgeometrie im Großen II: Über eine Abschätzung der Pickschen Invariante auf Affinsphären, *Math. Z.* **102** (1967), 1–8.

[314] R. Schneider: Translations- und Ähnlichkeitssätze für Eihyperflächen, Diploma Thesis Univ. Frankfurt/Main, 1964.

[315] R. Schneider: Über die Finslerräume mit S_{ijkl} = 0, *Arch. Math.* **19** (1968), 656–658.

[316] R. Schneider: Affine-invariant approximation by convex polytopes, *Stud. Sci. Math. Hungar.* **21** (1986), 401–408.

[317] R. Schneider: A characteristic property of the ellipsoid, *Amer. Math. Monthly* **74** (1967), 416–418.

[318] R. Schneider: On A. D. Aleksandrov's inequalities for mixed discriminants, *J. Math. Mechanics* **15** (1966), 285–290.

[319] R. Schneider: Equality in the Aleksandrov-Fenchel inequality – present state and new results. In: *Intuitive Geometry*, Conference Szeged 1991, Colloq. Math. Soc. János Bolyai 63, pp. 425–438, North Holland Publ. Co., Amsterdam, 1994.

[320] C. Schütt and E. Werner: The convex floating body, *Math. Scand.* **66** (1990), 275–290.

[321] A. Schwenk: Eigenwertprobleme des Laplaceoperators und Anwendungen auf Untermannigfaltigkeiten, Dissertation TU Berlin, 1984.

[322] A. Schwenk and U. Simon: Hypersurfaces with constant equiaffine mean curvature, *Arch. Math.* **46** (1986), 85–90.

[323] U. Simon: Integralformeln zur Kennzeichnung von Hyperflächen durch Krümmungen und Stützabstand, Dissertation FU Berlin, 1965.

[324] U. Simon: Minkowskische Integralformeln und ihre Anwendungen in der Differentialgeometrie im Großen, *Math. Ann.* **173** (1967), 307–321.

[325] U. Simon: Kennzeichnungen von Sphären, *Math. Ann.* **175** (1968), 81–88.

[326] U. Simon: Zur Relativgeometrie: Symmetrische Zusammenhänge auf Hyperflächen, *Math. Z.* **106** (1968), 36–46.

[327] U. Simon: The Pick invariant in equiaffine differential geometry, *Abh. Math. Sem. Hamburg* **535** (1983), 225–228.

[328] U. Simon: *Hypersurfaces in affine differential geometry and eigenvalue problems*, in: Proc. Conf. Diff. Geo. Appl., Nove Mésto na Morave (ČSSR) 1983, Part I, pp. 127–136, 1984.

[329] U. Simon: Hypersurfaces in equiaffine differential geometry, *Geom. Dedicata* **17** (1984), 157–168.

[330] U. Simon: *Dirichlet problems and the Laplacian in affine hypersurface theory*, Lecture Notes in Math. 1369, pp. 243–260, Springer, Berlin, 1989.

[331] U. Simon: Kongruenzsätze der affinen Differentialgeometrien, *Math. Z.* **120** (1971), 365–368.

[332] U. Simon: Charakterisierung von Relativsphären und Ellipsoiden, *Arch. Math.* **24** (1973), 100–104.

[333] U. Simon: Zur Entwicklung der affinen Differentialgeometrie nach Blaschke, in: [22], pp. 35–88, 1985.

[334] U. Simon: The fundamental theorem in affine hypersurface theory, *Geom. Dedicata* **26** (1988), 125–137.

[335] U. Simon: Connections and conformal structure in affine differential geometry, in: D. Krupka and A. Švec (eds.), *Differential Geometry and Its Applications*, Proceedings Conf. Aug. 1986, Brno, Czechoslovakia, pp. 315–327, D. Reidel Publ. Comp., Dordrecht, 1987.

[336] U. Simon: Global uniqueness for ovaloids in Euclidean and affine differential geometry, *Tôhoku Math. J.* **44** (1992), 327–334.

[337] U. Simon (ed.): Affine Differentialgeometrie, TU Berlin, 1988. (ISBN 3-7983-1192-7), 189–443. [Zbl.646.53001].

[338] U. Simon (ed.): Local classification of twodimensional affine spheres with constant curvature metric, *Diff. Geom. Appl.* **1** (1991), 123–132.

[339] U. Simon (ed.): Recent developments in affine differential geometry. Diff. Geom. and its Applications, in: Proc. Conf., Dubrovnik/Yugosl. 1988, pp. 327–347, 1989 [Zbl.681.53003].

[340] U. Simon (ed.): Affine hypersurface theory revisited: Gauge invariant structures, *Russian Math.* **48**(11) (2004), 48–73.

[341] U. Simon, A. Schwenk-Schellschmidt, and H. Viesel: *Introduction to the Affine Differential Geometry of Hypersurfaces*, Lecture Notes. Science University Tokyo. 1991. (ISBN 3 7983 15299).

[342] U. Simon and C.-P. Wang: Local theory of affine 2-spheres, in: R. E. Greene and S. T. Yau (eds.), Proceedings 1990, Summer Institute Diff. Geometry, *Proc. Symposia Pure Math.* **54** (1993), 585–598.

[343] U. Simon and W. Wendland: Zur Indexmethode in der Differentialgeometrie im Großen, *Math. Z.* **116** (1970), 242–246.

[344] U. Simon and H. Wissner: Geometry of the Laplace operator. Proc. Alg. Geom. Kuwait, 171–191(1981), Alden Press Oxford.

[345] M. Spivak: *A comprehensive introduction to differential geometry*, vol. 3, Publish or Perish, Boston, 1975.

[346] M. Spivak: *Calculus on manifolds. A modern approach to classical theorems of advanced calculus*. W. A. Benjamin, Inc., New York Amsterdam, 1965.

[347] B. Su: *Affine differential geometry*, Science Press, Beijing, and Gordon and Breach, New York, 1983.

[348] B. Su: Some intrinsic invariants of a parametric curve in affine hyperspace, *Chinese Ann. Math.* **1** (1980), 199–206.

[349] B. Su: *Selected mathematical papers*, Science Press, Beijing, and Gordon and Breach, New York, 1983.

[350] B. Su and D. Liu: An affine invariant theory and its application in computational geometry, *Scientia Sinica Ser. A* **26** (1983), 3, 259–272.

[351] B. Su and Y. Xin: Some intrinsic invariants of a parametric curve in higher dimensional affine space, *Acta Math. Appl. Sinica* **3** (1980), 139–146.

[352] W. Süss: Zur relativen Differentialgeometrie: I. Über Eilinien und Eiflächen in der elementaren und affinen Differentialgeometrie, *Jap. J. Math.* **4** (1927), 57–75.

[353] W. Süss: Zur relativen Differentialgeometrie V: Über Eihyperflächen im $\mathbb{R}^{n+1}$, *Tôhoku Math. J.* **31** (1929), 202–209.

[354] W. Süss: Eindeutigkeitssätze und ein Existenztheorem der Eiflächen im Großen, *Tôhoku Math. J.* **35** (1932), 290–293.

[355] W. Süss: Über Kennzeichnungen der Kugeln und Affinsphären durch Herrn K. P. Grotemeyer, *Arch. Math.* **3** (1952), 311–313.

[356] W. Süss: Ein affines Analogon zur Bestimmung einer Fläche aus einer Grundform und einer Krümmungsfunktion durch W. Scherrer, *Archiv Math.* **52** (1950), 698–702.

[357] A. Švec: Global differential geometry of surfaces in affine space, *Casopis Math.* **69** (1964), 340–346, [Zbl.124.143].

[358] A. Švec: Differential geometry of surfaces. Czech. Math. J. (Praha) 39(114), 303–322(1989).

[359] A. Švec: Infinitesimal rigidity of surfaces in A^3, *Czech. Math. J. (Praha)* **38**(113) (1988), 479–485.

[360] A. Švec: On equiaffine Weingarten surfaces, Czech. Math. J. (Praha) 37(112), 567–572(1985).

[361] A. Švec: On the affine normal, *Czech. Math. J. (Praha)* **40**(115) (1990), 332–342.

[362] A. Švec: On the Pick invariant, *Czech. Math. J. (Praha)* **38**(113) (1988), 493–497.

[363] A. Švec: Surfaces in general affine space, *Czech. Math. J. (Praha)* **39**(114) (1989), 280–287.

[364] C.-L. Terng: Affine minimal surfaces, *Ann. Math. Stud.* **103** (1983), 207–216.

[365] E. Teufel: Kinematische Berührung im Äquiaffinen, *Geom. Dedicata* **33** (1990), 317–326.

[366] G. Thomsen: AG XXXIX. Über Affinminimalflächen, die gleichzeitig Minimalflächen sind, *Abh. Math. Sem. Hamburg* **2** (1923), 71–73.

[367] N. S. Trudinger and X.-J. Wang: The Bernstein problem for affine maximal hypersurfaces, *Invent. Math.* **140** (2000), 399–422.

[368] N. S. Trudinger and X.-J. Wang: Affine complete locally convex hypersurfaces, *Invent. Math.* **150** (2002), 45–60.

[369] N. S. Trudinger and X.-J. Wang: The Monge–Ampère equation and its geometric applications, Handbook of Geometric Analysis, vol. I, 467–524(2008), Adv. Lect. Math. 7, Int. Press, Somerville, MA.

[370] G. Tzitzeica: Sur certaines surfaces reglées, *C. R. Acad. Paris* **146** (1907), 132–133.

[371] G. Tzitzeica: Sur une nouvelle classe de surfaces, *Rend. Circ. Math. Palermo* **25** (1908), 180–187.

[372] G. Tzitzeica: Sur les surfaces isothermiques. Proceedings fifth Int. Congress Math. Cambridge 1912, vol II, 88–92. Cambridge Univ. Press (1913).

[373] E. Vassiliou: Some aplications of conjugate connections, in: Proc. of the 3rd Congress of Geometry, May 1991. Thessaloniki, pp. 434–442, 1991.

[374] I. N. Vekua: Systems of differential equation of the first order of elliptic type and boundary value problems; Applications to the theory of shells, *Matem. Sbornik* **31**(73) (1952), 217–314.

[375] I. N. Vekua: *Generalized analytic functions*, International series of monographs on pure and applied math. 25, Pergamon Press, London Paris Frankfurt, 1962.

[376] L. Verstraelen and L. Vrancken: Affine variation formulas and affine minimal surfaces, *Michigan Math. J.* **36** (1989), 77–93.

[377] K. Voss: Einige Eindeutigkeitssätze in der affinen Differentialgeometrie, *Results Math.* **13** (1988), 379–385.

[378] K. Voss: Variation of curvature integrals, *Results Math.* **20** (1991), 789–796.

[379] G. Vránceanu: Invariants centro-affines d'une surface, *Rev. Roum. Math. Pures et Appl.* **24** (1979), 6, 979–982, [Zbl. 413. 53007].

[380] G. Vránceanu: Tzitzéica fondateur de la géométrie centroaffine, *Revue Roumaine* **24** (1979), 983–988.

[381] L. Vrancken: Affine surfaces with constant affine curvature, *Geom. Dedicata* **33** (1990), 177–194.

[382] L. Vrancken: Affine higher order parallel hypersurfaces. Ann. Fac. Sci. Toulouse, V. Ser., Math. 9, No. 3, 341–351(1988).

[383] L. Vrancken: Affine surfaces with higher order parallel cubic form, *Tôhoku Math. J.* **43** (1991), 127–139.

[384] L. Vrancken: Affine quasi umbilical hypersurfaces which are flat with respect to the affine metric, *Results Math.* **20** (1991), 756–776.

[385] L. Vrancken: Affine hypersurfaces with constant affine sectional curvature, in: F. Dillen and L. Verstraelen (eds.), *Geometry and Topology of Submanifolds*, IV. Proc. Conf. Diff. Geom. Vision, Leuven 1991, World Scientific, Singapore etc., 1992.

[386] L. Vrancken: The Magid-Ryan conjecture for equiaffine hyperspheres with constant sectional curvature, *J. Diff. Geom.* **54** (2000), 99–138.

[387] L. Vrancken, A.-M. Li, and U. Simon: Affine spheres with constant sectional curvature, *Math. Z.* **206** (1991), 651–658.

[388] R. Walter: Über zweidimensionale parabolische Flächen im vierdimensionalen affinen Raum. Teil I. Allgemeine Flächentheorie, *J. reine angew. Math.* **227** (1967), 178–208.

[389] R. Walter: Über zweidimensionale parabolische Flächen im vierdimensionalen affinen Raum. Teil II. Spezielle Flächen, *J. reine angew. Math.* **228** (1967), 71–92.

[390] R. Walter: Centroaffine differential geometry: submanifolds of codimension 2, *Results Math.* **13** (1988), 386–402.

[391] R. Walter: Compact centroaffine spheres of codimension 2, *Results Math.* **20** (1991), 777–788.

[392] R. Walter: Lineare Algebra und analytische Geometrie. Vieweg, Braunschweig etc., 1985.

[393] R. Walter: On Projective and Affine Differential Geometry, in: Proc. 3rd Congress of Geometry, May 1991, pp. 443, Thessaloniki, 1991.

[394] B. Wang: Some remarks on Euclidean boundary points of locally convex immersed hypersurfaces, *J. Sichuan Univ. (Nat. Sci. Ed.)* **51**(1) (2014), 16–20.

[395] C. P. Wang: Some examples of complete hyperbolic affine 2-spheres in $\mathbb{R}^3$, in: Proceedings Conf. Global Diff. Geometry Global Analysis, Berlin 1990, Lecture Notes Math. 1481, pp. 272–280, Springer, Berlin, 1991.

[396] C. P. Wang: Canonical equiaffine hypersurfaces in $\mathbb{R}^{n+1}$, *Math. Z.* **214** (1993), 579–592.

[397] C. P. Wang: Centroaffine minimal hypersurfaces in $\mathbb{R}^{n+1}$, *Geom. Dedicata* **51** (1994), 63–74.

[398] Y. Y. L. Wang: On the order of the Euler–Lagrange equations of the variational problem of the affine arc length., *J. reine angew. Math.* **245** (1970), 55–62.

[399] Y. Y. L. Wang: On the solution of the variational problem of the arc length in 4-dimensional affine space, *J. reine angew. Math.* **255** (1972), 99–103.

[400] B. Wegner: Über eine charakteristische Eigenschaft affiner Abbildungen, *Math.-Phys. Semesterber.* **19** (1972), 68–72.

[401] W. L. Wendland: *Elliptic systems in the plane*, Monographs and Studies in Mathematics 3, Pitman (Advanced Publishing Program), Boston (MA) London, 1979.

[402] H. Weyl: On the volume of tubes, *Amer. J. Math.* **61** (1939), 461–472.

[403] S. van Wilkinson: Characterizing Gauss maps, Ph. D. Thesis, Rice Univ., Houston, 1984.

[404] S. van Wilkinson: General affine differential geometry for low codimension immersions, *Math. Z.* **197** (1988), 583–594.

[405] T. Willmore: Riemann extensions and affine differential geometry, *Results Math* **13** (1988), 403–408.

[406] H. H. Wu: *The Equidistribution Theory of Holomorphic Curves*, Princeton Univ. Press, Princeton, and Univ. Tokyo Press, Tokyo, 1970.

[407] H. H. Wu: The spherical images of convex hypersurfaces, *J. Diff. Geom.* **9** (1974), 279–290.

[408] C.-M. Yau: Affine conormal of convex hypersurfaces, *Proc. Amer. Math. Soc.* **106** (1989), 465–470.

[409] S.-T. Yau and R. Schoen: *Differential Geometry*, Chinese Science Press, Beijing, 1988.

[410] I. Yokota: Exceptional Lie groups. arXiv: math.DG/0902. 0431v1.

[411] J. Yu: Affine spheres with constant sectional curvature in A^4, *J. Sichuan Univ. (Nat. Sci. Ed.)* **27**(4) (1990), 383–387.

[412] Z. Y. Zhang and F. Jia: Bernstein property of a relatively extremal hypersurface (in Chinese), *J. Sichuan Univ. (Nat. Sci. Ed.)* **49**(3) (2012), 489–493.

[413] G. Zhao: Ovaloids with parallel Ricci curvature in affine space (in Chinese), *J. Sichuan Univ. (Nat. Sci. Ed.)* **35**(2) (1998), 174–177.

Index

Abreu equation   259
adapted frame   46
affine
– isoperimetric inequality   303
– maximal hypersurfaces   217, 223
– Bernstein conjecture   272
– Bernstein problem   232, 236, 272
– completeness   131, 136, 151, 165
– conormal vector field   161
– hypersphere   93, 95
– invariance   35
– isoperimetric inequality   308
– mapping   10
– maximal surface   217
– mean curvature   55, 183, 310
– Minkowski formula   182
– normal   49, 56
– normal line   52
– principal curvature   55, 136, 310
– shape operator   40
– space   7
– support function   42, 62, 182, 294
– technique   4, 259
– transformation groups   10
– Weingarten surface   207
affine hyperspheres
– elliptic   96, 144
– hyperbolic   96, 154
– improper   95
– parabolic   96, 151, 229
– proper   95
affine hypersurface
– elliptic type   277
– hyperbolic type   277
Alexandrov–Fenchel inequality   310, 311
Alexandrov–Pogorelov-Heinz's theorem   257
apolarity condition   52, 54
argument principle   210

Blaschke
– completeness   290
– hypersurface   39
– metric   39, 40, 48
Bochner-Lichnerowicz formula   186, 187, 196
bootstrap method   250

boundary point   80, 328
– first class   81
– second class   81

Calabi
– completeness   131
– composition   100–103
– conjecture   154, 157, 159, 166, 240
– metric   90, 272
Cartan's lemma   46, 321
Cauchy–Riemann equation   206
center   95
central projection   281
centroaffine
– area   314
– hypersurface   278
– normalization   41
Cohn–Vossen's theorem   214
comparison principle of determinants   249
completeness
– affine   131, 165
– Calabi   131, 273
– Euclidean   131, 165, 289
– Weingarten   278, 289
conformal
– class   43, 172
– curvature tensor   144
– structure   172
conjecture
– E. Calabi   231, 240, 244
– S. S. Chern   231, 246, 250, 259
conjugate
– connection   31, 61
– triple   31
conormal   9, 25
– connection   30, 193
conormal field   25
– equiaffine   59
– p.d.e.   62
– regular   28
constant
– affine mean curvature   183
– mean curvature   104
– Pick invariant   68, 108, 188, 194

– scalar curvature 104, 108, 109, 186, 188
– sectional curvature 104, 112
contraction 318
convexity 48
covariant Hessian 63
cross product 9
cubic form 32, 53, 54
– parallel 116
curvature
– affine 54, 310
– functions, affine 54, 184, 312
– functions, constant 183, 184, 197, 198
– Gauß–Kronecker 55, 236, 277, 300
cut locus 341
cut point 341

decomposable net 203
determinant form 8, 9, 278
– dual 8
difference tensor 32, 102, 103
Dirichlet problems 195
distribution 202
divergence formula 339
dual 14

Einstein space 186, 335
ellipsoid 65, 96, 104, 306, 314
– characterization 183, 208, 222, 306
elliptic
– affine hypersphere 96
– paraboloid 63, 104, 246
embedding (imbedding) 324
equiaffine
– conormal 59
– frame 52, 220
– structure equations 56
Euclidean
– boundary point 80
– compatibility conditions 18
– completeness 131, 165, 232, 289
– space 7
exceptional Lie group 127
exponential mapping 337
extremal Kähler metrics 4, 259

Fubini-Pick form 54, 80, 103, 113, 173
fundamental theorem 76

gauge
– invariant 43
– surface 34
– transformation 43
Gauß basis 13, 324
Gauß structure equation 22, 26, 46, 54, 56, 72
geodesic 337
– distance function 341
global solutions of p.d.e.'s 190
globally convex 83
gradient map 88, 97, 156
graph hypersurface 85, 278
Green's formula 187, 200, 222, 339

Hadamard's theorem 180
Hadamard–Sacksteder–Wu's theorem 132, 155
Harnack inequality 247
Herglotz integral formula 216
Hessian metric 90
Hilbert normal form 206
Hilbert space 178
Hofer's Lemma 240
Hölder estimates 247
holomorphic parameters 172
Hopf and Rinow's theorem 338
hyperbolic affine hypersphere 96
hyperboloid 66, 96
hyperquadrics 63, 68
hypersurface
– immersion 20, 24
– nondegenerate 29
– with boundary 194

immersion 324
index 202, 205
– form 342
– method 201
induced connection 26, 52, 216
inequality
– Cauchy–Schwarz 135, 142, 237, 253, 260, 265, 269, 274, 296
– Gårding 184
– Hölder 309, 315
– Minkowski 305
– Moser 178
– Newton 183, 191, 200
– Young 253, 256
integrability conditions 72–76
interior point 328

isolated singularity 202, 205
isometry of linear mapping 16
isothermal coordinate 173, 232
isotropic 130

Jacobi
– equation 338
– field 338
Jordan multiplication 128
Jörgens–Calabi–Pogorelov's theorem 153

Klein geometry 20
Koszul formula 333

Laplace spectrum 186
Laplace-Beltrami operator 65, 177
Laplacian comparison theorem 341
Lax-Milgram lemma 179
Legendre
– transformation 93, 278
– transformation domain 88, 157, 257, 280,
    289
– transformation function 88, 99, 168, 254,
    258, 278, 280, 286
Lelieuvre formula 60
level set 132
Levi–Civita connection 333
Liouville's theorem 231
locally strongly convex 39

Maschke–Pick–Berwald's theorem 68
maximum modulus theorem 246
maximum principle 144, 283, 286, 292, 293,
    299
mean connection 61
mean curvature
– constant 104, 183, 197
minimal geodesic 143, 338
Minkowski
– integral formula 222
– uniqueness theorem 193
Minkowski's inequality 305
Minkowski's theorem 310
mixed volumes 310
Monge–Ampère equation 94, 99, 153, 156,
    166, 247
– linearization 247
Myers' theorem 146

net 203
nondegenerate hypersurface 29
normal coordinate system 337
normal field 24
normalization 29
– centroaffine 40
– convex set 250
– equiaffine 39, 85
– Euclidean 22, 38
– relative 38, 39, 90
normalized
– affine transformation 250
– convex domain 251, 263, 266
– convex set 250

ovaloid 183, 191, 303

parabolic affine hypersphere 96, 153
paraboloid 63, 96, 104, 112
parallel
– cubic form 116
– Fubini–Pick form 116
– vector field 336
– volume form 331
partition of unity 327
Pascali system 207
Peterson correspondence 34
Pick
– form 175
– invariant 37, 54, 144
Poincaré index formula 203, 210
polar cone 156
projectively flat manifold 63

Radon's theorem 79
Rayleigh's theorem 178
regular
– affine group 19
– conormal field 28
relative
– hypersurface 30
– metric 29, 39
– normal field 28, 29
– support function 42
– Tchebychev vector field 90
Ricci curvature 147, 151, 332
Ricci identity 152, 331
Ricci lemma 333

Riemann–Roch formula 176
rigidity theorem 191

scalar curvature 76
– constant 104, 109, 186, 188
Schauder interior estimate 246
section 142, 249, 251, 254, 296
sectional curvature 112
shadow boundary 199
shadow submanifold 194
singularity 202
specific curvature 257, 258
spectrum of the Laplacian 186
standard imbedding 125, 126, 129
Steiner symmetrization 303
Stokes' formula 146, 328, 339
structure equations 14, 26, 37, 40, 47, 54
– equiaffine 56
– Euclidean 22
– moving frames 73

Tchebychev
– vector field 32
– form 32, 43, 279
– function 279

theorema egregium 76, 145, 175, 229
transversal field 24, 26

unimodular
– affine frame 13
– group 10

variational formulas 218, 222
variational vector field 219
Vekua's system 206
volume form 8, 21–23, 27–29, 38, 40
– parallel 21, 40
– Riemannian 22, 39, 333

weak solution 147, 149
Weierstraß representation 226
Weingarten
– completeness 278, 289
– form 37, 40, 173, 186
– form, equiaffine 54
– form, relative 30
– metric 277, 281, 287
– operator 22, 41, 54
– structure equation 22, 26, 46, 56, 72, 280
– surface 208

Young's inequality 253, 256

De Gruyter Expositions in Mathematics

Volume 60
Benjamin Fine, Anthony Gaglione, Alexei Myasnikov, Gerhard Rosenberger, Dennis Spellman
The Elementary Theory of Groups, 2014
ISBN 978-3-11-034199-7, e-ISBN 978-3-11-034203-1, Set-ISBN 978-3-11-034204-8

Volume 59
Friedrich Haslinger
The d-bar Neumann Problem and Schrödinger Operators, 2014
ISBN 978-3-11-031530-1, e-ISBN 978-3-11-031535-6, Set-ISBN 978-3-11-031536-3

Volume 58
Oleg K. Sheinman
Current Algebras on Riemann Surfaces, 2012
ISBN 978-3-11-026452-4, e-ISBN 978-3-11-026452-4, Set-ISBN 978-3-11-916387-3

Volume 57
Helmut Strade
Simple Lie Algebras, Completion of the Classification, 2012
ISBN 978-3-11-026298-8, e-ISBN 978-3-11-026301-5, Set-ISBN 978-3-11-916682-9

Volume 56
Yakov Berkovich, Zvonimir Janko
Groups of Prime Power Order 3, 2011
ISBN 978-3-11-020717-0, e-ISBN 978-3-11-025448-8, Set-ISBN 978-3-11-218909-2

Volume 55
Rainer Picard, Des McGhee
Partial Differential Equations, 2011
ISBN 978-3-11-025026-8, e-ISBN 978-3-11-025027-5, Set-ISBN 978-3-11-218895-8

Volume 54
Edgar E. Enochs, Overtoun M. G. Jenda
Relative Homological Algebra, 2011
ISBN 978-3-11-021522-9, e-ISBN 978-3-11-021523-6, Set-ISBN 978-3-11-173442-2

Volume 53
Adolfo Ballester-Bolinches, Ramon Esteban-Romero, Mohamed Asaad
Products of Finite Groups, 2010
ISBN 978-3-11-020417-9, e-ISBN 978-3-11-022061-2, Set-ISBN 978-3-11-173407-1

www.degruyter.com